Advances in
ECOLOGICAL RESEARCH

VOLUME 30

Advances in
ECOLOGICAL RESEARCH

Edited by

A.H. FITTER

Department of Biology, University of York, UK

D.G. RAFFAELLI

Department of Zoology, University of Aberdeen, UK

VOLUME 30

ACADEMIC PRESS

A Harcourt Science and Technology Company

San Diego   London   Boston   New York
Sydney   Tokyo   Toronto

This book is printed on acid-free paper.

Copyright © 2000 by ACADEMIC PRESS

All Rights Reserved.
No part of this publication may be reproduced or transmitted in any form or by any means, electronic or mechanical, including photocopying, recording, or any information storage and retrieval system, without permission in writing from the publisher.

Academic Press
24–28 Oval Road, London NW1 7DX, UK
http://www.hbuk.co.uk/ap/

Academic Press
A Harcourt Science and Technology Company
525 B Street, Suite 1900, San Diego, California 92101–4495, USA
http://www.apnet.com

ISBN 0–12–013930–8

A catalogue for this book is available from the British Library

Typeset by Saxon Graphics Ltd, Derby, UK
Printed in Great Britain by Redwood Books Ltd, Trowbridge

00 01 02 03 04 05 MP 9 8 7 6 5 4 3 2 1

Contributors to Volume 30

R. AERTS, *Department of Systems Ecology, Vrije Universiteit Amsterdam, De Boelelaan 1087, NL-1081 HV Amsterdam, The Netherlands.*
M. AUBINET, *Faculty of Agricultural Science, Department of Physics, B-5030 Gembloux, Belgium.*
P. BERBIGIER, *INRA Bioclimatologie, BP81, F-33883 Villenave D'Ornon Cedex, France.*
CH. BERNHOFER, *Institute for Hydrology and Meteorology, Technical University of Dresden, D-01737 Tharandt, Germany.*
J.W.G. CAIRNEY, *Mycorrhiza Research Group, School of Science, University of Western Sydney, Nepean, PO Box 10, Kingswood, NSW 2747, Australia.*
F.S. CHAPIN III, *Institute of Arctic Biology, University of Alaska, Fairbanks, AK 99775, USA.*
R. CLEMENT, *Edinburgh University, Darwin Building, Mayfield Road, Edinburgh EH9 3JU, UK.*
J. ELBERS, *DLO Winand Staring Centre, NL-6700 AC, Wageningen, The Netherlands.*
T. FOKEN, *Department of Micrometeorology, University of Bayreuth, D-95440 Bayreuth, Germany.*
A. GRANIER, *INRA Unité d'Ecophysiologie Forestière, F-54280 Champenoux, France.*
A. GRELLE, *Department for Production Ecology, Swedish University of Agricultural Sciences, PO Box 7042, SE-750 07 Uppsala, Sweden.*
T. GRÜNWALD, *Institute for Hydrology and Meteorology, Technical University of Dresden, D-01737 Tharandt, Germany.*
A. IBROM, *Institute of Bioclimatology, Georg August University of Göttingen, D-37077 Göttingen, Germany.*
A.S. KOWALSKI, *Department of Biology, University of Antwerp (UIA), B-2610 Wilrijk, Belgium.*
P.H. MARTIN, *Institute for Systems, Informatics, and Safety, European Commission Joint Research Centre, TP650, I-21020 Ispra (VA), Italy.*
A.A. MEHARG, *Department of Plant and Soil Science; University of Aberdeen, Cruickshank Building, Aberdeen AB24 3UU, UK.*

H. MOLLER, *Department of Zoology, University of Otago, Private Bag, Dunedin, New Zealand.*

J. MONCRIEFF, *Darwin Building, Edinburgh University, Mayfield Road, Edinburgh EH9 3JU, UK.*

K. MORGENSTERN, *Institute of Bioclimatology, Georg August University of Göttingen, D-37077 Göttingen, Germany.*

K. PILEGAARD, *RISOE National Laboratory, DK-4000 Roskilde, Denmark.*

D. RAFFAELLI, *Culterty Field Station, University of Aberdeen, Newburgh, Ellon AB41 6AA, UK*

Ü. RANNIK, *Department of Physics, University of Helsinki, FIN-00014 Helsinki, Finland.*

C. REBMANN, *Max Planck Institute for Biogeochemistry, PO Box 10 01 64, D-07701 Jena, Germany.*

P.E. SCHMID, *School of Biological Sciences, Queen Mary and Westfield College, University of London, Mile End Road, London E1 4NS, UK.*

W. SNIJDERS, *DLO Winand Staring Centre, NL-6700 AC, Wageningen, The Netherlands.*

J.R. SPEAKMAN, *Aberdeen Centre for Energy Regulation and Obesity, Department of Zoology, University of Aberdeen, Aberdeen AB24 2TZ, UK.*

R. VALENTINI, *DISAFRI, University of Tuscia, I-01100 Viterbo, Italy.*

T. VESALA, *Department of Physics, University of Helsinki, FIN-00014 Helsinki, Finland.*

Preface

The six reviews in this latest volume of *Advances in Ecological Research* cover a broad spectrum of ecology, from micro-patterns and processes, to the ecophysiology of the individual organism, to forest-scale processes. Several of the papers deal with the possible evolutionary forces that have shaped particular strategies while others explore the potential and limitations for techniques in ecology, such as fractal geometry, field experiments and eddy covariance measures. Despite this diversity of topics, there are plenty of points of contact and cross-reference.

The first article by Aerts and Chapin explores the adaptive strategies behind the interactions between plants and nutrients, focusing on plant growth, nutrient acquisition, nutrient usage and nutrient cycling. They argue that the type of nutrient limitation is determined by a combination of nutrient biogeochemistry, disturbance and ecosystem management, with foliar N:P mass ratio being a good indicator of limitation type. Biomass allocation patterns seem to play a more important role than uptake kinetics in nutrient uptake capacity, with various strategies employed according to habitat nutrient status. Reducing nutrient loss may be more important than increasing acquisition in nutrient-poor habitats, and clear patterns with respect to nutrient use efficiency of plants in soils with different fertilities are difficult to discern. Finally, the relationships between litter chemistry, evergreen versus deciduous strategies and ecological scale are discussed.

Meharg and Cairney explore the co-evolution of mycorrhizal symbionts and their plant hosts in metal-contaminated soils. While there is an extensive literature on metal resistance in higher plants, mycorrhizal fungi have been neglected in this respect. The authors identify three main evolutionary strategies: mycorrhiza confer metal resistance on their host; the host does not rely on the fungal symbiont for resistance; and the combined resistance of both host and mycorrhizal symbiont confers enhanced resistance to the plant. The factors that promote co-evolution under these conditions are discussed and gaps in our knowledge identified.

Moving to a much larger ecological scale—the forest—Aubinet *et al.* describe the approach behind the EUROFLUX programme for estimating annual net carbon and water exchange over large spatial scales. The project was established to help fill gaps in our knowledge of fluxes and to formulate better predictive capabilities, a need highlighted by the "missing sink" problem, amongst others. The authors describe the rationale behind the use of eddy covariance for flux measurements which they have carried out across a range of latitudes, from the Mediterranean to the Arctic. The relevance of this approach for international protocols aimed at addressing the problem of increased carbon emissions is discussed.

In Speakman's review, we move on to the problems involved in estimating the energy expenditure of small mammals in the field. He reviews the findings of 185 field measurements made on 73 species using the doubly labelled water method, which he himself has pioneered. Field metabolic rate of small mammals was found to be dependent on body mass, ambient temperature and latitude of the study site, and these relationships are explored further, particularly with respect to phylogeny and diet.

The review by Raffaelli and Moller deals with the thorny question of the limitations of manipulative field experiments for detecting the effects of species interactions. The problems seem to be particularly acute for terrestrial systems, where the spatial and temporal scales demanded are often at odds with the resources available. These constraints are discussed in comparison with aquatic systems, where resource limitations are potentially less of a problem, but where the scales chosen to date may still be inappropriate. Finally, the potential of alternative approaches when resources are limiting, such as meta-analysis, are discussed.

Defining and describing aspects of habitat that are important to the associated biota is a difficult and challenging process, in large part because of the likely scale-dependency of the measurements. Fractal geometry provides a quantitative approach to this problem and is the subject of Schmid's review. The potential of the method is explored in relation to freshwater benthic substrates, but serves as a generic model for many other types of system. Different variants of the technique are discussed and data are presented linking these measures to organism abundance, with particular respect to patch structure.

Dave Raffaelli
Alastair Fitter

Contents

Contributors to Volume 30 v
Preface vii

The Mineral Nutrition of Wild Plants Revisited: A Re-evaluation of Processes and Patterns

R. AERTS AND F.S. CHAPIN III

I.	Summary 2	
II.	Introduction 3	
III.	Nutrient-limited Plant Growth: Which Nutrient is Limiting? 4	
	A. Differential Limitation by N and P 4	
	B. How to Detect N- or P-limited Plant Growth 7	
IV.	Nutrient Acquisition 9	
	A. Root Uptake 9	
	B. Mycorrhizal Uptake 11	
	C. Leaf Uptake and Loss 13	
	D. Symbiotic N Fixation 14	
	E. Rhizosphere Effects 14	
V.	Biomass Allocation in Relation to Nutrient Acquisition 15	
	A. Patterns in Biomass Allocation 16	
	B. Dependence on Nutrient Availability 18	
VI.	Nutrient Storage 19	
	A. The Concept of Storage 19	
	B. Variation among Growth-forms 19	
	C. Nutrient Dependency 20	
VII.	Nutritional Aspects of Leaf Traits 21	
	A. Leaf Nutrient Concentrations 21	
	B. Nutrient Resorption from Senescing Leaves 24	
	C. Leaf Lifespan 26	
VIII.	Nutrient Use for Biomass Production 28	
	A. Nutrient Use Efficiency (NUE) 28	
	B. Patterns in NUE 31	
IX.	Patterns in Leaf-level Nutrient Use Efficiency and its Components 34	
	A. Leaf Traits and their Contribution to Leaf-level NUE: Theoretical Considerations 34	

		B.	Actual patterns in Leaf-level Nutrient Use Efficiency and the Relation with Underlying Leaf Traits 37
		C.	Physiological Constraints on Maximization of Leaf-level NUE 42
		D.	Ecological Consequences 44
	X.	Litter Decomposition 45
		A.	Nutrient Leaching from Senesced Leaves 45
		B.	Climatological and Chemical Controls on Litter Decomposition ...47
		C.	Variation in Litter Decomposition among Growth-forms 49
	XI.	Trade-off Between Nutrient Use Efficiency and Litter Decomposability? 51
	XII.	Conclusions: Plant Strategies 53
References 55

Co-evolution of Mycorrhizal Symbionts and their Hosts to Metal-contaminated Environments

A.A. MEHARG AND J.W.G. CAIRNEY

I.	Summary 70
II.	Introduction 71
	A. Role of Mycorrhiza on Metal-contaminated Soils 71
	B. Review Outline 72
III.	Arbuscular Mycorrhizas 73
	A. Arbuscular Mycorrhiza and Metal-contaminated Sites 73
	B. Species Diversity 76
	C. Spore Germination 76
	D. Germ Tube Growth 79
	E. Hyphal Penetration 80
	F. Root Colonization 81
	G. Spore Production 82
	H. Metal Assimilation in Plant Tissues 82
	I. Plant Sensitivity 85
	J. Nutritional Aspects 86
	K. Do Plants Benefit from being Colonized with Resistant Arbuscular Mycorrhizal Strains? 87
IV.	Ericoid Mycorrhizas 88
	A. The Ericaceae and Metal-contaminated Environments 88
	B. Copper 89
	C. Zinc 93
	D. Role of Mycorrhizas in Metal Resistance of the Ericaceae 95
V.	Ectomycorrhizal Fungi 95
	A. Introduction 95

		B.	General Response of Ectomycorrhizal Associations to Raised Levels of Metals 95
		C.	Genetic Adaptation of Trees to Metal Contaminants 98
		D.	Intraspecific Variation in Metal Resistance in Ectomycorrhizal Fungi..99
		E.	Studies Conducted with Hosts Colonized with Resistant and Sensitive Strains 102
		F.	Role of Ectomycorrhizal Fungi in Facilitating Host Metal Resistance 103
	VI.	Conclusions..104	
		A.	Current State of Knowledge 104
		B.	Future Research Requirements 105
References 107			

Estimates of the Annual Net Carbon and Water Exchange of Forests: The EUROFLUX Methodology

M. AUBINET, A. GRELLE, A. IBROM, Ü. RANNIK, J. MONCRIEFF,
T. FOKEN, A.S. KOWALSKI, P.H. MARTIN, P. BERBIGIER, Ch. BERNHOFER,
R. CLEMENT, J. ELBERS, A. GRANIER, T. GRÜNWALD,
K. MORGENSTERN, K. PILEGAARD, C. REBMANN, W. SNIJDERS,
R. VALENTINI AND T. VESALA

I.	Introduction 114	
II.	Theory 116	
III.	The Eddy Covariance System 119	
	A.	Sonic Anemometer 119
	B.	Temperature Fluctuation Measurements 120
	C.	Infrared Gas Analyser 120
	D.	Air Transport System 123
	E.	Tower Instrumentation 126
IV.	Additional Measurements 127	
V.	Data Acquisition: Computation and Correction 127	
	A.	General Procedure 127
	B.	Half-hourly Means, (Co-)variances and Uncorrected Fluxes 130
	C.	Intercomparison of Software 134
	D.	Correction for Frequency Response Losses 136
VI.	Quality control 144	
	A.	Raw Data Analysis 145
	B.	Stationarity Test 145
	C.	Integral Turbulence Test 146
	D.	Energy Balance Closure 147
VII.	Spatial Representativeness of Measured Fluxes 154	

VIII.	Summation Procedure		156
IX.	Data Gap Filling		158
	A.	Interpolation and Parameterization	158
	B.	Neural Networks	158
X.	Corrections to Night-time Data		162
XI.	Error Estimation		164
XII.	Conclusions		167

Acknowledgements 168
References 168
Appendix A 173
Appendix B 175

The Cost of Living: Field Metabolic Rates of Small Mammals

J.R. SPEAKMAN

I.	Summary		178
II.	Introduction		179
	A.	The Importance of Energy in Living Systems	179
	B.	Limitations on Animal Energy Expenditure	181
	C.	The Extrinsic Limitation Hypothesis	189
	D.	The Intrinsic Limitation Hypothesis	192
	E.	Experimental Studies of the Limitation Hypotheses	197
	F.	The Central Limitation Hypothesis and Links between FMR and RMR	200
	G.	Interspecific Reviews of the Link between DEE and RMR	202
	H.	Summary and Aims	204
III.	Methods		205
	A.	Measuring Energy Expenditure by Indirect Calorimetry	205
	B.	Basal and Resting Energy Expenditure	206
	C.	Time and Energy Budget Estimates of Daily Energy Expenditure	209
	D.	Direct Measurements of Free-living Energy Expenditure	212
	E.	Summary and Data Inclusion Criteria for the Present Review	217
IV.	Results		218
	A.	Overview of the Database	218
	B.	Factors Influencing Daily Energy Expenditure	221
V.	Discussion		242
	A.	Links between FMR and RMR	242
	B.	Sustainable Metabolic Scope	255

Acknowledgements 264
References 264
Appendix A 283
Appendix B 289
Appendix C 294

Manipulative Field Experiments in Animal Ecology: Do They Promise More Than They Can Deliver?

D. RAFFAELLI AND H. MOLLER

I.	Summary	299
II.	Introduction	300
III.	Designs of Published Field Experiments	302
	A. Plot Sizes, Replication and Duration	302
	B. Why Are Some Experimental Designs So Weak?	312
	C. For How Long Should Experiments Run?	320
	D. Issues of Scale	324
	E. Meta-analysis: A Way Forward?	327
IV.	Conclusions	328
Acknowledgements		330
References		330

Fractal Properties of Habitat and Patch Structure in Benthic Ecosystems

P.E. SCHMID

I.	Summary	339
II.	Introduction	341
III.	Fractal Dimension and its Measurement	342
	A. Fractal Dimension of Boundary and Profile Lines	342
	B. Fractal Dimension of Deposition Processes	348
	C. Fractal Dimension of Surfaces	349
IV.	Fractal Properties in Riverine Ecosystems	358
	A. Large Spatial Scales	358
	B. Intermediate to Small Spatial Scales	363
	C. Fractal Dimension Across Several Spatial Scales	373
V.	Fractal Random Walks	380
VI.	Fractal Properties of Biotic Structures	387
	A. Coral Reefs	387
	B. Mussel Beds	388
	C. Colonization of Artificial Pond Weeds	388
VII.	Fractal Coexistence and Species Abundance Distribution	390
VIII.	Concluding Remarks	395
Acknowledgements		396
References		397
Appendix A		401
Appendix B		401
Cumulative List of Titles		403
Index		407

The Mineral Nutrition of Wild Plants Revisited: A Re-evaluation of Processes and Patterns

R. AERTS AND F.S. CHAPIN III

I.	Summary	2
II.	Introduction	3
III.	Nutrient-limited Plant Growth: Which Nutrient is Limiting?	4
	A. Differential Limitation by N and P	4
	B. How to Detect N- or P-limited Plant Growth	7
IV.	Nutrient Acquisition	9
	A. Root Uptake	9
	B. Mycorrhizal Uptake	11
	C. Leaf Uptake and Loss	13
	D. Symbiotic N Fixation	14
	E. Rhizosphere Effects	14
V.	Biomass Allocation in Relation to Nutrient Acquisition	15
	A. Patterns in Biomass Allocation	16
	B. Dependence on Nutrient Availability	18
VI.	Nutrient Storage	19
	A. The Concept of Storage	19
	B. Variation among Growth-forms	19
	C. Nutrient Dependency	20
VII.	Nutritional Aspects of Leaf Traits	21
	A. Leaf Nutrient Concentrations	21
	B. Nutrient Resorption from Senescing Leaves	24
	C. Leaf Lifespan	26
VIII.	Nutrient Use for Biomass Production	28
	A. Nutrient Use Efficiency (NUE)	28
	B. Patterns in NUE	31
IX.	Patterns in Leaf-level Nutrient Use Efficiency and its Components	34
	A. Leaf Traits and their Contribution to Leaf-level NUE: Theoretical Considerations	34
	B. Actual Patterns in Leaf-level Nutrient Use Efficiency and the Relation with Underlying Leaf Traits	37
	C. Physiological Constraints on Maximization of Leaf-level NUE	42
	D. Ecological Consequences	44
X.	Litter Decomposition	45

	A. Nutrient Leaching from Senesced Leaves 45
	B. Climatological and Chemical Controls on Litter Decomposition 47
	C. Variation in Litter Decomposition among Growth-forms 49
XI.	Trade-off Between Nutrient Use Efficiency and Litter Decomposability? .51
XII.	Conclusions: Plant Strategies 53
References 55

I. SUMMARY

This review considers four main topics in the nutritional ecology of wild plants: nutrient-limited plant growth, nutrient acquisition, nutrient use efficiency, and nutrient recycling through litter decomposition. Within each of these four areas, plants have evolved major adaptive strategies with respect to their nutritional ecology.

In most terrestrial ecosystems, plant growth is N-limited but P-limitation also occurs frequently. The type of limitation is determined by differences in biogeochemistry between N and P, by anthropogenic disturbances such as increased atmospheric N deposition and other forms of eutrophication, and by the effects of ecosystem management on N and P cycles. The foliar N : P mass ratio is a good indicator of the type of limitation: plant growth is N-limited at N : P ratios < 14, P-limited at N : P ratios > 16, and co-limited by N and P at intermediate values.

Variation in nutrient uptake capacity is, especially for immobile ions such as phosphate, determined more by biomass allocation patterns than by uptake kinetics. However, species from nutrient-poor habitats do not necessarily allocate more biomass to their roots than species from more fertile environments. These species may compensate for low biomass allocation to the roots by having a high specific root length (SRL: root length per unit root mass). The uptake of organic nitrogen compounds by both mycorrhizal and non-mycorrhizal plants is a novel aspect of the terrestrial N cycle. The ability of plants to use this 'short-cut' of the N cycle may be of great adaptive significance in nitrogen-poor habitats, because it gives some plants access to a nitrogen source of which other species are deprived.

Whole-plant nutrient use efficiency (NUE) is generally measured as productivity per unit nutrient uptake or loss. It is shown that in nutrient-poor environments selection is on traits that reduce nutrient losses rather than on traits conferring a high NUE *per se*. Low tissue nutrient concentrations and low tissue turnover rates, characteristic traits of evergreens, are the most important determinants of high nutrient retention in nutrient-poor environments. High nutrient resorption efficiency, however, is important in all species and does not differ consistently between species from nutrient-poor and nutrient-rich environments. Due to selection on the components of NUE rather than on NUE itself, there are no clear patterns in whole-plant NUE when comparing species from environments differing in soil fertility. At the phenotypic level, however, NUE decreases with increasing soil fertility.

A theoretical and experimental analysis of the relation between leaf-level NUE and underlying leaf traits showed that, for woody species, leaf-level nitrogen use efficiency (NUE_N) is most strongly determined by variation in mature leaf N concentration. For herbaceous species, however, N resorption efficiency is the most important determinant of NUE_N. For phosphorus use efficiency (NUE_P), P resorption efficiency contributes most strongly to maximization of NUE_P in all growth-forms. This occurs because maximum P resorption efficiency is higher than maximum N resorption efficiency. In addition, at high resorption efficiencies (r) NUE is disproportionally increased by small increases of r. In all growth-forms, leaf lifespan is only a minor contributor to variation in both leaf-level NUE_N and NUE_P. This is not in agreement with the pattern at the whole-plant level. Evergreen species have higher leaf-level NUE_N and NUE_P than other growth-forms.

Litter decomposition is a key process in the nutrient cycles of most terrestrial ecosystems. At large geographical scales, litter decomposition is determined mainly by climatic factors. At a regional scale, however, litter chemistry is the most important determinant of litter decomposability. In most climatic regions, the leaf litter of evergreen shrubs and trees decomposes slower than that of deciduous shrubs and trees. This is due mainly to the low nutrient concentrations in evergreen leaves and the high concentrations of secondary compounds. This implies that the plant characteristics of evergreens do not only lead to high NUE, but also keep soil fertility low and thereby influence the competitive balance with deciduous species in their favour.

II. INTRODUCTION

The mineral nutrition of plants is a central topic of plant ecology. For decades, ecologists have investigated the various adaptations of plant species to different levels of nutrient availability in their natural habitat (e.g. Grime, 1979; Chapin, 1980; Grime *et al.*, 1997). The adaptations to low levels of soil fertility have received particular attention. In 1980, the well-known, and extensively cited paper 'The mineral nutrition of wild plants' was published in the *Annual Review of Ecology and Systematics* (Chapin, 1980). Since then, almost two decades have elapsed, and our ecological knowledge about most of the processes and patterns described in that paper has increased steadily. However, at present there is no paper in which the ideas and hypotheses put forward in that paper (which had a small factual basis) have been critically re-evaluated and subjected to rigorous statistical testing. Here, we review the advances that have been made in understanding the ecology of the mineral nutrition of wild plants from terrestrial ecosystems.

This review is based largely on an analysis of the extensive databases we have collected on various aspects of the mineral nutrition of wild plants. These databases are used to study (quantitatively and qualitatively) the interrelations among various parameters related to mineral nutrition. We will do this both with respect to variation among growth-forms and to variation in soil nutrient availability. Thus, a substantial part of the review consists of a quantitative evaluation of various aspects of the mineral nutrition of perennials originating from habitats differing in soil fertility with special emphasis on the 'strategies' of different growth-forms and their ecological implications. We will approach the concept of strategies with regard to mineral nutrition by means of the nutrient use efficiency (NUE) of species (see Vitousek, 1982; Berendse and Aerts, 1987). This review focuses on N and P, because these nutrients are the main growth-limiting nutrients for plants in natural environments (Vitousek and Howarth, 1991; Koerselman and Meuleman, 1996; Verhoeven *et al.*, 1996).

Our review is organized along three lines. First, we treat the issues of nutrient-limited plant growth and nutrient uptake, with special emphasis on the importance of the uptake of nutrients in organic form (both by mycorrhizal and by non-mycorrhizal plants) and the importance of symbiotic nitrogen fixation. In addition, we describe the influence of allocation patterns on mineral nutrient uptake. Next, we explore some of the nutritional aspects of leaf functioning and how nutrients are used for biomass production by the plant. We do that by studying the NUE of plants and the various components of NUE. Finally, we investigate the feedback of plant species to soil nutrient availability by reviewing patterns in litter decomposition and nutrient mineralization. The review concludes with a synthesis of the various aspects of the mineral nutrition of wild plants. To that end, we present a conceptual description of plant strategies with respect to mineral nutrition.

III. NUTRIENT-LIMITED PLANT GROWTH: WHICH NUTRIENT IS LIMITING?

A. Differential Limitation by N and P

By definition, mineral nutrients have specific and essential functions in plant metabolism: nitrogen is an important constituent of proteins and thus plays an essential role in all enzymatic activity, whereas phosphorus is involved in the energy transfer in the cell (adenosine triphosphate (ATP), reduced nicotinamide adenine dinucleotide phosphate (NADPH)) and, together with nitrogen, is an important structural element in nucleic acids (Marschner, 1995). Both elements are required for plant growth in relatively large quantities and are therefore classified as macronutrients. As potassium, the other macronutrient, limits plant growth only infrequently (Vitousek and Howarth, 1991; Koerselman and Meuleman, 1996), this element will not be discussed here. Empirical data shown that a N : P mass ratio in leaves of about 10 is optimal for plant growth (Van den Driessche, 1974; Ingestad, 1979; Lajtha and Klein, 1988), although this ratio varies considerably among species. This suggests that plants should absorb (on a mass basis) about 10 times more N than P to promote balanced plant growth. Deviations from this ratio should, therefore, lead to N- or P-limited plant growth (see below). Nutrient limitation can be evaluated at the level of individual species or at the community level. It is important to distinguish between these levels, because in multispecies communities plant species may coexist that are differentially limited by N and P (DiTomasso and Aarssen, 1989; Koerselman and Meuleman, 1996), although the cause of differential nutrient limitation (N versus P) is not well understood.

At first sight, it seems surprising that plant growth in most non-tropical terrestrial ecosystems is N-limited (DiTomasso and Aarssen, 1989; Bridgham *et al.*, 1995; Shaver and Chapin, 1995; Wassen *et al.*, 1995) despite the widespread occurrence of biological N_2- fixation (Vitousek and Howarth, 1991). Vitousek and Howarth (1991) explain the relatively minor contribution of N_2-fixation to N supply in temperate terrestrial ecosystems by energetic constraints on the colonization or activity of nitrogen fixers or by limitation of N- fixers by another nutrient (phosphorus, molybdenum, iron) which would then represent the ultimate limiting factor to net primary production. Ecosystem N budget studies do indeed show that N_2- fixation is only a minor contributor to the total N supply for plant growth in late-successional ecosystems (Hemond, 1983; Morris, 1991; Koerselman and Verhoeven, 1992).

The reasons for differential nutrient limitation include differences in biogeochemistry between N and P (Vitousek and Howarth, 1991), anthropogenic disturbances such as increased atmospheric N deposition and other forms of eutrophication (e.g. Aerts *et al.*, 1992a), and effects of ecosystem management on N and P cycles (e.g. Koerselman and Verhoeven, 1992).

The biogeochemical cycles of N and P show many differences. First, they differ in their ultimate source. Nitrogen derives primarily from the atmosphere and phosphorus from rock weathering. As a result, nitrogen is nearly absent from new soils and does then limit Net Primary Production (NPP), whereas it gradually accumulates in the soil during later succession (Tilman, 1986; Vitousek *et al.*, 1987; Berendse, 1990). In contrast, the amount and availability of P decline during long-term soil development and eventually lead to extremely P-deficient soils, as in many tropical soils (Vitousek and Sanford, 1986), soils under Australian and South African–Mediterranean-type vegetation (Specht and Rundel, 1990; Cowling, 1993), and glacial and aeolean sandy soils under north-west European heathlands (Aerts and Heil, 1993). These opposing trends suggest that during long-term ecosystem development, plant growth should shift from N to P limitation. Second, nitrogen (especially nitrate) is much more mobile in the soil, which promotes nitrogen leaching.

Moreover, nitrogen can easily move across ecosystem boundaries in gaseous form, especially in systems where substantial denitrification occurs or in terrestrial ecosystems with frequent fires. Depending on the magnitude of the N losses due to leaching, denitrification and fire, ecosystems may remain N-limited over long periods of time. Third, detrital N is mostly carbon-bonded whereas detrital P is mostly ester-bonded and often soluble. Due to the production of extracellular phosphatases, this phosphorus may quickly become available for plant uptake again (Hunt *et al.*, 1983; Howarth, 1988) whereas carbon-bonded N may be immobilized for a long time. This would promote N-limitation.

Differential nutrient limitation is caused not only by inherent differences in the biogeochemistry of N and P, but also by differences or changes in external inputs. Atmospheric nitrogen deposition is an important N input to many terrestrial ecosystems in north-west Europe and has a substantial impact on nutrient cycling in these ecosystems (Aerts and Heil, 1993). Human activities have now doubled the annual amount of N sequestered by terrestrial and aquatic ecosystems (Vitousek, 1994). Long-term exposure of ecosystems to high loads of atmospheric nitrogen will lead to an increase in the N : P mass ratio in plant tissues, and this may lead to changes from N-limited to P-limited plant growth (Aerts and Bobbink, 1999). This has been investigated in ombrotrophic raised bogs in Sweden, where strong temporal and spatial gradients in N deposition exist (Malmer, 1988, 1990; Aerts *et al.*, 1992a). These bogs, which are dominated by *Sphagnum* mosses, receive their mineral nutrients solely by atmospheric deposition and are therefore very suitable for monitoring the effects of N deposition on plant growth (Woodin and Lee, 1987; Malmer, 1988; Aerts *et al.*, 1992a). The increase of atmospheric N deposition in southern Sweden which has occurred during the past few decades has indeed resulted in strong increases in the N : P mass ratio in *Sphagnum* mosses (Malmer, 1990). Moreover, fertilization experiments by Aerts *et al.* (1992a) in *Sphagnum*-dominated bogs in Swedish Lapland (low N deposition) and in southern Sweden (high N deposition) showed that plant growth in the low N deposition area was N-limited, whereas in the high N deposition area it had become P-limited. Such a change of the primary element limitation on plant growth from N to P has also been observed in Dutch and Danish heathlands (Aerts and Berendse, 1988; Riis-Nielsen, 1997) and in mesotrophic fens in the Netherlands (Verhoeven and Schmitz, 1991).

Differential limitation by N and P may also be caused by ecosystem management. In north-western Europe, many species-rich fens are managed to preserve the high plant species diversity. The management usually consists of hay-making in summer and removal of the mown biomass (and the nutrients contained therein). Nutrient-budget studies of Dutch fens with a harvest regime have shown that this results in a substantial net P loss from these systems and a relatively small loss of N (Koerselman *et al.*, 1990). The overall result is that plant growth in fens with a long history of mowing is P-limited, whereas it is N-limited in fens that are very infrequently mown or where mowing has started only recently (Verhoeven and Schmitz, 1991).

B. How to Detect N- or P-limited Plant Growth

It is common practice to study the nature of nutrient limitation in natural plant communities by factorial fertilizer experiments with the macronutrients N, P and K (Chapin *et al.*, 1986b; Boeye *et al.*, 1997). This is a straightforward method for studying nutrient limitation, but there are some associated problems. First, these experiments are time-consuming, laborious and impose some type of disturbance on the study site. Furthermore, due to specific site conditions, interpretation of the results may be difficult. For example, in chalk grasslands, most of the supplied P will be bound to calcium in the soil and will not be taken up by the vegetation. For this reason no growth response may be recorded after P addition, despite the fact that plant growth in such a community may be severely P-limited (Bobbink, 1991, 1992). Another example is provided by fertilization experiments by Shaver and Chapin (1995) in Alaskan tundra. Despite high N and P supply rates, recovery of the supplied N and P in plant biomass was extremely low. The authors suggested that immobilization in the upper moss layer and microbial immobilization played a major role in controlling the fertilizer response of vascular arctic tundra vegetation. Thus, chemical adsorption and microbial immobilization of the supplied nutrients may strongly interfere with the plant responses to nutrient additions.

Recently, Koerselman and Meuleman (1996) proposed to use the N : P mass ratio in plant tissues as an indicator of the type of nutrient limitation. Although this idea is certainly not new (see Redfield, 1958; Van den Driessche, 1974; Ingestad, 1979; Aerts *et al.*, 1992a), a strong factual basis for this idea was provided. Koerselman and Meuleman (1996) collected data from 40 fertilization studies conducted in European wetland ecosystems (bogs, fens, wet heathlands, dune slacks, wet grasslands). They plotted the P-concentration against the N-concentration in above-ground bulk samples from the unfertilized controls of these experiments and indicated which type of limitation was found in the fertilization studies (Figure 1). The vegetation N : P ratio clearly discriminates between N- and P-limited sites. At N : P ratios > 16, the community biomass production was P-limited, whereas at N : P ratios < 14 there was N-limited plant growth in all but one study. At N : P ratios between 14 and 16, co-limitation by N and P occurred. In their analysis N concentrations varied more than 3-fold, whereas P concentrations varied more than 16-fold. Koerselman and Meuleman argued that this huge variation probably reflects differences in the supply ratio of N and P rather than differences in absolute N and P availability. Figure 1 also shows that there is no clear

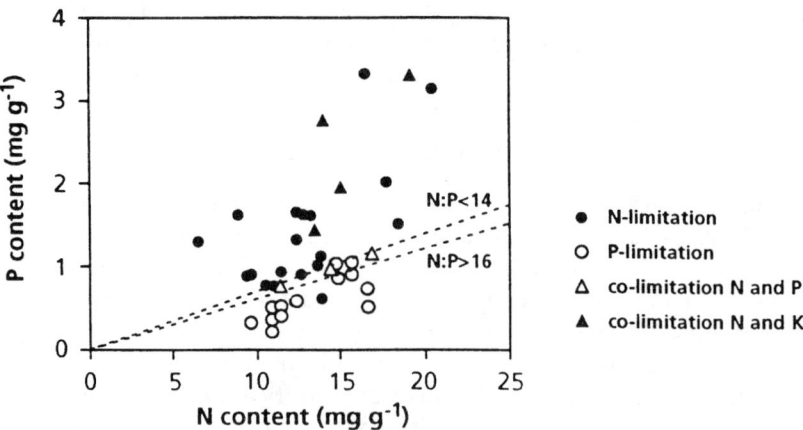

Fig. 1. Relationship between vegetation N and P content and the nature of nutrient limitation in 40 European wetlands (bogs, poor fens, rich fens, wet heathlands, wet grasslands and dune slacks). Data points are from studies in which fertilization experiments showed which type of nutrient limitation occurred. The nutrient contents shown were determined in unfertilized control plots. Dashed lines depict N : P mass ratios of 14 and 16, respectively. Redrawn from Koerselman and Meuleman (1996).

relationship between the nutrient concentration in plant tissue and the nature of nutrient limitation. For example, in wet grasslands nitrogen was limiting plant growth at relatively high concentrations of about 20 mg N g^{-1}, whereas it did not limit plant growth in wet heathlands and rich fens at concentrations of around only 10 mg N g^{-1}. Thus, in the data analysed by Koerselman and Meuleman (1996), it really is the ratio between N and P that controls plant growth, not the absolute concentration in plant tissue. However, in some cases very low concentrations of N and/or P may limit plant growth despite the fact that the N : P ratio may be favourable.

Use of the N : P ratio is a meaningful and easy first approach to the determination of nutrient limitation in plant species and plant communities. However, use of this parameter does not solve all problems involved in the determination of the type of nutrient limitation. The first problem with the analysis by Koerselman and Meuleman (1996) is that they apparently assume that the growth of plant species or vegetation is always limited by N and/or P. However, when both nutrients are in ample supply, the N : P mass ratio will still indicate a type of nutrient limitation, despite the fact that this is clearly not the case. A second point is that it is necessary to differentiate between nutrient limitation at the community level and limitation at the species level. Koerselman and Meuleman (1996) present a nice example in which it is shown that in several communities the N : P mass ratio of individual species is so different from that of the community as a whole that it must be concluded that the growth of these species is controlled by an element that does not control community biomass production. This conclusion is supported by experimental data reviewed by DiTomasso and Aarssen (1989).

IV. NUTRIENT ACQUISITION

A. Root Uptake

The rate of nutrient acquisition by a plant is a function of the rate of uptake per gram of biomass, its total biomass, and the proportion of biomass allocated to nutrient-absorbing tissues. In this section we consider only the first of these factors. Allocation will be treated in the next section. Nutrient uptake is a complex process that involves (1) nutrient movement to the root surface, (2) the transport of nutrients from the root surface into the plant, (3) the capacity to form symbiotic linkages with mycorrhizal fungi or nitrogen-fixing bacteria and (4) modification of the soil environment to enhance nutrient supply. Each of these processes is strongly influenced by the physical relationship between roots and soil, making it difficult to determine which processes and traits actually control nutrient acquisition under natural conditions. We arbitrarily separate the plant effects on these processes in terms of physiological uptake by roots or leaves, rhizosphere effects, and mycorrhizal and N-fixing symbioses.

1. Patterns of Nutrient Uptake

Plants adapted to infertile soils typically have a high capacity (high V_{max}) to absorb mobile ions such as potassium (Veerkamp and Kuiper, 1982) but a relatively lower capacity to absorb immobile ions like phosphate (Chapin et al., 1986a; Raab et al., 1998). Presumably a low uptake capacity for immobile ions is not strongly disadvantageous in a low-nutrient environment because diffusion so strongly limits uptake in this situation that uptake capacity has little influence on nutrient uptake, and because mycorrhizal transfer may be more important than root uptake. Diffusion is less limiting to mobile ions, so there is selection for a high capacity to absorb these ions in plants adapted to infertile soils (Chapin, 1988).

Nitrogen differs from other nutrients in that it occurs as multiple forms in the soil: nitrate, ammonium and organic N. Most plants absorb any form of soluble nitrogen that is available in the soil, especially if acclimated to its presence (Atkin, 1996). However, since plants must synthesize different carrier proteins to absorb each N form, they differ in their relative preference for different N forms, depending on the relative supply of ions in the soil. For example, arctic and alpine plants, which experience high amino acid concentrations in soil and where N mineralization and nitrification are strongly constrained by low temperature, preferentially absorb and grow on amino acids, whereas barley preferentially absorbs inorganic nitrogen (Chapin et al., 1993; Kielland, 1994; Raab et al., 1996). Spruce, which also grows on acid soils, preferentially absorbs ammonium over nitrate (Kronzucker et al., 1997), with untested capacity to absorb amino acids. Boreal forest herbs and shrubs also preferentially absorb glycine (an amino acid) over inorganic N in field experiments in Sweden, regardless of mycorrhizal status (Näsholm et al., 1998). In general, species from habitats with high nitrate availability (e.g. calcareous grasslands) preferentially absorb nitrate and have higher capacities to reduce nitrate than do species from low-nitrate habitats. Conversely, many plant species from habitats where ammonium is the dominant form of available nitrogen preferentially absorb ammonium relative to nitrate. Until recently, few studies have examined the capacity of plants to absorb organic N because it was assumed that microbes would outcompete plants for this N form. However, since N is typically transformed from insoluble organic N to soluble organic N to ammonium to nitrate, with some uptake of these forms by plants and/or microbes at each step, the supply rate in any soil must be in the order: soluble organic N $\geq$ ammonium-N $\geq$ nitrate-N (Eviner and Chapin, 1997). Thus the potential of plants to absorb soluble organic N may be much more important than previously appreciated (see section IV.B).

Plants can also tap organic P through production of phosphatases by roots. Phosphatase production is enhanced by a low P supply to the plants. The capacity to use organic P varies among species and also depends on soil conditions. It may range from almost zero to a capacity sufficient to explain most of the P acquisition by the plant (Kroehler and Linkins, 1991; Hübel and Beck, 1993). However, at present it is not clear whether this high potential uptake of organic P sources is also realized under field conditions.

2. Dependence on Nutrient Availability

Any factor that increases plant demand for a particular nutrient increases the plant's capacity to absorb that nutrient. Thus, high availability of light, warm temperature, and favourable moisture all lead to a high uptake capacity by a given genotype. With respect to nutrients, this adjustment in uptake capacity is specific to the nutrient that limits growth: nitrogen stress increases the potential to absorb nitrate or ammonium but decreases the potential to absorb other non-limiting nutrients, whereas phosphorus stress increases the capacity to absorb phosphate but reduces the capacity to absorb other ions (Lee, 1982) (Table 1). The situation with nitrate uptake is more complex than that of other ions, because plants must both absorb and assimilate nitrate (i.e. convert it to ammonium) before it becomes nutritionally useful. Plants exhibit a low constitutive level of nitrate uptake when grown in the complete

Table 1
Effect of environmental stresses on nutrient uptake rate (from Chapin, 1991).

Stress	Ion absorbed	Uptake rate by stressed plant (% of control)
Nitrogen	Ammonium	209
	Nitrate	206
	Phosphate	56
	Sulfate	56
Phosphorus	Phosphate	400
	Nitrate	35
	Sulfate	70
Sulfur	Sulfate	895
	Nitrate	69
	Phosphate	32
Water	Phosphate	13
Light	Nitrate	73

Values are for barley, except for water stress (tomato).

absence of nitrate; this uptake rate increases upon exposure to nitrate as a result of induction of synthesis of more carriers and of nitrate reductase (Huang et al., 1996).

Plants not only increase the maximum rate at which they acquire a nutrient under conditions of low supply but also reduce the leakage of nutrients out of the root, enabling them to acquire nutrients at lower external concentrations (Kronzucker et al., 1997). Efflux may actually be more important than influx in explaining differences in net nutrient uptake by plants (Kronzucker et al., 1997). Thus, plants are able to reduce external concentrations to lower levels under conditions of nutrient stress. This is consistent with Tilman's (1988) R^* concept of competition, in which plants are presumed to compete effectively with their neighbours by reducing the concentration of a resource to concentrations lower than can be acquired by their neighbours. The capacity of plants to alter uptake capacity in response to growth at different nutrient levels is best developed in plants with high relative growth rate (Veerkamp and Kuiper, 1982; Chapin et al., 1986a), which is consistent with Grime's (1979) concept that rapidly growing plants have a high capacity to acquire nutrients.

B. Mycorrhizal Uptake

Most plants in both natural ecosystems and croplands are mycorrhizal and probably acquire a large proportion of their nutrients via mycorrhizas. However, a few plant species, especially those that predominate in disturbed or fertile soils (e.g. Brassicaceae), do not form mycorrhizal associations (Allen, 1991; Smith and Read, 1997). Mycorrhizas are thought to acquire nutrients at a lower carbon cost than roots because of their smaller diameter and greater surface : volume ratio. Arbuscular mycorrhizas (AMs), which are generally associated with herbaceous plants or tropical trees, are critically important for the uptake of nutrients that diffuse slowly in soil, especially P, although mycorrhizas have also been implicated in absorption of NH_4^+ and other nutrients (Smith and Read, 1997). When mycorrhizal associations are experimentally suppressed in N-limited grasslands, plant production can become P-limited, suggesting that effective P uptake by mycorrhizas may be an important factor explaining the widespread N-limitation of temperate ecosystems (Grogan and Chapin, in press).

In recent years, it has become clear that the uptake of organic nitrogen compounds by both mycorrhizal and non-mycorrhizal plants is an important pathway in the terrestrial nitrogen cycle (Read, 1991; Chapin et al., 1993; Kielland, 1994; Northup et al., 1995; Turnbull et al., 1995). The ability of plants to use this 'short-cut' of the N cycle may be of great adaptive significance in nutrient-poor habitats, because it gives plants access to a nitrogen source of which other species are deprived. However, in temperate ecosystems the ability to take up organic N sources is restricted mainly to plants with ericoid mycorrhizas (EMs) and ectomycorrhizas (ECMs), and hardly occurs in species with arbuscular mycorrhizas (AMs) and in non-mycorrhizal plants (Smith and Read, 1997). The litter of EM plants usually has higher concentrations of secondary compounds than litter from AM and non-mycorrhizal plants, which may retard N mineralization and thus decrease the availability or inorganic N in the soil (Aerts, 1997b). It has been hypothesized that the use of differential nitrogen sources by the different mycorrhizal types may create positive feedbacks between plant species dominance, litter chemistry and mycorrhizal type. However, until now there has been hardly any field evidence for this hypothesis. Heathlands are very suitable ecosystems for investigating the ecological significance of differential uptake of organic and inorganic nitrogen sources. In nutrient-poor heathlands, ericoid species (in north-west Europe: *Erica tetralix, Calluna vulgaris* and *Empetrum nigrum*) predominate (Aerts and Heil, 1993). These ericoid mycorrhizal species have the ability to use organic N sources for their mineral nutrition, thus making them less dependent on mineralization of organic matter (Read, 1991). In these heathlands, the vegetation also contains grasses such as *Deschampsia flexuosa* and *Molinia caerulea* which have a high competitive ability only at high levels of nutrient availability. These species, with AMs, have no or only a very limited capacity to utilize organic N sources. This is a strong disadvantage under nutrient-poor conditions. This fascinating mechanism of species coexistence as a result of differential use of soil N sources may be disrupted as a result of increased levels of atmospheric N deposition (Aerts and Bobbink, 1999). This results in both higher availability of inorganic N and in an increase in the ratio between inorganic and organic N in the soil. This may affect the degree of EM colonization, thus depriving the ericoid species of their relative advantage in nutrient-poor soils, and may increase the competitive ability of the grasses, because they can now utilize the inorganic N sources. Clearly, investigation of this type of species interaction may significantly contribute to our understanding of the regulation of species distribution over soil fertility gradients.

Plants reduce their degree of mycorrhizal infection under conditions of high nutrient supply (reviewed in Aerts and Bobbink, 1999), because under these circumstances plants can readily meet their nutrient requirements through direct uptake by roots without the additional carbon expenditure required to support mycorrhizal fungi. This explains why non-mycorrhizal genotypes grow more rapidly then mycorrhizal ones in fertile soils but grow quite slowly in low-P soils (Koide, 1991).

C. Leaf Uptake and Loss

Leaves, like roots, can both absorb and lose nutrients to solutions on the leaf surface, primarily by movement through the stomata (Sutton *et al.*, 1993). Therefore, uptake rates by leaves are greatest in species that have high stomatal conductance (generally rapidly growing species with a high Specific Leaf Area—leaf area per unit leaf mass). Environmental conditions that favour high stomatal conductance (e.g. high light and water supply) also favour high uptake rates by leaves (Sutton *et al.*, 1993). The effect of nitrogen status of leaves is variable among studies. Leaves with high N status have high stomatal conductance (promoting uptake) but also tend to lose NH_3 more readily through stomata (see below).

Leaves with high nutrient concentrations lose more nutrients than those with low tissue concentrations, in either gaseous form (NH_3) or as solutions. The susceptibility to leaching loss is greatest for monovalent cations (Na, K), less for divalent cations, and least for organically bound nutrients such as N and P (Tukey, 1970; Chapin and Moilanen, 1991). Nutrient loss by leaching is greatest when water first contacts a leaf (Tukey, 1970). N and P loss is most pronounced during autumn, when nutrients are mobilized by the senescence process. Deciduous and evergreen trees both lose approximately 15% of their annual above-ground nutrient return to the soil via leaching (Table 2). The higher rates of nutrient loss by deciduous species are balanced by the greater time during which evergreen leaves are available to be leached.

Nutrient absorption by leaves can be a net source of nutrients to plants when concentrations are high in rain water or when there are high concentrations of

Table 2
Nutrients leached from the canopy (throughfall) as a percentage of the total aboveground nutrient return from plants to the soil for 12 deciduous and 12 evergreen forests (from Chapin, 1991)

Nutrient	Throughfall (% of annual return)	
	Evergreen forests	Deciduous forests
N	14 ± 3	15 ± 3
P	15 ± 3	15 ± 3
K	59 ± 6	48 ± 4
Ca	27 ± 6	24 ± 5
Mg	33 ± 6	38 ± 5

Values are mean ± SD.

NH_3 in the air. The most dramatic examples of N uptake by leaves occur downwind from industrial sources of NO_x or agricultural sources of volatilized NH_3.

D. Symbiotic N Fixation

There is a vast literature on symbiotic N fixation, so only those patterns related to plant nutrition will be summarized here. As with root uptake and mycorrhizal association, species differ in their capacity to acquire N by symbiotic fixation, with this capacity restricted to eight families among vascular plants, including Fabaceae (with about 3000 N-fixing species) and Betulaceae (e.g. *Alnus*) (Lambers *et al.*, 1998). Among species with the capacity to fix N, there are strong differences in the N fixation rate among genotypes of both the plant and microbial symbiont. The N fixation rate is sensitive to plant nutrient status. Under conditions of low N supply, symbiotically fixed N can account for more than 75% of the nitrogen acquired by clover (Boller and Nösberger, 1987) and often accounts for half or more of plant nitrogen in other species with N-fixing symbionts (Gault *et al.*, 1995; Vance, 1995; Sprent *et al.*, 1996; Lambers *et al.*, 1998). N-fixing plants reduce the N fixation rate when grown in soils with a high N supply, presumably to reduce the high energetic cost of N fixation (Marschner, 1995). Conversely, soils that are low in P support low fixation rates because of the important role of P in the energetics of N fixation.

E. Rhizosphere Effects

Plants strongly influence rates of N and P uptake from soil by modifying the physical and chemical nature of the rhizosphere. Mass flow of soil solution to the root caused by transpiration carries soluble N and P to the root surface, where it is available for uptake. In crops, mass flow supplies an estimated

80% of total N delivery to the root but only 5% of P delivery, because of the greater mobility of N in soils (Lambers et al., 1998). In infertile tundra soils, where the concentration of N and P in the soil solution are much lower, mass flow supplies less than 1% of N and P delivery to the root surface. Diffusion and transport by mycorrhizal fungi are the major mechanisms, other than mass flow, that move nutrients to the root surface.

When plants absorb an excess of cations over anions, as commonly occurs when N is absorbed as an organic acid or ammonium, roots secrete H^+ to maintain charge balance. The resulting decline in rhizosphere pH can reduce P availability in acidic soils (Marschner, 1995).

In organic soils, most soluble P is organic. Root surface phosphatases can account for most of the P acquired by plants in these soils (Kroehler and Linkins, 1991), as described above. Some plants also secrete organic acids which can increase P availability both by acidifying the rhizosphere (in calcareous soils) and by chelating cations that form insoluble phosphates. By solubilizing these compounds, chelates enhance P diffusion to the root surface. Iron phosphate is a common form of insoluble phosphate in acid soils, and plants typical of these soils secrete pisidic acid which chelates iron and solubilizes phosphate. Plants typical of calcareous soils produce citrate, which increases P availability by chelating calcium (Marschner, 1995). Secretion of these organic acids is greatest under P-limiting conditions (Johnson et al., 1996).

Plants also strongly influence nutrient availability in the rhizosphere through the exudation of organic acids, carbohydrates and amino acids, which stimulate microbial activity. Root exudates appear to stimulate net N mineralization under relatively fertile conditions, where soil microbes are carbon-limited (Zak et al., 1994; Hungate, 1998), but cause microbial immobilization of N in infertile soils (Diaz et al., 1993). An additional mechanism by which exudation can enhance nutrient supply is through the stimulation of microbial grazers such as amoebae and nematodes, which feed on bacteria and excrete excess N; this N is then available for absorption by plants (Clarholm, 1985). Annual N uptake by vegetation is often twice the N mineralization estimated from incubation of soils in the absence of roots (Chapin et al.,1988). Much of this discrepancy could involve more rapid nutrient cycling that occurs in the rhizosphere, as fuelled by root exudation. Root exudation rates differ several-fold among species (V.T. Eviner, unpublished data), but the differences in exudation rates among plant growth forms and the nutritional controls on exudation are poorly known.

V. BIOMASS ALLOCATION IN RELATION TO NUTRIENT ACQUISITION

Because the bulk of the nutrients required for plant growth usually enters the plant by means of root and/or mycorrhizal uptake (section IV), the allocation of biomass to roots is an important determinant of nutrient acquisition. Studies on the allocation of biomass began early in this century (e.g. Turner, 1922; Crist and Stout, 1929). Important progress was made with the classical papers by Brouwer (1962a,b) on the effect of light intensity and nutrient supply on the partitioning of biomass between shoots and roots. These papers were the basis for the 'functional equilibrium' concept. Since then, Brouwer's findings have been corroborated by numerous other studies (e.g. Chapin, 1980; Robinson and Rorison, 1983; Boot and Den Dubbelden, 1990; Boot and Mensink, 1990; Garnier, 1991; Aerts et al. 1991, 1992b). Allocation to the root system can be studied at the between-organ level (root weight ratio (RWR)—root mass as a fraction of total plant mass) or at the within-organ level (specific root length (SRL)—root length per unit root mass) (Trewavas, 1986). Moreover, root hair density is an important determinant of nutrient acquisition, particularly for nutrients with a low mobility in soil, such as phosphate (Clarkson, 1985; Hofer, 1991). Finally, root architecture is an important determinant of the efficiency of exploration and exploitation of mobile soil resources (Fitter, 1987, 1991).

An important question is whether there are inherent differences in biomass allocation patterns between species from nutrient-poor and nutrient-rich habitats. Studies on this topic are often complicated by the fact that allocation patterns are very plastic and are strongly influenced by light intensity and nutrient levels (Brouwer, 1962a,b; Robinson, 1986; Robinson and Rorison, 1988; Aerts and De Caluwe, 1989; Boot and Den Dubbelden, 1990; Olff et al., 1990, Olff, 1992). Therefore, the adaptive significance of allocation patterns has to be studied at several levels of resource availability. In this review, we will confine ourselves to the relation between allocation patterns and nutrient availability.

A. Patterns in Biomass Allocation

It has long been postulated that plants from infertile sites allocate more biomass to roots than do plants from more fertile sites (Grime, 1979; Chapin, 1980; Tilman, 1985, 1988; Tilman and Wedin, 1991). This contention was mostly based on relatively short-term studies (several weeks) in growth cabinets with juvenile plants grown at optimal levels of resource availability without interference with other plants. For example, Poorter and Remkes (1990) and Poorter et al. (1990) studied relative growth rate (RGR—growth rate per unit plant mass) and allocation in 24 wild plant species and found that biomass allocation to the leaves (leaf weight ratio (LWR)) and the ratio between leaf area and leaf mass (specific leaf area (SLA)) were positively correlated with RGR. Moreover, differences in potential RGR were habitat related: fast-growing species are found in environments with high levels of resource availability, whereas slow-growing species occur in all kinds of adverse environments. These observations agree with the hypothesized higher root allocation of slow-growing species. Garnier (1991, 1992) showed that the trend of increasing RGR with increasing biomass allocation to leaves, as reported by Poorter and

Remkes (1990), only holds for dicotyledonous species and not for grasses. Monocotyledonous herbaceous species allocate relatively more biomass to the roots and less to leaves, compared with dicotyledonous herbaceous species with the same inherent RGR (Garnier, 1991). Van der Werf *et al.* (1993) also found that there were only minor interspecific differences in allocation patterns in a growth analysis among five monocotyledonous species from low- and high-productive habitats at high N, whereas at low N supply the fast-growing species allocated more biomass to their roots than did the slow-growing species. Thus, these growth cabinet studies indicate that allocation patterns are quite variable and depend strongly on the growth-form under consideration.

These growth analysis data, obtained from studies with juvenile plants under artificial conditions, cannot easily be extrapolated to field situations (Garnier and Freijsen, 1994). Field studies conducted by Berendse and Elberse (1989), Olff *et al.* (1990), Tilman and Cowan (1989), Gleeson and Tilman (1990) and Aerts (1993) showed that fast-growing species from nutrient-rich sites had higher biomass allocation to the roots than slow-growing species from nutrient-poor sites. This suggests that these species have a higher capacity for nutrient uptake. However, nutrient uptake is often better correlated with root length than with root mass (Fitter, 1991). Thus, species may compensate for low biomass allocation to the roots by having a high SRL. This has indeed been found in some studies with species from nutrient-poor sites (Aerts *et al.*, 1991; Elberse and Berendse, 1993). This strongly suggests that there are different evolutionary solutions to the allocation problem: plants may have a high nutrient uptake capacity as a result of high biomass allocation to the roots, or they may have high uptake capacity by having a lower root allocation but a higher SRL. These results are not consistent with theories of evolutionary trade-offs in allocation between shoot and root.

Nutrient uptake depends not only on specific absorption rate (SAR–uptake rate per gram of root), SRL and proportional allocation to roots, but also on plant biomass. At the level of the ecosystem, nutrient supply should be considered as an absolute resource (expressed, for example, as the amount of nutrients per unit of ground area). So nutrient acquisition depends on the absolute amount of nutrient-absorbing tissues (expressed, for example, as root mass or root length per unit of ground area). Root biomass depends on both the biomass of plants and the proportion of that biomass allocated to roots. This implies that allocation ratios are only poor indicators of resource capture. This is illustrated by a study with experimental populations of *Carex diandra, C. rostrata* and *C. lasiocarpa* (species from mesotrophic fens) and *C. acutiformis* (a species from eutrophic fens). It was found that percentage biomass allocation to below-ground parts (roots + rhizomes) was higher in the low-productive species (Aerts *et al.*, 1992b). However, because of the higher total biomass of the high-productive species at high N supply, their absolute below-ground tissue mass significantly exceeded that of the low-productive species. This implies that these high-productive species have a lower RWR, but a higher root mass per unit of ground area, and can therefore acquire more nutrients.

From these results, we conclude that allocation patterns found in growth cabinet studies cannot be extrapolated directly to field situations, especially when species with different biomasses and/or different growth-forms are compared. Moreover, the adaptive significance of root biomass allocation patterns is probably less important than root morphology (i.e. SRL) in explaining species adaptations to habitats with different levels of nutrient availability.

B. Dependence on Nutrient Availability

The most frequently observed phenotypic response to suboptimal nitrogen supply is an increase in biomass allocation to the roots, especially in fast-growing species (e.g. Aerts and De Caluwe, 1989; Boot and Den Dubbelden, 1990; Shipley and Peters, 1990; Tilman and Wedin, 1991; Lambers and Poorter, 1992; Van der Werf *et al.*, 1993; Wilson and Tilman, 1995). This effect is observed both in growth cabinet and field studies. This raises the issue of the relative importance of phenotypic responses of root allocation compared with inherent interspecific differences. Olff (1992) studied the effects of light and nutrient availability on dry matter and N allocation in six successional grass species. He found that phenotypic responses to variation in light and nutrient supply were much larger than interspecific differences. This emphasizes once again the very plastic nature of resource allocation in plants.

Nutrient supply varies spatially as well as temporally (Campbell and Grime, 1989, 1992). Many species are capable of rapid root proliferation into nutrient-rich patches (De Jager, 1982; Crick and Grime, 1987; Eissenstat and Caldwell, 1987; Campbell *et al.*, 1991; Berntson *et al.*, 1995; Grime *et al.*, 1997). This high plasticity in the spatial arrangement of the root system may be of great adaptive significance in soils with strong spatial differences in nutrient availability.

Nutrient supply effects on SRL are variable. In some studies, SRL does not change with changes in nutrient supply (Fitter, 1991; Aerts *et al.*, 1992b). In other studies, SRL is higher under conditions of low nutrient availability (Fitter, 1985; Boot and Mensink, 1990; Hetrick *et al.*, 1991; Berntson *et al.*, 1995). This increases the uptake capacity of the root system.

Root hair formation also depends on nutrient availability. In agricultural plants root hair formation increased at low levels of phosphate and nitrate supply (Föhse and Jungk, 1983). Also in grass species an increased root hair length was observed in response to low nutrient supply, especially in slow-growing species (Boot and Mensink, 1990; Liljeroth *et al.*, 1990).

In conclusion, most plant species are very plastic in adjusting various aspects of root allocation to changes in mineral nutrient supply. Among herbaceous species, fast-growing species show a higher phenotypic plasticity than slow-growing species (Lambers and Poorter, 1992). This high degree of plasticity contributes to an ability for high rates of resource acquisition in productive habitats and as such confers to the success of fast-growing species in these environments (Grime *et al.*, 1997). However, among broader comparisons of growth-forms there is no relationship between growth rate and plasticity of root : shoot ratio (Reynolds and D'Antonio, 1996). Often plasticity in some traits (e.g. root : shoot ratio) leads directly to low plasticity in other traits (e.g. nutrient concentration), so that it is unlikely that there is a general relationship between plasticity and relative growth rate.

VI. NUTRIENT STORAGE

A. The Concept of Storage

Storage constitutes the resources that build up in a plant and can be mobilized in the future to support biosynthesis. There are three general categories of storage (Chapin *et al.*, 1990):

(1) Accumulation is the increase in compounds that do not directly promote growth. Accumulation occurs when resource acquisition exceeds demands for growth and maintenance (Millard, 1988). Accumulation accounts for much of the short-term fluctuation in chemical composition of plants, for example the daily fluctuation of leaf carbohydrates or the nitrogen accumulation, also termed 'luxury consumption', that occurs following pulses of nitrogen availability.

(2) Reserve formation involves the metabolically regulated synthesis of storage compounds that might otherwise directly promote growth. Reserve formation directly competes for resources with growth and defence (Rappoport and Loomis, 1985). For example many plants divert nitrogen to storage organs in autumn, despite strong limitation of production by N supply (Shaver *et al.*, 1986).

(3) Recycling is the reutilization of compounds whose immediate physiological function contributes to growth or defence but which can subsequently be broken down to support future growth (Chapin *et al.*, 1990). Recycling of nutrients following leaf senescence allows reutilization of about half of the nitrogen and phosphorus originally contained in the leaf (section VII.B). These stored nutrients are then a nutrient source for new growth at a later time.

B. Variation among Growth-forms

Most plants depend strongly on storage, but the time-scale of this dependence and the type of storage differ among growth-forms. Annual plants invest most nutrients directly into growth with minimal reserve storage. At the initiation of reproduction, roots and leaves begin to senesce, with nutrients being recycled from vegetative to reproductive tissues. Typically 50–90% of N and P but less than 5% of the carbon is recycled from vegetative to reproductive tissues (Chapin and Wardlaw, 1988). Annuals also show relatively modest short-term nutrient accumulation in response to pulses of nutrient supply, because their rapid growth enables pulses of growth to match pulses of supply (Chapin *et al.*, 1990).

Most perennial plants depend strongly on storage, but the type of storage differs among species. Biennials and species adapted to frequent disturbances develop large nutrient storage reserves, even under conditions of nutrient limitation. For example, during the first year of growth biennials such as *Arctium tomentosum* develop a taproot that stores nutrients and carbohydrates, but form only a rosette of leaves above ground (Heilmeier *et al.*, 1986; Heilmeier and Monson, 1994). These reserves in the taproot support rapid growth the following year. Similarly, many mediterranean shrubs develop large N and P reserves that can support regrowth after fire, despite strong nutrient limitation in these habitats; these storage reserves are typically best developed in 'sprouter' species with low RGR rather than in more rapidly growing 'seeder' species (Keeley and Zedler, 1978).

Evergreen species are thought to depend less on reserve storage than deciduous species, because much of the leaf senescence coincides with new growth, allowing direct recycling of nutrients from old to new leaves (Chapin and Shaver, 1989; Nambiar and Fife, 1991). Jonasson (1989) tested this hypothesis by removing old leaves of several evergreen species from different ecosystems but found that this had no effect on the nutrient pools in new leaves. This discrepancy demonstrates clearly that it is impossible to make conclusions about the importance of storage simply by comparing the patterns of nutrient loss from old tissues with nutrient gain in new tissues. Often the patterns of nutrient stores in below-ground parts and nutrient uptake are unknown and assumed to be unaffected. Clearly, experiments such as those of Jonasson (1989) are required to demonstrate the situations when storage is important.

C. Nutrient Dependency

Increased nutrient supply directly augments nutrient accumulation in all plants (i.e. the accumulation of nutrients in excess of immediate needs for growth) (Schulze *et al.*, 1985). This nutrient accumulation is most pronounced in plants that grow slowly, because these species typically have lower plasticity in their capacity to invest newly acquired nutrients in growth (Chapin *et al.*, 1990). Seasonal reserve storage is much more pronounced (and increases more strongly in response to nutrient addition) in perennial forbs, graminoids and deciduous shrubs than in evergreens (Chapin and Shaver, 1989).

VII. NUTRITIONAL ASPECTS OF LEAF TRAITS

In the coming sections we explore how various leaf traits (leaf nutrient concentration, nutrient resorption from senescing leaves, and leaf lifespan) vary among growth-forms and how they are affected by changes in mineral nutrient supply. Next, we explore how these leaf traits contribute to leaf-level nutrient use efficiency (NUE). In perennials, NUE is generally measured as productivity per unit nutrient uptake or loss (cf. Vitousek, 1982). A commonly used index for NUE is the inverse of litter nutrient concentrations.

The data used in these sections mainly originate from the database presented in Aerts (1996) with minor additions from the literature about leaf lifespan and related traits listed in Reich *et al.* (1992). The calculation of the NUE index of Vitousek (1982) requires data on both nutrient concentration in mature leaves and on nutrient resorption from senescing leaves. As a result, about 15% of the data of Aerts (1996) could not be used, because those data originated from papers in which only one of the two parameters was presented. Nevertheless, the data set used here is about five times as large as the 'NITROGEN' data set of Reich *et al.* (1992), and three times as large as the data set of Killingbeck (1996) on resorption efficiency, and therefore provides a strong basis for evaluating general patterns in leaf-level NUE and its underlying components.

A. Leaf Nutrient Concentrations

In a recent analysis of a wide array of plant traits, Grime *et al.* (1997) suggested that mineral nutrients should be considered as the fundamental currency of vegetation processes at scales ranging from the individual to ecosystems. They found a marked correlation between foliar concentrations of N, P, K, Ca and Mg, high concentrations of which coincided with the capacity for rapid growth in productive conditions and an inability to sustain yield under limiting supplies of nutrients. Thus, leaf nutrient concentrations are important determinants of the functioning of plant species in their habitat.

1. Variation among Growth-forms

Nitrogen concentrations in mature leaves of perennials show clear differences between growth-forms (Figure 2A). Evergreens have significantly lower N concentrations than forbs and graminoids, and deciduous species have the highest N concentrations. The large data set ($n = 77$) of Thompson *et al.* (1997) on N and P concentrations in temperate herbaceous species was not included in the analysis because it did not meet the criterion that leaf-level NUE could be calculated from the basic data. If those data had been included, then forbs and graminoids would have had the highest N and P concentrations in mature

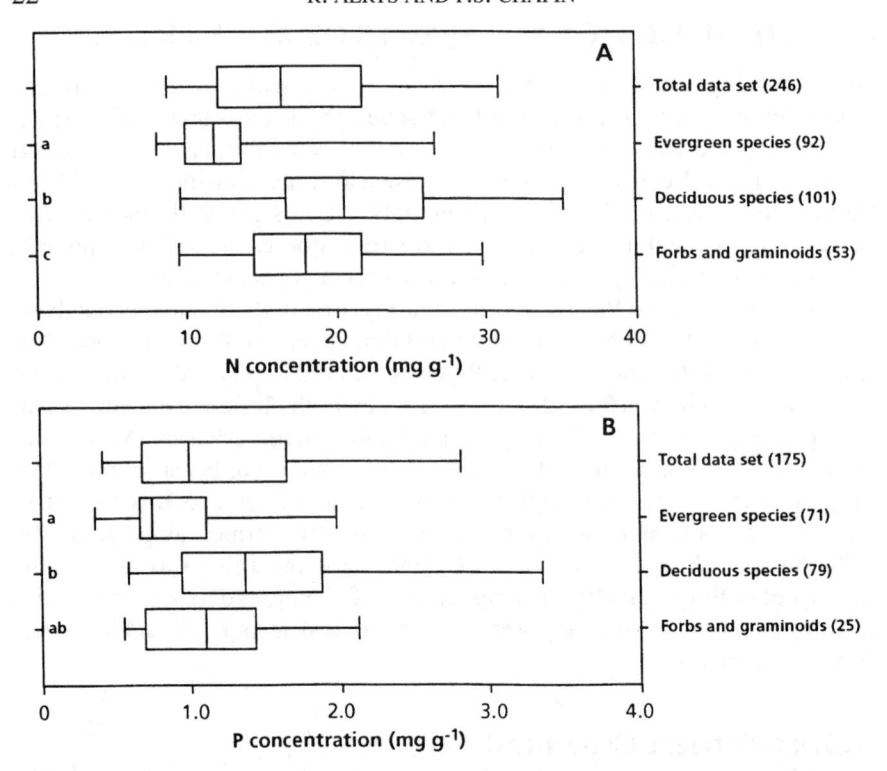

Fig. 2. Box plots showing the distribution of (A) leaf N and (B) leaf P concentrations. Results are presented for the entire data set, evergreen shrubs and trees, deciduous shrubs and trees, and forbs and graminoids. The middle line in each box indicates the geometric mean of the observed distribution, the left and right parts the 25th and 75th percentiles, and the left and right 'error bars' the 5th and 95th percentiles. Values in parentheses indicate number of observations. Different letters indicate statistical difference between growth-forms (Tukey test after analysis of variance, $P < 0.05$).

leaves. However, as is clearly shown in Figure 2, there is considerable overlap among the various growth-forms. Thus, growth-form comparisons within small subsets of the data may lead to another pattern. As the carbon assimilation of a leaf is linearly related to the nitrogen content of that leaf (Field, 1983; Hunt *et al.*, 1985; Hirose and Werger, 1987; Evans, 1989), this implies that the patterns of N concentration in mature leaves have clear consequences for the carbon gain of these growth-forms. Another ecological factor that might be correlated with variation in leaf N concentration is chemical defence against herbivory. According to the carbon-nutrient balance theory of Bryant *et al.* (1983), chemical defence against herbivory should be carbon based in nutrient-poor habitats and nitrogen based in more carbon-limited habitats. At present, it is not clear how much of the nitrogen present in leaves should be considered as 'defence N'. Thus, it is not clear what the contribution is of N-based defence to the variation in leaf N concentrations as shown in Figure 2.

The pattern of phosphorus concentrations is more or less similar to that of N concentrations, but in general the variation is larger (Figure 2B). Phosphorus concentrations in leaves of evergreen species are lower than in deciduous species. Forbs and graminoids occupy an intermediate position (but see the comment above). As is the case with nitrogen, these patterns may have direct consequences for the rate of photosynthesis in the leaves. It has been shown that the rate of photosynthesis is dependent on the leaf P concentration (Herold, 1980) and that P deficiency inhibits photosynthesis (Terry and Ulrich, 1973). However, in most cases photosynthetic rate correlates more closely with N than with P concentration.

The larger variation in leaf P concentration compared with leaf N concentration was also observed by Fitter *et al.* (1998) for temperate and tropical plants. They found that leaf P concentration is not correlated simply with leaf N concentration, resulting in an even larger variation in leaf N : P ratios compared with the variation in concentrations of N and P (cf. section III). They concluded that leaf N concentrations are more closely regulated than leaf P concentrations and that, despite the widespread occurrence of mycorrhizal symbiosis, the variation in leaf P concentrations most likely reflects the variation in P availability in various habitats. The question that remains then is how plants respond to variation in leaf P concentration. Fitter *et al.* (1998) showed that leaf P concentration probably plays a crucial role in the regulation of resource allocation between vegetative and reproductive development.

2. Dependence on Nutrient Availability

Leaf nutrient concentrations as determined in field studies reflect both genotypic and phenotypic sources of variation. Nutrient availability has two different effects on leaf nutrient concentrations. First, it may increase concentrations within species, especially when another nutrient becomes limiting (e.g. Aerts, 1989a,b; Millard and Proe, 1993; Bowman, 1994). Second, it may also lead to changes in community species composition (e.g. Tilman and Cowan, 1989; Gleeson and Tilman, 1990; Tilman and Wedin, 1991; Verhoeven and Schmitz, 1991; Bobbink, 1992; Aerts and Bobbink, 1999), which may have important consequences for leaf nutrient concentrations at the community level. In general, nutrient-poor sites are dominated by slow-growing species with low leaf nutrient concentrations, notably evergreens (Monk, 1966; Aerts, 1995). At increasing levels of nutrient availability these species are replaced by forbs and graminoids and/or deciduous shrubs and trees (e.g. Aerts and Berendse, 1988), which have higher leaf nutrient concentrations (Aerts, 1995). Both genetic differences and phenotypic responses of plants to nutrient supply contribute to the low N and P concentrations of evergreen species observed in the field (Figure 2) (Craine and Mack, 1998). Thus, increased nutrient availability leads to higher leaf nutrient concentrations at the community level due to both phenotypic responses and species replacements.

B. Nutrient Resorption from Senescing Leaves

Resorption of nutrients from senescing leaves is of great adaptive significance, because it enables plants to reuse these nutrients and thereby can lead to a higher nutrient retention (Chapin, 1980; Chabot and Hicks, 1982; Aerts, 1990). This process has important implications at both the population level (see section VIII) and the ecosystem level (sections X and XI). At the ecosystem level, nutrient resorption from senescing leaves has a profound influence on element cycling. The nutrients that are resorbed during senescence are directly available for further plant growth, which makes a species less dependent on current nutrient uptake. Nutrients that are not resorbed, however, will be circulated through litterfall. The litter must be decomposed and the nutrients contained in that litter must be remineralized to become available for plant uptake again. Compared with the resorption pathway, this dependency of plants on the decomposition pathway has the disadvantages that each plant must compete for the mineralized N with micro-organisms (Kaye and Hart, 1997) and with neighbouring plants, and that part of the nitrogen can be incorporated in stable soil organic N pools and become unavailable for plant uptake (Aerts, 1997a). However, low molecular organic N compounds can be taken up by mycorrhizal (Read, 1991) and non-mycorrhizal plants and provide a large part of the annual N requirement (Kielland, 1994).

A repeated suggestion of early studies based on comparison of a few plant species was that plants from nutrient-poor environments have a higher nutrient resorption efficiency than those from more nutrient-rich environments. Is this true?

1. Variation among Growth-forms

As evergreen shrubs and trees are the dominant growth-form in nutrient-poor environments, it seems reasonable to assume that they have higher resorption efficiency than other growth-forms. However, an analysis of the literature shows that there is no difference in N and P resorption between growth-forms (Figure 3A and B). In all growth-forms, median N and P resorption is somewhere around 50%. This implies that, although nutrient resorption is an important nutrient conservation mechanism at the species level, it apparently does not explain the distribution of growth-forms over habitats differing in soil

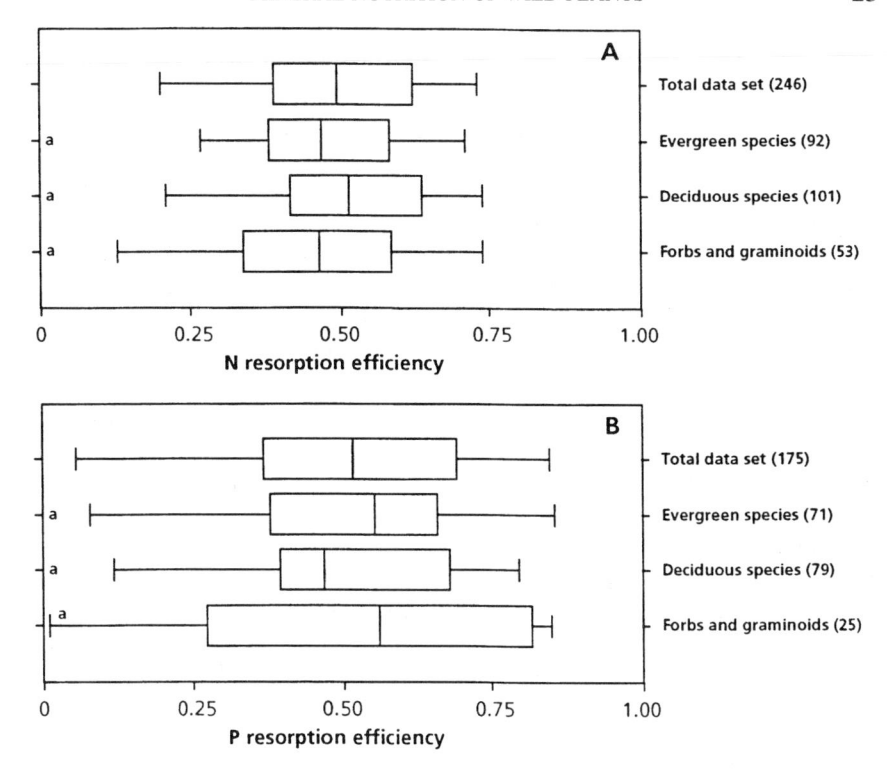

Fig. 3. Box plots showing the distribution of (A) N and (B) P resorption efficiency from senescing leaves. See Figure 2 for further explanation.

fertility (Chapin and Kedrowski, 1983; Aerts, 1996; Killingbeck, 1996). However, note that the variation in P resorption efficiency is much larger than that in N resorption efficiency. Moreover, maximum values of P resorption efficiency are higher than those for N and the minimum levels of P resorption are lower than those for N. These differences between N and P resorption efficiency have important consequences for leaf-level N and P use efficiency (section IX).

2. Nutritional Controls

Although the distribution of certain growth-forms is clearly related to soil fertility, this cannot be explained by differences in nutrient resorption efficiency (Aerts, 1996). This raises the question of whether there are consistent differences at the phenotypic level in nutrient resorption between high- and low-fertility conditions. The answer is negative: in an analysis of published fertilization experiments covering 60 species (Aerts, 1996) there was no response of N resorption in 63% of the experiments analysed, whereas in 32% there was a decrease in N resorption in response to increased nutrient availability. Also, P resorption (37 species analysed) showed no response in 57% of cases, and in 35% P resorption decreased upon enhanced nutrient supply. Nutrient resorption in evergreen shrubs and trees showed especially low responsiveness to changed nutrient availability.

Thus, if there are nutritional controls on nutrient resorption efficiency at the phenotypic level, these must be less important than other factors that have yet to be identified. The most clear nutritional control on nutrient resorption is found when plants are grown at abnormally high nutrient concentrations, where nutrient resorption efficiency is low (Chapin and Moilanen, 1991). Under natural nutrient supplies there is no relationship between resorption efficiency and the lability of N- and P-containing chemical fractions in plants (Chapin and Kedrowski, 1983).

The lack of nutritional controls on nutrient resorption raises the question of which other factors control nutrient resorption. Several possible controls have been proposed, including the relative sink strength of plant organs (Nambiar and Fife, 1991), the rate of phloem transport (source–sink interactions) (Chapin and Moilanen, 1991) and soil moisture availability (Boerner, 1985; del Arco et al., 1991; Escudero et al., 1992; Pugnaire and Chapin, 1993). However, as Nambiar and Fife (1991) have emphasized, there is unlikely to be a single explanation for variation in nutrient resorption efficiency.

If there are indeed multiple controls over nutrient resorption efficiency, most resorption efficiencies observed in the field and reported in the literature are unlikely to be the maximum resorption efficiency of which a species is capable (Killingbeck, 1996). A second possible explanation for the large range in resorption efficiency reported in the literature could be that plants control the minimum N and P concentration of senesced leaves (resorption proficiency) rather than the proportion of nutrients withdrawn during senescence (resorption efficiency) (Killingbeck, 1996). Killingbeck (1996) has argued that resorption proficiency is the trait on which selection is most likely to have acted. We discuss the nutritional significance of resorption efficiency in section IX.

C. Leaf Lifespan

Although leaf lifespan is not a direct nutritional leaf trait, variation in leaf lifespan is of great importance for a wide variety of ecological processes, including those related to mineral nutrition. These issues have been considered extensively by Reich et al. (1992), so we will confine ourselves to a synthesis of the most relevant results for this review.

1. Variation among Growth-forms

By definition, leaf lifespan of sclerophyllous evergreen species is much longer than that of other growth-forms, which do not differ consistently in lifespan (Figure 4.) Reich *et al.* (1992) showed that long leaf lifespans, as found in evergreen species, are negatively correlated with the maximum rate of photosynthesis (both per unit leaf area and per unit leaf mass), the mass-based N concentration in the leaf, and the specific leaf area (SLA). They found that species with the longest leaf lifespan, the lowest SLA, the lowest leaf nitrogen concentration, and the lowest maximum rate of photosynthesis are either conifers in low-temperature, dry and/or nutrient-poor environments, or evergreen broad-leaf species inhabiting nutrient-poor environments. The low maximum rate of photosynthesis (per unit leaf mass or leaf area) of evergreens is significantly correlated with the low maximum RGR of seedlings of evergreens.

In addition to the earlier review, Reich *et al.* (1997) have demonstrated that similar interspecific relationships exist among leaf structure and function and plant growth in a wide variety of biomes, ranging from tropical and temperate forests to alpine tundra and desert. For a data set of 280 species they found that photosynthesis and respiration increase in similar proportion with decreasing leaf lifespan, increasing leaf N concentration and increasing SLA. The productivity of individual plants and of leaves in vegetation canopies also changes in constant proportion to leaf lifespan and SLA. This global-scale convergent evolution can provide a quantitative basis for evaluating interspecific and intraspecific species differences and for comparing among ecosystems and biomes.

From the analysis of Reich *et al.* (1992) it appeared that the largest variability in leaf trait occurs among species with short lifespans (< 1 year), and the lowest

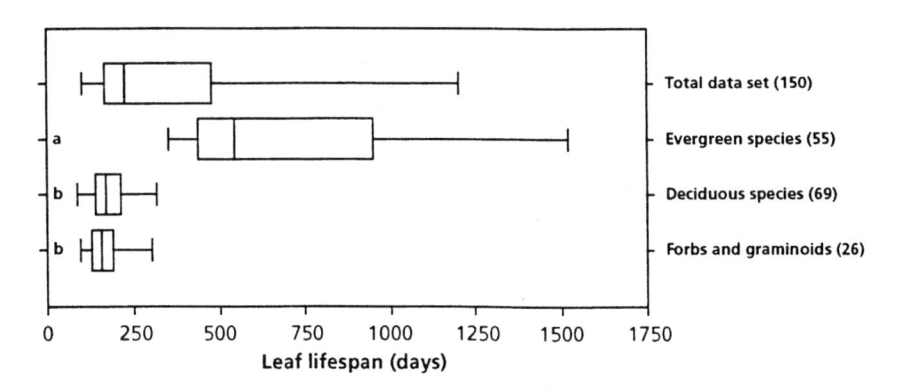

Fig. 4. Box plots showing the distribution of leaf lifespan. See Figure 2 for further explanation.

variability among species with leaf lifespans of more than 2 years. As a result, deciduous species with leaves that persist for 9–10 months are likely to share traits more closely with evergreen species that retain foliage for 2–3 years than with deciduous species that keep their leaves for 2–3 months. Similarly, the nutrient concentrations of senesced leaves differ strikingly between species with leaf longevities greater or lesser than 1.5 years (Craine and Mack, 1998). Thus, the tendency to 'lump' together all temperate evergreens with leaf lifetimes > 1 year and to compare them with deciduous plants with lifetimes of < 1 year may be misleading in certain instances.

2. Dependence on Nutrient Availability

At the community level, there is a clear relation between leaf lifespan and nutrient availability. Nutrient-poor communities are generally dominated by slow-growing (mostly) evergreen species with long leaf lifespans, whereas nutrient-rich communities are dominated by fast-growing deciduous species with high rates of leaf turnover.

The effect of increased nutrient availability on leaf lifespan at the phenotypic level is, however, rather variable. No effect was found by Aerts (1989a) with the evergreen shrub *Calluna vulgaris*, and by Reader (1980) with the evergreen shrubs *Kalmia polifolia* and *Ledum groenlandicum*. Similarly, in a study on leaf lifespan with four *Carex* species from the fens differing in nutrient availability Aerts and De Caluwe (1995) found that leaf lifespan was not significantly affected by enhanced N supply, except in *C. diandra*, where leaf lifespan decreased upon enhanced N supply. Decreasing leaf lifespan upon enhanced nutrient supply was also found by Reader (1980) with the evergreen shrub *Chamaedaphne calyculata*, by Aerts (1989a) with the evergreen shrub *Erica tetralix*, and by Shaver (1981, 1983) with the evergreen shrub *Ledum palustre*. Bazzaz and Harper (1977), on the other hand, found an increased leaf lifespan upon enhanced nutrient supply with the herb *Linum usitatissimum*. Apparently, the leaf longevity of individual species is not unambiguously controlled by nutrient availability.

VIII. NUTRIENT USE FOR BIOMASS PRODUCTION

A. Nutrient Use Efficiency (NUE)

The nutrients taken up by plants are generally used for the production of biomass. The relative amounts of nutrients taken up and the relative amounts of carbon fixation determine whether a species can persist in natural plant communities (Tilman, 1988; Aerts *et al.*, 1990; 1991; Huston, 1994). Due to its prime importance in the explanation of the distribution of plant species over environmental gradients, the interrelation between the carbon and nutrient economy of plant species has become one of the major topics in plant ecology. This issue has been investigated with respect to both variation in resource use efficiency among species and the dependence of resource use efficiency on environmental circumstances, notably nutrient availability. Major progress has been achieved by studying the nutrient use efficiency (NUE) of plant species (e.g. Vitousek, 1982; Boerner, 1984; Pastor *et al.*, 1984; Shaver and Melillo, 1984; Birk and Vitousek, 1986; Berendse and Aerts, 1987; Aerts, 1990; Shaver and Chapin, 1991; Escudero *et al.*, 1992; Aerts and Van der Peijl, 1993; Bridgham *et al.*, 1995). In perennials, NUE is generally measured as productivity per unit nutrient uptake or loss (cf. Vitousek, 1982). As such, this parameter integrates a wide variety of physiological processes, including the relation between the net carbon assimilation rate of plants and leaf nutrient content and the partitioning of nutrients between resorption and decomposition pathways (Aerts, 1997a). In the paragraphs below we refer to nitrogen use efficiency as NUE_N and to phosphorus efficiency as NUE_P.

High NUE is considered to be advantageous under conditions of low soil fertility, because it entails a high biomass production per unit of nutrient taken up. The problem with most indices of NUE is that they are ratios without a time dimension and therefore do not take into account the dynamic behaviour of plants in response to nutrient supply. Thus, these indices of NUE do not directly show why a high NUE is beneficial in low-nutrient habitats. This problem was addressed by Berendse and Aerts (1987), who postulated that selection in low-nutrient habitats is not necessarily on a high NUE but rather on plant traits that reduce nutrient losses (see Grime, 1979), whereas in high-nutrient habitats selection will be on characteristics that lead to a high rate of dry matter production. Moreover, they hypothesized that, as a result of evolutionary trade-offs, genotypically determined plant characteristics that lead to a high growth rate are inversely correlated with those that reduce nutrient losses. Thus, nutrient-poor habitats will be dominated by slow-growing species with low nutrient turnover rates, and nutrient-rich habitats by fast-growing species with high rates of nutrient turnover. Berendse and Aerts (1987) proposed to analyse the adaptive strategies of perennials with respect to nitrogen availability by distinguishing two components of NUE: the mean residence time of nitrogen in the plant (MRT) and the nitrogen productivity (A_{NP}). In this concept NUE (g biomass g^{-1} N) equals the product of MRT and A_{NP}, which in turn equals total productivity divided by the total loss of the growth-limiting nutrient in litter (both above and below ground). The mean residence time (MRT, years) measures how long a unit of nitrogen is present in the population. The nitrogen productivity (A_{NP}, g biomass g^{-1} N $year^{-1}$) is defined as annual productivity divided by the annual average of the amount of the growth-limiting nutrient present in the population.

Growth analysis with juvenile plants has shown that a high nitrogen productivity is strongly correlated with high leaf nitrogen concentrations and a high specific leaf area (Ågren, 1985, 1988; Poorter and Remkes, 1990; Grime *et al.*, 1997; Hunt and Cornelissen, 1997). The MRT of nutrients in the plant is determined by a variety of plant traits, such as tissue lifespan, tissue nutrient concentration and nutrient resorption efficiency from senescing tissues (Aerts, 1990). It is common knowledge that species from nutrient-poor habitats have low tissue turnover rates (e.g. Aerts, 1990; Escudero *et al.*, 1992; Reich *et al.*, 1992; Ryser and Lambers, 1995; Schläpfer and Ryser, 1996; Eckstein and Karlsson, 1997; Eissenstat and Yanai, 1997). Recently, it has been postulated that tissue density is a possible link between tissue lifespan, nutrient retention and growth rate. High maximum relative growth rate is correlated with low tissue density (Garnier and Laurent, 1994; Ryser and Lambers, 1995; Schläpfer and Ryser, 1996). Tissue lifespan might also be determined by tissue density, as the high amount of sclerified structures in tissues with a high density increases its resistance to environmental hazards and thus increases lifespan (Garnier and Laurent, 1994; Van Arendonk and Poorter, 1994). Thus, fast-growing species will have high resource acquisition capacity due to high areas of resource-absorbing surfaces, but due to low tissue density the lifespan of leaves and roots is short, leading to low nutrient retention. Slow-growing species, on the other hand, have a lower resource acquisition capacity, but the high tissue density leads to low tissue turnover rates and thus to a high nutrient retention.

Two recently developed models, which are based on the NUE concept of Berendse and Aerts (1987), show that a high MRT leads to clear advantages in low-nutrient habitats (Aerts and Van der Peijl, 1993; Berendse, 1994a). Both models explicitly assume that there is a trade-off between plant traits that lead to low nutrient loss rates and those that lead to a high dry matter production. In the model of Aerts and Van der Peijl (1993), two species are considered which have an equal NUE, but one species is nutrient conserving (high MRT) and has a low nitrogen productivity, whereas the other has a high nutrient turnover rate (low MRT) and a high nitrogen productivity. The model clearly demonstrates that the species with a high MRT attains a higher equilibrium biomass than that with the low MRT, but that the species with a high nitrogen productivity attains the equilibrium biomass at a faster rate (Figure 5). Thus, the model shows that: (1) a nutrient-conserving strategy leads to a direct advantage in nutrient-poor environments; (2) NUE alone is not a good predictor for the success of a species in nutrient-poor environments; and (3) high-productivity species initially have a higher net biomass increase than low-productivity species even in nutrient-poor environments. Using a similar approach, Berendse (1994a) also showed that under conditions of nutrient-limited growth species with low nutrient loss rates can outcompete species with high nutrient loss rates, even when these species have a higher competitive ability for nutrient uptake.

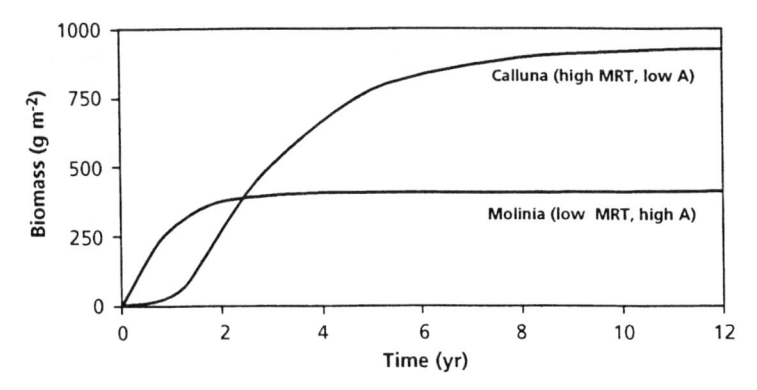

Fig. 5. Simulated long-term biomass dynamics of two species with an equal nutrient use efficiency (NUE), but which differ in their components of NUE. *Calluna vulgaris* has a high mean residence time of nutrients (MRT), but a low nitrogen productivity (A_{NP}), whereas *Molinia caerulea* has the opposite characteristics. Redrawn from Aerts and Van der Peijl (1993).

B. Patterns in NUE

Numerous recent studies on patterns in NUE have provided a basis for evaluating the ideas put forward in the previous sections. At the phenotypic level, NUE generally decreases with increasing soil fertility or fertilization (e.g. Pastor *et al.*, 1984; Birk and Vitousek, 1986; Chapin and Shaver, 1989; Aerts and De Caluwe, 1994; Bowman, 1994; Bridgham *et al.*, 1995; van Oorschot *et al.*, 1997), although at very low soil fertility levels NUE may also decrease (Bridgham *et al.*, 1995). However, species from high-fertility sites do not necessarily have lower NUE than species from low-fertility sites (Aerts, 1990; Son and Gower, 1991; Aerts and De Caluwe, 1994). This emphasizes that it is important to differentiate between phenotypic and genotypic responses of NUE to changes in soil fertility. In addition, many of these patterns are based on study of a single organ (e.g. leaves) or on above-ground parts only, which may be misleading.

A proper evaluation of NUE requires data at the whole-plant level, because patterns of above-ground NUE are not necessarily similar to whole-plant NUE (Aerts, 1990). Unfortunately, few whole-plant field studies have been conducted owing to the difficulty of measuring below-ground nutrient dynamics in natural ecosystems. Therefore, we will consider interspecific differences in NUE at three levels: whole plant, above ground, and leaves (see Reich *et al.*, 1992).

1. Whole-plant NUE

To our knowledge, the only field study on whole-plant NUE was performed by Aerts (1990). In that study, which was performed with species from nutrient-poor and nutrient-rich heathlands, it was found that the nutrient-conserving ericaceous species *Erica tetralix* and *Calluna vulgaris* (dominant in nutrient-poor heathlands) had a high MRT and a low A_{NP}, whereas the high-productivity grass *Molinia caerulea* (dominant in nutrient-rich heathlands) had a low MRT and a high A_{NP} (Aerts, 1990). The NUE of these species was equal. However, when only above-ground NUE was considered, the grass species had a much higher NUE than the ericaceous species.

2. Above-ground NUE

Contrary to general ecological belief, there are no clear patterns in above-ground NUE at the interspecific level. Gray (1983) studied NUE of the evergreen shrub *Ceanothus megacarpus* in Californian chaparral and the drought-deciduous shrubs *Salvia leucophylla* and *Artemisia californica* in coastal sage. NUE was higher in the evergreen species for all major elements. This was due mainly to a higher nutrient retention compared with the coastal sage. However, Aerts (1990) reported lower above-ground NUE for two ericaceous shrubs compared with a co-occurring grass species. A similar pattern was observed by Son and Gower (1991) in a study of evergreen and deciduous plantation trees.

An interesting case of species-specific patterns in N and P use efficiency in response to fertilization was reported by Bowman (1994). In two alpine tundra communities differing in nutrient limitation, the N-limited dry meadow had a higher N use efficiency, whereas the N–P co-limited wet meadow had a higher P use efficiency. Fertilization with N, P or N + P reduced the NUE of the supplied nutrients at the community level. A large part of this change was mediated through changes in species and growth-form composition in response to enhanced nutrient supply.

Eckstein and Karlsson (1997) studied above-ground growth and NUE of 14 plant species in subarctic Sweden. These 14 wild species represented four life-forms (woody evergreen, woody deciduous, graminoid and herb) and were growing in two contrasting habitats: a nutrient-rich meadow birch forest and a nutrient-poor bog. Their data showed that above-ground NUE was not affected by habitat, life-form or species. However, they did find an inverse relation between MRT and A_{NP}, as predicted by Berendse and Aerts (1987): species dominant in the low-nutrient bog (woody evergreens) had the highest MRT and the lowest A_{NP}, whereas species dominant in the more nutrient-rich forest (herbs and graminoids) were characterized by a low MRT and a high A_{NP}. Deciduous woody species occupied an intermediate position (Figure 6). Their results, as well as the community level patterns observed by Bridgham *et al.* (1995), indicate that NUE varies within rather narrow limits (by a factor of less than 3) compared to its components (variation by a factor 10 or more). This strongly supports the hypotheses of

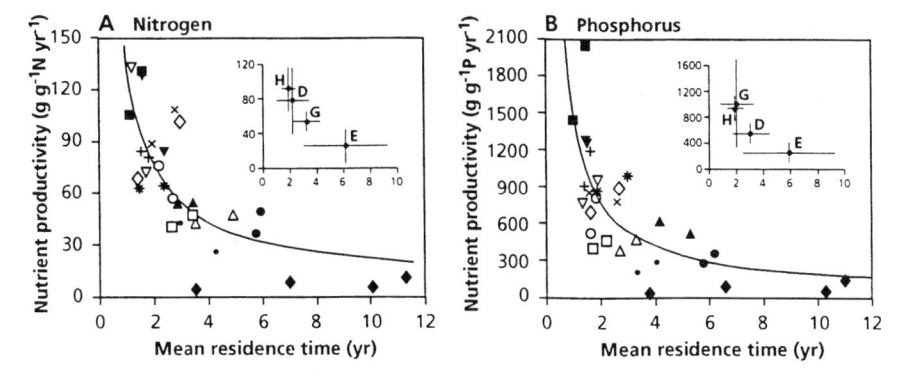

Fig. 6. Relationship between above-ground nutrient mean residence time (MRT) and nutrient productivity (A_{NP}) for 14 species from subarctic Sweden. Regression statistics: (A) $r^2 = 0.60$, $P < 0.001$ for nitrogen (B) $r^2 = 0.54$, $P < 0.001$ for phosphorus. The inset figures indicate the mean ± SD of MRT and A_{NP} for the respective growth-forms: D, deciduous shrubs; E, evergreen shrubs; G, graminoids; H, herbs. Redrawn from Eckstein and Karlsson (1997).

Berendse and Aerts (1987) which state that selection in nutrient-poor habitats is not necessarily on a high NUE, but rather on a high nutrient retention and that, as a result of evolutionary trade-offs, there is an inverse relation between nutrient retention and nutrient productivity.

These hypotheses are further supported by a re-analysis of the large data set that Vitousek (1982) used in his seminal paper on NUE of forest ecosystems (Knops *et al.*, 1997). In contrast with Vitousek's original hypothesis that the amount of nutrients in litterfall increases with increasing soil fertility, Knops and co-workers found that the relationship between the fertility of ecosystems and the amount of nutrients in the litterfall was inconclusive. They concluded that there is no indication of higher NUE in nutrient-poor ecosystems. Rather, they suggested that there might be a more general relationship between the mean residence time of nutrients or nutrient productivity and ecosystem fertility.

3. Leaf-level NUE

In contrast with data on whole-plant and above-ground NUE, the data on leaf-level NUE are very abundant. This makes it possible to investigate the relative contribution of various leaf traits (leaf nutrient concentrations, nutrient resorption from senescing leaves, and leaf lifespan) to leaf-level NUE in different growth-forms. Given the large variation in, for example, leaf lifespan between evergreen and deciduous species (Reich *et al.*, 1992; Aerts, 1995), it is likely that growth-forms will differ in determinants of leaf-level NUE. These issues will be treated in the next section.

IX. PATTERNS IN LEAF-LEVEL NUTRIENT USE EFFICIENCY AND ITS COMPONENTS

A. Leaf Traits and their Contribution to Leaf-level NUE: Theoretical Considerations

Leaf-level nutrient use efficiency expresses how much biomass is produced by a leaf during the retention time of a unit of nutrient that has been taken up (Small, 1972; Reich *et al.* 1992). However, it should be noticed that this organ-level efficiency ignores the carbon that is fixed by the leaf and supports production or respiration elsewhere in the plant. First, we explore theoretically how various leaf traits may contribute to NUE. We will illustrate this for nitrogen, but a similar reasoning can be developed for phosphorus.

Small (1972) developed a simple approach for calculating an index of NUE for perennial species. The mean residence time (MRT, years) of a unit of N in the leaf was estimated according to the formula:

$$\text{MRT} = \text{leaf lifespan} * 1/(1-r) \tag{1}$$

where r is the nutrient resorption efficiency (fraction of the total leaf nutrient pool that is resorbed during senescence: $0 < r < 1$) and leaf lifespan is expressed in years.

Based on the rate of net photosynthesis of individual leaves on a dry mass basis (A_{mass}, mmol CO_2 g^{-1} h^{-1}) and the N content of the mature leaf ([N], mg N g^{-1}) the NUE index of Small (1972), which he named 'Potential photosynthate g^{-1} N', can then be calculated as:

$$\text{NUE} = A_{mass} * \text{MRT}/[\text{N}] \tag{2}$$

which equals

$$\text{NUE} = (A_{mass} * \text{leaf lifespan} * 1/(1-r))/[\text{N}] \tag{3}$$

This formula does not, of course, yield a direct estimate of the amount of biomass that is produced during the retention time of a unit of nutrient. To calculate this, information must be available for the conversion efficiency of photosynthate into biomass of the plant species under study. Moreover, Small assumed that A_{mass} does not change during the lifespan of a leaf, that no self-shading occurs during canopy development, and that there is no leaf mass loss during senescence. Despite these simplifying assumptions, the approach of Small (1972) clearly shows which parameters are important as determinants of leaf-level NUE and how they are related to NUE. In addition, the conversion efficiency of photosynthate into biomass is nearly constant for all plant parts and species (Chapin, 1989; Poorter, 1994). In the remainder of this review, we will concentrate on a discussion of the importance of leaf lifespan, resorption efficiency and leaf N concentrations as determinants of NUE. Unfortunately, there are hardly any literature sources in which data on both A_{mass}, leaf lifespan, nutrient resorption efficiency and N concentrations are provided. There is, however, a substantial amount of data on both leaf lifespan, r and [N]. Moreover, A_{mass} and leaf N concentrations are highly correlated (r^2 = 0.74–0.85; Reich et al., 1992), which might obscure possible direct relationships between these two parameters and NUE. As a result, the analysis will be restricted to the three parameters mentioned above.

Thus, the relation between NUE and leaf traits can be described by the following equation for NUE:

$$\text{NUE} = (1/[N]) * (1/(1-r)) * \text{lifespan} * c \qquad (4)$$

in which c is a constant which takes A_{mass} and the conversion efficiency into account.

Thus, leaf lifespan shows a linear relation with NUE, but both [N] and r are inversely related to NUE. This equation shows that NUE approaches infinity when [N] approaches zero, and/or r approaches 1, and/or leaf lifespan approaches infinity. These are, of course, biologically unrealistic situations, but they indicate which values the underlying leaf traits of NUE should approach in order to maximize NUE. The response of NUE to changes in the underlying leaf traits can conveniently be studied by partial differentiation of NUE with respect to these traits. This partial derivative shows how NUE changes per unit change of the trait under consideration.

1. Nutrient Concentration

Partial differentiation of NUE with respect to the leaf N concentration yields:

$$d(\text{NUE})/d([N]) = (-1/[N]^2) * (1/(1-r)) * \text{lifespan} * c \qquad (5)$$

Equation (5) shows that the greatest change in NUE per unit change in [N] takes place at low N concentrations. The minus sign indicates that NUE decreases with increasing N concentration. These implications are illustrated graphically in Figure 7.

2. Resorption Efficiency

Partial differentiation of NUE with respect to resorption efficiency yields a different type of relationship:

$$d(\text{NUE})/d(r) = (1/(1-r)^2) * (1/[N]) * \text{lifespan} * c \ (0 < r < 1) \qquad (6)$$

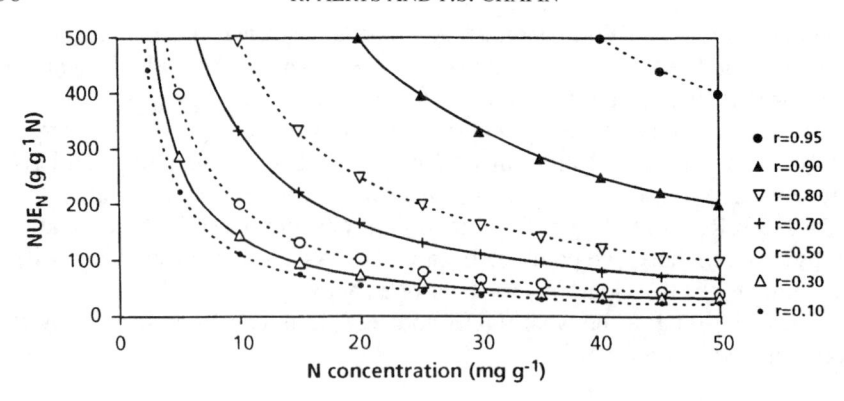

Fig. 7. Theoretical values of leaf-level nitrogen use efficiency (NUE$_N$) as a function of N concentration in mature leaves. Isoclines of N resorption efficiency (r) are shown.

In this case, the greatest changes in NUE per unit change in r are obtained at high values of r (Figure 8). NUE increases with increasing resorption efficiency.

3. Leaf Lifespan

When NUE is partially differentiated with respect to leaf lifespan, the following equation is obtained:

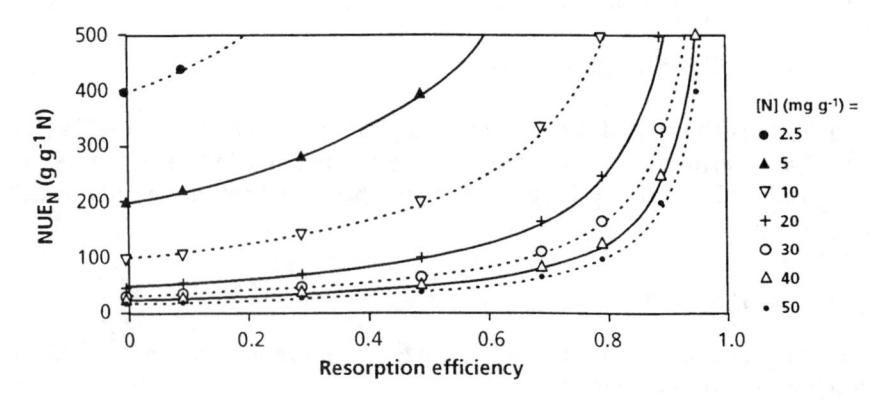

$$d(\text{NUE})/d(\text{lifespan}) = (1/[\text{N}]) * (1/(1-r)) * c \tag{7}$$

Fig. 8. Theoretical values of leaf-level nitrogen use efficiency (NUE$_N$) as a function of N resorption efficiency (r). Isoclines of N concentration in mature leaves are shown.

Equation (4) shows that NUE increases with increasing leaf lifespan, but equation (7) shows that changes in NUE per unit change in leaf lifespan are independent of the actual value of leaf lifespan (Figure 9).

These equations and their graphic illustrations show that the largest changes in NUE per unit change in the leaf trait under consideration are obtained when N resorption efficiency increases above values of about 0.75 and/or when leaf N concentrations decrease below values of about 15 mg N g^{-1}. For leaf lifespan, the change in NUE per unit change in lifespan is independent of the actual value of the leaf lifespan. These results imply that maximization of leaf-level NUE requires natural selection to favour traits that lead to N resorption efficiencies higher than 0.75 and leaf N concentrations lower than 15 mg N g^{-1}. For phosphorus, for which the concentrations are about a factor 10 lower, these values are 0.75 and 1.5 mg P g^{-1}, respectively (data not illustrated graphically). In the coming paragraphs, we explore whether and how these findings are realized in plant species belonging to different growth-forms.

B. Actual Patterns in Leaf-level Nutrient Use Efficiency and the Relation with Underlying Leaf Traits

Vitousek (1982) and Birk and Vitousek (1986) have proposed a more simple index of NUE, the inverse of leaf litter N concentration:

$$\text{NUE} = 1/([N] * (1-r)) \tag{8}$$

This index is similar to that of Small (1972) and has the advantages that only two parameters are needed for its calculation and that it gives a direct estimate

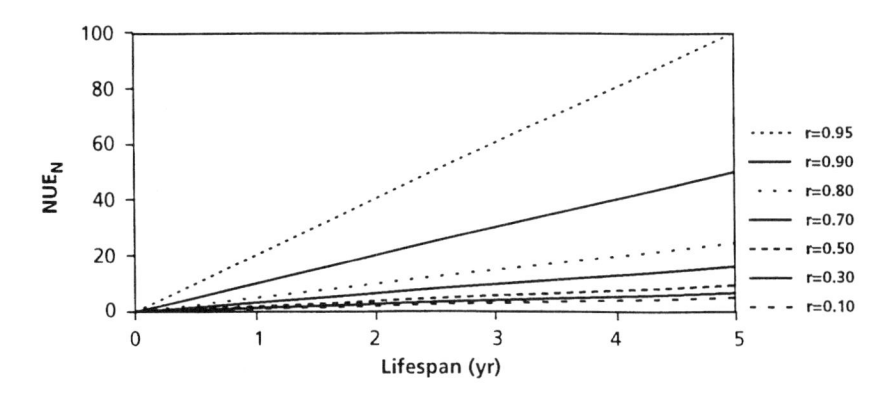

Fig. 9. Theoretical values of leaf-level nitrogen use efficiency (NUE$_N$) as a function of leaf lifespan. Isoclines of N resorption efficiency (r) are shown. Note that NUE values are on a relative scale (see text for explanation).

of the amount of biomass produced per unit of N taken up. Therefore, this index will be used for the calculation of NUE.

1. Nitrogen Use Efficiency

Leaf-level nitrogen use efficiency of evergreens is higher than that of the other growth-forms. However, there is considerable overlap (Figure 10). This implies that, within small subsets of this large data set, the differences between growth-forms may deviate from the patterns presented here.

We used path analyses to unravel the contribution of various leaf traits to NUE_N. When the entire data set was considered, leaf N concentration showed the highest (negative) correlation with NUE_N (Figure 11A). Moreover, a large part of the correlation of leaf lifespan with NUE_N was due to an indirect effect of leaf N concentration on leaf lifespan. A similar picture emerged when evergreen shrubs and trees were considered (Figure 11B). In this growth-form, leaf

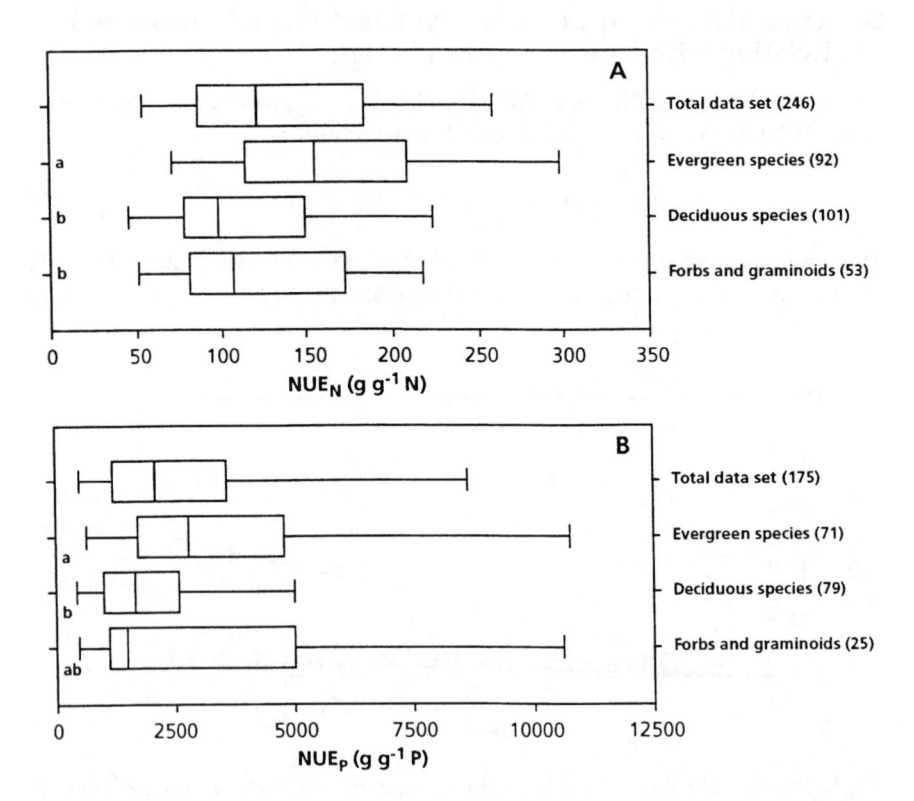

Fig. 10. Box plots showing the distribution of nitrogen use efficiency (NUE_N) and phosphorus use efficiency (NUE_P). See Figure 2 for further explanation.

(A) Total data set (n=148)

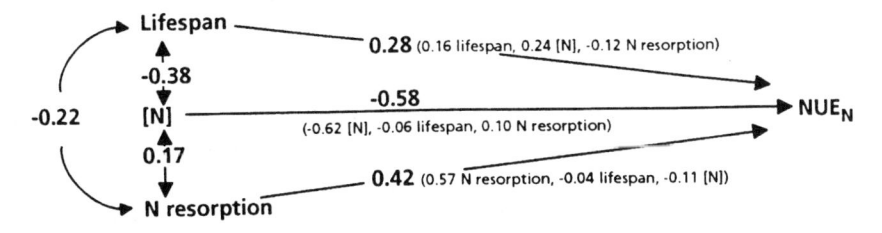

```
       ┌──► Lifespan ──────────── 0.28 (0.16 lifespan, 0.24 [N], -0.12 N resorption)
       │       ▲
       │     -0.38
       │       ▼                    -0.58
  -0.22│      [N] ─────────────────────────────────────────────────► NUE_N
       │       ▲       (-0.62 [N], -0.06 lifespan, 0.10 N resorption)
       │     0.17
       │       ▼
       └──► N resorption ────────── 0.42 (0.57 N resorption, -0.04 lifespan, -0.11 [N])
```

(B) Evergreen species (n=54)

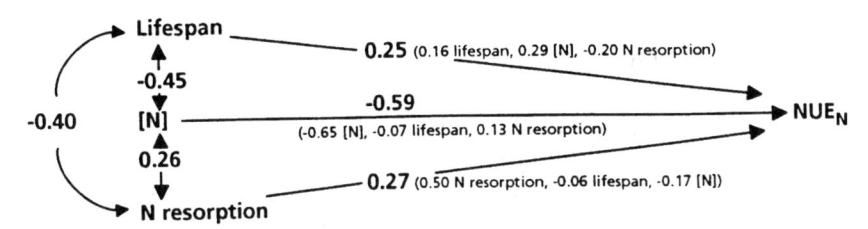

```
       ┌──► Lifespan ──────────── 0.25 (0.16 lifespan, 0.29 [N], -0.20 N resorption)
       │       ▲
       │     -0.45
       │       ▼                    -0.59
  -0.40│      [N] ─────────────────────────────────────────────────► NUE_N
       │       ▲       (-0.65 [N], -0.07 lifespan, 0.13 N resorption)
       │     0.26
       │       ▼
       └──► N resorption ────────── 0.27 (0.50 N resorption, -0.06 lifespan, -0.17 [N])
```

(C) Deciduous species (n=68)

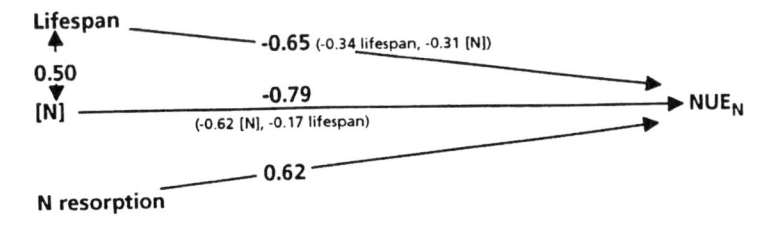

```
           Lifespan ──────────── -0.65 (-0.34 lifespan, -0.31 [N])
              ▲
            0.50
              ▼                    -0.79
             [N] ──────────────────────────────────────────► NUE_N
                         (-0.62 [N], -0.17 lifespan)

           N resorption ────────── 0.62
```

(D) Forbs and graminoids (n=26)

```
           [N] ──────────── -0.43
                                      ──► NUE_N
           N resorption ──── 0.81
```

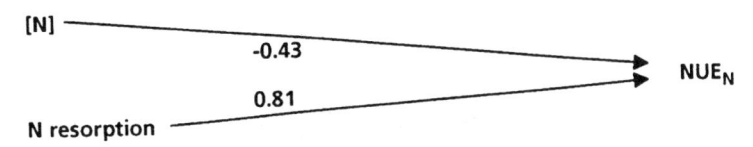

Fig. 11. Path diagrams describing the structure of the relationship between leaf-level nitrogen use efficiency (NUE$_N$) and various leaf characteristics for (A) the total data set, (B) evergreen woody species, (C) deciduous woody species and (D) forbs and graminoids. Numbers in bold type show the Pearson correlation coefficients among the variables, whereas values in parentheses partition the Pearson correlation coefficients between NUE$_N$ and the predictor variables into direct and indirect (i.e. attributable to indirect relationships with the other predictor variables) effects.

N concentration has an overriding effect on NUE_N, both directly and indirectly through effects on both other leaf traits. For deciduous shrubs and trees, N resorption efficiency from senescing leaves is more important as a determinant of NUE_N than in the evergreens (Figure 11C). However, in the deciduous shrubs and trees, leaf lifespan shows a negative correlation with NUE_N. Thus, within this growth-form a short leaf lifespan is associated with higher NUE_N. This is probably due to the fact that leaf lifespan is positively correlated with leaf N concentration. In the other growth-form with short-lived leaves, the forbs and graminoids, N resorption efficiency is the most important determinant of NUE_N (Figure 11D). In these herbaceous species, there is no significant effect of leaf lifespan on NUE_N. Thus, variation in the NUE_N of evergreen and deciduous shrubs and trees is determined mainly by variation in N concentration in mature leaves. In forbs and graminoids, however, N resorption efficiency is the most important predictor of variation in NUE_N. For all growth-forms, leaf lifespan is only of minor importance in explaining variation in NUE_N.

Theoretical analysis indicated that NUE_N shows the greatest increase per unit change in N concentration at values of N below 15 mg g^{-1} (equation (5); Figure 7) and also that the greatest increase per unit change in resorption efficiency was obtained at values of r above 0.75 (equation (6); Figure 8). This raises the question of how plants in real life maximize NUE_N and whether there are differences among growth-forms. To investigate this, we plotted the N concentration in mature leaves against N resorption efficiency from senescing leaves, and have indicated various NUE_N isoclines (Figure 12). This figure shows a number of interesting

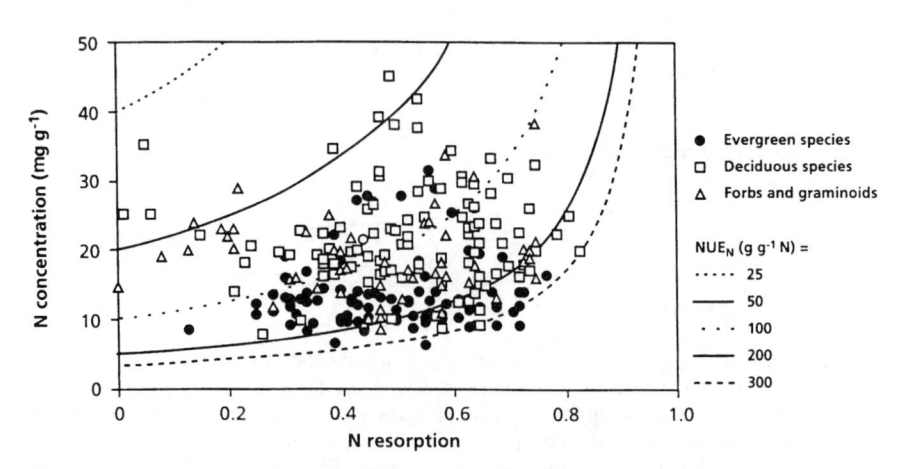

Fig. 12. N concentration in mature leaves as a function of N resorption from senescing leaves of species from various ecosystems ($n = 249$). Various NUE_N isoclines are shown. The relation between N concentration and N resorption was not significant.

phenomena. First, it appears that leaf-level NUE_N has a maximum value of about 300 g g^{-1} N (see Figure 10), which corresponds to a minimum N concentration in leaf litter of about 3 mg g^{-1}. Second, of the 47 cases where NUE_N exceeded 200 g g^{-1} N, in only four cases was the resorption efficiency greater than 0.75, whereas in 36 cases the N concentration in mature leaves was lower than 15 mg g^{-1}. Third, in 13 of 18 cases where NUE_N was greater than 250 g g^{-1} N, the species were evergreen shrubs and trees. From these observations, and from the relations presented in Figures 7–11, it can be concluded that plant species maximize their NUE_N more by the synthesis of leaves with low N concentrations than by having leaves with high N resorption efficiency and/or long lifespans. The species with the highest values of NUE_N are mainly evergreen shrubs and trees.

2. Phosphorus Use Efficiency

In evergreen shrubs and trees leaf-level phosphorus use efficiency was higher than in deciduous shrubs and trees, whereas the NUE_P of forbs and graminoids did not differ significantly from that of the other growth-forms (Figure 10). As was the case with NUE_N, there is large variation in NUE_P within growth-forms.

The path analyses showed that P resorption from senescing leaves was the most important determinant of leaf-level NUE_P (Figure 13). In evergreen species, it was the only determinant of NUE_P. This contrasts with the pattern for NUE_N, in which N concentration in mature leaves was the most important determinant of NUE_N. In general, the correlation between leaf traits and NUE_P was higher than that between leaf traits and NUE_N (see Figures 11 and 13). As was the case with NUE_N, leaf lifespan showed only a weak correlation (both directly and indirectly) with NUE_P.

According to the theoretical analysis, the greatest increase of NUE_P per unit change in P concentration would occur at P concentrations lower than 1.5 mg g^{-1} and/or at P resorption efficiencies higher than 0.75. An analysis similar to that for NUE_N (Figure 12) showed an interesting pattern. In contrast with the pattern for NUE_N, NUE_P was most apparently maximized due to very high P resorption efficiencies (Figure 14), with a maximum value of 0.98 in *Pinus strobus* (Small, 1972). This agrees with the findings of the path analyses. Apparently, P resorption efficiencies can be higher than N resorption effiencies. This has also been noted by Walbridge (1991), DeLucia and Schlesinger (1995) and Aerts (1996). This suggests that leaf P fractions are more readily broken up or retranslocatable than N fractions. The maximum value of NUE_P is about 15 000 g g^{-1} P, which corresponds to a minimum P concentration in senesced leaves of 0.07 mg g^{-1}. This extremely low P concentration is achieved mainly by the very high maximum P resorption efficiency (cf. equation (4)).

(A) Total data set (n=101)

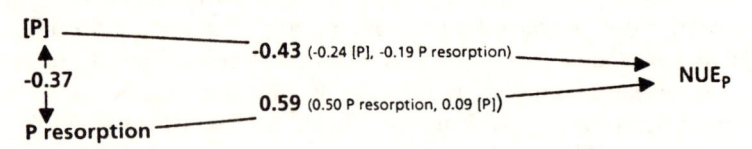

[P] ————————————— -0.43 (-0.24 [P], -0.19 P resorption) —————————→ NUE_P
↑↓ -0.37
P resorption ———— 0.59 (0.50 P resorption, 0.09 [P]) ————————————→

(B) Evergreen species (n=37)

P resorption ———— 0.82 ————————————————————→ NUE_P

(C) Deciduous species (n=46)

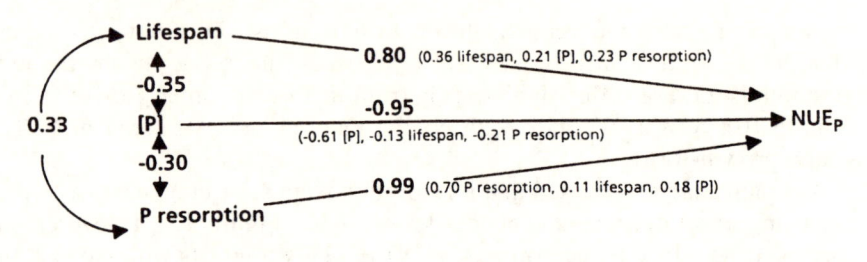

Lifespan ———— 0.80 (0.36 lifespan, 0.21 [P], 0.23 P resorption) ————→
↑ -0.35
0.33 [P] ———— -0.95 ————————————————————→ NUE_P
↑ -0.30   (-0.61 [P], -0.13 lifespan, -0.21 P resorption)
P resorption ———— 0.99 (0.70 P resorption, 0.11 lifespan, 0.18 [P]) ——→

(D) Forbs and graminoids (n=17)

[P] ————————————— -0.62 ————————————————→ NUE_P
P resorption ———— 0.86 ————————————————————→

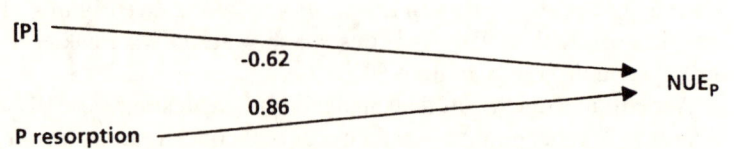

Fig. 13. Path diagrams describing the structure of the relationship between leaf-level phosphorus use efficiency (NUE_p) and various leaf characteristics for (A) the total data set, (B) evergreen woody species, (C) deciduous woody species and (D) forbs and graminoids. For further explanation see Figure 11.

C. Physiological Constraints on Maximization of Leaf-level NUE

It can be concluded that for woody species leaf-level nitrogen use efficiency is most strongly determined by variation in mature leaf N concentration. For herbaceous species, however, N resorption efficiency is the most important determinant of nitrogen use efficiency. For phosphorus use efficiency, P resorption efficiency contributes most strongly to maximization of NUE_p in all

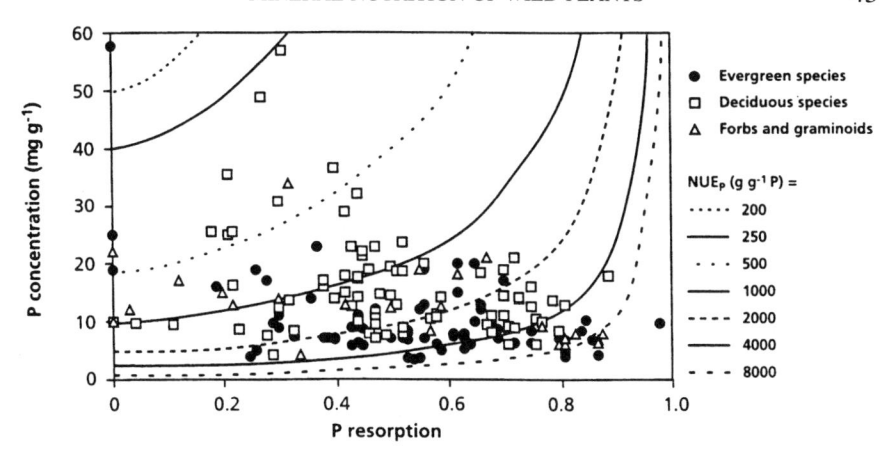

Fig. 14. P concentration in mature leaves as a function of P resorption from senescing leaves of species from various ecosystems ($n = 175$). Various NUE_p isoclines are shown. $[P] = 2.059 - 0.0145 \text{ PRESORPTION}$; $r^2 = 0.14$, $P < 0.01$.

growth-forms. It appears that, for all growth-forms, leaf lifespan is only a minor contributor to variation in both NUE_N and NUE_P.

An intriguing question arising from these findings is why N resorption does not contribute strongly to maximization of NUE_N, whereas P resorption does this very strongly. The most likely explanation is that N resorption has an upper limit of about 0.80, whereas P resorption has an upper limit of about 0.90 (see Walbridge, 1991; Aerts, 1996; Killingbeck, 1996). It is clearly shown that above resorption efficiencies of 0.75 NUE shows a strongly disproportional increase, owing to the inversely quadratic nature of the partial derivative of NUE against r (see equation (6)). Thus, this relatively small difference in maximum resorption efficiencies has disproportionate consequences for maximization of NUE. From our analysis it appears that the minimum level to which the N concentration can be reduced in senesced leaves is about 3 mg N g^{-1}; for P this is about 0.07 mg P g^{-1}. These concentrations appear to be the biochemical lower limits for N and P in senesced leaves and are similar to the values (3 mg N g^{-1} and 0.1 mg P g^{-1}) reported by Killingbeck (1996).

At first sight, it may seem strange that leaf lifespan is not the most important contributor to maximization of NUE, because lifespan data vary more than 20-fold (Figure 4), whereas both nutrient resorption efficiencies and nutrient concentrations vary much less (Figures 4 and 5). The weak relation between NUE_N and leaf lifespan was also found by Reich *et al.* (1992) in the analysis of their 'NITROGEN' data set ($r^2 = 0.27$). For nitrogen, the path analysis provides some clues to the solution of this problem. It is obvious that most of the variation in leaf lifespan is present within the evergreens (Figure 4). In this growth-form, there is a strong negative correlation between leaf lifespan and the N concentration of the leaf (see Reich et al., 1992) and N resorption efficiency (Figure 11). Thus, evergreen species with long leaf lifespans have low N concentrations in mature leaves and low N resorption efficiency. As a result of this pattern, and because of the different response of NUE to changes in leaf lifespan on the one hand and nutrient concentration and nutrient resorption efficiency on the other (see equations 4–7; Figures 7–9), leaf N concentration is the most important contributor to NUE_N in evergreen species. For the growth-forms with short-lived leaves (< 1 year), leaf lifespan is negatively related to NUE_N (deciduous shrubs and trees) or does not show a relation with NUE_N (forbs and graminoids). For phosphorus, this explanation does not hold. An important reason is that P resorption efficiency shows in all growth forms a very large variation (Figure 3B) and has very high maximum values. Thus, variation in P resorption efficiency can contribute more to maximization of NUE_P than leaf lifespan (see equations (7) and (8)).

D. Ecological Consequences

The theoretical analysis and the analysis of actual patterns in NUE and its underlying components made clear that in perennials NUE_N is maximised by the synthesis of low-nitrogen leaves. This is most apparent in evergreen shrubs and trees, species that are confined mainly to nutrient-poor sites (Monk, 1966; Aerts, 1995). However, this strategy has some important ecological consequences. First, the penalty for having leaves with low N concentrations is a low carbon assimilation rate and thus a low growth rate (Chapin, 1980; Hirose and Werger, 1987; Reich et al., 1992). As a result, evergreen species show a low responsiveness to environmental changes, in terms of both biomass increase and reproductive output (Chapin et al., 1983; Aerts, 1995). This low responsiveness of evergreens to environmental change is probably due to the trade-off between plant traits that reduce nutrient losses and those that lead to high rates of dry matter production (Aerts, 1990; Shaver and Chapin, 1991; Reich et al., 1992; Grime et al., 1997). This implies that changes in nutrient availability may lead to a decrease of the relative abundance of evergreen species. Such a decrease has been observed in heathlands in north-west Europe after anthropogenic increases of atmospheric nitrogen deposition (Aerts and Heil, 1993).

Compared with the evergreens, the deciduous shrubs and trees and the forbs and graminoids have higher leaf N concentrations, comparable N resorption efficiency, higher litter N concentrations, much shorter leaf lifespans, and lower values of NUE_N (Figures 4–6, 10). Furthermore, these growth-forms are characterized by another type of relationship between NUE_N and the underlying leaf traits (Figure 11). In these growth-forms, and especially in the forbs and graminoids, NUE_N depends more on resorption efficiency than in the evergreens. In addition, these growth-forms have a higher SLA than evergreens (Reich et al., 1991, 1992; Reich, 1993). This suite of leaf traits (high N concentrations, high SLA and high leaf turnover rates) is the main determinant of a high potential productivity of plants and makes these growth-forms more plastic in their response to environmental changes (Poorter and Remkes, 1990; Poorter et al., 1990; Reich et al., 1992; Grime et al., 1997; Hunt and Cornelissen, 1997). As a result, species with these traits dominate the more fertile habitats and replace evergreen species when nutrient availability in a particular habitat is increased (Aerts and Berendse, 1988; Aerts et al., 1990, 1991; Fox, 1992; Jonasson, 1992; Havström et al., 1993; Wookey et al., 1993; Parsons et al., 1994, 1995; Aerts and Bobbink, 1999).

The ecological consequences of traits leading to high NUE_P (mainly high P resorption efficiency) are of a different nature. High P resorption efficiency leads to very low P concentrations in litter, which may affect the decomposability of that litter (Aerts and De Caluwe, 1997). This suggests that NUE and leaf litter decomposability are strongly interrelated, as discussed in sections X and XI.

X. LITTER DECOMPOSITION

Decomposition of plant litter is a key process in the nutrient cycle of terrestrial ecosystems (Meentemeyer, 1978; Vitousek, 1982; Van Vuuren et al., 1993; Vitousek et al., 1994; Aerts and De Caluwe, 1997). Litter decomposition rates are controlled by environmental conditions, by the chemical composition of the litter and by soil organisms (Swift et al. 1979; Blair et al. 1990; Beare et al. 1992). Mineralization of nutrients contained in litter is often referred to a three-stage process: first, nutrients in soluble form are leached from the litter; second, nutrient immobilization occurs; and, finally, net nutrient mineralization from the litter takes place, thereby making nutrients available for plant uptake again.

A. Nutrient Leaching from Senesced Leaves

There is a marked difference between the leachability of different ions: K and Mg are very mobile and easily leached from fresh litter, whereas N, P and Ca show a much lower rate of leaching (Swift et al., 1979). Although many authors have emphasized the importance of nutrient leaching from fresh litter, there are relatively few quantitative estimates of this process.

Most authors estimate how much N and P are leached from senesced leaves by expressing it as a percentage of the initial N and P content of those leaves (Table 3). From these data it is clear that nutrient leaching is extremely variable.

Table 3
Potential N and P leaching from senescing leaves, expressed as a percentage of initial leaf N and P pool in senesced leaves, as reported in different studies

Reference	Species	N leaching (%)	P leaching (%)
Morton (1977)	*Molinia caerulea* in England	0	0
Chapin and Kedrowski (1983)	Four deciduous and evergreen trees in Alaska	0.1–0.3	0.04–0.6
Boerner (1984)	Four deciduous tree species in Ohio	0–8	0–10
Ganzert and Pfadenhauer (1986)	*Schoenus ferrugineus* in Germany	0	0
Pastor et al. (1987)	*Schizachyrium scoparium* in Minnesota	4–57*	N.D.
Chapin and Moilanen (1991)	*Betula papyrifera* in Alaska	31	31
Ibrahima et al. (1995)	Seven evergreen and deciduous Mediterranean species	0–15†	N.D.
Aerts and De Caluwe (1997)	Four *Carex* species in fens in the Netherlands	8–12	3–14

*Four levels of N supply; positive relation between initial N content and N leaching.
†In four species 0%, in three other 9–15%.
N.D., not determined.

Site fertility effects on leaching have been investigated by several authors. Boerner (1984) studied N leaching in relation to site fertility in four deciduous tree species. He found lower N and P leaching at low-fertility sites in 10 of 16 cases. Pastor et al. (1987) found no consistent effect of fertilization on initial N content of leaf litter of *Schizachyrium scoparium*, but they did find a strong positive correlation between N leaching and initial N content of the leaves. Aerts and De Caluwe (1997) studied nutrient supply effects on N and P leaching from senesced leaves of four *Carex* species. They found that potential N leaching from field litter differed among species and amounted to values between 8% and 12% of the initial N content of the leaves. However, there was no consistent difference between N leaching of field litter and of litter produced under high-fertility growth conditions. Potential P leaching from field litter varied between 3% and 14%, and was strongly increased when plants were grown under high P conditions. In that case, a maximum value of 65% was observed in *Carex lasiocarpa*. Rustad

(1994) also found such high P leaching from tree leaf litter from a red spruce ecosystem in Maine, USA, which had initial C : P ratios similar to those in the experimental litters from the study of Aerts and De Caluwe (1997). These data indicate that litter with high N and P concentrations generally shows relatively higher N and P leaching losses than low-nutrient litter.

B. Climatological and Chemical Controls on Litter Decomposition

There are three main levels of litter decomposition control which operate in the following order: climate > litter chemistry > soil organisms (Swift *et al.*, 1979; Lavelle *et al.*, 1993). Climate has a direct effect on litter decomposition due to the effects of temperature and moisture. However, as a result of the climatic control of soil formation and nutrient cycling (Vitousek and Sanford, 1986; Lavelle *et al.*, 1993), climate must also have an indirect effect through the climatic impact on litter chemistry (Swift and Anderson, 1989).

Aerts (1997b) analysed the effects of climate and litter chemistry on first-year leaf litter decomposition using literature data from 44 locations, ranging from cool temperate sites to humid tropical sites. Actual evapotranspiration (AET) was used as an index for the climatic control of decomposition, and litter chemistry was characterized by N and P concentrations, C : N and C : P ratios, lignin concentrations, and lignin : N and lignin : P ratios. The decomposability of the litter was expressed by the decomposition constants (k values) which are based on the single exponential decay model proposed by Olson (1963):

$$W_t = W_0 \, e^{-kt} \quad (9)$$

in which W_t and W_0 are litter masses at time t and time 0; k is the decomposition constant (year^{-1}) and t is time (years).

At a global scale, climate (expressed as AET) is the best predictor for the decomposition constants of the litter (Figure 15), whereas litter chemistry parameters have much lower predictive values (Table 4). This is in agreement with work of Berg *et al.* (1993) who studied litter decomposition of *Pinus* species using 39 experimental sites spanning climatic regions from subarctic to subtropical and Mediterranean sites. Also in their experimental study, climate (expressed as AET) exerted the strongest influence on litter decomposition over broad geographical scales, whereas litter chemistry was important at local scales only. Using path analysis, Aerts (1997b) showed that the climatic control on litter decomposability is mediated partly through an indirect effect of climate on litter chemistry. This is due to the fact that climate determines to a large extent the distribution of soil types. Indirectly, the soil type may influence decomposition via the composition of the decomposer

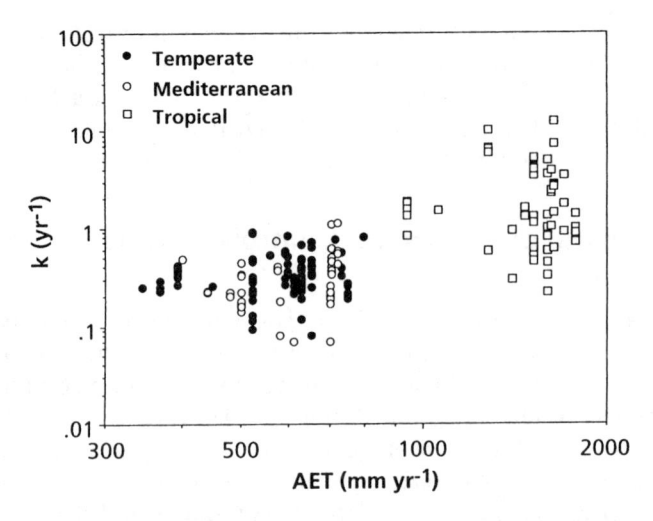

Fig. 15. Relationship between first-year leaf-litter decomposition constants (k) and actual evapotranspiration (AET) for temperate, Mediterranean and tropical sites. Redrawn from Aerts (1997b).

community and the resource quality of the plant litter input (Swift and Anderson, 1989). Vitousek and Sanford (1986) found that, in comparison with rates in temperate forests, N mineralization rates and nitrification rates are high on fertile tropical lowland soils. These results are consistent with the high N concentrations in litter from tropical plant species (Aerts, 1997b) and rapid circulation of N in vegetation in most lowland tropical sites (except on spodosols).

Within a particular climatic region, however, litter chemistry parameters exert the strongest control on litter decomposability, especially in the tropics (Table 4). In general, litters from the tropical sites have higher N concentrations and lower lignin : N ratios than litters from other climatic regions, which results in a higher decomposability of the litter. In both the tropics and the Mediterranean region, the lignin : N ratio is the best chemical predictor of litter decomposability. In the temperate region, however, there appeared to be no good chemical predictor of decomposability. This is in contrast to the findings of many field studies conducted in the temperate zone, in which limited sets of species were studied within a particular site (e.g. Melillo *et al.*, 1982; Meentemeyer and Berg, 1986; Blair, 1988; Taylor *et al.*, 1991). In these studies, a relationship was found between litter decomposability and litter chemistry. A possible explanation for this discrepancy might be that within a specific site there is a match between the local climate, soil characteristics, soil fauna and litter chemistry. As a result, the microbial activity is closely related to the resource quality, but in a way that is specific for the site. When data from

Table 4
Summary of significant regressions of decomposition constants (k: year^{-1}) against actual evapotranspiration (AET: mm year^{-1}) and various litter chemistry parameters (percentage concentrations) in different climatic regions (from Aerts, 1997b)

Parameter	d.f.	Regression (k)	r^2	P
All climatic zones				
AET	192	0.000057 (AET)$^{1.371}$	0.46	< 0.0001
C : N ratio	62	20.184 (CNRATIO)$^{-0.966}$	0.26	< 0.0001
N concentration	187	0.555 (NCONC)$^{0.888}$	0.24	< 0.0001
Lignin : N ratio	130	4.612 (LIGNIN/N)$^{-0.718}$	0.24	< 0.0001
P concentration	133	1.601 (PCONC)$^{0.369}$	0.10	< 0.0001
C : P ratio	57	4.345 (CPRATIO)$^{-0.350}$	0.06	< 0.05
Lignin : P ratio	94	0.993 (LIGNIN/P)$^{-0.133}$	0.04	< 0.05
Temperate zone				
N concentration	89	0.895 (NCONC)$^{0.174}$	0.05	< 0.05
Mediterranean zone				
Lignin : N ratio	29	3.165 (LIGNIN/N)$^{-0.714}$	0.24	< 0.005
C : N ratio	23	7.830 (CNRATIO)$^{-0.781}$	0.16	< 0.05
AET	37	0.00041 (AET)$^{1.024}$	0.08	< 0.05
N concentration	37	0.901 (NCONC)$^{0.546}$	0.08	< 0.05
Tropical zone				
Lignin : N ratio	34	20.26 (LIGNIN/N)$^{-0.907}$	0.57	< 0.0001
Lignin : P ratio	28	26.99 (LIGNIN/P)$^{-0.525}$	0.44	< 0.0001
C : P ratio	16	3231 (CPRATIO)$^{-1.175}$	0.39	< 0.005
C : N ratio	16	19.82 (CNRATIO)$^{-0.729}$	0.26	< 0.05
Lignin concentration	34	24.07 (LIGNIN)$^{-0.957}$	0.26	< 0.001
N concentration	57	1.291 (NCONC)$^{0.744}$	0.24	< 0.0001
P concentration	51	5.590 (PCONC)$^{0.534}$	0.23	< 0.0001

many sites are combined and litter decomposition is related to litter chemistry, as in the analysis of Aerts (1997b), the variation at the intermediate level (local climate, soil characteristics, soil fauna) is ignored. As a result, the relationship between litter decomposition and litter chemistry can be masked by this unaccounted for variation.

C. Variation in Litter Decomposition among Growth-forms

Leaves of evergreens usually contain higher concentrations (on a mass basis) of lignin and other secondary compounds than leaves of deciduous species (Chapin, 1989; Aerts and Heil, 1993; Van Vuuren *et al.*, 1993). These high amounts of secondary compounds in leaves of low-productivity species are usually explained as a defence mechanism against herbivory (Coley *et al.*, 1985; Grime *et al.*, 1996) or as structural components that enable evergreen leaves to withstand unfavourable conditions in the non-growing season

(Poorter and Bergkotte, 1992). This may have serious implications for the decomposability of those leaves (Aerts, 1995). This effect can be reinforced by the lower N concentrations in the leaves of evergreens compared with leaves from other growth-forms (Figure 2). The rate of litter decomposition is in many cases negatively correlated with the lignin concentration or the lignin : N ratio, and positively correlated with the N concentration (Aerts, 1997b). This suggests that the leaf decomposition rate of evergreen species is lower than that of deciduous species. We investigated this hypothesis by a reanalysis of the leaf litter decomposition data from Aerts (1997b). As litter decomposition is controlled most strongly by climate (see above), the data for each climatic region were analysed separately. In both temperate and tropical regions leaf litter decomposition rates of evergreen shrubs and trees and of forbs and graminoids are significantly lower than those of deciduous shrubs and trees (Table 5). In the Mediterranean region, however, there was no difference in leaf litter decomposability between evergreen and deciduous species. A possible reason might be that slow-growing deciduous species, as found in this region, also contain high concentrations of (hemi)cellulose, insoluble sugars and lignin (Niemann et al., 1992; Poorter and Bergkotte, 1992), which may lead to low decomposition rates.

A low decomposability of plant litter in low-nutrient environments is less detrimental to evergreen than to deciduous species (Hobbie, 1992; Chapin, 1993; Van Breemen, 1993; Berendse, 1994b). Using a simulation model in which nutrient cycling and plant competition between an evergreen and a deciduous species were included, Berendse (1994b) demonstrated that the plant traits of evergreens (long tissue lifespans, low nutrient concentrations and low litter decomposition rates) can be favourable under nutrient-limited growth conditions. Low litter decomposability and the resulting low rate of nutrient release from that litter, as observed in evergreen species, leads to longer dominance of the evergreen species. This implies that the plant characteristics of evergreens do not only lead to high NUE, but also keep soil fertility

Table 5
First-year decomposition constants (k: year^{-1}) for leaf litter in different climatic zones, according to growth-form.

Zone	Evergreen species	Deciduous species	Forbs and graminoids
Temperate	0.31 ± 0.09 (33)	0.45 ± 0.20* (32)	0.29 ± 0.11 (22)
Mediterranean	0.37 ± 0.26 (20)	0.40 ± 0.24 (13)	N.D.
Tropical	1.37 ± 1.16 (21)	3.20 ± 1.45* (10)	1.66 ± 2.10 (13)

Values are mean ± SD with number of observations shown in parentheses.
*$P < 0.05$ versus other two growth-forms in that zone.

low and thereby influence the competitive balance with deciduous species in their favour (see Hobbie, 1992; Van Breemen, 1993).

XI. TRADE-OFF BETWEEN NUTRIENT USE EFFICIENCY AND LITTER DECOMPOSABILITY?

In section IX it was shown that high nutrient resorption contributes to high NUE but leads to low litter nutrient concentrations, and vice versa. This has direct implications for litter decomposition rates and nutrient release, because decomposition and nutrient release from litter are often positively related to the nutrient concentrations in the litter and negatively related to the C : nutrient ratio or the lignin : nutrient ratio (Coulson and Butterfield, 1978; Taylor *et al.*, 1989; Table 4). These observations suggest that there may be a trade-off between leaf-level NUE and litter decomposability. Aerts (1997a) investigated this hypothesis by a reanalysis of the data presented in section X.B. This analysis shows that there is indeed a trade-off between both NUE_N and NUE_P and litter decomposability: first-year decomposition constants are negatively related to NUE_N and NUE_P (Figure 16). However, the percentage of variance explained by the regression models is relatively low (24% for NUE_N, 10% for NUE_P). This is due mainly to the fact that in many low-productivity species both NUE and *k* are low (Figure 16), i.e. the rate of decomposition is low even if the N or P concentration of the litter is relatively high.

Owing to the strong climatic control on litter decomposition rates, part of the large scatter might be explained by the different climatic origin of the data points. However, a reanalysis of the data for each climatic zone separately did not result in an improved relationship between *k* values and NUE (data not shown). This was also the case when a separation was made between growth-forms (data not shown). A more likely explanation for the relatively weak relationship between NUE and litter decomposability lies in the secondary chemistry of the species considered. The chemistry of secondary compounds in plants is directly related to both the growth rate of plant species and to the decomposability of litter. Although both aspects have received considerable attention over the past few years, only a few attempts have been made to search for a casual connection between them (see Grime *et al.*, 1996; Cornelissen and Thompson, 1997). In general, low productivity species have higher amounts of secondary compounds than high-productivity species (Ellis, 1972; Bryant *et al.*, 1983; Niemann *et al.*, 1992; Poorter and Bergkotte, 1992).

These differences in secondary chemistry have a profound effect on the productivity of plant species per unit N in the plant and thereby on their NUE. On a daily basis, low-productivity species produce less biomass per unit N in the plant than high-productivity species (Lambers and Poorter, 1992). This difference is partly compensated by the longer lifespan of plant tissues in low-productivity species, but it may still lead to equal or lower plant-level NUE

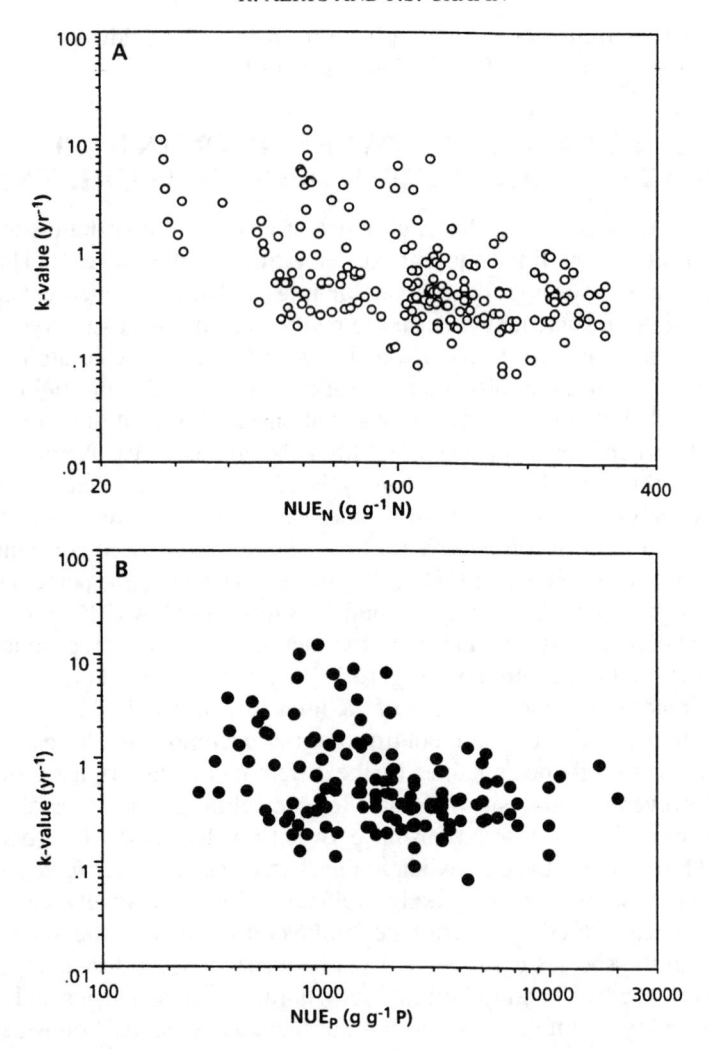

Fig. 16. Relationship between first-year leaf-litter decomposition constants (k) and (A) leaf-level nitrogen use efficiency (NUE_N) and (B) phosphorus use efficiency (NUE_P). Data were obtained from Aerts (1997b). Regression models: $k = 33.11\ (NUE_N)^{-0.888}$, $r^2 = 0.24$, 187 = d.f., $P < 0.0001$; $k = 8.69\ (NUE_P)^{-0.368}$, $r^2 = 0.10$, 133 = d.f., $P < 0.001$. (A) is redrawn from Aerts (1997a).

compared with that in high-productivity species (Aerts, 1990, 1995; Son and Gower, 1991). Moreover, high concentrations of phenolics in leaves, as observed in low-productivity species growing at infertile sites (Ellis, 1972; Haukioja et al., 1985; Nicolai, 1988), may lead to precipitation of proteins before protein hydrolysis, which reduces nutrient resorption (Chapin and Kedrowski, 1983) and thereby NUE. Thus, low-productivity species do not necessarily have a high NUE (see Aerts, 1990, 1995; Son and Gower, 1991).

The secondary chemistry also affects litter decomposability. Litter decay is in many cases negatively related to the lignin concentration or the lignin : nutrient ratio of the litter (Berg, 1984; Van Vuuren *et al.*, 1993; Aerts, 1997b; Aerts and De Caluwe, 1997). Also the phenolics : nutrient ratio may have a strongly retarding effect on litter decay (Thomas Asakawa, 1993; Constantinides and Fownes, 1994; Aerts and De Caluwe, 1997).

These observations show that, in addition to their significant role in defence against herbivory, the presence of secondary compounds in plants may lead both to low NUE and to low litter decomposability, and alter the hypothesized trade-off between NUE and litter decomposability. This may explain the large number of data points in Figure 16 in which values of both NUE and k are low. This combination of low productivity (and thus low litter production), low NUE and low litter decomposability may lead to a low rate of ecosystem N cycling (Chapin, 1993; Van Breemen, 1993). This may prevent the invasion of highly competitive species that are dependent on high N availability (Aerts and Van der Peijl, 1993; Berendse, 1994a,b). However, the decomposition data presented here refer to first-year mass loss and not to net N mineralization from the litter. Especially during the initial stages of decomposition, considerable N immobilization may occur (Blair, 1988; Aerts and De Caluwe, 1997). Moreover, element dynamics in decomposing litter may differ between the early and late stages of decomposition (Rustad, 1994).

XII. CONCLUSIONS: PLANT STRATEGIES

In this review, we have distinguished four main topics in the nutritional ecology of plants: nutrient-limited plant growth, nutrient acquisition, nutrient use efficiency, and nutrient recycling through litter decomposition. Within each of these areas, plants have evolved sets of adaptive traits ('strategies') with respect to their nutritional ecology. Species from nutrient-poor habitats often have the ability to take up organic nitrogen sources, they have a root allocation pattern directed towards the acquisition of nutrients that diffuse slowly to the roots, and they possess traits that lead to high nutrient retention such as tissues with slow turnover rates, low concentrations of mineral nutrients and high concentrations of secondary compounds, which serve amongst others as a defence against herbivory. All these traits lead to a low growth rate and/or to a low potential of resource capture. High nutrient resorption efficiency, however, is important in all species and does not differ consistently between species from nutrient-poor and nutrient-rich environments. Species from nutrient-rich habitats have traits that lead to rapid growth and quick capture of both above- and below-ground resources. The fact that this differentiation does occur between species from habitats differing in soil fertility strongly suggests that there is a trade-off between their respective traits. This trade-off is adequately described by the NUE model of Berendse and Aerts (1987) in which two components of NUE are distinguished: the mean residence time (MRT) of nitrogen which reflects the 'nutrient retention strategy', and nitrogen productivity (A_{NP}) which reflects the 'rapid growth strategy'. Because of selection on the components of NUE rather than on NUE itself, there are no clear patterns in whole-plant NUE when comparing species from environments differing in soil fertility. At the phenotypic level, however, NUE decreases with increasing soil fertility.

Our analysis shows that the underlying leaf traits associated with high leaf-level NUE differ both among growth forms and among the nutrients under consideration. For woody species, leaf-level nitrogen use efficiency (NUE_N) is most strongly determined by variation in mature leaf N concentration. For herbaceous species, however, N resorption efficiency is the most important determinant of NUE_N. For phosphorus use efficiency (NUE_P), the situation is different: P resorption efficiency contributes most strongly to maximization of NUE_P in all growth-forms. This is due to the fact that maximum P resorption efficiency is higher than maximum N resorption efficiency and that at high resorption efficiencies (r) NUE is disproportionally increased by small increases of r. It appears that, for all growth-forms, leaf lifespan is only a minor contributor to variation in both leaf-level NUE_N and NUE_P. This is not in agreement with the pattern at the whole-plant level. Evergreen species have higher leaf-level NUE_N and NUE_P than other growth-forms.

It is important to note that the traits associated with competitive dominance in habitats differing in soil fertility may also have effects on ecosystem nutrient cycling. In nutrient-poor environments, species produce relatively small amounts of litter due to low productivity and long lifespans of the various tissues. This litter generally has low nutrient concentrations and high concentrations of secondary compounds such as lignin and phenolics. Thus, species from nutrient-poor environments produce litter that decomposes slowly and from which only low amounts of nutrients are released. The opposite holds for species from fertile environments. Owing to their high productivity and high tissue turnover rates, they produce relatively large amounts of litter. Moreover, this litter contains relatively high concentrations of mineral nutrients and low concentrations of secondary compounds. As a result, this litter decomposes relatively quickly and releases high amounts of nutrients. Thus, in nutrient-poor ecosystems the combination of low productivity (and thus low litter production) and low litter decomposability may lead to a low rate of ecosystem nutrient cycling. This may prevent the invasion of highly competitive species which are dependent on high nutrient availability. On the other hand, the traits of species from fertile environments lead to a high rate of ecosystem nutrient cycling and this excludes slow-growing and nutrient-conserving species from these habitats. From these patterns, it can be concluded that the strategies of species from nutrient-poor and nutrient-rich habitats promote long-term ecosystem stability. This is an important evolutionary consequence of these strategies, although it is questionable whether long-term ecosystem stability is prone to natural selection.

Within the nutritional ecology of plants, several major issues still have to be resolved. A fascinating issue is how the differential uptake of inorganic and organic N sources contributes to plant coexistence and how this is related to ecosystem nitrogen cycling and ecosystem stability. A major gap in our knowledge of patterns in NUE along soil fertility gradients is that there are hardly any data on whole-plant NUE in natural ecosystems. Moreover, we still do not know which factors control nutrient resorption from senescing tissues, how this is related to the dynamics of secondary compounds in senescing tissues, and how these processes are reflected in litter decomposition and nutrient release from litter. Moreover, the fact that both mycorrhizal and non-mycorrhizal plants are capable of taking up various organic N compounds requires that studies of plant nutrition in natural ecosystems focus more on the dynamics of these compounds in the soil.

REFERENCES

Aerts, R. (1989a). The effect of increased nutrient availability on leaf turnover and above-ground productivity of two evergreen ericaceous shrubs. *Oecologia* **78**, 115–120.

Aerts, R. (1989b). Aboveground biomass and nutrient dynamics of *Calluna vulgaris* and *Molinia caerulea* in a dry heathland. *Oikos* **56**, 31–38.

Aerts, R. (1990). Nutrient use efficiency in evergreen and deciduous species from heathlands. *Oecologia* **84**, 391–397.

Aerts, R. (1993). Biomass and nutrient dynamics of dominant plant species in heathlands. In: *Heathlands, Patterns and Processes in a Changing Environment* (Ed. by R. Aerts and G.W. Heil), pp. 51–84. Kluwer Academic Publishers, Dordrecht.

Aerts, R. (1995). The advantages of being evergreen. *Trends Ecol. Evol.* **10**, 402–407.

Aerts, R. (1996). Nutrient resorption from senescing leaves of perennials: are there general patterns? *J. Ecol.* **84**, 597–608.

Aerts, R. (1997a). Nitrogen partitioning between resorption and decomposition pathways: a trade-off between nitrogen use efficiency and litter decomposability? *Oikos* **80**, 603–606.

Aerts, R. (1997b). Climate, leaf litter chemistry and leaf litter decomposition in terrestrial ecosystems: a triangular relationship. *Oikos* **79**, 439–449.

Aerts, R. and Berendse, F. (1988). The effect of increased nutrient availability on vegetation dynamics in wet heathlands. *Vegetatio* **76**, 63–69.

Aerts, R. and Bobbink, R. (1999). The impact of atmospheric nitrogen deposition on vegetation processes in terrestrial non-forest ecosystems. In: *The Impact of Nitrogen Deposition on Natural and Semi-Natural Ecosystems* (Ed. by S. Langan Kluwer, Dordrecht, pp. 85–122.

Aerts, R. and De Caluwe, H. (1989). Aboveground productivity and nutrient turnover of *Molinia caerulea* along an experimental gradient of nutrient availability. *Oikos* **54**, 320–324.

Aerts, R. and De Caluwe, H. (1994). Nitrogen use efficiency of *Carex* species in relation to nitrogen supply. *Ecology* **75**, 2362–2372.
Aerts, R. and De Caluwe, H. (1995). Interspecific and intraspecific differences in shoot and leaf lifespan of four *Carex* species which differ in maximum dry matter production. *Oecologia* **102**, 467–477.
Aerts, R. and De Caluwe, H. (1997). Nutritional and plant-mediated controls on leaf litter decomposition of *Carex* species. *Ecology* **78**, 244–260.
Aerts, R. and Heil, G.W. (Editors). (1993). *Heathlands, Patterns and Processes in a Changing Environment*. Kluwer Academic Publishers, Dordrecht.
Aerts, R. and Van der Peijl, M.J. (1993). A simple model to explain the dominance of low-productive perennials in nutrient-poor habitats. *Oikos* **66**, 144–147.
Aerts, R., Boots, R.G.A. and Van der Aart, P.J.M. (1991). The relation between above- and belowground biomass allocation patterns and competitive ability. *Oecologia* **87**, 551–559.
Aerts, R., Berendse, F., de Caluwe, H. and Schmitz, M. (1990). Competition in heathland along an experimental gradient of nutrient availability. *Oikos* **57**, 310–318.
Aerts, R., Wallén, B. and Malmer, N. (1992a). Growth-limiting nutrients in *Sphagnum*-dominated bogs subject to low and high atmospheric nitrogen supply. *J. Ecol.* **80**, 131–140.
Aerts, R., De Caluwe, H. and Konings, H. (1992b). Seasonal allocation of biomass and nitrogen in four *Carex* species from mesotrophic and eutrophic fens as affected by nitrogen supply. *J. Ecol.* **80**, 653–664.
Ågren, G.I. (1985). Theory for growth of plants derived from the nitrogen productivity concept. *Physiol. Plantarum* **64**, 17–28.
Ågren, G.I. (1988). The ideal nutrient productivities and nutrient proportions. *Plant Cell Environ.* **11**, 613–620.
Allen, M.F. (1991). *The Ecology of Mycorrhizae*. Cambridge University Press, Cambridge.
Atkin, O.K. (1996). Reassessing the nitrogen relations of arctic plants: a mini-review. *Plant Cell Environ.* **19**, 695–704.
del Arco, J.M., Escudero, A. and Garrido, M.V. (1991). Effects of site characteristics on nitrogen retranslocation from senescing leaves. *Ecology* **72**, 701–708.
Bazzaz, F.A. and Harper, J.L. (1977). Demographic analysis of the growth of *Linum usitatissimum*. *New Phytol* **78**, 193–208.
Beare, M.H., Parmelee, R.W., Hendrix, P.F., Cheng, W., Coleman, D.C. and Crossley, D.A. (1992). Microbial and faunal interactions and effects on litter nitrogen and on decomposition in agroecosystems. *Ecol. Monogr.* **62**, 569–591.
Berendse, F. (1990). Organic matter accumulation and nitrogen mineralization during secondary succession in heathland ecosystems. *J. Ecol.* **78**, 413–427.
Berendse, F. (1994a). Competition between plant populations at low and high nutrient supplies. *Oikos* **71**, 253–260.
Berendse, F. (1994b). Litter decomposability—a neglected component of plant fitness. *J. Ecol.* **82**, 187–190.
Berendse, F. and Aerts, R. (1987). Nitrogen use efficiency: a biologically meaningful definition? *Funct. Ecol.* **1**, 293–296.
Berendse, F. and Elberse, W.T. (1989). Competition and nutrient losses from the plant. In: *Causes and Consequences of Variation in Growth Rate and Productivity of Higher Plants* (Ed. by H. Lambers, H. Konings, M.L. Cambridge and T.L. Pons), pp. 269–284. SPB Academic Publishing, The Hague.
Berg, B. (1984). Decomposition of moss litter in a mature Scots pine forest. *Pedobiologia* **26**, 301–308.

Berg, B., Berg, M.P., Bottner, P., Box, E., Breymeyer, A., Couteaux, M. *et al.* (1993). Litter mass loss rates in pine forests of Europe and eastern United States: some relationships with climate and litter quality. *Biogeochemistry* **20**, 127–159.
Berntson, G.M., Farnsworth, E.J. and Bazzaz, F.A. (1995). Allocation, within and between organs, and the dynamic of root length changes in two birch species. *Oecologia* **101**, 439–447.
Birk, E.M. and Vitousek, P.M. (1986). Nitrogen availability and nitrogen use efficiency in loblolly pine stands. *Ecology* **67**, 69–79.
Blair, J.M. (1988). Nitrogen, sulfur and phosphorus dynamics in decomposing deciduous leaf litter in the southern Apalachians. *Soil Biol. Biochem.* **20**, 693–701.
Blair, J.M., Parmelee, R.W. and Beare, M.H. (1990). Decay rates, nitrogen fluxes, and decomposer communities of single- and mixed species foliar litter. *Ecology* **71**, 1976-1985.
Bobbink, R. (1991). Effects on nutrient enrichment in Dutch chalk grassland. *J. Appl. Ecol.* **28**, 28–41.
Bobbink, R. (1992). Critical loads for nitrogen eutrophication of terrestrial and wetland ecosystems based upon changes in vegetation and fauna. In: *Critical Loads for Nitrogen* (Ed. by P. Grennfelt and R. Thornelof), pp. 111–159. Nord Miljrapport 41, Nordic Council of Ministers, Copenhagen.
Boerner, R.E.J. (1984). Foliar nutrient dynamics and nitrogen use efficiency of four deciduous tree species in relation to site fertility. *J. Appl. Ecol.* **21**, 1029–1040.
Boerner, R.E.J. (1985). Foliar nutrient dynamics, growth, and nutrient use efficiency of *Hamamelis virginiana* in three forest microsites. *Can. J. Bot.* **63**, 1476–1481.
Boeye, D., Verhagen, B., Van Haesebroeck, V. and Verheyen, R.F. (1997). Nutrient limitation in species-rich lowland fens. *J. Veg. Sci.* **8**, 415–424.
Boller, B.C. and Nösberger, J. (1987). Symbiotically fixed nitrogen from field-grown white and red clover mixed with ryegrass at low levels of ^{15}N-fertilization. *Plant Soil* **104**, 219–226.
Boot, R.G.A. and Den Dubbelden, K.C. (1990). Effects of nitrogen supply on growth, allocation and gas exchange characteristics of two perennial grasses from inland dunes. *Oecologia* **85**, 15–121.
Boot, R.G.A. and Mensink, M. (1990). Size and morphology of root systems of perennial grasses from contrasting habitats as affected by nitrogen supply. *Plant Soil* **129**, 291–299.
Bowman, W.D. (1994). Accumulation and use of nitrogen and phosphorus following fertilization in two alpine tundra communities. *Oikos* **70**, 261–270.
Bridgham, S.D., Pastor, J., McClaugherty, C.A. and Richardson, C.J. (1995). Nutrient-use efficiency: a litterfall index, a model, and a test along a nutrient-availability gradient in North Carolina peatlands. *Am. Nat.* **145**, 1–21.
Brouwer, R. (1962a). Distribution of dry matter in the plant. *Neth. J. Agric. Sci.* **10**, 361–376.
Brouwer, R. (1962b). Nutritive influences on the distribution of dry matter in the plant. *Neth. J. Agric. Sci.* **10**, 399–408.
Bryant, J.P., Chapin, F.S. and Klein, D.R. (1983). Carbon/nutrient balance of boreal plants in relation to vertebrate herbivory. *Oikos* **40**, 357–368.
Campbell, B.D. and Grime, J.P. (1989). A comparative study on plant responsiveness to the duration of episodes of mineral nutrient enrichment. *New Phytol.* **112**, 261–267.
Campbell, B.D. and Grime, J.P. (1992). An experimental test of plant strategy theory. *Ecology* **73**, 15–29.
Campbell, B.D., Grime, J.P. and Mackey, J.M.L. (1991). A trade-off between scale and precision in resource foraging. *Oecologia* **87**, 532–538.

Chabot, B.F. and Hicks, D.J. (1982). The ecology of leaf life spans. *Annu. Rev. Ecol. Syst.* **13**, 229–259.

Chapin, F.S. III (1980). The mineral nutrition of wild plants. *Annu. Rev. Ecol. Syst.* **11**. 233–260.

Chapin, F.S. III (1988). Ecological aspects of plant mineral nutrition. *Adv. Miner. Nutr.* **3**, 61–191.

Chapin, F.S. III (1989). The cost of tundra plant structures: evaluation of concepts and currencies. *Am. Nat.* **133**, 1–19.

Chapin, F.S. III (1991). Effects of multiple stresses on nutrient availability and use. In: *Response of Plants to Multiple Stresses* (Ed. by H.A. Mooney, W.E. Winner and E.J. Pell), pp. 67–88. Academic Press, San Diego.

Chapin, F.S. III (1993). The evolutionary basis of biogeochemical soil development. *Geoderma* **7**, 223–227.

Chapin, F.S. III and Kedrowski, R.A. (1983). Seasonal changes in nitrogen and phosphorus fractions and autumn retranslocation in evergreen and deciduous taiga trees. *Ecology* **64**, 376–391.

Chapin, F.S. III and Moilanen, L. (1991). Nutritional controls over nitrogen and phosphorus resorption from Alaskan birch leaves. *Ecology* **72**, 709–715.

Chapin, F.S. III and Shaver, G.R. (1989). Differences in growth and nutrient use among arctic plant growth forms. *Funct. Ecol.* **3**, 73–80.

Chapin, F.S. III and Wardlaw, I.F. (1988). Effect of phosphorus deficiency on source–sink interactions between the flag leaf and developing grain in barley. *J. Exp. Bot.* **39**, 165–177.

Chapin, F.S. III, Van Cleve, K. and Tryon, P.R. (1983). Influence of phosphorus on the growth and biomass allocation of Alaskan taiga tree seedlings. *Can. J. Forest Res.* **13**, 1092–1098.

Chapin, F.S. III, Shaver, G.R. and Kedrowski, R.A. (1986a). Environmental controls over carbon, nitrogen and phosphorus chemical fractions in *Eriophorum vaginatum* L. in Alaskan tussock tundra. *J. Ecol.* **74**, 167–195.

Chapin, F.S., Vitousek, P.M. and Van Cleve, K. (1986b). The nature of nutrient limitation in plant communities. *Am. Nat.* **127**, 48–58.

Chapin, F.S. III, Fetcher, N., Kielland, K., Everett, A.R. and Linkins, A.E. (1988). Productivity and nutrient cycling of Alaskan tundra: enhancement by flowing soil water. *Ecology* **69**, 693–702.

Chapin, F.S. III, Schulze, E.-D., Mooney H.A., (1990). The ecology and economics of storage in plants. *Annu. Rev. Ecol. Syst.* **21**: 423–447.

Chapin, F.S. III, Moilanen, L. and Kielland, K. (1993). Preferential use of organic nitrogen for growth by a non-mycorrhizal arctic sedge. *Nature* **361**, 150–153.

Clarholm, M. (1985). Interactions of bacteria, protozoa and plants leading to mineralization of soil nitrogen. *Soil Biol. Biochem.* **17**, 181–187.

Clarkson, D.T. (1985). Factors affecting mineral nutrient acquisition by plants. *Annu. Rev. Plant Physiol.* **26**, 77–115.

Coley, P.D., Bryant, J.P. and Chapin, F.S. III (1985). Resource availability and plant anti-herbivore defense. *Science* **230**, 895–899.

Constantinides, M. and Fownes, J.H. (1994). Nitrogen mineralization from leaves and litter of tropical plants: relationship to nitrogen, lignin and soluble polyphenol concentrations. *Soil Biol. Biochem.* **26**, 49–55.

Cornelissen, J.H.C. and Thompson, K. (1997). Functional leaf attributes predict litter decomposition rate in herbaceous plants. *New Phytol.* **135**, 109–114.

Coulson, J.C. and Butterfield, J. (1978). An investigation of the biotic factors determining the rates of plant decomposition on blanket bog. *J. Ecol.* **66**, 631–650.

Cowling, R.M. (1993). *The Ecology of Fynbos. Nutrients, Fire and Diversity.* Oxford University Press, Oxford.
Craine, J.M. and Mack, M.C. (1998). Detection of nutrient resorption proficiency: comment. *Ecology* **79**: 1818–1820.
Crick, J.C. and Grime, J.P. (1987). Morphological plasticity and mineral nutrients capture in two herbaceous species of contrasting ecology. *New Phytol.* **107**, 403–414.
Crist, J.W. and Stout, G.J. (1929). Relation between top and root size in herbaceous plants. *Plant Physiol.* **4**, 63–85.
de Jager, A. (1982). Effects of a localized supply of H_2PO_4, NO_3, SO_4, Ca and K on the production and distribution of dry matter in young maize plants. *Neth. J. Agric. Sci.* **30**, 193–203.
DeLucia, E.H. and Schlesinger, W.H. (1995). Photosynthetic rates and nutrient-use efficiency among evergreen and deciduous shrubs in Okefenokee swamp. *Int. J. Plant Sci.* **156**, 19–28.
Diaz, S.A., Grime, J.P., Harris, J. and McPherson, E. (1993). Evidence of a feedback mechanism limiting plant response to elevated carbon dioxide. *Nature* **364**, 616–617.
DiTomasso, A. and Aarssen, L.W. (1989). Resource manipulations in natural vegetation: a review. *Vegetatio* **84**, 9–29.
Eckstein, R.L. and Karlsson, P.S. (1997). Above-ground growth and nutrient use by plants in a subarctic environment: effects of habitat, life-form and species. *Oikos* **79**, 311–324.
Eissenstat, D.M. and Caldwell, M.M. (1987). Characterization of successful competitors: an evaluation of potential growth rate in two cold desert tussock grasses, *Oecologia* **71**, 167–173.
Eissenstat, D.M. and Yanai, R.D. (1997). The ecology of root lifespan. *Adv. Ecol. Res.* **27**, 1–60.
Elberse, W.Th. and Berendse, F. (1993). A comparative study of the growth and morphology of eight grass species with different nutrient availabilities. *Funct. Ecol.* **7**, 223–229.
Ellis, R.C. (1972). The mobilization of iron by extracts of *Eucalyptus* leaf litter. *J. Soil Sci.* **22**, 8–22.
Escudero, A., del Arco, J.M., Sanz, I.C. and Ayala, J. (1992). Effects of leaf longevity and retranslocation efficiency on the retention time of nutrients in the leaf biomass of different woody species. *Oecologia* **90**, 80–87.
Evans, J.R. (1989). Photosynthesis and nitrogen relationships in leaves of C_3 plants. *Oecologia* **78**, 9–19.
Eviner, V.T. and Chapin F.S. III (1997). Plant–microbial interactions. *Nature* **385**, 26–27.
Field, C. (1983). Allocating leaf nitrogen for the maximization of carbon gain: leaf age as a control on the allocation program. *Oecologia* **56**, 341–347.
Fitter, A.H. (1985). Functional significance of root morphology and root system architecture. In: *Ecological Interactions in Soil. Plants, Microbes and Animals* (Ed. by A.H. Fitter, D. Atkinson and D.J. Read), pp. 87–106. Blackwell, Oxford.
Fitter, A.H. (1987). An architectural approach to the comparative ecology of plant root systems. *New Phytol.* **106**, 61–77.
Fitter, A.H. (1991). Characteristics and functions of root systems. In: *Plant Roots: The Hidden Half* (Ed. by A.E.Y. Waisel, A. Eshel and U. Kafkafi), pp. 3–25. Marcel Dekker, New York.
Fitter, A.H., Wright, W.J., Williamson, L., Belshaw, M., Fairclough, J. and Meharg, A.A. (1998). *The Phosphorus Nutrition of Wild Plants and the Paradox of Arsenate Tolerance: Does Leaf Phosphate Concentration Control Flowering?* Penn State University, New York.

Föhse, D. and Jungk, A. (1983). Influence of phosphate and nitrate supply on root hair formation of rape, spinach and tomato plants. *Plant Soil* **74**, 359–368.
Fox, J.F. (1992). Responses of diversity and growth-form dominance to fertility in Alaskan tundra fellfield communities. *Arctic Alp. Res.* **24**, 233–237.
Ganzert, C. and Pfadenhauer, J. (1986) Seasonal dynamics of shoot nutrients in *Schoenus ferrugineus* (Cyperaceae). *Holarct. Ecol.* **9**, 137–142.
Garnier, E. (1991). Resource capture, biomass allocation and growth in herbaceous plants. *Trends Ecol. Evol.* **6**, 126–131.
Garnier, E. (1992). Growth analysis of congeneric annual and perennial grass species. *J. Ecol.* **80**, 665–675.
Garnier, E. and Freijsen, A.H.J. (1994). On ecological inference from laboratory experiments conducted under optimum conditions. In: *A Whole Plant Perspective on Carbon–Nitrogen Interactions* (Ed. by J. Roy and E. Garnier), pp. 267–292. SPB Academic Publishing, The Hague.
Garnier, E. and Laurent, G. (1994). Leaf anatomy, specific mass and water content in congeneric annual and perennial grass species. *New Phytol.* **128**, 725–736.
Gault, R.R., Peoples, M.B., Turner, G.L., Lilley, D.M., Brockwell, J. and Bergersen, F.J. (1995). Nitrogen fixation by irrigated lucerne during the first three years after establishment. *Aust. J. Agric. Res.* **56**, 1401–1425.
Gleeson, S.K. and Tilman, D. (1990). Allocation and the transient dynamics of succession on poor soils. *Ecology* **71**, 1144–1155.
Gray, J.T. (1983). Nutrient use by evergreen and deciduous shrubs in southern California. I. Community nutrient cycling and nutrient-use efficiency. *J. Ecol.* **71**, 21–41.
Grime, J.P. (1979). *Plant Strategies and Vegetation Processes*. Wiley, Chichester, UK.
Grime, J.P., Cornelissen, J.H.C., Thompson, K. and Hodgson, J.G. (1996). Evidence of a causal connection between anti-herbivore defence and the decomposition rate of leaves. *Oikos* **77**, 489–494.
Grime, J.P., Thompson, K., Hunt, R., Hodgson, J.G., Cornelissen, J.H.C., Rorison, I.H. et al. (1997). Integrated screening validates primary axes of specialisation in plants. *Oikos* **79**, 259–281.
Grogan, P. and Chapin F.S. III. (1999). Nitrogen limitation of production in a Californian annual grassland: the contribution of arbuscular mycorrhizae. *Oecologia* (submitted).
Haukioja, E., Niemelä, P. and Sirén, S. (1985). Foliage phenols and nitrogen in relation to growth, insect damage, and ability to recover after defoliation, in the mountain birch *Betula pubescens* ssp. *tortuosa*. *Oecologia* **65**, 214–222.
Havström, M., Callaghan, T.V. and Jonasson, S. (1993). Differential growth responses of *Cassiope tetragona*, an arctic dwarf-shrub, to environmental perturbations among three contrasting high- and subarctic sites. *Oikos* **66**, 389–402.
Heilmeier, H. and Monson, R.K. (1994). Carbon and nitrogen storage in herbaceous plants. In: *A Whole Plant Perspective on Carbon–Nitrogen Interactions*. (Ed. by J. Roy and E. Garnier), pp. 149–171. SPB Academic Publishing, The Hague.
Heilmeier, H., Schulze E.-D. and Whale, D.M. (1986). Carbon and nitrogen partitioning in the biennial monocarp *Arctium tomentosum* Mill. *Oecologia* **70**, 466–474.
Hemond, H.F. (1983). The nitrogen budget of Thoreau's bog. *Ecology* **64**, 99–109.
Herold, A. (1980). Regulation of photosynthesis by sink activity—the missing link. *New Phytol.* **86**, 131–144.
Hetrick, B.A.D., Wilson, G.W.T. and Leslie, J.F. (1991). Root architecture of warm- and cool-season grasses: relationship to mycorrhizal dependence. *Can. J. Bot.* **69**, 112–118.

Hirose, T. and Werger, M.J.A. (1987). Nitrogen use efficiency in instantaneous and daily photosynthesis of leaves in the canopy of a *Solidago altissima* stand. *Physiol. Plant.* **70**, 215–222.
Hobbie, S.E. (1992). Effects of plant species on nutrient cycling. *Trends Ecol. Evol.* **7**, 336–339.
Hofer, R.-M. (1991). Root Hairs. In: *Plant Roots: The Hidden Half.* (Ed. by A.E.Y. Waisel, A. Eshel and U. Kafkafi), pp. 129–148. Marcel Dekker, New York.
Howarth, R.W. (1988). Nutrient limitation of net primary production in marine ecosystems. *Annu. Rev. Ecol. Syst.* **19**, 89–110.
Huang, N.-C., Chiang C.-S., Crawford N.M. and Tsay, Y.F. (1996). *Chl1* encodes a component of the low-affinity nitrate uptake system in *Arabidopsis* and shows cell type-specific expression in roots. *Plant Cell* **8**, 2183–2191.
Hübel, F. and Beck, F. (1993). *In-situ* determination of the P-relations around the primary root of maize with respect to inorganic and phytate-P. *Plant Soil* **157**, 1–9.
Hungate, B.A. (1998). Ecosystem responses to rising atmospheric CO_2 feedbacks through the nitrogen cycle. In: *Interactions of Elevated CO_2 and environmental stress*. (Ed. by J. Seeman and Y. Luo). Academic Press, San Diego.
Hungate, B.A. (1999) Ecosystem responses to rising atmospheric CO_2: feedbacks through the nitrogen cycle. In: Luo Y. and Mooney H.A. (eds) *CO_2 and environmental stress*. (265–285) Academic Press, San Diego.
Hunt, E.R., Weber, J.A. and Gates, D.M. (1985). Effects of nitrate application on *Amaranthus powellii* Wats. I. Changes in photosynthesis, growth rates, and leaf area. *Plant Physiol.* **79**, 609–613.
Hunt, H.W., Stewart, J.B. and Cole, C.V. (1983). A conceptual model for interactions of carbon, nitrogen, phosphorus, and sulphur in grasslands. In: *The Major Biogeochemical Cycles and their Interactions* (Ed. by B. Bolin and R.B. Cook), pp. 303–326. Wiley, New York.
Hunt, R. and Cornelissen, J.H.C. (1997). Components of relative growth rate and their interrelations in 59 temperate plant species. *New Phytol.* **135**, 395–417.
Huston, M.A. (1994). *Biological Diversity. The Coexistence of Species on Changing Landscapes*. Cambridge University Press, Cambridge.
Ibrahima, A., Joffre, R. and Gillon, D. (1995). Changes in litter during the initial leaching phase: an experiment on the leaf litter of Mediterranean species. *Soil Boil. Biochem.* **27**, 931–939.
Ingestad, T. (1979). Nitrogen stress in birch seedlings. II. N, K, P, Ca, and Mg nutrition. *Physiol. Plant.* **45**, 149–157.
Johnson, F.F., Allan, D.L., Vance C.P. and Weiblen, G. (1996). Root carbon dioxide fixation by phosphorus-deficient *Lupinus albus*. Contribution to organic acid exudation by proteoid roots. *Plant Physiol.* **112**, 19–30.
Jonasson, S. (1989). Implications of leaf longevity, leaf nutrient re-absorption and translocation for the resource economy of five evergreen plant species. *Oikos* **56**, 121–131.
Jonasson, S. (1992). Plant responses to fertilization and species removal in tundra related to community structure and clonality. *Oikos* **63**, 420–429.
Kaye, J.P. and Hart, S.C. (1997). Competition for nitrogen between plants and soil microorganisms. *Trends Ecol. Evol.* **12**, 139–143.
Keeley, J.E. and Zedler P.H. (1978). Reproduction in chaparral shrubs after fire: a comparison of sprouting and seeding strategies. *Am. Midl. Nat.* **99**, 142–161.
Kielland, K. (1994). Amino acid absorption by arctic plants: implications for plant nutrition and nitrogen cycling. *Ecology* **75**, 2373–2383.

Killingbeck, K.T. (1996). Nutrients in senesced leaves: keys to the search for potential resorption and resorption proficiency. *Ecology* **77**, 1716–1727.

Knops, J.H.M., Koenig, W.D. and Nash, T.H. (1997). On the relationship between nutrient use efficiency and fertility in forest ecosystems. *Oecologia* **110**, 550–556.

Koerselman, W. and Meuleman, A.F.M. (1996). The vegetation N : P ratio: a new tool to detect the nature of nutrient limitation. *J. Appl. Ecol.* **33**, 1441–1450.

Koerselman, W. and Verhoeven, J.T.A. (1992). Nutrient dynamics in mires of various trophic status: nutrient inputs and outputs and the internal nutrient cycle. In: *Fens and Bogs in the Netherlands: Vegetation, History, Nutrient Dynamics and Conservation* (Ed. by J.T.A. Verhoeven), pp. 397–432. Kluwer, Dordrecht.

Koerselman, W., Bakker, S.A. and Blom, M. (1990). Nitrogen, phosphorus and potassium budgets for two small fens surrounded by heavily fertilized pastures. *J. Ecol.* **78**, 428–442.

Koide, R. T. (1991). Nutrient supply, nutrient demand and plant response to mycorrhizal infection. *New Phytol.* **117**, 365–386.

Kroehler, C.J. and Linkins, A.E. (1991). The absorption of inorganic phosphate from ^{32}P-labeled inositol hexaphosphate by *Eriophorum vaginatum*. *Oecologia* **85**, 424–428.

Kronzucker, H.J., Siddiqi, M.Y. and Glass, A.M. (1997). Conifer root discrimination against soil nitrate and the ecology of forest succession. *Nature* **385**: 59–61.

Lajtha, K. and Klein, M. (1988). The effect of varying nitrogen and phosphorus availability on nutrient use by *Larrea tridentata*, a desert evergreen shrub. *Oecologia* **75**, 348–353.

Lambers, H. and Poorter, H. (1992). Inherent variation in growth rate between higher plants: a search for physiological causes and ecological consequences. *Adv. Ecol. Res.* **22**, 187–261.

Lambers, H., Chapin F.S. III and Pons, T. (1998). *Plant Physiological Ecology.* Springer, Berlin.

Lavelle, P., Blanchart, E., Martin, A., Spain, A., Toutain, F., Barois, I. and Schaefer, R. (1993). A hierarchical model for decomposition in terrestrial ecosystems: application to soils of the humid tropics. *Biotropica* **25**, 130–150.

Lee, R.B. (1982). Selectivity and kinetics of ion uptake by barley plant following nutrient deficiency. *Ann. Bot.* **50**, 429–449.

Liljeroth, E., Bååth, E., Mathiason, I. and Lundborg, T. (1990). Root exudation and rhizopane bacterial abundance on barley (*Hordeum vulgare* L.) in relation to nitrogen fertilization and root growth. *Plant Soil* **127**, 81–89.

Malmer, N. (1988). Patterns in the growth and the accumulation of inorganic constituents in the *Sphagnum* cover on ombrotrophic bogs in Scandinavia. *Oikos* **53**, 105–120.

Malmer, N. (1990). Constant or increasing nitrogen concentrations in *Sphagnum* mosses in mires in southern Sweden during the last few decades. *Aquilo Ser. Bot.* **28**, 57–65.

Marschner, H. (1995). *Mineral Nutrition of Higher Plants.* Academic Press, London.

Meentemeyer, V. (1978). Macroclimate and lignin control of litter decomposition rates. *Ecology*, **59**, 465–472.

Meentemeyer, V. and Berg, B. (1986). Regional variation in rate of mass loss of *Pinus sylvestris* needle litter in Swedish pine forests as influenced by climate and litter quality. *Scand. J. Forest Res.* **1**, 167–180.

Melillo, J.M., Aber, J.D. and Muratore, J.F. (1982). Nitrogen and lignin control of hardwood leaf litter decomposition dynamics. *Ecology* **63**, 621–626.

Millard, P. (1988). The accumulation and storage of nitrogen by herbaceous plants. *Plant Cell Env.* **11**, 1–8.

Millard, P. and Proe, M.F. (1993). Nitrogen uptake, partitioning and internal cycling in *Picea sitchensis* (Bong.) Carr. as influenced by nitrogen supply. *New Phytol.* **125**, 113–119.
Monk, C.D. (1966). An ecological significance of evergreenness. *Ecology* **47**, 504–505.
Morris, J.T. (1991). Effects of nitrogen loading on wetland ecosystems with particular reference to atmospheric deposition. *Annu. Rev. Ecol. Syst.* **22**, 257–279.
Morton, A.J. (1977). Mineral nutrient pathways in a *Molinietum* in autumn and winter. *J. Ecol.* **65**, 993–999.
Nambiar, E.K.S. and Fife, D.N. (1991). Nutrient retranslocation in temperate conifers. *Tree Physiol.* **9**, 185–207.
Näsholm, T., Ekblad, A., Nordin, A., Gieslr, R., Högberg, M. and Högberg, P. (1998). Boreal forest plants take up organic nitrogen. *Nature* **392**, 914–916.
Nicolai, V. (1988). Phenolic and mineral content of leaves influences decomposition in European forest ecosystems. *Oecologia* **75**, 575–579.
Niemann, G.J., Pureveen, J.B.M., Eijkel, G.B.H., Poorter, H. and Boon, J.J. (1992). Differences in relative growth rate in 11 grasses correlate with differences in chemical composition as determined by pyrolysis mass spectrometry. *Oecologia* **89**, 567–573.
Northup, R.R., Yu, Z., Dahlgren, R.A. and Vogt, K.A. (1995). Polyphenol control of nitrogen release from pine litter. *Nature* **377**, 227–229.
Olff, H. (1992). Effects of light and nutrient availability on dry matter and N allocation in six successional grass species. Testing for resource ratio effects. *Oecologia* **89**, 412–421.
Olff, H., Van Andel, J. and Bakker, J.P. (1990). Biomass and shoot/root allocation of five species from a grassland succession series at different combinations of light and nutrient supply. *Funct. Ecol.* **4**, 193–200.
Olson, J.S. (1963). Energy storage and the balance of producers and decomposers in ecological systems. *Ecology* **44**, 322–331.
Parsons, A.N., Welker, J.M., Wookey, P.A., Press, M.C., Callaghan, T.V. and Lee, J.A. (1994). Growth-response of four sub-arctic dwarf shrubs to simulated environmental change. *J. Ecol.* **82**, 307–318.
Parsons, A.N., Press, M.C., Wookey, P.A., Welker, J.M., Robinson, C.H., Callaghan, T.V. and Lee, J.A. (1995). Growth responses of *Calamagrostis lapponica* to simulated environmental change in the sub-arctic. *Oikos* **72**, 61–66.
Pastor, J., Aber, J.D., McClaugherty, C.A. and Melillo, J.M. (1984). Aboveground production and N and P cycling along a nitrogen mineralization gradient on Blackhawk Island, Wisconsin. *Ecology* **65**, 256–268.
Pastor, J., Stillwell, M.A. and Tilman, D. (1987). Little bluestem litter dynamics in Minnesota old fields. *Oecologia* **72**, 327–330.
Poorter, H. (1994). Construction costs and payback time of biomass: a whole plant perspective. In: *A Whole Plant Perspective on Carbon–Nitrogen Interactions* (Ed. by J. Roy and E. Garnier), pp. 111–127. SPB Academic Publishing, The Hague.
Poorter, H. and Bergkotte, M. (1992). Chemical composition of 24 wild species differing in relative growth rate. *Plant Cell Env.* **15**, 221–229.
Poorter, H. and Remkes, C. (1990). Leaf area ratio and net assimilation rate of 24 wild species differing in relative growth rate. *Oecologia* **83**, 553–559.
Poorter, H., Remkes, C. and Lambers, H. (1990). Carbon and nitrogen economy of 24 wild species differing in relative growth rate. *Plant Physiol.* **94**, 621–627.
Pugnaire, F.I. and Chapin, F.S. III (1993). Controls over nutrient resorption from leaves of evergreen Mediterranean species. *Ecology* **74**, 124–129.

Raab, T.K., Lipson, D.A. and Monson, R.K. (1996). Non-mycorrhizal uptake of organic N by the alpine dry meadow sedge, *Kobresia myosuroides*. Implications for the alpine nitrogen cycle. *Oecologia* **108**, 488–496.

Raab, T.K., Lipson, D.A. and Monson, R.K. (1998). Soil amino acid utilization among species of the Cyperaceae: plant and soil processes. *Ecology* (in press).

Raab, T.K., Lipson, D.A. and Monson, R.K. (1999). Soil amino acid utilization among species of the Cyperaceae: plant and soil processes. *Ecology* (in press).

Rappoport, H.F. and Loomis, R.S. (1985). Interaction of storage root and shoot in grafted sugarbeet and chard. *Crop Sci.* **25**, 1079–1084.

Read, D.J. (1991). Mycorrhizas in ecosystems. *Experientia* **47**, 376–391.

Reader, R.J. (1980). Effects of nitrogen fertilizer, shade and removal of new growth on longevity of overwintering bog ericad leaves. *Can. J. Bot.* **58**, 1737–1743.

Redfield, A.C. (1958). The biological control of chemical factors in the environment. *Am. Sci.* **46**, 206–226.

Reich, P.B. (1993). Reconciling apparent discrepancies among studies relating life span, structure and function of leaves in contrasting plant life forms and climates: "the blind men and the elephant retold". *Funct. Ecol.* **7**, 721–725.

Reich, P.B., Uhl, C., Walters, M.B. and Ellsworth, D.S. (1991). Leaf lifespan as a determinant of leaf structure and function among 23 tree species in Amazonian forest communities. *Oecologia* **86**, 16–24.

Reich, P.B., Walters, M.B. and Ellsworth, D.S. (1992). Leaf life-span in relation to leaf, plant, and stand characteristics among diverse ecosystems. *Ecol. Monogr.* **62**, 365–392.

Reich, P.B., Walters, M.B. and Ellsworth, D.S. (1997). From tropics to tundra: global convergence in plant functioning. *Proc. Natl. Acad. Sci. U.S.A.* **94**, 13730–13734.

Reynolds, H.L. and D'Antonio, C. (1996). The ecological significance of plasticity in root weight ratio in response to nitrogen. Opinion. *Plant Soil* **185**, 75–97.

Riis-Nielsen, T. (1997). *Effects of Nitrogen on the Stability and Dynamics of Danish Heathland Vegetation*. Thesis, University of Copenhagen.

Robinson, D. (1986). Compensatory changes in the partitioning of dry matter in relation to nitrogen uptake and optimal variations in growth. *Ann. Bot.* **86**, 841–848.

Robinson, D. and Rorison, I.H. (1983). A comparison of the responses of *Lolium perenne* L., *Holcus lanatus* L. and *Deschampsia flexuosa* L. Trin. to a localized supply of nitrogen. *New Phytol.* **94**, 263–273.

Robinson, D. and Rorison, I.H. (1988). Plasticity in grass species in relation to nitrogen supply. *Funct. Ecol.* **2**, 249–257.

Rustad, L.E. (1994). Element dynamics along a decay continuum in a red spruce ecosystem in Maine, USA. *Ecology* **75**, 867–879.

Ryser, P. and Lambers, H. (1995). Root and leaf attributes accounting for the performance of fast- and slow-growing grasses at different nutrient supply. *Plant Soil* **170**, 251–265.

Schläpfer, B. and Ryser, P. (1996). Leaf and root turnover of three ecologically contrasting grass species in relation to their performance along a productivity gradient. *Oikos* **75**, 398–406.

Schulze, E.-D., Koch, G.W., Percival, F., Mooney, H.A. and Chu, C. (1985). The nitrogen balance of *Raphanus sativus* × *raphanistrum* plants. I. Daily nitrogen use under high nitrate supply. *Plant Cell Env.* **8**, 713–720.

Shaver, G.R. (1981). Mineral nutrition and leaf longevity in an evergreen shrub, *Ledum palustre* ssp. *decumbens*. *Oecologia* **49**, 362–365.

Shaver, G.R. (1983). Mineral nutrition and leaf longevity in *Ledum palustre*: the role of individual nutrients and the timing of leaf mortality. *Oecologia* **56**, 160–165.

Shaver, G.R. and Chapin, F.S. III (1991). Production : biomass relationships and element cycling in contrasting arctic vegetation types. *Ecol. Monogr.* **61**, 1–31.

Shaver, G.R. and Chapin, F.S. III (1995). Long-term responses to factorial NPK fertilizer treatment by Alaskan wet and moist tundra sedge species. *Ecography* **18**, 259–275.

Shaver, G.R. and Melillo, J.M. (1984). Nutrient budgets of marsh plants: efficiency concepts and relation to availability. *Ecology* **65**, 1491–1510.

Shaver, G.R., Chapin, F.S. III and Gartner, B.L. (1986). Factors limiting seasonal growth and peak biomass accumulation in *Eriophorum vaginatum* in Alaskan tussock tundra. *J. Ecol.* **74**, 257–278.

Shipley, B. and Peters, R.H. (1990). The allometry of seed weight and seedling relative growth rate. *Funct. Ecol.* **4**, 523–529.

Small, E. (1972). Photosynthetic rates in relation to nitrogen recycling as an adaptation to nutrient deficiency in peat bog plants. *Can. J. Bot.* **50**, 2227–2233.

Smith, S.E. and Read, D.J. (1997). *Mycorrhizal Symbiosis*. Academic Press, London.

Son, Y. and Gower, S.T. (1991). Aboveground nitrogen and phosphorus use by five plantation-grown trees with different leaf longevities. *Biogeochemistry* **14**, 167–191.

Specht, R.L. and Rundel, P.W. (1990). Sclerophylly and foliar nutrient status of Mediterranean-climate plant communities in southern Australia. *Aust. J. Bot.* **38**, 459–474.

Sprent, J.I., Geoghagan, I.E., Whitty, P.W. and James, E.K. (1996). Natural abundance of ^{15}N and ^{13}C in nodulated legumes and other plants in the cerrado and neighbouring regions of Brazil. *Oecologia* **105**, 440–446.

Sutton, M.A., Pitcairn, C.E.R. and Fowler, D. (1993). The exchange of ammonia between the atmosphere and plant communities. *Adv. Ecol. Res.* **24**, 301–393.

Swift, M.J. and Anderson, J.M. (1989). Decomposition. In: *Ecosystems of the World, 14B. Tropical Rain Forest Ecosystems, Biogeographical and Ecological Studies* (Ed. by H. Lieth and M.J.A. Werger), pp. 547–569. Elsevier, Amsterdam.

Swift, M.J., Heal, O.W. and Anderson, J.M. (1979). *Decomposition in Terrestrial Ecosystems*. University of California Press, Berkeley.

Taylor, B.R., Parkinson, D. and Parsons, W.F.J. (1989). Nitrogen and lignin content as predictors of litter decay rates: a microcosm test. *Ecology* **70**, 97–104.

Taylor, B.R., Prescott, C.E., Parsons, W.J.F. and Parkinson, D. (1991). Substrate control of litter decomposition in four Rocky Mountain coniferous forests. *Can. J. Bot.* **69**, 2242–2250.

Terry, N. and Ulrich, A. (1973). Effects of phosphorus deficiency on the photosynthesis and respiration of leaves in sugar beet. *Plant Physiol.* **51**, 43–47.

Thomas, R.J. and Asakawa, N.M. (1993). Decomposition of leaf litter from tropical forage grasses and legumes. *Soil Biol. Biochem.* **25**, 1351–1361.

Thompson, K., Parkinson, J.A., Band, S.R. and Spencer, R.E. (1997). A comparative study of leaf nutrient concentrations in a regional herbaceous flora. *New Phytol.* **136**, 679–689.

Tilman, D. (1985). The resource-ratio hypothesis of plant succession. *Am. Nat.* **125**, 827–852.

Tilman, D. (1986). Nitrogen-limited growth in plants from different successional stages. *Ecology* **67**, 555–563.

Tilman, D. (1988). *Plant Strategies and the Dynamics and Structure of Plant Communities*. Princeton University Press, Princeton, NJ.

Tilman, D. and Cowan, M.L. (1989). Growth of old field herbs on a nitrogen gradient. *Funct. Ecol.* **3**, 425–438.

Tilman, D. and Wedin, D. (1991). Plant traits and resource reduction for five grasses growing on a nitrogen gradient. *Ecology* **72**, 685–700.

Trewavas, A.J. (1986). Resource allocation under poor growth conditions: a major role for growth substances in developmental plasticity. In: *Plasticity in Plants* (Ed. by D.A. Jennings and A.J. Trewavas), pp. 31–76. The Company of Biologists Limited, Cambridge.

Tukey, H.B. Jr. (1970). The leaching of substances from plants. *Annu. Rev. Plant Physiol.* **21**, 305–324.

Turnbull, M.H., Goodall, R. and Stewart, G.R. (1995). The impact of mycorrhizal colonization upon nitrogen source utilization and metabolism in seedlings of *Eucalyptus grandis* Hill ex Maiden and *Eucalyptus maculata* Hook. *Plant Cell Env.* **18**, 1386–1394.

Turner, T.W. (1922). Studies of the mechanism of the physiological effects of certain mineral salts in altering the ratio of top growth to root growth in seed plants. *Am. J. Bot.* **9**, 415–445.

Van Oorschot, M., Robbemont, E., Boerstal, M., van Shrien, I. and van Kerkhoven-Schmitz, M. (1997). Effects of enhanced nutrient availability on plant and soil nutrient dynamics in two English riverine ecosystems. *J. Ecol.* **85**: 167–179.

Vance, C.P. (1995). Root–bacteria interactions. Symbiotic nitrogen fixation. In: *Plant Roots: The Hidden Half* (Ed. by Y. Waisel, A. Eshel and U. Kafkaki), pp. 723–755. Marcel Dekker, New York.

Veerkamp, M.T. and Kuiper, P.J.C. (1982). The uptake of potassium by *Carex* species from swamp habitats varying from oligotrophic to eutrophic. *Physiol. Plant.* **55**, 237–241.

Verhoeven, J.T.A. and Schmitz, M.B. (1991). Control of plant growth by nitrogen and phosphorus in mesotrophic fens. *Biogeochemistry* **12**, 135–148.

Verhoeven, J.T.A., Koerselman, W. and Meuleman, A.F.M. (1996). Nitrogen- or phosphorus-limited growth in herbaceous, wet vegetation: relations with atmospheric inputs and management regimes. *Trends Ecol. Evol.* **11**, 494–497.

Vitousek, P.M. (1982). Nutrient cycling and nutrient use efficiency. *Am. Nat.* **119**, 553–572.

Vitousek, P.M. (1994). Beyond global warming. Ecology and global change. *Ecology* **75**, 1861–1876.

Vitousek, P.M. and Howarth, R.W. (1991). Nitrogen limitation on land and in the sea: how can it occur? *Biogeochemistry* **13**, 87–115.

Vitousek, P.M. and Sanford, R.L. (1986). Nutrient cycling in moist tropical forest. *Annu. Rev. Ecol. Syst.* **17**, 137–167.

Vitousek, P.M., Walker, L.R., Whiteaker, L.D., Mueller-Dombois, D. and Matson, P.A. (1987). Biological invasion by *Myrica faya* alters ecosystem development in Hawaii. *Science* **238**, 802–804.

Vitousek, P.M., Turner, D.R., Parton, W.J. and Sanford, R.L. (1994). Litter decomposition on the Mauna Loa environmental matrix, Hawaii: patterns, mechanisms, and models. *Ecology* **75**, 418–429.

van Arendonk, J.J.C.M. and Poorter, H. (1994). The chemical composition and anatomical structure of leaves of grass species differing in relative growth rate. *Plant Cell Environ.* **17**, 963–970.

van Breemen, N. (1993). Soils as biotic constructs favouring net primary productivity. *Geoderma* **57**, 183–211.

van den Driessche, R. (1974). Prediction of mineral nutrient status of trees by foliar analysis. *Bot. Rev.* **40**, 347–394.

van der Werf, A., van Nuenen, M., Visser, A.J. and Lambers, H. (1993). Contribution of physiological and morphological plant traits to species' competitive ability at high and low nitrogen supply. *Oecologia* **94**, 434–440.

van Vuuren, M.M.I., Berendse, F. and De Visser, W. (1993). Species and site differences in the decomposition of litters and roots from wet heathlands. *Can. J. Bot.* **71**, 167–173.

Walbridge, M. (1991). Phosphorus availability in acid organic soils of the lower North Carolina coastal plain. *Ecology* **72**, 2083–2100.

Wassen, M.J., Olde Venterink, H.G.M. and de Swart, E.O.A.M. (1995). Nutrient concentrations in mire vegetation as a measure of nutrient limitation in mire ecosystems. *J. Veg. Sci.* **6**, 5–16.

Wilson, S.D. and Tilman, D. (1995). Competitive responses of eight old-field plant species in four environments. *Ecology* **76**, 1169–1180.

Woodin, S.J. and Lee, J.A. (1987). The fate of some components of acidic deposition in ombrotrophic mires. *Environ. Pollut.* **45**, 61–72.

Wookey, P.A., Parsons, A.N., Welker, J.M., Potter, J.A., Callaghan, T.V., Lee, J.A. and Press, M.C. (1993). Comparative responses of phenology and reproductive development to simulated environmental change in sub-arctic and high arctic plants. *Oikos* **67**, 490–502.

Zak, D.R., Pregitzer, K.S., Curtis, P.S., Teeri, J.A., Fogel, R. and Randlett, D.A. (1994). Elevated atmospheric CO_2 and feedback between carbon and nitrogen cycles. *Plant Soil* **151**, 105–117.

Co-evolution of Mycorrhizal Symbionts and their Hosts to Metal-contaminated Environments

A.A. MEHARG AND J.W.G. CAIRNEY

I.	Summary	70
II.	Introduction	71
	A. Role of Mycorrhiza on Metal-contaminated Soils	71
	B. Review Outline	72
III.	Arbuscular Mycorrhizas	73
	A. Arbuscular Mycorrhiza and Metal-contaminated Sites	73
	B. Species Diversity	76
	C. Spore Germination	76
	D. Germ Tube Growth	79
	E. Hyphal Penetration	80
	F. Root Colonization	81
	G. Spore Production	82
	H. Metal Assimilation in Plant Tissues	82
	I. Plant Sensitivity	85
	J. Nutritional Aspects	86
	K. Do Plants Benefit from being Colonized with Resistant Arbuscular Mycorrhizal Strains?	87
IV.	Ericoid Mycorrhizas	88
	A. The Ericaceae and Metal-contaminated Environments	88
	B. Copper	89
	C. Zinc	93
	D. Role of Mycorrhizas in Metal Resistance of the Ericaceae	95
V.	Ectomycorrhizal Fungi	95
	A. Introduction	95
	B. General Response of Ectomycorrhizal Associations to Raised Levels of Metals	95
	C. Genetic Adaptation of Trees to Metal Contaminants	98
	D. Intraspecific Variation in Metal Resistance in Ectomycorrhizal Fungi	99
	E. Studies Conducted with Hosts Colonized with Resistant and Sensitive Strains	102

F. Role of Ectomycorrhizal Fungi in Facilitating Host Metal
   Resistance 103
VI. Conclusions 104
    A. Current State of Knowledge 104
    B. Future Research Requirements 105
References 107

I. SUMMARY

Mycorrhizas provide the interface between roots and soil and are very effective at assimilating essential metals (copper, zinc and nickel) and their analogues, many of which are at toxic concentrations on contaminated soils. Higher plants established on highly metalliferous soils are generally mycorrhizal. However, metal resistances in mycorrhizal fungi have been investigated in much less detail than their hosts. This is despite the fact that they may be key to plant survival on contaminated soils, not only for potential conferred or enhanced metal resistances, but also for their role in nutrient and water acquisition.

This review of the co-evolution of plant and fungal symbionts to metalliferous soils aims to establish the evolutionary strategies that enable them to survive in soils with toxic levels of metals. Constitutive and adaptive mycorrhizal resistance to a number of metals has been demonstrated. Three basic strategies have been identified: (1) in some associations (principally ericoid) the mycorrhiza are essential to plant survival as they endow metal resistance on their host; (2) plants that have adaptive or constitutive metal resistances do not require mycorrhizas to provide or enhance resistance; and (3) there is some evidence that, in ericoid associations, resistances have evolved in both symbionts for some metals, and that the combined resistances in this association confer enhanced resistance to the plant.

As ectomycorrhizal (EcM) and arbuscular mycorrhizal (AM) fungi have evolved resistances to metal-contaminated soils and colonize their metal-resistant hosts, yet do not confer added resistance, why have these mycorrhizas co-evolved with their hosts? Symbiosis on metal-contaminated sites may benefit the host by providing the same ecological functions (nutrient acquisition, water relationships, defence against pathogens) that mycorrhiza provide to plants in uncontaminated environments. Hence both metal-resistant hosts and mycorrhizal fungi have co-evolved to the extreme ecological niches provided by metal-contaminated sites.

Whilst it has been established that mycorrhizas can be constitutively or adaptively resistant to high metal concentrations, their ecological role on metal-contaminated sites (both metal resistance and nutrient and water acquisition) is still uncertain in all associations, with the exception of copper resistances in the Ericaceae where ericoid mycorrhiza confer metal resistance. For EcM, AM and ericoid mycorrhiza (other than copper resistance), experimental models are presented to address this fundamental gap in knowledge. It is concluded that much research is required before an understanding of the co-evolution of symbionts to metal-contaminated environments is established.

II. INTRODUCTION
A. Role of Mycorrhiza on Metal-contaminated Soils

The majority of higher plants that grow on highly metal-contaminated soils are mycorrhizal. Mycorrhizal fungi can exhibit both constitutive and adaptive resistance to metals. The ability of ericoid mycorrhizal associations to establish on soil contaminated with copper, for example, appears to be constitutive for all ericoid mycorrhizal fungal populations from both contaminated and uncontaminated soils (Bradley *et al.*, 1981, 1982). The plant is entirely dependent on the fungus for survival at toxic metal concentrations (Bradley *et al.*, 1982). The situation for arbuscular mycorrhizal (AM) associations is quite different. In this instance, host plant taxa that grow on very contaminated sites are always resistant to the metals concerned and do not necessarily require colonization by AM fungi for survival on such soils (Griffioen *et al.*, 1994). However, most potentially AM plants found growing on metal-contaminated soils are colonized by AM fungi, suggesting that the fungi may confer a degree of benefit to the plant, either in the form of increased metal resistance or by other mechanisms such as enhanced nutrient supply or decreased water stress. Populations of AM fungi isolated from metal-contaminated sites are often more resistant to metals than those collected from uncontaminated soils (Weissenhorn *et al.*, 1993, 1994; Griffioen, 1994; Griffioen *et al.*, 1994). Adaptation to metal-contaminated soils by ectomycorrhizal (EcM) associations is less well resolved. Conflicting studies suggest either that EcM fungi possess constitutive resistance to a wide range of toxic metals (regardless of the metal status of their soils of origin) or that EcM fungi isolated from metal-contaminated sites have evolved resistance to the metals concerned (Hartley *et al.*, 1997). Only a limited number of tree species has been screened for metal resistance (for a review see Wilkinson and Dickinson, 1995), but from these studies it appears that tree species colonizing contaminated soils generally display constitutive resistances.

Mycorrhizal associations appear to have evolved a number of strategies for colonizing highly metal-contaminated soils:

(1) Constitutively low sensitivity to metals in all populations of certain mycorrhizal fungal taxa, with limited or no requirement for adaptation for resistance in the host.
(2) Constitutively low sensitivity to metals in all populations of certain mycorrhizal fungal taxa, with either adaptation of resistance, or constitutive resistance in the host.

(3) Resistance adaptation by the host plant alone with no requirement for mycorrhizal colonization.
(4) Adaptive resistance by plant and fungal symbionts.

Mycorrhizas are efficient at acquiring both micro-nutrient metals that are often toxic at high concentrations, such as copper, nickel and zinc (Smith and Read, 1997), and non-essential, toxic metals and semi-metals, such as arsenic, cadmium and lead. Being mycorrhizal may therefore be detrimental to the plant unless the mycorrhizas can effectively down regulate metal acquisition. Even if mycorrhizas have evolved metal resistances, it may be obligatory for the co-evolution of symbiont and host that the symbiont has evolved mechanisms for reducing exposure of itself and its host to metals. It has often been postulated that mycorrhizas provide a barrier, restricting metal transport to host tissues through metal binding and/or sequestration in both external and internal fungal tissues (Wilkins, 1991; Wilkinson and Dickinson, 1995; Leyval et al., 1997). Whether this is the case, and whether there is selection for such properties, needs to be ascertained.

Mycorrhizas may have a more general role in plant nutrition and water relationships on metal-contaminated soils, which are often nutrient deficient, have very poor soil structure and are freely drained. Therefore, mycorrhizal fungi and their host plants may have co-evolved metal resistances because the association is generally beneficial in metal-contaminated environments, even if the mycorrhizas are not intrinsically required for enhanced or conferred metal resistance. Furthermore, mycorrhizal fungi have complex cost–benefit relationships with their host plants, and there may be a cost–benefit surface regulated by severity of metal contamination, nutrient deficiency, water stress, etc. That is, it may be beneficial in some metal-contaminated habitats to be mycorrhizal, but detrimental in others. As metal-contaminated sites are normally very heterogeneous, mycorrhizal benefits and costs may have considerable spatial heterogeneity within such habitats.

B. Review Outline

In this review we have attempted to synthesize the available literature on mycorrhizal fungi–host plant adaptation to metal-contaminated soils in order to draw conclusions on the co-evolutionary strategies for the establishment of mycorrhizal associations on contaminated sites. The review considers the major classes of associations found on metal-contaminated sites (ericoid, EcM and AM), to determine whether systematic differences or similarities occur for common polluting metals and whether constitutive or adaptive resistance is observed in the fungi and/or their hosts.

We further consider why some associations show constitutive resistance whereas others require adaptive resistance, as this may reveal differing underlying ecophysiological strategies. Associations that colonize acidic soils (the ericoid association and coniferous EcM, for example) will be exposed to high levels of toxic metals (aluminium, iron, manganese) in most environments which they colonize naturally and, therefore, they may have evolved mechanisms for coping with these metals. Mechanisms for coping with high levels of aluminium, iron and manganese may also confer constitutive resistance to other metals such as copper, cadmium, lead and zinc as the modes of toxic action and entry into cells are similar.

Where genetic adaptation of mycorrhiza to metal contaminants does occur, it must be considered whether the fungus benefits the host with respect to increased metal resistance or whether the benefit is due to generally enhanced plant fitness from being mycorrhizal. Mine soils are normally nutrient-deficient and freely drained, and mycorrhizal associations may generally benefit the plant in such environments. Conversely, there is little interspecies competition, ground cover is normally limited and plants are very slow growing. Hence, there may be limited benefit, or even detriment, to being mycorrhizal in such environments.

Known physiological attributes associated with enhanced resistance in mycorrhizal fungi will be discussed, as will the likely mechanisms of genetic adaptation. Mechanisms by which mycorrhizas adapt to high levels of common polluting metals will be considered and placed in context with other microbial and higher plant studies.

III. ARBUSCULAR MYCORRHIZAS

A. Arbuscular Mycorrhiza and Metal-contaminated Sites

Populations of AM fungi generally colonize metal-contaminated sites once plant cover has begun to establish. Thus, even on highly contaminated sites, most potentially AM plant taxa are colonized by AM fungi (Table 1), with more than one AM fungal species normally present (Griffioen, 1994; Weissenhorn *et al.*, 1994; Shetty *et al.*, 1995). Colonization levels (percentage of root length) of AM plants on mine spoils may be high. For example, 96% colonization has been reported for the legume *Lotus corniculatus* L. on a calamine spoil heap that had 49 g kg^{-1} zinc, 4.5 g kg^{-1} lead and 180 mg kg^{-1} cadmium (Pawlowska *et al.*, 1996). Similarly, Ietswaart *et al.* (1992) found up to 73% colonization of metal-resistant *Agrostis capillaris* L. on a site highly contaminated with lead and zinc.

Studies of AM colonization of metal-contaminated soils have concentrated on three habitat types: sites receiving metal-contaminated sewage sludge (Boyle and Paul, 1988; Weissenhorn *et al.*, 1995c; Weissenhorn and Leyval, 1996), soil impacted by metal smelting (Gildon and Tinker, 1981, 1983; Dueck *et al.*, 1986; Ietswaart *et al.*, 1992; Griffioen *et al.*, 1994; Weissenhorn

Table 1
Presence of arbuscular mycorrhizas at metal-contaminated sites

Soil matrix	As	Cd	Cu	Pb	Zn	Root (%) colonization*	Vegetation cover	Plant adaptation	Reference
Zn smelter		2.5	30	103	575		Wheat	No	Leyval et al. (1995)
Zn smelter		131	1630	6060	24 410		Shrubs and trees	Not tested	
Non-polluted		1	23	58	184		Oats	No	
Natural alum shales		3	91	49	215		Potato	No	
Non-polluted		0.2	13	39	77		Wheat	No	
Non-polluted		0.2	13	35	70		Wheat	No	
Non-polluted		0.2	18	28	60		Oat	No	
Non-polluted		0.1	9	18	23		Oat	No	
Zn spiked		18	95	895	1 220	38	Maize	No	Weissenhorn et al. (1993)
Cd spiked		53	18	33	44	12	Ryegrass	No	
Copper mine spoil		0.5	779		294	1	Agrostis capillaris	Yes	Griffioen et al. (1994)
Zn smelter		13.5	170		1 020	41	Agrostis capillaris	Yes	
Non-polluted		0.4	0.6		5	58	Agrostis capillaris	Yes	
Zn and Cd contaminated		863			1 400	35	Clover	Not tested	Gildon and Tinker (1981, 1983)
Zn and Cd contaminated		488			1 300	51	Clover	Not tested	
Zn and Cd contaminated		5			600	65	Clover	Not tested	
Cd spiked		53	18	33	44	12	Ryegrass	No	Weissenhorn et al. (1994)
Unspiked Cd control		0.6	21	32	47	12	Ryegrass	No	
Zn sewage sludge		6	67	189	1 074	0	Maize	No	
Sewage sludge control		0.5	23	43	51	0	Maize	No	
As mine soil	550					22	Holcus lanatus	Yes	A.A. Meharg (unpublished data)
As mine soil	486					48	Holcus lanatus	Yes	
Non-polluted	20					38	Holcus lanatus	Yes	
Unvegetated mine tailings		4	18	13	824		None	Not tested	Shetty et al. (1994)
Vegetated mine tailings		1		10	43		Not given	Not tested	
Non-polluted		0.4		5	6		Not given		
Calamine mine spoil (disturbed)		180	18	4560	49 000	0–90	40 species	Metal-resistant ecotypes present	Pawlowska et al. (1996)
Calamine mine spoil (undisturbed)		180	18	4560	49 000	0–96	25 species		
Zn sewage sludge soil 1					11.4	41	Barley	No	Boyle and Paul (1988)
Zn sewage sludge soil 1					206	33	Barley	No	
Control					61	42	Barley	No	
Zn sewage sludge soil 2					246	4	Barley	No	
Zn sewage sludge soil 2					422	5	Barley	No	
Control					114	28	Barley	No	

*Where no colonization percentage is given, AM fungal presence was identified from spores isolated from the soil.

et al., 1994, 1995a,b; Weissenhorn and Leyval, 1995, 1996) and mine spoils (Griffioen and Ernst, 1990; Griffioen *et al.*, 1994; Shetty *et al.*, 1994, 1995). Sewage sludge investigations have generally focused on agronomic plant species that are not adapted to metal contaminants (Boyle and Paul, 1988; Weissenhorn *et al.*, 1995c; Weissenhorn and Leyval, 1996). Soils contaminated by metals as a result of their proximity to smelters may be either moderately contaminated (and still support agronomic plants not adapted to raised metal levels (Gildon and Tinker, 1981, 1983; Weissenhorn *et al.*, 1994, 1995a,b; Weissenhorn and Leyval, 1995, 1996)) or highly contaminated, supporting only ecotypes with resistance to the contaminants present (Dueck *et al.*, 1986; Ietswaart *et al.*, 1992; Griffioen *et al.*, 1994). Mine spoils represent extreme metalliferous environments, where, generally, only metal-resistant higher plant populations that have evolved resistance to the contaminants present can establish (Macnair, 1993). There is abundant evidence that AM fungi are generally present in soils that support metal-resistant plant ecotypes (Dueck *et al.*, 1986; Griffioen and Ernst, 1990; Ietswaart *et al.*, 1992; Griffioen *et al.*, 1994); hence it may be concluded that AM fungi have also evolved strategies that allow their colonization of highly metalliferous soils. Evidence for this hypothesis is expanded below.

The first conclusive studies on evolution of metal resistance in AM fungi were conducted by Gildon and Tinker (1981, 1983). These workers demonstrated that strains of *Glomus mosseae* isolated from soils that were heavily contaminated with cadmium and zinc could sustain high colonization levels in clover in the presence of zinc and cadmium, while greatly reduced colonization was observed for strains isolated from an uncontaminated habitat. In a series of experiments on AM fungal spores isolated from zinc- and cadmium-contaminated soils, Leyval, Weissenhorn and their associates (Weissenhorn *et al.*, 1993, 1994; Weissenhorn and Leyval, 1995) were the first to demonstrate that other stages of the AM fungal life cycle, besides colonization, demonstrate intraspecific variation in response to toxic metals.

Most studies on AM metal sensitivity and colonization of contaminated sites by potentially AM plants and their fungi have concentrated on cadmium and zinc (Table 1). It is only for these two metals that conclusive evidence exists for evolution of metal resistance in AM fungi. The majority of the copper-resistant grasses present on highly copper-contaminated sites are free of AM colonization, and those that are colonized have very low levels of colonization (Griffioen *et al.*, 1994). Griffioen *et al.* (1994) postulated that copper is fungicidal to AM fungi and that decreased sensitivity to the element has not evolved in this group. Griffioen and Ernst (1990), however, conversely detailed a study in which they inoculated *Agrostis capillaris* from the same copper mine population with its indigenous AM.

Metals may potentially inhibit AM function at any point in their life cycle. Host–symbiont interactions with metals has a crucial role as a number of feedback loops regulating the symbiosis may be perturbed by soil contamination. If the plant is unhealthy from exposure to the metal, this may affect colonization by, and regulation of, the symbiosis (Gildon and Tinker, 1983).

Consideration of the following parameters needs to be considered when assessing adaptation of AM fungi to metal-contaminated sites: AM diversity, spore germination, germ tube growth, plant infection, colonization, plant growth and differential host responses to colonization.

B. Species Diversity

A number of studies have shown that considerable AM fungal species diversity can exist in metal-contaminated soils; see Griffioen (1994) for a summary of literature data, and Pawlowska et al. (1996) and Shetty et al. (1994, 1995) for subsequently published studies. In each of these studies, the genera *Acaulospora, Gigaspora, Glomus* and *Sclerocystis* were all represented, with a number of species within each genus present. Perhaps more importantly, while the extent to which 'functional redundancy' exists within AM (and indeed other mycorrhizal fungal) taxa is currently unclear (see Allen et al. (1995) for a full discussion), it seems likely that considerable functional heterogeneity exists within such diverse communities. Thus adaptation of a diverse fungal community to metal contamination may be important in enhanced host fitness on contaminated sites.

C. Spore Germination

Although germination of AM fungal spores may be inhibited at high metal concentrations (Hepper and Smith, 1976), that spores of metal-resistant ecotypes can germinate under such conditions is implicit in the observations that AM fungi colonize plants at metal-contaminated sites (Table 1). In a series of experiments with AM fungal spores isolated from metal-contaminated sites, metal-resistant *G. mosseae* strains were isolated from zinc- and cadmium-contaminated soils (Weissenhorn et al., 1993, 1994; Weissenhorn and Leyval, 1995) (Table 2). Spores were placed in a membrane sandwich device and incubated in sand culture modified with a range of metal concentrations, and germination was assayed after a suitable time period. Spores isolated from contaminated sites demonstrated considerably decreased sensitivity towards cadmium and zinc compared with those from uncontaminated soils (Table 2). Concentrations causing a 50% decline in germination (EC_{50}) were 4–9-fold higher in populations from contaminated soils. However, at the highest zinc and cadmium concentrations (10 and 9.8 mg l^{-1} cadmium and zinc respectively in sand culture) germination was markedly inhibited in all spore populations, regardless of their soil of origin (Weissenhorn et al., 1994).

Table 2

EC$_{50}$ values for germination, hyphal growth and colonization levels for cadmium and zinc sensitive and resistant phenotypes of *Glomus mosseae*

Soil of origin	Cadmium in soil (mg kg^{-1}) Total	Cadmium in soil (mg kg^{-1}) EDTA extracted	Zinc in soil (mg kg^{-1}) Total	Zinc in soil (mg kg^{-1}) EDTA extracted	EC$_{50}$ for cadmium (μmol l^{-1}) Spore germination	EC$_{50}$ for cadmium (μmol l^{-1}) Hyphal growth	EC$_{50}$ for cadmium (μmol l^{-1}) Colonization levels
Not described, uncontaminated	–	–	–	–	7.1	58	17.8
Silty clay loam, contaminated by a zinc smelter	17.7	13.6	1220	350	62.5	50.8	47.3
Loamy sand, uncontaminated	0.6	0.3	47	3.2	18.5	78	–
Loamy sand, amended with cadmium nitrate	53	46	44	2.2	72.5	66	–
Farmyard manure modified, uncontaminated	0.5	0.4	51	32	8.8	11	–
Sewage sludge modified, zinc contaminated	5.7	3.1	1074	390	7.5	9	–

EDTA, ethylene diamine tetra-acetic acid.
Data from Weissenhorn *et al.* (1993, 1994) and Weissenhorn and Leyval (1995).

It appears that evolution of resistance is rapid, with cadmium and zinc resistance developing in a population after 1 year of field exposure to soil containing cadmium nitrate at a level of 40 mg cadmium kg^{-1} (Weissenhorn et al., 1994). When cultured through one reproduction cycle under uncontaminated conditions, the insensitivity was lost from the population. Weissenhorn et al. (1994) explained this rapid assimilation and then loss of metal resistance in terms of the polynuclear nature of AM fungal spores, with each spore known to contain 1000–3850 nuclei (Smith and Read, 1997). Thus rapid evolution may be due to phenotypic plasticity endowed by being polynucleic. If this was the case, then phenotypic plasticity may be expected in all populations, including those from uncontaminated environments. The reproductive biology of AM fungi is not understood. It may be that spores containing nuclei with genes encoding for metal resistance are rapidly transferred through AM populations in the presence of selection pressures (such as toxic metals). Metal resistances in higher plants from uncontaminated populations are generally maintained by the stochastic mutation rate (Macnair, 1993), and this may be the case in AM populations. The reproductive biology of AM may facilitate rapid transfer of these genes through the population when high metal burdens are imposed. When the selection pressure is removed, genes coding for metal resistance may be rapidly lost and return to low frequencies as metal resistance is thought to generally carry a high metabolic cost (Macnair, 1993).

A further observation in support of this hypothesis is that the metal dose–response curves for germination of spores of resistant *G. mosseae* strains are not steep and, although germination rates in sensitive strains are decreased compared with those in resistant strains, germination is not totally inhibited, even at the highest exposure concentrations (Weissenhorn et al., 1993, 1994). Indeed, at the highest concentrations tested, sensitive and resistant strains have very similar germination rates.

In spores from populations from uncontaminated soils that do germinate in high metal concentrations, there is little difference in hyphal growth at high metal concentrations between metal-sensitive and metal-resistant ecotypes (Weissenhorn et al., 1994). This lends support to the hypothesis of Weissenhorn et al. (1994) that, if plasticity (as a function of the polynucleic nature of spores) is responsible for the rapid appearance and disappearance of metal resistance in AM fungal populations, then spores that germinate under high metal concentrations appear to grow equally well from both resistant and sensitive populations (Table 2).

Comparisons of spore germination in soils with differing degrees of aluminium saturation (6–100%) suggest that intraspecific and interspecific variation in sensitivity exists in AM fungi (Bartolome-Esteban and Schenck, 1994). *Gigaspora* and *Scutellospora* species exhibit relatively high resistance to aluminium, whereas, all *Glomus* species are extremely sensitive to the metal, with the exception of a population of *G. manihotis* isolated from an acid soil (pH 4). Unfortunately the degree of aluminium saturation of the soils from which the fungi were isolated was not given, and pH of only a limited number of soils was reported, hindering interpretation of the results with the respect to evolution of aluminium resistance in AM fungi.

In general, it appears that selection for AM fungi that are resistant to inhibition of spore germination by cadmium, zinc and potentially aluminium does occur in metal-contaminated environments. However, the ecological significance of spore germination insensitivity is somewhat clouded by the fact that studies conducted in soil show that other edaphic factors (besides the metal contaminants) are the principal regulators of spore germination (Leyval *et al.*, 1995; Weissenhorn and Leyval, 1996). It may be the case in such studies that the metal levels in the soil solution to which spores were exposed were insufficiently high to be a significant determining factor in regulating germination. Soils with very high metal burdens (such as mine spoil soils) may be required before metal concentrations regulate germination rates. However, under extreme selection pressures, resistant and sensitive strains have been shown to display similar germination rates (Weissenhorn *et al.*, 1993, 1994), and it is only at intermediate concentrations that resistant strains fare better than sensitive ones with respect to cadmium and zinc exposure.

D. Germ Tube Growth

Germ tube growth was ascertained in cadmium and zinc resistant and sensitive strains of *G. mosseae* isolated from soil impacted by a zinc smelter, soil modified with cadmium nitrate, and soil modified with zinc-contaminated sewage sludge (Weissenhorn *et al.*, 1993, 1994; Weissenhorn and Leyval, 1995). Spores of germinated 'sensitive' strains of AM fungi showed equivalent growth in the presence of cadmium and zinc to 'resistant' strains (Table 2). Germ tube growth by itself may not be a good criterion for assessing sensitivity towards metals because of the effects of energy reserves contained within spores, which may facilitate germ tube growth under adverse conditions (Bartolome-Esteban and Schenck, 1994). However, this argument for resistance implies (a) that metals do not exert a toxic stress on the plasma membrane of the extending germ tube and (b) that the metals are not being assimilated by germ tubes and subsequently causing intracellular disruption. Plasma membrane processes are known to be sensitive to toxic metals (Meharg, 1993). It seems unlikely that germ tubes should be constitutively resistant to cadmium and zinc since they possess the same array of physiological processes (carriers, channels, adenosine triphosphatases, etc.) in the plasma membrane (e.g. Thompson *et al.*, 1990; Lei *et al.*, 1991) as other plant and fungal plasma membranes.

An alternative hypothesis has been outlined above in the section on spore germination, principally that the genetic make-up of AM fungal spores facilitates a high stochastic mutation rate. In essence, nuclei that encode the genes for metal resistance should be at higher frequencies in contaminated soils, leading to higher spore germination rates. However, the stochastic mutation rate may mean that a proportion of spores from uncontaminated soils carries genes for metal resistance, and that they will display hyphal extension rates in the presence of metals that are consistent with ecotypes from contaminated environments. The plasticity incurred by being polynuclear may explain the ecological advantage of spores carrying such a high amount of genetic diversity—not just with respect to adapting to metal-contaminated environments, but also to a wide range of other edaphic factors.

E. Hyphal Penetration

It is thought that the plant initiates differential morphogenesis in AM fungal hyphae leading to colonization, probably determined by chemical signals released from the host (Koide and Schreiner, 1992; Giovannetti *et al.*, 1994). The presence of toxic metals in plant tissues may affect signalling and the response of AM fungi to potential plant hosts. There is evidence from the study by Gildon and Tinker (1983) that, in metal-stressed plants, penetration and/or colonization is suppressed. In this study, maize (*Zea mays*) was grown in a split-root experiment where half the root system was exposed to soil modified with 100 mg kg^{-1} zinc and the other half of the root system grown in unmodified soil. Colonization in both root chambers was identical and was suppressed compared with that in a control split-root treatment were no zinc was added to the soil. This suggested that penetration and/or colonization was being suppressed by the roots in conditions were there was no zinc inhibition of spore germination or spore hyphal growth. Zinc levels in the unmodified half the split-root experiment were increased 3-fold above those of the control treatment, but were at half the concentration of roots in the modified part of the microcosm. Even though zinc levels in roots were different, colonization was equal in both parts of the microcosm. Plant health was not affected by zinc treatment in this experiment as the zinc-treated plant grew as well as the control.

Barkdoll and Schenck (1986) investigated the regulation of spore germination and hyphal penetration of bean roots by aluminium for 10 fungal isolates from four species (*G. mosseae, G. manihotis, Gi. pellucida* and *Acaulospora longula*). Although responses of the fungi to aluminium were observed, root penetration (number of fungal penetration points) was affected more by aluminium than was spore germination.

F. Root Colonization

It is a general phenomenon in both sensitive and resistant AM fungal strains that increasing the metal concentration in the growth matrix reduces the percentage of root colonized (Gildon and Tinker, 1981, 1983; Shetty et al., 1994, 1995; Weissenhorn and Leyval, 1995). However, the factors underlying this decrease in percentage colonization are not clear. Inhibition of spore germination, extension of free-living hyphae, root hyphal penetration and disruption of the internal hyphae may all result in decreased colonization levels.

Exposure of extramatrical mycelium to metals notwithstanding, once the fungus is inside the host it receives relatively little exposure to the contaminating metals, since free metal activity within host tissue should be low. It is possible that the host may transfer metals, either as free ions or as complexes to the fungus, or that the fungus transfers metals from the soils via its hyphal network to its own tissues within the plant. The former hypothesis has not been tested. Experiments with cadmium indicate that the latter may be the case, as cadmium absorbed by external hyphae is transferred to the plant, but transfer from fungus to the plant is restricted due to fungal immobilization (Joner and Leyval, 1997). It may be that development of AM fungal structures within the plant is affected by disruption of both plant and fungal physiology by metals. Fungal growth, unchecked, would carry a high cost to the host, and colonization is highly regulated by the host (Koide and Schreiner, 1992). Regulation is thought to involve combined mechanisms that enhance or limit colonization, dependent on environmental conditions and nutrient supply. Disruption of these regulatory processes may occur due to stress caused by plant exposure to high metal levels. Although the mechanistic basis of the interactions that regulate the symbiosis are not well understood, it is clear that exposure of the symbiosis to metals causes severe disruption of AM fungal morphology in roots. Experiments with onion indicate that, in the presence of concentrations of zinc and copper that inhibit the extent of colonization, internal hyphae are less dense, shorter and more irregular than those colonizing non-exposed plants (Gildon and Tinker, 1983). However, these morphological changes may be caused by changes to host plant cell anatomy.

Root colonization was more severely impeded by zinc in onion than in clover, with colonization by a sensitive *G. mosseae* isolate being totally eliminated at a zinc level of 100 mg kg^{-1}, while clover roots remained colonized (though greatly reduced) at zinc concentrations of 1000 mg kg^{-1} (Gildon and Tinker, 1983). Onion (shoot) biomass was not greatly affected by zinc concentrations that severely inhibited colonization. No data for clover was presented. It was unclear whether the clover strain used was more resistant to zinc than the onion. In any case, the data indicate that AM fungal strains can respond differently with respect to their colonization of different host roots in the presence of an imposed metal stress. Whether this was due to a decrease in the number of hyphal penetration points and/or disruption of plant regulation of colonization was uncertain.

Experiments with maize (Weissenhorn and Leyval, 1995; Weissenhorn et al., 1995b) and clover (Gildon and Tinker, 1981, 1983) using cadmium and zinc sensitive and resistant strains of *G. mosseae* showed that the resistant strains colonized the roots more effectively than the sensitive strains. Plant suppression of colonization by the physiological disruption caused by the metal could not explain why resistant AM fungal ecotypes fared so well with respect to colonization in comparison with sensitive AM fungi. In the clover experiments (Gildon and Tinker, 1981, 1983), colonization by the resistant strain was maintained at high levels (40–50% versus 50–70% in unexposed plants) at very high cadmium and zinc concentrations. The sensitive ecotype still colonized roots at the highest concentrations tested, although colonization was reduced by 10–20%. Differences in colonization rates between sensitive and resistant strains are regulated by the initial density of viable inoculum, spore germination success, growth of free-living hyphae, hyphal penetration of roots and subsequent colonization of the roots. Reduced colonization of roots by sensitive strains may simply reflect decreased germination and decreased hyphal penetration by this strain, rather than decreased fitness within its host. Alternatively, the resistant strain may be able to regulate transfer and/or detoxification of metals, thus reducing exposure of internal tissues to the metals, resulting in more effective colonization of the host root. No plant growth parameters were given in the studies of Gildon and Tinker (1981, 1983), making interpretation with respect to plant health difficult. In the studies of Weissenhorn and Leyval (1995), root and shoot biomass was inhibited by only about 50% at cadmium concentrations that totally inhibited colonization by cadmium-sensitive and -resistant *G. mosseae*.

G. Spore Production

Spore production is an essential stage in the AM life cycle. There is limited research into the effects of toxic metals and spore production. Leyval et al. (1997) reported a study where spore production was not affected by cadmium concentrations of up to 100 mg kg^{-1}.

H. Metal Assimilation in Plant Tissues

AM fungi are generally thought to facilitate the assimilation of micronutrient divalent cations (principally zinc and copper) by the host (Smith and Read, 1997). Studies using radio-isotopes of the essential element zinc (^{65}Zn) (Cooper and Tinker, 1978; Burkert and Robson, 1994) and the non-essential element cadmium (^{109}Cd) (Joner and Leyval, 1997) have shown that extra-radical hyphae can accumulate and transfer these metals to roots.

Essential divalent cations such as zinc and copper are toxic at high concentrations, as are non-essential transition metal cations. It is possible, in some cases, that AM fungi impose enhanced toxic stress on the plant as they may facilitate increased assimilation in contaminated soils. One mechanism that may be required for decreased sensitivity in AM fungi or in the AM fungus–host association is that the adapted AM fungus reduces assimilation of metals and consequently decreases the toxic stress to the host. This may be through binding of metals to external hyphae, restricted transport of metals to internal AM and plant tissues, or sequestration with AM tissues (Leyval *et al.*, 1997). Being AM may also lead to the down regulation of metal assimilation by host tissues.

AM fungi tend to cause enhanced assimilation of micro-nutrient metals from deficient or unenriched soils (Weissenhorn *et al.*, 1995b). On metal-enriched soils the situation is more complicated, with reports showing that AM fungi can enhance metal uptake, and indeed lead to enhanced toxicity (Killham and Firestone, 1983; Weissenhorn *et al.*, 1995b). Other studies have shown that AM fungi decrease metal uptake on highly contaminated soils (Hetrick *et al.*, 1994). AM fungi may also have no effect on plant uptake of toxic metals (copper and zinc) across a wide range of concentrations (Gildon and Tinker, 1993).

Zinc assimilation by *Festuca arundina* and *Andropogon gerardii* inoculated with AM fungi from uncontaminated and contaminated sites was determined in soils where zinc concentrations were sufficiently high to limit the growth of *A. gerardii* but not *F. arundina* (Shetty *et al.*, 1994). The AM fungi were probably mixed populations and no attempt was made to characterize sensitivity. AM fungi from the contaminated site enhanced the uptake of zinc to a greater extent than the AM inoculum from the uncontaminated site by 25–50% per plant for both grass species, and both AM fungal populations increased the assimilation compared with the non-mycorrhizal treatments. For *A. gerardii* enhanced assimilation was observed in both roots and shoots, but for *F. arundina* assimilation was enhanced only in roots.

Hetrick *et al.* (1994) grew *A. gerardii* and *F. arundinacea* on zinc-contaminated mine tailings (750 mg Zn kg^{-1}) and compared their zinc accumulation to that in plants grown in uncontaminated soils. Plants were inoculated with AM fungi (mixed population) from the uncontaminated soils. The fungi colonized the roots of both species well on the contaminated mine spoil (approximately 50% for both species). The non-mycorrhizal treatment had considerably higher zinc tissue concentrations (100% for *A. gerardii* and 2700% for *F. arundinacea*) in the mine tailing/clay treatment, with a similar, although less dramatic, pattern on the mine spoil alone treatment. Various fertilization treatments greatly enhanced uptake of zinc, and generally resulted in greatly enhanced (up to 50-fold) increases in plant yield. The conclusions that can be drawn from this study are that AM colonization reduces zinc uptake in extremely contaminated and nutrient-deficient soils.

A. gerardii was grown on soil (a silt loam) modified with a range of zinc concentrations (up to 1000 mg kg^{-1}) and inoculated with either an AM fungal population from a mine tailing rich in zinc or from an uncontaminated soil (Shetty *et al.*, 1995). Mycorrhizal colonization ranged from 15% to 29% and there was no significant effect of zinc or origin of inocula on colonization. At low (< 200 mg kg^{-1} zinc) and at high (1000 mg kg^{-1}) zinc concentrations there was little difference in metal assimilation in shoots between mine and non-mine colonized and uncolonized controls. At 500 mg kg^{-1} the mine mycorrhizal population assimilated significantly less zinc than the non-mine population, with the uncolonized treated having intermediate zinc levels.

In experiments with maize growing on lead–zinc smelter contaminated soil (1220 mg kg^{-1} zinc, 895 mg kg^{-1} lead), inoculation with zinc and cadmium sensitive and resistant strains of *G. mosseae* resulted in no difference in zinc or lead assimilation (both roots and shoots) in comparison with each other and in comparison with an uninoculated control, even though colonization levels differed greatly (0% for uninoculated, 19% for the sensitive strain and 32% for the resistant strain; Weissenhorn *et al.*, 1995b). In experiments using the same *G. mosseae* isolates, Weissenhorn and Leyval (1995) showed that, over a range of cadmium concentrations in sand culture, the resistant and sensitive AM isolates behaved similarly with respect to cadmium accumulation. The non-mycorrhizal control behaved in a similar fashion to the mycorrhizal treatments at high and low cadmium treatments, but at intermediate levels grew much better and assimilated less cadmium. Colonization levels were much higher with the resistant strain at intermediate cadmium concentrations, yet there were no differences in cadmium assimilation.

Griffioen and Ernst (1990) grew copper-sensitive and -resistant populations of *Agrostis capillaris*, which were either non-mycorrhizal or colonized with inocula from soil of origin of the metal-resistant plants. AM colonization decreased the copper content of sensitive plants but significantly increased the copper content of resistant plants. In a similar experiment, Dueck *et al.* (1986) collected *Festuca rubra* L. and *Calamagrostis epigejos* from zinc-contaminated sand dunes downwind of a blast furnace and grew them in zinc-contaminated soil, either inoculated or uninoculated with AM fungi. It was assumed, although not stated explicitly, that these plants were zinc tolerant. The origin of the inoculum was not given and colonization levels were not stated. In zinc-contaminated sand (458 mg kg^{-1} zinc), AM-colonized plants assimilated slightly less zinc in their roots (8% for both species), although shoot levels were identical for both species. AM colonization did not greatly ameliorate zinc toxicity.

Evidence to date is inconclusive as to whether colonization with resistant AM strains decreases metal exposure to resistant or sensitive hosts. The effects of AM colonization in the studies discussed in this subsection were variable, dependent on the metal, substrate and nutrition. Therefore, being AM on highly contaminated sites could be either detrimental, beneficial of neutral with respect to metal assimilation, and local conditions within contaminated sites could regulate the cost–benefit relationship of being AM.

I. Plant Sensitivity

A. gerardii and *F. arundina* colonized with AM fungi from a zinc-contaminated site produced less biomass than plants colonized with the AM fungi from an uncontaminated site, but this was true for both the uncontaminated and the contaminated soil (Shetty *et al.*, 1994). Hetrick *et al.* (1994) grew *A. gerardii* and *F. arundina* on zinc-contaminated mine tailings (750 mg kg^{-1} zinc) and compared growth to that on uncontaminated soils. Plants were inoculated with AM fungi (mixed population) from the uncontaminated soils. The fungi colonized the roots of both species well on the contaminated mine spoil (approximately 50% for both species), but conferred no growth advantage to the host on this spoil. The mycorrhizal treatment was disadvantageous on mine spoil amended with calcined montmorillonite clay (3 : 1 mixture of mine spoil to clay), reducing growth of *A. gerardii* by 50% and that of *F. arundinacea* by 30% compared with non-mycorrhizal treatments. *A gerardii* was grown on soil (a silt loam) modified with a range of zinc concentrations (up to 1000 mg kg^{-1}) and inoculated with an AM fungal population either from a mine tailing rich in zinc or from an uncontaminated soil (Shetty *et al.*, 1995). Mycorrhizal colonization ranged from 15% to 29% and there was no significant effect of zinc or origin of inocula on colonization. Both mycorrhizal treatments greatly enhanced plant biomass production, with the inoculum from the uncontaminated soil being much more effective than that from the contaminated soil (Shetty *et al.*, 1995). At intermediate concentrations (100 and 500 mg kg^{-1} zinc), plant biomass with the inoculum from the contaminated source increased 3-fold, but decreased 5-fold over this concentration range with AM from the uncontaminated source. However, only at 500 mg kg^{-1} did the fungi from contaminated soil out perform mycorrhiza from uncontaminated soil, the concentration at which it significantly reduced zinc assimilation. Inoculation (from both sources) increased biomass production at all zinc treatments compared with uninoculated plants. At 1000 mg kg^{-1} zinc, biomass was suppressed in all treatments, although the inoculated plants fared slightly better than the uninoculated plants, with biomass production equivalent for both inoculum treatments.

For maize growing on lead–zinc smelter contaminated soil (1220 mg kg^{-1} zinc, 895 mg kg^{-1} lead), inoculation with zinc and cadmium sensitive and resistant strains of *G. mosseae* resulted in no difference in biomass (both roots and shoots) in comparison with each other and with an uninoculated control (Weissenhorn *et al.*, 1995b). Colonization levels were zero for uninoculated, 19% for the sensitive strain and 32% for the resistant strain. Over a range of cadmium concentrations in sand culture, resistant and sensitive *G. mosseae* isolates behaved very similarly with respect to biomass production (Weissenhorn and Leyval, 1995). The non-mycorrhizal control behaved in a similar fashion to the mycorrhizal treatments at high and low cadmium treatments, but at intermediate levels grew much better and assimilated less cadmium. Colonization levels were much higher with the resistant strain at intermediate cadmium concentrations, yet there were no differences in biomass production. From the experiments of Weissenhorn and Leyval (1995) and Weissenhorn *et al.* (1995b), it must be concluded that resistant AM fungi do not confer any benefit to the host with respect to biomass production.

J. Nutritional Aspects

Andropogon gerardii colonized with AM fungi from a zinc-contaminated and -uncontaminated site had similar amounts of phosphorus in their shoots across a wide range of zinc concentrations (up to 1000 mg kg^{-1}) (Shetty *et al.*, 1995). P concentration was 2–3 times lower in the uncolonized plants. Besides an initial decrease in zinc assimilation at 100 mg kg^{-1} in both inoculated treatments, there was little change in all AM and non-AM fungal treatments at all zinc concentrations. Mycorrhizal colonization ranged from 15% to 29%, and there was no significant effect of zinc or origin of inocula on colonization. The situation was quite different in the study of Gildon and Tinker (1983), in which shoot P concentrations decreased markedly in clover inoculated for a zinc-sensitive *G. mosseae* isolate and the uncolonized control across a range of zinc soil concentrations. For a zinc-resistant strain, there was no effect on shoot P levels at the maximum concentration tested (1000 mg kg^{-1}). Colonization was greatly impeded with the sensitive isolate, but reduced only slightly for the resistant isolate, which may explain the differences in P status.

P concentrations were determined in roots and shoots of maize, either uninoculated or inoculated with cadmium-sensitive and -resistant strains of *G. mosseae* grown over a range of cadmium concentrations (Weissenhorn and Leyval, 1995). The non-mycorrhizal treatment generally maintained lower root and shoot P concentrations, especially at intermediate cadmium levels, than the inoculated treatments. There was little or no difference between inoculation treatments. However, mycorrhizal colonization did not greatly enhance P root and shoot concentrations, with a maximum difference of about 40% between inoculated and uninoculated treatments. At high cadmium exposure, P concentrations were the same in all treatments, concurrent with almost total inhibition of colonization. P tissue levels were relatively stable across a range of cadmium concentrations in the uninoculated controls. Gildon and Tinker (1983) found a similar response in P concentrations of clover inoculated with a cadmium-resistant and -sensitive strain of *G. mosseae,* with inoculated plants having slightly higher P shoot levels than uninoculated plants across a wide range of cadmium concentrations, with little or no effect of cadmium. Resistant and sensitive isolates had very similar P levels. P levels were not greatly affected in any of the plant treatments by increasing cadmium soil concentrations.

From the limited studies conducted on nutritional aspects of mycorrhizal–host associations on contaminated sites, it appears that cadmium- and zinc-resistant AM can maintain plant P status at high soil cadmium and zinc levels, benefiting the plant in comparison to sensitive AM strains and uncolonized controls. This seems principally to be a function of the resistant strains maintaining AM colonization at high levels of metal exposure.

K. Do Plants Benefit from being Colonized with Resistant Arbuscular Mycorrhizal Strains?

AM fungi can evolve resistance to a toxic metals (Gildon and Tinker ,1981, 1983; Dueck *et al.,* 1986; Griffioen and Ernst, 1990; Ietswaart *et al.*, 1992; Weissenhorn *et al.,* 1993, 1994; Bartoleme-Esteban and Schenck, 1994; Griffioen, 1994; Griffioen *et al.,* 1994; Weissenhorn and Leyval, 1995). The definitive experiments to study co-evolution in host and symbiont have not been conducted (reciprocal inoculation of sensitive and resistant host and sensitive and resistant AM fungus over a range (dose–response) of metal concentrations). However, the evidence from experiments conducted to date with resistant and sensitive hosts with (presumed) resistant AM fungi (Griffioen and Ernst, 1990), resistant hosts with resistant AM fungal strains or populations (Dueck *et al.,* 1986) and sensitive hosts with sensitive and resistant AM fungal strains or populations (Gildon and Tinker, 1981, 1983; Weissenhorn *et al.,* 1993, 1994; Hetrick *et al.,* 1994; Shetty *et al.,* 1994, 1995; Weissenhorn and Leyval, 1995) indicates that AM fungi, whether sensitive or resistant, confer little or no enhanced metal resistance to their hosts. There was no systematic evidence that AM fungi protected their hosts from exposure to metal contaminants by reduced assimilation of metals, with a wide range of metal assimilation responses observed.

AM presence on contaminated sites then raises the question as to why potentially AM plants on highly contaminated soils are colonized by AM fungi if they do not confer added resistance? The likely answer is that AM colonization is maintained on mine plant populations for the same reasons that they are maintained on uncontaminated environments: AM fungi confer benefit to their host besides metal resistance, such as maintaining phosphorus acquisition in resistant strains observed at high metal exposures (section III.J). Therefore, AM fungi and their hosts have co-evolved to survive in metal-contaminated environments to exploit such ecological niches. This conclusion is preliminary, and studies need to be conducted with AM fungal strains from highly contaminated environments in combination with resistant hosts from similar habitats to determine whether or not AM fungi facilitate enhanced metal resistance.

IV. ERICOID MYCORRHIZAS

A. The Ericaceae and Metal-contaminated Environments

Investigations into mycorrhizal endophytes of the Ericaceae provide the most clear and conclusive roles for mycorrhiza in providing metal resistance to their hosts (Bradley et al., 1981, 1982). Species of the family Ericaceae colonize soils with naturally raised metal concentrations such as serpentine soils (Proctor and Woodell, 1971; Shewry and Peterson, 1976; Marrs and Bannister, 1978; Marrs and Proctor, 1978) and metalliferous soils contaminated as a result of mining and smelting activity (Porter and Peterson, 1975; Marrs and Bannister, 1978; Oxbrow and Moffat, 1979; Freedman and Hutchinson, 1980; Benson et al., 1981; Bradley et al., 1981, 1982; Bagatto and Shorthouse, 1991; Eltrop et al., 1991; Brown, 1994; Aparicio, 1995). Members of the Ericaceae are often the dominant vegetation cover on such soils (Proctor and Woodell, 1971; Shewry and Peterson, 1976; Marrs and Bannister, 1978; Marrs and Proctor, 1978; Oxbrow and Moffat, 1979; Freedman and Hutchinson, 1980; Bradley et al., 1982). The metals and semi-metals contaminating soil colonized by Ericaceae include arsenic, cadmium, chromium, cobalt, copper, lead, molybdenum, nickel and zinc (Table 3).

Soils colonized naturally by Ericaceae are generally acidic and nutrient deficient (Marrs and Bannister, 1978; Bradley et al., 1982; Shaw and Read, 1989; Shaw et al., 1990; Yang et al., 1996), and are often boggy and exposed to routine flooding (Jones and Etherington, 1970; Jones, 1971). Both low pH and anaerobic soil conditions facilitate the mobilization of transition metals. Many of the metals mobilized under such conditions are micro-nutrients such as copper, iron, manganese, nickel and zinc, which are toxic at high concentrations. It is thought the Ericaceae in general have adapted their physiology to cope with high external concentrations of these metals (Shaw and Read, 1989; Leake et al., 1990; Shaw et al., 1990). Non-essential metals such as aluminium, cadmium and lead are also phytotoxic and are bioavailable at low pH. Aluminium, iron and manganese are problematic in most habitats inhabited by the Ericaceae, although it should be noted that some members of what is considered to be a calcifuge family may colonize calcareous soils (Leake et al., 1990). Since the Ericaceae are generally adapted to grow in soils where metal availability is high, they have an intrinsic ability to tolerate high levels of aluminium (Yang et al., 1996), iron (Shaw et al., 1990; Hashem, 1995a) and manganese (Hashem, 1995b). The chemistry of metal ions such as cadmium, copper, nickel and zinc ions is somewhat analogous to

Table 3
Ericaceae colonization of metal-contaminated soils

Metal	Ericoid species	Reference
Arsenic	*Calluna vulgaris*	Benson *et al.* (1981)
		Porter and Peterson (1975)
Chromium	*Calluna vulgaris*	Marrs and Bannister (1978)
		Marrs and Proctor (1978)
	Erica vagans	Marrs and Proctor (1978)
	Erica cinera	Marrs and Proctor (1978)
Copper	*Calluna vulgaris*	Benson *et al.* (1981)
		Bradley *et al.* (1981, 1982)
		Marrs and Bannister (1978)
		Porter and Peterson (1975)
	Vaccinium angustifolium	Freedman and Hutchinson (1980)
		Bagatto and Shorthouse (1991)
Lead	*Calluna vulgaris*	Marrs and Bannister (1978)
		Eltrop *et al.* (1991)
		Oxbrow and Moffat (1979)
Nickel	*Calluna vulgaris*	Marrs and Bannister (1978)
		Marrs and Proctor (1978)
	Erica vagans	Marrs and Proctor (1978)
	Erica cinera	Marrs and Proctor (1978)
	Vaccinium angustifolium	Freedman and Hutchinson (1980)
		Bagatto and Shorthouse (1991)

that of aluminium, iron and manganese, and they tend to exert similar modes of toxic action in both higher and lower plants (Clarkson and Luttge, 1989; Gadd, 1993; Ross, 1993; Welsh *et al.*, 1993; Fox and Guerinot, 1998). Adaptation of Ericaceae to high levels of available aluminium, iron and manganese may thus endow resistance to other metals (Bradley *et al.*, 1981, 1982). The Ericaceae are, however, mycotrophic in all habitats, and Bradley *et al.* (1981, 1982) have shown that the fungal symbiont confers copper and zinc resistance to the plant, since ericoid plants are much more sensitive to copper and zinc when they are non-mycorrhizal.

B. Copper

Bradley *et al.* (1981, 1982) conducted a series of experiments to investigate the role of mycorrhizas in ameliorating copper and zinc toxicity in the Ericaceae. They compared two races of *C. vulgaris,* one from a copper mine and the other from an uncontaminated site, each inoculated with endophytes from their site of origin. The *C. vulgaris* responses to copper and zinc were also compared with responses to these metals by *Vaccinium macrocarpon* and *Rhododendron ponticum* (Bradley *et al.*, 1982) (Figure 1).

There were no differences in copper toxicity between the two *C. vulgaris* races, regardless of their mycorrhizal status (Figure 1; Table 4), although the endophyte-colonized plants were considerably more resistant to copper than

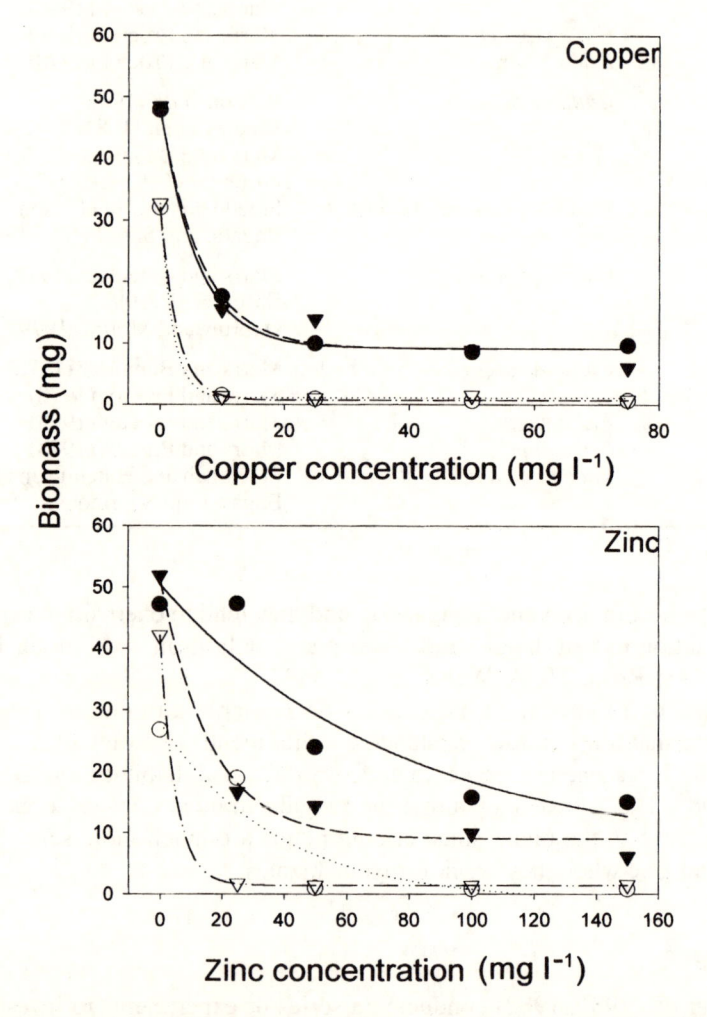

Fig. 1. Dose–response of copper mine and non-mine *Calluna vulgaris*, either uninoculated or inoculated with endophytes from their soil of origin, to copper and zinc. Mycorrhizal treatments are denoted by filled symbols and non-mycorrhizal treatments by open symbols: ●, ○, copper mine population; ▼, ▽, control population. Data from Bradley *et al.* (1982).

Table 4
Effective concentration inhibiting growth (biomass production) by 50% (EC_{50}) for ericoid species inoculated with mycorrhiza from their soil of origin (data from Bradley et al., 1982)

Species	Copper (mg l⁻¹)		Zinc (mg l⁻¹)	
	Non-mycorrhizal	Mycorrhizal	Non-mycorrhizal	Mycorrhizal
Calluna vulgaris (copper mine)	2	8	28	74
C. vulgaris (non-mine)	2	6	4	14
Vaccinium macrocarpon	3	12	68	98
Rhododendron ponticum	3	8	17	33

uncolonized plants. The EC_{50} values for copper, calculated from Figure 1 and given in Table 4, increased 3–4-fold when colonized and uncolonized plants were compared. Growth of uncolonized plants was almost entirely inhibited at a copper concentration of 10 mg l⁻¹, while in mycorrhizal plants the maximum decrease in biomass (to 20–30% of initial weight) occurred at 25 mg l⁻¹ and was sustained at this value at copper concentrations of up to 75 mg l⁻¹.

Similarly, in reciprocal transplant experiments, Marrs and Bannister (1978) showed that when races of *C. vulgaris* from soil not contaminated with copper were transplanted to copper-contaminated mine spoil soil they grew as well as a *C. vulgaris* race indigenous to the copper mine spoil. As in the later work of Bradley *et al.* (1981, 1982), Marrs and Bannister concluded that *C. vulgaris* (endophyte associations) were constitutively adapted to copper-contaminated soil.

Endophyte-mediated resistance was also observed for *V. macrocarpon* and *R. ponticum*: uncolonized plants were shown to be much more sensitive to copper than colonized plants (Bradley *et al.*, 1982). Constitutive copper resistance is, in all species examined, endowed upon the Ericaceae by their fungal endophytes.

When the toxicity of mycorrhizal fungal endophytes isolated from *C. vulgaris, V. macrocarpon* and *R. ponticum* was assayed in solution culture in the absence of the host they exhibited similar dose responses to copper, being resistant at concentrations of up to 50 mg l⁻¹, but inhibited at concentrations of 100 mg l⁻¹ (Bradley *et al.*, 1982). Intraspecific responses of the endophytes were not, however, assayed so it could not be determined whether endophytes showed differential response depending on their site of origin in the absence of their hosts.

Mycorrhizal colonization greatly reduced copper concentrations in the shoots of the Ericaceae over a wide range of external copper concentrations (Bradley *et al.*, 1981, 1982). Bradley *et al.* (1981, 1982) postulated that the mycorrhizal endophytes bind or sequester copper in external hyphae and in hyphal coils within colonized root cells, restricting translocation to the host tissues. While this seems the likely mechanism of resistance, a number of questions remain:

(1) Ericoid mycorrhizal fungal endophytes produce relatively little mycelium external to the root (Read, 1992), so that hyphal tissue will offer a finite and limited number of binding or sequestering sites. What happens when these sites are saturated and hyphae are no longer capable of buffering the host plant from copper?
(2) Copper, because of its redox properties, initiates free radical damage of membrane lipids and cytoplasmic components, which then induces considerable cellular stress (Gadd, 1993). How do mycorrhizal fungal hyphae and uncolonized root cells protect themselves from free radical attack?

A possible answer to the first of these questions might be that hyphae do reach their buffering capacity for copper, but that, once saturated, the hyphae cease to function or become senescent and new hyphae are generated to act as a barrier between the plant and copper, and to continue the other roles of symbiotic association such as nutrient acquisition. Ericoid mycorrhizal colonization appears to turn over relatively quickly even in the absence of toxic metals, with 5–6 weeks regarded as the mean longevity of hyphal coils in individual epidermal cells (Smith and Read, 1997). Alternatively, the fungus may be able to exclude copper from its cells through reduced uptake or a copper efflux mechanism. Copper exclusion from cells has been demonstrated for other eukaryotic micro-organisms such as algae (Foster, 1977; Twiss *et al.*, 1993), filamentous fungi (Gadd and White, 1985) and yeasts (Gadd *et al.*, 1984). Cellular sequestration (Silver and Misra, 1988) and production of extracellular copper-binding protein complexes (Harwood-Sears and Gordon, 1990) have been postulated as mechanisms of copper resistance in some bacteria.

Wainwright and Woolhouse (1977) were able to show that tolerant ecotypes of the grass *Agrostis capillaris* were resistant to peroxidative damage of cell membranes by copper, whereas membrane damage occurred in sensitive ecotypes. This mechanism of resistance was also demonstrated in copper-tolerant *Mimulus guttatus* (Strange and Macnair, 1991). A similar protective mechanism, preventing lipid peroxidation by copper, may have evolved in ericoid mycorrhizal fungi to enable them to survive in copper-contaminated environments. It is noteworthy in this context that the ericoid mycorrhizal endophyte *Hymenoscyphus ericae* is known to release hydrogen peroxide (via carbohydrate oxidase activity) into the extracellular medium during growth in axenic culture media (Burke and Cairney, 1998), implying an ability to withstand peroxidative damage to its hyphae. Host root cells may also have evolved such protective mechanisms, but may require the presence of fungal endophytes to prevent copper assimilation before copper resistance is conferred. Data on root copper levels were presented only for mycorrhizal plants in the study of Bradley *et al.* (1982), and conclusions cannot be made regarding the role of mycorrhizal fungal sequestration of copper in the roots compared with non-mycorrhizal plants.

Iron resistance in the Ericaceae shows some similarities to mycorrhiza-mediated copper resistance; however, the growth benefit and restriction of translocation to the shoot in mycorrhizal plants were substantially less for iron (Hashem, 1995a) than for copper (Bradley *et al.*, 1981, 1982). It therefore appears that constitutive copper resistance observed in ericoid mycorrhizal associations is not endowed solely by resistance to iron. However, underlying similarities in copper and iron metabolism may mean that some common mechanisms in assimilation, metabolism and resistance are also involved (Welsh *et al.*, 1993).

C. Zinc

C. vulgaris from a copper mine spoil population was more resistant to zinc than a population from an uncontaminated soil (Bradley *et al.*, 1982; see Figure 1 and Table 4). This mine site, Parys Mountain in Wales, UK, is contaminated with zinc, and *C. vulgaris* on this site has raised shoot zinc levels (Marrs and Bannister, 1978). Enhanced sensitivity was conferred in these plants by endophyte colonization, considerably increasing the EC_{50} values (Table 4) and enabling the plant to grow at concentrations that completely inhibit growth of uncolonized plants. This selection for host resistance meant that endophyte-colonized plants from the mine population grew much better than colonized plants from the non-mine population (EC_{50} values of 78 and 14 mg l^{-1} zinc respectively). Reciprocal experiments with mine and non-mine plants and mine and non-mine fungal endophytes were not performed (i.e. non-mine fungi were not inoculated on to mine plants and vice versa), so it is not possible to gauge whether the endophytes had also evolved zinc resistance on the mine soils. However, the non-mine endophytes greatly enhanced the resistance of the non-mine *C. vulgaris* to zinc.

Notwithstanding the fact that zinc resistance appears to have evolved in *C. vulgaris,* and that *C. vulgaris* endophytes alleviate zinc toxicity, the contaminated soils from which the mine populations were collected also had raised copper levels as the selective pressure (Marrs and Bannister, 1978; Bradley *et al.*, 1981, 1982). It is interesting that uncolonized *C. vulgaris* from copper mines does not exhibit copper resistance but does show zinc resistance. Both mine and non-mine mycorrhizal endophytes can decrease host sensitivity to copper and zinc, suggesting that the resistance is conferred by a common constitutive mechanism. Pleiotropically conferred resistance to different metals (co-resistance) has been observed in higher plants (Verkleij and Prast, 1989; Schat and Vooijs, 1997;

Tilson and Macnair, 1997), but no examples of copper or zinc co-resistance have been identified (Walley *et al.,* 1974; Wu and Antonovics, 1975). As discussed above, *C. vulgaris* and its fungal endophytes are adapted to conditions where transition metal ions are highly available, at levels that would be toxic to many plant species. Therefore, the selective pressures on *C. vulgaris* and its endophytes are quite different to those of most plant species, and mechanisms evolved for living in such environments may receive further selection in metalliferous environments, leading to evolution of co-resistances not observed for other higher and lower plants. This hypothesis requires further testing. As with *C. vulgaris,* endophyte colonization of both *V. macrocarpon* and *R. ponticum* enhanced zinc resistance compared with that in uncolonized plants (Bradley *et al.,* 1982), (Table 4).

In the experiments of Bradley *et al.* (1982), root zinc concentrations were available only for colonized plants, so that it could not be ascertained whether colonization altered root zinc assimilation patterns. As for copper, zinc resistance is expressed in colonized plants (mine populations of *C. vulgaris* as well as *V. macrocarpon* and *R. ponticum*) by considerably reduced zinc assimilation in the shoots (Bradley *et al.,* 1981, 1982). However, the *C. vulgaris* population from the non-mine population had very similar assimilation patterns in both colonized and uncolonized plants. It appears that mycorrhizal endophytes from the non-mine environment did not restrict zinc transport to the shoots, unlike the situation with the mine endophytes where restriction in zinc translocation was observed. Endophyte associations from the non-mine environment were much more sensitive to zinc than the associations from mine environments (Table 4; Figure 1). This may reflect both to the greater inherent sensitivity of the non-mine *C. vulgaris* and lack of inhibition of zinc transport to the shoots by the mycorrhiza for the non-mine population. It appears that copper resistance is constitutive in *C. vulgaris* associations but that there are phenotypic differences in zinc sensitivity, selected for by the metal contamination status of the environment from which the plants are derived.

Using a similar argument as for copper (section IV.B), it appears that zinc resistance conferred by ericoid endophytes is not constitutively endowed by iron resistance, as assimilation and resistance patterns differ greatly for zinc (Bradley *et al.,* 1981, 1982) and iron (Hashem, 1995a). That zinc resistance varies intraspecifically for hosts (Bradley *et al.,* 1982) and for the ericoid mycorrhizal endophyte *H. ericae* (Denny and Ridge, 1995) further argues against zinc resistance being linked solely to iron resistance.

The dose response of *C. vulgaris, V. macrocarpon* and *R. ponticum* endophytes to zinc has been ascertained in solution culture (Bradley *et al.,* 1982). These experiments were conducted with only a single *C. vulgaris* endophyte, and the origin of this isolate was not given. Work with defined isolates of *H. ericae* showed that there is an intraspecific response to zinc by this fungal species with one isolate only showing 25% inhibition of growth at a zinc concentration of 392 mg l^{-1} while another isolate was inhibited by 65% at this concentration

(Denny and Ridge, 1995). The zinc status of the soils from which these isolates were obtained is not known. Denny and Ridge (1995) also determined zinc uptake in solution culture of these two isolates of *H. ericae*. Uptake of zinc at one concentration (1.5 mmol l^{-1}) in the two *H. ericae* isolates differed significantly, one assimilating approximately 9 and the other about 14 mmol per g tissue. There appear to be intraspecific differences in zinc assimilation for *H. ericae*.

D. Role of Mycorrhizas in Metal Resistance of the Ericaceae

For copper, metal resistance is conferred on sensitive hosts by the mycorrhiza, and mycorrhiza from studied populations are constitutively resistant. In the case of zinc, ericoid mycorrhizas from both contaminated and uncontaminated environments confer resistance. However, intraspecific variation for zinc resistance has been demonstrated in both hosts and mycorrhizas.

Ericoid associations colonize a wide range of metal-contaminated soils (Table 4). It is interesting to postulate whether the phenomena observed for copper resistance are repeated for other metals, or whether, as the data for zinc suggest, there is selection for metal resistance in either the host or symbiont.

Nutrient relationships on contaminated sites have not been determined for ericoid associations. However, the Ericaceae, in general, are adapted to nutrient-deficient soils (Smith and Read, 1997) and it is likely that the ericoid mycorrhizas present on contaminated sites are effective in acquiring the nutrients required by the plants for sustenance.

V. ECTOMYCORRHIZAL FUNGI

A. Introduction

It is widely perceived that ectomycorrhizal (EcM) fungi constitutively endow metal resistance to trees (Wilkins, 1991; Wilkinson and Dickinson, 1995; Leyval *et al.*, 1997), and that resistance is achieved by the fungi reducing metal burdens in shoots, through binding of metals to the surface of extramatrical hyphae and on to the surface of mycorrhizal sheaths (Denny and Wilkins, 1987c; Wilkins, 1991; Leyval *et al.*, 1997). In highly metalliferous soils, however, the adsorptive capacity of the mycelium will be rapidly saturated (Colpaert and van Assche, 1993), and it is argued in this section that there is little or no convincing evidence that EcM fungi generally endow their hosts with metal resistance.

B. General Response of Ectomycorrhizal Associations to Raised Levels of Metals

To illustrate the lack of an apparent role of EcM fungi in metal resistance to trees, the study of Dixon and Buschena (1988) will be considered in some detail. They grew *Pinus banksiana* and *Picea glauca* seedlings, either inoculated with

Suillus luteus or uninoculated, in a sterilized sandy loam soil dosed with a range of cadmium, copper, nickel, lead and zinc concentrations. Figure 2 shows the response of the ratio of EcM : non-EcM biomass, the ratio of EcM : non-EcM metal burden in needles and the percentage of EcM short lateral roots. Total biomass in both EcM and non-EcM plants was decreased to a greater or lesser extent in both species for all metals with the exception of zinc, but the ratio of

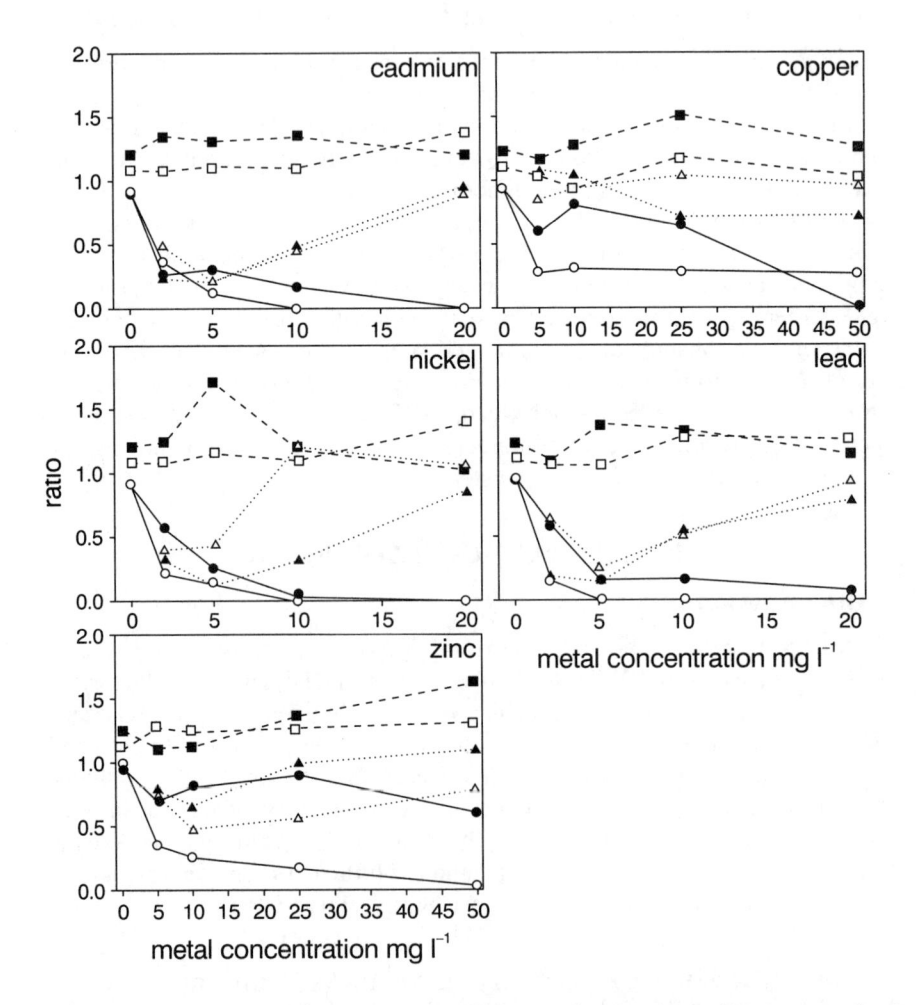

Fig. 2. Response of *Pinus banksiana* and *Picea glauca* to cadmium, copper, lead, nickel and zinc: ratio of colonized : uncolonized biomass (■, □), ratio of colonized : uncolonized needle metal burden (▲, △) and percentage colonization (●, ○) with *Suillus varigatus*. Filled symbols are for *P. banksiana* and open symbols for *P. glauca*. Data from Dixon and Buschena (1988).

biomass production in EcM : non-EcM plants changed very little in most dose responses (Figure 2). Even with treatments that did produce a shift in the relative ratios, these shifts were not substantial. The lack of a marked shift in biomass ratio over a wide range of metal concentrations suggests that EcM colonization does not benefit the host by decreasing metal toxicity. Thus colonized and uncolonized plants appear to have identical responses to the metals. Indeed, the constant ratio was maintained over concentrations at which 93% of EcM plant root tips were colonized, to concentrations where colonization was totally inhibited (Figure 2). EcM fungi did not attenuate the toxicity of the hosts to metals in their growth medium.

The ability of EcM to decrease shoot metal burdens does not lead to decreased toxicity of the metals to the host (Figure 2). The response of EcM and non-EcM plants to metal toxicity was identical for a range of metals, regardless of the fact that non-EcM plants had higher shoot burdens than EcM plants. As metal concentrations rose in the medium for all metals with the exception of copper, non-EcM plants had much higher burdens at low metal exposure, but as exposure increased the ratio of metals in EcM : non-EcM plants approached unity, concurrent with the inhibition of colonization. EcM fungi do not protect plants at high metal exposures. This is principally because EcM colonization is itself inhibited at these concentrations, while plant biomass never decreased to less than 50% of the control treatments, and often plant biomass production was decreased to a much lesser extent (Dixon and Buschena, 1988).

It must be stressed that the study of Dixon and Buschena (1988) is not an isolated example. Studies on *Quercus rubra* inoculated with a single EcM fungal isolate and exposed to cadmium, nickel and lead (Dixon, 1988), *P. sylvestris* with six EcM fungal isolates exposed to zinc (Colpaert and van Assche, 1992a), *P. sylvestris* with three EcM fungal isolates exposed to zinc (Buckling and Heyser, 1994) and *P. sylvestris* with nine EcM fungal isolates exposed to cadmium (Colpaert and van Assche, 1993) all showed the same trends and confirm that EcM fungi do not confer enhanced resistance on their hosts. The majority of fungal isolates used in these studies would, however, not be considered as adaptively metal resistant. Only the experiments of Colpaert and van Assche (1992a, 1993) used several isolates of a single fungal species which were shown to be differentially sensitive to metals. The role of resistant and sensitive isolates in metal resistance of the host will be discussed in detail in section V.d.

An important fact to note from the study of Dixon and Buschena (1988) is that, in terms of the number of short laterals colonized with *S. luteus,* the *P. banksiana* association was more resistant to all metals than the *P. glauca* association (Figure 2). These differences were most notable for zinc. Given that the same isolate of *S. luteus* was used to infect both host plant taxa, it must be concluded that differential sensitivity of the association partially regulates colonization. A host that can maintain an association in metal-contaminated environments will be at a competitive advantage over host species that cannot. Also, EcM fungi can be active at high metal concentrations in their substrate, but may not successfully form associations with some hosts. Similar differential sensitivity of hosts in maintaining associations was demonstrated for AM associations (Gildon and Tinker, 1983).

The studies of Dixon and Buschena (1988) and Dixon (1988) showed that there was no interaction between mycorrhizal and non-mycorrhizal plants and metal levels in media with respect to nutrient status (N, P, K, Ca and Mg). Mycorrhizal plants did have higher tissue P concentrations. The other studies discussed in this section did not report nutritional status.

C. Genetic Adaptation of Trees to Metal Contaminants

A range of tree species can colonize metal-contaminated soils that select for evolutionary adaptation in grasses and non-woody herbs (Jones and Hutchinson, 1986; Borgegard and Rydin, 1989; Eltrop *et al.*, 1991; Dickinson *et al.*, 1992; Turner and Dickinson, 1993; Watmough and Dickinson, 1996). There are reports of metal-resistant and sensitive ecotypes of *Betula* spp. (Brown and Wilkins, 1985a, 1986; Denny and Wilkins, 1987a). However, Wilkinson and Dickson (1995) have suggested that there is little evidence that tree species have adapted genetically to grow on metal-contaminated soils. This hypothesis is based largely on the tenet that in slow-growing tree species resistance to metals cannot be selected for quickly, and that the large genome of tree species endows phenotypic plasticity which may allow the colonization of extreme environments such as metalliferous soils (Wilkinson and Dickson, 1995). These authors further suggested that mycorrhizal fungi (AM and EcM) associated with trees may adapt more rapidly, due to their shorter life cycles, and, therefore, that selection of metal-resistant mycorrhizal fungi may be involved in tree adaptation to contaminated soils. It is pertinent in this context, however, that short lateral root-tip colonization by EcM fungi generally decreases with increasing metal burden of the growth medium (Dixon, 1988; Dixon and Buschena, 1988). Furthermore, extramatrical mycelium is also greatly impacted by metal contamination (Colpaert and van Assche, 1992a). It has often been argued that EcM fungi protect their hosts from metal by the hyphal mantle acting as a physical barrier between the fungus and the host, and that the mycelial network in soils acts as a buffer, adsorbing metals and protecting the host (Wilkins, 1991). However, if metal contamination decreases EcM colonization, in terms of both root tips and mycelial density in soil, this argument can only partially explain any potential role that EcM fungi may have in mediating host metal resistance.

An alternative hypothesis is that host trees are more resistant to the metal contaminants than the EcM fungi. A number of studies have shown that *P. sylvestris* (Colpaert and van Assche, 1992a, 1993; Buckling and Heyser,

1994), *P. banksia* (Dixon and Buschena, 1988), *P. glauca* (Dixon and Buschena, 1988), *Q. rubra* (Dixon, 1988) and *P. sylvestris* can be considerably more metal resistant than their EcM associations. Even where decreased EcM colonizations was observed, mycorrhizal trees generally produced more biomass than uncolonized plants. Both colonized and uncolonized plants were generally affected by increasing metal contribution to the same relative extent (Figure 2). This all points to a higher intrinsic resistance in the host compared with the EcM fungus, but the host still benefits from being colonized. As the fungus does not appear to confer decreased metal sensitivity, it is likely to be benefiting the plant in other ways such as enhanced nutrient absorption, phytohormone production and/or water acquisition. Therefore, while EcM colonization may play a vital role in revegetating contaminated sites, this may simply reflect their general benefit to plant health and have little to do with any metal resistance conferred to their hosts. Many tree species are adapted to nutrient-deficient soils with low pH. As argued in section IV.A for ericoid mycorrhizas, which have similar habitats to such tree species, trees and their associated mycorrhizal fungi may have acquired general resistances to metal ions that are available in soils under low pH conditions (such as aluminium, iron and manganese). These mechanisms of adaptation may confer resistances to analogous ions such as cadmium, copper, lead and zinc.

D. Intraspecific Variation in Metal Resistance in Ectomycorrhizal Fungi

EcM fungi have been widely studied with respect to their role in metal resistance of their host, metal uptake and distribution within the association, and their toxicity in isolated culture (see Hartley *et al.* (1997) for a comprehensive review). They show considerable interspecific response to toxic metals (Hartley *et al.*, 1997), yet the extent to which adaptive (intraspecific) resistance exists is still unclear.

1. Zinc

Most studies investigating intraspecific differences in metal sensitivity in EcM fungi in *in vitro* culture and in symbiosis with a host plant have concentrated on zinc. The results of these studies are somewhat contradictory, with Colpaert and van Assche (1987, 1992b) showing that a range of EcM species had adapted to zinc-contaminated environments, whilst the studies of Denny and Wilkins (1987b,c) and Brown and Wilkins (1985b) showed no adaptation. Colpaert and van Assche (1987) explained this disparity as being due to differences in levels of zinc contamination of the environment from which they collected the fungi for their experiments (3166 mg kg^{-1} Zn compared with 379 mg kg^{-1} in the Brown and Wilkins (1985b) study). This is borne out in plots of data from Colpaert and van Assche (1987) and Brown and Wilkins (1985b) (Figure 3) for the zinc response of the EcM fungus *Amanita muscaria*. All isolates from the Brown and Wilkins' (1985b) study fell within the same range of dose response as the sensitive ecotype of Colpaert and van Assche (1987). The zinc-resistant ecotype used by Colpaert and van Assche (1987) was, however, much more resistant to zinc than all other genotypes.

Colpaert and van Assche (1987) showed that strains of *S. luteus, Suillus bovinus, Thelephora terrestris* and *A. muscaria* from a highly zinc-contaminated environment all had much higher resistances to zinc than isolates of the corresponding species from an uncontaminated habitat (see Figure 3 for the resistance of the *A. muscaria* isolate to zinc). In *in vitro* studies, radial growth was inhibited by a minimum of only 44% at 1000 mg l^{-1} Zn in the resistant isolates, a concentration that totally inhibited growth in all but one of the isolates from an uncontaminated site.

Denny and Wilkins (1987b,c) conducted similar experiments using *Paxillus involutus* strains from the same zinc-contaminated site as the *A. muscaria* strain used by Brown and Wilkins (1985b). As in the former study, the *P. involutus* strains also showed no adaptation to zinc contamination, again suggesting that selection pressure at the site is low. The studies of Denny and Wilkins (1987b,c) and Brown and Wilkins (1985b) are often cited as showing that there may be no relationship between the soil of origin and metal resistance of EcM fungi isolated from such habitats. It is unfortunate that they appear to have chosen sites where the selection pressure (degree of zinc contamination) was not strong

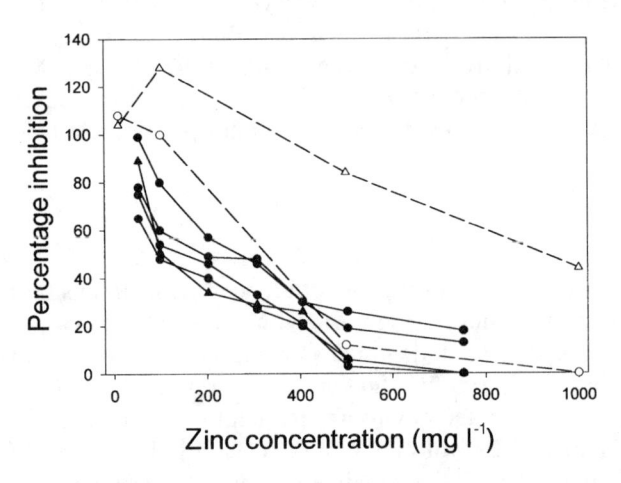

Fig. 3. Dose–response to zinc for *Amanita muscaria* isolates collected from contaminated (▲, △) and uncontaminated (●, ○) sites by Brown and Wilkins (1985b) (filled symbols and solid lines) and Colpaert and van Assche (1987) (open symbols with dashed lines).

enough to cause evolutionary change in zinc resistance. The data of Colpaert and van Assche (1987) suggest that, where selection pressure is high enough, selection for resistant ecotypes can indeed occur.

Colpaert and van Assche (1987) found that an isolate of *P. involutus* from the contaminated site did not show any selection for zinc resistance. It may be that this strain was isolated from a relatively uncontaminated micro-site within the contaminated area. Taken together with the data of Denny and Wilkins (1987b,c), however, it may be that *P. involutus* simply does not evolve resistances. While this hypothesis needs testing, it is noteworthy that only a limited number of higher plant species exhibit evolutionary resistance to metals, and these species can normally evolve multiple resistances (Macnair, 1993). The same may be true of EcM fungi, where only a subset of species has the genetic code for metal resistances. Alternatively, *P. involutus* may be constitutively resistant to zinc.

2. Cadmium

A. muscaria, S. luteus and *S. bovinus* isolates originating from a cadmium-polluted site were found to be more resistant to cadmium than strains of the same species from uncontaminated sites (Colpaert and van Assche, 1992b). Although there were intraspecific differences according to the soil or origin, there were also considerable interspecific differences in resistance, with some isolates from uncontaminated environments being more resistant than isolates from mine habitats.

3. Aluminium

Many tree species are adapted to soils with low pH and it is possible that these species and their associated EcM fungi may have evolved mechanisms for living in environments where the availability of toxic metals such as aluminium, iron and manganese is high. Both interspecific (Thompson and Medve, 1984; Jongbloed and Borst-Pauwels, 1992; Jones and Muehlchen, 1994; Tam, 1995) and intraspecific (Thompson and Medve, 1984; Egerton-Warburton and Griffin, 1995; Leski *et al.*, 1995) variation are known to exist in EcM fungal responses to aluminium *in vitro*.

P. tinctorius isolates from sites of unknown origin showed considerable differences in their resistance to aluminium in agar culture (Thompson and Medve, 1984). One isolate was inhibited by 25% at 500 mg l^{-1} aluminium compared with the zero aluminium exposure, while the growth of another isolate was only 2.8% that of the control treatment. In a more detailed study of *P. tinctorius*, Egerton-Warburton and Griffin (1995) investigated aluminium sensitivity in 11 isolates from a coal mine site (pH 4.0, ammonium acetate extractable Al 327 mg l^{-1} soil dry mass), six isolates from a rehabilitated mine site (pH 4.9, Al 22 mg l^{-1}) and four isolates from an uncontaminated forest site (pH 5.3, Al 6 mg l^{-1}). Isolates from the coal mine site had EC$_{50}$ values ranging from 4 to 2000 mg l^{-1}, while those from the rehabilitated and forest sites ranged from 0.5 to 22 mg l^{-1}. Only one of the rehabilitated and forest site isolates had an EC$_{50}$ value higher than that of two of the mine site isolates. This study provides strong evidence that *P. tinctorius* ecotypes that are resistant to aluminium have been selected for in soils with high aluminium availability. However, within this fungal population, there is considerable variation in response to aluminium, presumably reflecting the heterogeneity in soil aluminium availability within the coal mine site.

Similar variation was observed in *S. luteus* isolates from sites with high and low aluminium availability (Leski *et al.*, 1995). Isolates from contaminated sites all showed considerable resistance to aluminium, with the most sensitive isolate showing only 67% inhibition of growth (on a biomass basis) at an aluminium concentration of 1000 mg l^{-1} in agar culture, and three isolates (of a total of six) showing no inhibition of growth at this concentration. Five of six isolates from the uncontaminated site showed much greater sensitivity than all the isolates from the contaminated site. It is important to note, however, that one of the six isolates from the uncontaminated site was resistant to aluminium in the culture medium. Furthermore, in a more recent study of 18 *P. involutus* isolates from the same aluminium-contaminated sites, Rudawska and Leski (1998) found evidence for constitutive, rather than site-specific, adaptive resistance to aluminium. This observation suggests that interspecific differences may exist in EcM fungal taxa with respect to the evolution of aluminium resistance, and strengthens the argument above that *P. involutus* does not evolve adaptive resistance to metals in the environment. None the less, aluminium-resistant *S. luteus* and *P. tinctorius* ecotypes appear to be selected for at contaminated sites, and resistant ecotypes can also be observed (at a lower frequency) on uncontaminated sites. There may be a great deal of spatial variability in aluminium exposure in environments inhabited by EcM fungi, as reflected in the observed range of sensitivities at 'uncontaminated sites'. Where aluminium concentrations in a habitat are universally high, selection for resistant ecotypes of some EcM taxa appears to occur.

E. Studies Conducted with Hosts Colonized with Resistant and Sensitive Strains

To date, the only studies to have considered the effects of metals on plants colonized with metal-resistant and -sensitive strains of the same species of EcM fungus are those of Colpaert and van Assche (1992b, 1993). In the first experiment, Colpaert and van Assche (1992b) studied the response of *P. sylvestris,* inoculated with either a zinc-sensitive or zinc-resistant *S. bovinus* isolate, to raised levels of zinc in the growth medium compared with an uninoculated control. Zinc concentrations of up to 916 mg l^{-1} in sand–vermiculite culture did not impede the growth of the uninoculated control plant (or the inoculated plants), yet mycelial biomass production was reduced in both sensitive and resistant isolates of *S. bovinus*. From these data it can be concluded that the host was more resistant to zinc than either isolate. The zinc-resistant isolate of *S. bovinus* produced considerably more biomass at 916 mg l^{-1} Zn (83% colonization of substrate) than the sensitive strain (47% of substrate). This enhanced biomass in the resistant strain did not, however, lead to reduced zinc levels in the shoot compared with those in the sensitive strain, but both did reduce *in planta* zinc levels compared with the uninoculated control. Similarly, it has been postulated that enhanced extramatrical mycelial biomass depletes metal levels in the soil solution (through adsorption of the metal to hyphal surfaces), leading to decreased exposure of the plant to the metals (Wilkins 1991). This was not the case in the study of Colpaert and van Assche (1992b), where metal levels in shoots were very similar, despite the resistant isolate of *S. bovinus* producing twice as much biomass. Furthermore, in this experiment, being mycorrhizal did not confer any benefit to the host. Indeed, with respect to zinc toxicity, inoculation with the resistant isolate actually reduced plant biomass production. The advantage to the fungus in being resistant was that its growth was impeded to a lesser extent than that of the sensitive strain.

Colpaert and van Assche (1993) found that shoot biomass production of non-mycorrhizal *P. sylvestris* did not differ from control treatments when grown in the presence of 5 mg l^{-1} cadmium. Inoculation with cadmium-sensitive and -resistant isolates of *S. luteus* and *S. bovinus* considerably reduced cadmium levels in shoots, although, again, there was no difference in cadmium levels between sensitive and resistant isolates. All mycorrhizal treatments, in both control and cadmium treatments, decreased biomass production in *P. sylvestris* compared with that in non-mycorrhizal plants.

F. Role of Ectomycorrhizal Fungi in Facilitating Host Metal Resistance

From the evidence currently available, it appears that EcM fungi do not endow their hosts with increased metal resistance. It must be stressed, however, that this tenative conclusion is based on data from a small number of investigations, which have employed a limited number of fungal and host taxa and strains. The possibility that some EcM fungal taxa may confer a degree of resistance to their hosts under some circumstances therefore remains, but this has yet to be demonstrated. Tree species so far investigated have tended to be more resistant to metals than either EcM fungi or the host–EcM fungus associations, although there is interspecific variation with respect to the host species in maintaining associations as metal availability increases (Dixon and Buschena, 1988).

It appears that intraspecific variation in metal resistance does occur in a range of EcM fungal species and that this variation may, in the case of aluminium at least, be linked to the metal status of the soils of origin (Thompson and Medve, 1984; Colpaert and van Assche, 1987, 1992b; Egerton-Warburton and Griffin, 1995; Leski *et al.*, 1995). We must be mindful, however, that, with the exception of the study of Egerton-Warburton and Griffin (1995) which screened 19 *P. tinctorius* isolates, these data have been derived from only a handful of isolates of particular fungal taxa. As already outlined elsewhere (Hartley *et al.*, 1997), screening of a large number of isolates from contaminated and uncontaminated habitats is required to determine adaptive metal resistance in EcM fungal populations. Other studies, such as those of Brown and Wilkins (1985b) and Denny and Wilkins (1987b), screened EcM fungi isolates from soils that were only moderately contaminated with zinc, the data suggesting no relationship between metal resistance and soil of origin. We argue that these data are an insufficient basis on which to draw such a conclusion, and that further screening using isolates from more highly zinc-contaminated sites is required.

In conclusion, it appears that EcM fungi can adapt to extreme metalliferous soils and colonize their hosts at such sites, but do not confer added metal resistances to their hosts. As with AM fungi (section III), host species establishing on metal-contaminated soils benefit from being mycorrhizal but this appears to have little to do with the association conferring metal resistance to the host, and probably reflects a more general benefit conferred by the EcM fungi. Experiments on nutritional aspects need to be conducted on contaminated soils before the role of resistant strains in such environments is elucidated.

VI. CONCLUSIONS

A. Current State of Knowledge

There is great interest in both the evolutionary and physiological aspects of metal resistances in plants (Macnair, 1993). The role of their symbiosis with mycorrhizal fungi must be considered when interpreting metal resistance in plants as mycorrhizas are the interface between the plant and soil and have a crucial role in regulating metal acquisition. Although there is a considerable literature on metal toxicity and assimilation relationships in mycorrhizal associations, studies that address co-evolution of host and symbiont are sparse.

This review started by considering a range of possibilities with respect to co-evolution of metal resistance in host–mycorrhiza associations. Three basic ecological strategies have emerged for co-evolution on highly metalliferous sites:

(1) Both EcM and AM fungi can evolve metal resistance and colonize hosts on metal-contaminated sites, but their hosts are either constitutively metal resistant (tree species) or have evolved resistance themselves (grasses, herbs and possibly trees). The mycorrhizal association performs its usual ecological role in such situations (nutrient acquisition, water relations, pathogenic resistance, phytohormone production, etc.), but does not confer enhanced metal resistance to the host in the examples studied to date.
(2) In ericoid associations, at least for copper, metal resistance is conferred on sensitive hosts by the mycorrhiza, and mycorrhiza from all studied populations are constitutively resistant.
(3) For zinc, ericoid mycorrhiza from both contaminated and uncontaminated environments confer resistance, but there is some evidence that there may be some selection for resistance in both the host and fungi on metal-contaminated sites. At least intraspecific variation appears to occur in both host and mycorrhiza. The role of co-evolution for zinc of ericoid associations requires further investigation if this scenario is to be verified.

However, the above ecological strategies are based on only a limited number of investigations. The studies of Bradley *et al.* (1981, 1982) on ericoid mycorrhiza with copper and zinc, of Gildon and Tinker (1981, 1983) on AM fungi with cadmium and zinc, and of Colpaert and van Assche (1992a, 1993) on EcM fungi with cadmium and zinc are the only investigations that have considered co-evolution. It is only for copper resistance in ericoid associations that the co-evolutionary strategy of the association has been demonstrated unequivocally. Definitive experiments must be conducted to elucidate co-evolutionary strategies in other associations and for other metals.

B. Future Research Requirements

What is required for all relevant associations is that resistant and sensitive plant populations need to be identified. Alternatively, if the host species is constitutively sensitive (such as *C. vulgaris* for copper resistance) or constitutively resistant (as may be the case for a number of tree species), populations established on highly contaminated and uncontaminated environments should be identified. Mycorrhizal fungi from the same extreme contaminated sites as the plant populations need to be identified (to species level), and the same mycorrhizal species identified and cultured in the plant populations from the uncontaminated environments. Dose responses need to be performed on the plants and fungi individually to determine whether selection for resistance in populations from contaminated environments has occurred.

Factorial experiments with plants and fungi from the contaminated and uncontaminated habitats (i.e. reciprocal cross-inoculation of contaminated with uncontaminated plants and fungi) would then be conducted, with the associations grown over a concentration range that would discriminate between sensitive and resistant populations. The highest dose should represent plant exposure for the most extreme environments which associations can survive on metalliferous soils. Clonal plant material is preferable, but not always practicable. For EcM and ericoid mycorrhizas, clonal mycorrhizal material should be used. The genetic composition of AM fungal spores does not facilitate clonal culture and populations of defined species should be used, preferably cultured from single spores. Plant responses (biomass, metal assimilation and nutritional status of differing tissues) and colonization levels should then be ascertained at suitable time points. Only this experimental design will determine whether (and how) mycorrhizas from either uncontaminated or contaminated habitats confer additional metal resistances to their resistant and insensitive hosts. This experimental model will also ascertain whether mycorrhizas decrease sensitivity to their hosts, which may be a possibility if they are facilitating enhanced metal assimilation.

Metalliferous soils normally have very poor soil structure and are generally nutrient deficient and freely drained, as such environments are usually either mine spoils or soils impacted by smelter activity. The high metal burdens inhibit normal microbial nutrient cycles in soils, driving down soil quality and sustainability. Given that much of the evidence presented in this review suggests that EcM and AM associations are not essential for metal resistance, it is important to establish cost–benefits for metal-resistant (constitutive or adaptive) plants maintaining metal-resistant (constitutive or adaptive) AM and EcM fungi on metalliferous soils. Plant populations from contaminated soils should be grown on a range of soils (varying in the degree of contamination and nutritional status) of origin with or without mycorrhizas (cultured and identified from the same habitat). A sufficiently large number of soils is required to enable the natural heterogeneity (in both nutrients and metal levels) found at mine sites to be investigated to determine the factors regulating mycorrhizal benefits and costs. Biomass production, metal assimilation, nutrient acquisition, transpiration, water acquisition and pathogen infection should all be determined over a suitable time course. Experiments in defined media where nutrients and metal concentrations can be manipulated factorially also need to be conducted to complement field soil experiments. Dose–response approaches are essential for both nutrient and metal additions. Experiments with metal-contaminated media should be carried out using defined matric potentials to determine whether the role of mycorrhizas on highly contaminated sites is regulating water relationships in habitats that often have poor soil structure and are often very freely drained. These sets of experiments would ascertain the costs and benefits to being mycorrhizal in contaminated environments. At the core of this approach is the dose–response relationship, as it is likely that the benefits of being mycorrhizal vary with the degree of metal contamination. The higher the concentrations, the more stressed and slower growing the plant will be. Under high stress conditions, mycorrhizal colonization may impose additional stress in terms of carbon economy and enhanced metal acquisition, especially if nutrient supply is sufficient. Alternatively, mycorrhizas may decrease metal exposure at high metal concentrations, leading to enhanced resistances and maintaining enhanced nutrient acquisition in environments that are nutrient deficient.

The principle that certain mycorrhizal species have evolved either constitutive or adaptive resistance to a range of metals has been established. The reasons for co-evolution of mycorrhizas and their hosts are, however, poorly understood. It is imperative that the experimental models outlined here be applied to determine the driving forces for co-evolution of mycorrhizas and their hosts to metalliferous soils.

REFERENCES

Allen, E., Allen, M.F., Helm, D.J., Trappe, J.M., Molina, R. and Rincon, E. (1995). Patterns and regulation of mycorrhizal fungal communities. *Plant Soil* **170**, 47–62.

Aparicio, A. (1995). Seed germination of *Erica andevalensis* Cabezudo and Rivera (Ericaceae), and endangered edaphic endemic in southwestern Spain. *Seed Sci. Technol.* **23**, 705–713.

Bagatto, G. and Shorthouse, J.D. (1991). Accumulation of copper and nickel in plant-tissues and an insect gall of lowbrush blueberry, *Vaccinium angustifolium*, near an ore smelter at Sudbury, Ontario, Canada. *Can. J. Bot.* **69**, 1483–1490.

Barkdoll, A.W. and Schenck, N.C. (1986). Effects of aluminium on spore germination of vesicular arbuscular mycorrhizal fungi and hyphal penetration of bean roots. *Phytopathology* **76**, 1063–1064.

Bartolome-Esteban, H. and Schenck, N.C. (1994). Spore germination and hyphal growth of arbuscular mycorrhizal fungi in relation to soil aluminium saturation. *Mycologia* **86**, 217–226.

Benson, L.M., Porter, E.K. and Peterson, P.J. (1981). Arsenic accumulation, resistance and genotypic variation in plants on arsenical mine wastes in S.W. England. *J. Plant Nutr.* **3**, 655–666.

Borgegard, S.O. and Rydin, H. (1989). Biomass, root penetration and heavy metal uptake in birch, in a soil cover over copper tailings. *Plant Soil* **26**, 585–595.

Boyle, M. and Paul, E.A. (1988). Vesicular–arbuscular mycorrhizal associations with barley on sewage-amended plots. *Soil Biol. Biochem.* **20**, 945–948.

Bradley, B., Burt, A.J. and Read, D.J. (1981). Mycorrhizal infection and resistance to heavy metal toxicity in *Calluna vulgaris. Nature* **292**, 335–357.

Bradley, R., Burt, A.J. and Read, D.J. (1982). The biology of mycorrhiza in the ericaceae. VIII. The role of mycorrhizal infection in heavy metal resistance. *New Phytol.* **91**, 197–209.

Brown, G. (1994). Soil factors affecting patchiness in community composition of heavy metal-contaminated areas of western Europe. *Vegetatio* **115**, 77–90.

Brown, M.T. and Wilkins, D.A. (1985a). Zinc resistance in *Betula*. *New Phytol.* **99**, 91–100.

Brown, M.T. and Wilkins, D.A. (1985b). Zinc resistance of *Amanita* and *Paxillus*. *Trans. Br. Mycol. Soc.* **84**, 367–369.

Brown, M.T. and Wilkins, D.A. (1986). The effects of zinc on germination, survival and growth of *Betula* seed. *Environ. Pollut. Series A* **41**, 53–61.

Buckling, H. and Heyser, W. (1994). The effect of ectomycorrhizal fungi on Zn uptake and distribution in seedlings of *Pinus sylvestris* L. *Plant Soil* **167**, 203–212.

Burke, R.M. and Cairney, J.W.G. (1998). Carbohydrate oxidases in ericoid and ectomycorrhizal fungi: a possible source of Fenton radicals during the degradation of lignocellulose. *New Phytol.* **139**, 637–645.

Burkert, B. and Robson, A. (1994). ^{65}Zn uptake in subterranean clover (*Trifolium subterraneum* L.) by three vesicular–arbuscular mycorrhizal fungi in root-free sandy soil. *Soil Biol. Biochem.* **26**, 1117–1124.

Clarkson, D.T. and Luttge, U. (1989). III. Mineral nutrition: divalent cations, transport and compartmentation. *Prog. Bot.* **51**, 93–112.

Colpaert, J.V. and van Assche, J.A. (1987). Heavy metal resistance is some ectomycorrhizal fungi. *Funct. Ecol.* **1**, 415–421.

Colpaert, J.V. and van Assche, J.A. (1992a). Zinc toxicity in ectomycorrhizal *Pinus sylvestris*. *Plant Soil* **143**, 201–211.

Colpaert, J.V. and van Assche, J.A. (1992b). The effects of cadmium and the cadmium–zinc interaction on the axenic growth of ectomycorrhizal fungi. *Plant Soil* **145**, 237–243.

Colpaert, J.V. and van Assche, J.A. (1993). The effects of cadmium on ectomycorrhizal *Pinus sylvestris* L. *New Phytol.* **123**, 325–333.

Cooper, K.M. and Tinker, P.B. (1978). Translocation and transfer of nutrients in vesicular–arbuscular mycorrhizas. II. Uptake and translocation of phosphorus, zinc and sulphur. *New Phytol.* **81**, 43–52.

Denny, H.J. and Ridge, I. (1995). Fungal slime and its role in the mycorrhizal amelioration of zinc toxicity to higher plants. *New Phytol.* **130**, 251–257.

Denny, H.J. and Wilkins, D.A. (1987a). Zinc resistance in *Betula* spp. I. Effect of external concentration of zinc on growth and uptake. *New Phytol.* **106**, 517–524.

Denny, H.J. and Wilkins, D.A. (1987b). Zinc resistance in *Betula* spp. III. Variation in response to zinc among ectomycorrhizal associates. *New Phytol.* **106**, 535–544.

Denny, H.J. and Wilkins, D.A. (1987c). Zinc resistance in *Betula* spp. IV. The mechanism of ectomycorrhizal amelioration of zinc toxicity. *New Phytol.* **106**, 545–553.

Dickinson, M.S., Turner, A.P., Watmough, S.A. and Lepp, N.W. (1992). Acclimation of trees to pollution stress: cellular metal resistance traits. *Ann. Bot.* **70**, 569–572.

Dixon, R.K. (1988). Response of ectomycorrhizal *Quercus rubra* to soil cadmium, nickel and lead. *Soil Biol. Biochem.* **20**, 555–559.

Dixon, R.K. and Buschena, C.A. (1988). Response of ectomycorrhizal *Pinus banksia* and *Picea glauca* to heavy metals in soil. *Plant Soil* **105**, 265–271.

Dueck, T.H., Visser, P., Ernst, W.H.O. and Schat, H. (1986). Vesicular–arbuscular mycorrhizae decrease zinc-toxicity to grasses growing on zinc-polluted soils. *Soil Biol. Biochem.* **18**, 331–333.

Egerton-Warburton, L.M. and Griffin, B.J. (1995). Differential responses of *Pisolithus tinctorius* isolates to aluminium *in vitro*. *Can. J. Bot.* **73**, 1229–1233.

Eltrop, L., Brown, G., Joachim, O. and Brinkmann, K. (1991). Lead resistance of *Betula* and *Salix* in the mining area of Mechernich/Germany. *Plant Soil* **131**, 275–285.

Foster, P.L. (1977). Copper exclusion as a mechanism of heavy metal resistance in a green alga. *Nature* **269**, 322–323.

Fox, T.C. and Guerinot, M.L. (1998). Molecular biology of cation transport in plants. *Annu. Rev. Plant Physiol. Plant Mol. Biol.* **49**, 669–696.
Freedman, B. and Hutchinson, T.C. (1980). Pollutant inputs from the atmosphere and accumulations in soils and vegetation near a nickel–copper smelter at Sudbury, Ontario, Canada. *Can. J. Bot.* **58**, 108–132.
Gadd, G.M. (1993). Interactions of fungi with toxic metals. *New Phytol.* **124**, 25–60.
Gadd, G.M. and White, C. (1984). Copper uptake by *Penicillium ochrochloron*: influence of pH on toxicity and demonstration of energy-dependent copper influx using chloroplasts. *J. Gen. Microbiol.* **131**, 1875–1879.
Gadd, G.M., Stewart, A., White, C. and Mowll, J.L. (1984). Copper uptake by whole cells and protoplasts of a wild-type and copper-resistant strain of *Saccharomyces cerevisiae*. *FEMS Microbiol.* **24**, 231–234.
Gildon, A. and Tinker, P.B. (1981). A heavy-metal tolerant strain of a mycorrhizal fungus. *Trans. Br. Mycol. Soc.* **77**, 648–649.
Gildon, A. and Tinker, P.B. (1983). Interactions of vesicular–arbuscular mycorrhizal infection and heavy metals in plants. I. The effects of heavy metals on the development of vesicular–arbuscular mycorrhizas. *New Phytol.* **95**, 247–261.
Giovannetti, M., Sbrana, C. and Logi, C. (1994). Early processes involved in host recognition by arbuscular mycorrhizal fungi. *New Phytol.* **127**, 703–709.
Griffioen, W.A.J. (1994). Characterization of a heavy metal-tolerant endomycorrhizal fungus from the surroundings of a zinc refinery. *Mycorrhiza* **4**, 197–200.
Griffioen, W.A.J. and Ernst, W.H.O. (1990). The role of VA mycorrhizae in heavy-metal resistance of *Agrostis capillaris* L. *Agric. Ecosyst. Environ.* **29**, 173–177.
Griffioen, W.A.J., Ietswaart, J.H. and Ernst, W.H.O. (1994). Mycorrhizal infection of an *Agrostis capillaris* population on a copper contaminated soil. *Plant Soil* **158**, 83–89.
Hartley, J., Cairney, J.W.G. and Meharg, A.A. (1997). Do ectomycorrhizal fungi exhibit adaptive resistance to potentially toxic metals in the environment? *Plant Soil* **189**, 303–319.
Harwood-Sears, V. and Gordon, A.S. (1990). Copper-induced production of copper-binding supernatant proteins by the marine bacterium *Vibrio alginolyticus*. *Appl. Environ. Microbiol.* **56**, 1327–1332.
Hashem, A.R. (1995a). The role of mycorrhizal infection in the resistance of *Vaccinium macrocarpon* to iron. *Mycorrhiza* **5**, 451–454.
Hashem, A.R. (1995b). The role of mycorrhizal infection in the resistance of *Vaccinium macrocarpon* to manganese. *Mycorrhiza* **5**, 289–291.
Hepper, C.M. and Smith, G.A. (1976). Observations on the germination of *Endogene* spores. *Trans. Br. Mycol. Soc.* **66**, 189–194.
Hetrick, B.A.D., Wilson, G.W.T. and Figge, D.A.H. (1994). The influence of mycorrhizal symbiosis and fertilizer amendments on establishment of vegetation in heavy metal mine spoil. *Environ. Pollut.* **86**, 171–179.
Ietswaart, J.H., Griffioen, W.A.J. and Ernst, W.H.O. (1992). Seasonality of VAM infection in three populations of *Agrostis capillaris* (Gramineae) on soil with or without heavy metal enrichment. *Plant Soil* **139**, 67–73.
Joner, E.J. and Leyval, C. (1997). Uptake of ^{109}Cd by the roots and hyphae of a *Glomus mosseae/Trifolium subterraneum* mycorrhiza from soil amended with high and low concentrations of cadmium. *New Phytol.* **135**, 353–360.
Jones, D. and Muehlchen, A. (1994). Effects of the potentially toxic metals, aluminium, zinc and copper on ectomycorrhizal fungi. *J. Environ. Sci. Health* **A29**, 949–966.
Jones, H.E. (1971). Comparative studies on plant growth and distribution in relation to waterlogging. III. The response of *Erica cinerea* L. to waterlogging in peat soils of different iron content. *J. Ecol.* **59**, 583–591.

Jones, H.E. and Etherington, J.R. (1970). Comparative studies on plant growth and distribution in relation to waterlogging. I. The survival of *Erica cinerea* L. and *E. tetralix* L. and its apparent relationship to iron and manganese uptake in waterlogged soil. *J. Ecol.* **58**, 487–496.

Jones, M.D. and Hutchinson, T.C. (1986). The effect of mycorrhizal infection on the response of *Betula papyrifera* to nickel and copper. *New Phytol.* **102**, 429–442.

Jongbloed, R.H. and Borst-Pauwels, G.W.F.H. (1992). Effects of aluminium and pH on growth and potassium uptake by three ectomycorrhizal fungi in liquid culture. *Plant Soil* **140**, 157–165.

Killham, K. and Firestone, M.K. (1983). Vesicular arbuscular mycorrhizal mediation of grass response to acidic and heavy metal depositions. *Plant Soil* **72**, 39–48.

Koide, R.T. and Schreiner, R.P. (1992). Regulation of the vesicular–arbuscular mycorrhizal symbiosis. *Ann. Rev. Plant Physiol. Plant Mol. Biol.* **43**, 557–581.

Leake, J.R., Shaw, J.R. and Read, D.J. (1990). The biology of mycorrhiza in the Ericaceae. XVI. Mycorrhiza and iron uptake in *Calluna vulgaris* (L.) Hull in the presence of two calcium salts. *New Phytol.* **114**, 651–657.

Lei, J., Becard, G., Catford, J.G. and Piche, Y. (1991). Root factors stimulate ^{32}P uptake and plasmalemma ATPase activity in vesicular–arbuscular mycorrhizal fungus, *Gigaspora margarita*. *New Phytol.* **118**, 289–294.

Leski, T., Rudawska, M. and Kieliszewska-Rokicka, B. (1995). Intra-specific aluminium response in *Suillus luteus* (L.) S.F. Gray, and ectomycorrhizal symbiont of Scots pine. *Acta Soc. Botan. Pol.* **64**, 97–105.

Leyval, C., Singh, B.R. and Joner, E.J. (1995). Occurrence and infectivity of arbuscular mycorrhizal fungi in some Norwegian soils influenced by heavy metals and soil properties. *Water Air Soil Pollut.* **84**, 203–216.

Leyval, C., Turnau, K. and Haselwandter, K. (1997). Effect of heavy metal pollution on mycorrhizal colonization and function: physiological, ecological and applied aspects. *Mycorrhiza* **7**, 139–153.

Macnair, M.R. (1993). The genetics of metal resistance in vascular plants. *New Phytol.* **124**, 541–559.

Marrs, R.H. and Bannister, P. (1978). The adaptation of *Culluna vulgaris* (L.) Hull to contrasting soil types. *New Phytol.* **81**, 753–761.

Marrs, R.H. and Proctor, J. (1978). Chemical and ecological studies of heath plants and soils of the Lizard Peninsula, Cornwall. *J. Ecol.* **66**, 417–432.

Meharg, A.A. (1993). The role of the plasmalemma in metal resistance in angiosperms. *Physiol. Plant.* **88**, 191–198.

Oxbrow, A. and Moffat, J. (1979). Plant frequency and distribution on high lead soil near Leadhills, Lanarkshire. *Plant Soil* **52**, 127–130.

Pawlowska, T.E., Blaszkowski, J. and Ruhling, A. (1996). The mycorrhizal status of plants colonizing a calamine spoil mound in southern Poland. *Mycorrhiza* **6**, 499–505.

Porter, E.K. and Peterson, P.J. (1975). Arsenic accumulation by plants on mine waste (United Kingdom). *Sci. Total Environ.* **4**, 365–371.

Proctor, J. and Woodell, S.R.J. (1971). The plant ecology of serpentine. I. Serpentine vegetation of England and Scotland. *J. Ecol.* **59**, 375–395.

Read, D.J. (1992). The mycorrhizal mycelium. In: *Mycorrhizal Functioning* (Ed. by M.F. Allen), pp. 102–133. Chapman and Hall, New York.

Ross, I.S. (1993). Membrane transport processes and responses to exposure to heavy metals. In: *Stress Resistance of Fungi* (Ed. by D.H. Jennings), pp. 97–126. Dekker, New York.

Rudawska, M. and Leski, T. (1998). Aluminium resistance of different *Paxillus involutus* Fr. strains originating from polluted and nonpolluted sites. *Acta Soc. Bot. Pol.* **67**, 115–122.
Schat, H. and Vooijs, R. (1987). Multiple resistance and coresistance to heavy metals in *Silene vulgaris*: a co-segregation analysis. *New Phytol.* **136**, 489–496.
Shaw, G. and Read, D.J. (1989). The biology of mycorrhiza in the Ericaceae. XIV. Effects of iron and aluminium on the activity of acid phosphatase in the ericoid endophyte *Hymenoscyphus ericae* (Read) Kork and Kernan. *New Phytol.* **113**, 529–533.
Shaw, G., Leake, J.R. and Read, D.J. (1990). The biology of mycorrhiza in the Ericaceae. XXII. The role of mycorrhizal infection in the regulation of iron uptake by ericaceous plants. *New Phytol.* **115**, 251–258.
Shetty, K.G., Hetrick, B.A.D., Figge, D.A.H. and Schwab, A.P. (1994). Effects of mycorrhizae and other soil microbes on revegetation of heavy metal contaminated mine spoil. *Environ. Pollut.* **86**, 181–188.
Shetty, K.G., Hetrick, B.A.D. and Schwab, A.P. (1995). Effects of mycorrhizae and fertilizer amendments on zinc resistance of plants. *Environ. Poll.* **88**, 307–314.
Shewry, P.R. and Peterson, P.J. (1976). Distribution of chromium and nickel in plants and soil from serpentine and other sites. *J. Ecol.* **64**, 195–212.
Silver, S. and Misra, T.K. (1988). Plasmid-mediated heavy metal resistances. *Ann. Rev. Microbiol.* **42**, 717–743.
Smith, S.E. and Read, D.J. (1997). *Mycorrhizal Symbiosis,* 2nd edn. Academic Press, San Diego.
Strange, J. and Macnair, M.R. (1991). Evidence for a role for the cell membrane in copper resistance of *Mimulus guttatus* Fischer ex DC. *New Phytol.* **119**, 383–388.
Tam, P.C. (1995). Heavy metal resistance by ectomycorrhizal fungi and metal amelioration by *Pisolithus tinctorius*. Mycorrhiza **5,** 181–187.
Thompson, B.D., Clarkson, D.T. and Brain, P. (1990). Kinetics of phosphorus uptake by the germ-tubes of the vesicular–arbuscular mycorrhizal fungus, *Gigaspora margarita*. *New Phytol.* **116**, 647–653.
Thompson, G.W. and Medve, R.J. (1984). Effects of aluminium and manganese on the growth of ectomycorrhizal fungi. *Appl. Environ. Microbiol.* **48**, 556–560.
Tilson, G.H. and Macnair, M.R. (1997). Nickel resistance and copper–nickel co-resistance in *Mimulus guttatus* from a copper mine and serpentine habitats. *Plant Soil* **191**, 173–180.
Turner, A.P. and Dickinson, N.M. (1993). Survival of *Acer pseudoplatanus* L. (sycamore) seedlings on metalliferous soils. *New Phytol.* **123**, 509–521.
Twiss, M.R., Welborn, P.M. and Schwartzel, E. (1993). Laboratory selection for copper resistance in *Scenedesmus acutus* (Chlorophyceae). *Can. J. Bot.* **71**, 333–338.
Verkleij, J.A.C. and Prast, J.E. (1989). Cadmium resistance and co-resistance in *Silene vulgaris* (Moench) Garcke [= *S. cucubalus* (L) Wib]. *New Phytol.* **111**, 637–645.
Wainwright, S.J. and Woolhouse, H.W. (1977). Some physiological aspects of copper and zinc resistance in *Agrostis tenuis* Sibth: cell elongation and membrane damage. *J. Exp. Bot.* **28**, 1029–1036.
Walley, K.A., Khan, M.S.I. and Bradshaw, A.D. (1974). The potential for evolution of heavy metal resistance in plants. I. Copper and zinc resistance in *Agrostis tenuis*. *Heredity* **32**, 309–319.
Watmough, S.A. and Dickinson, N.M. (1996). Variability of metal resistance in *Acer pseudoplatanus* L. (sycamore) callus tissue of different origins. *Environ. Exp. Bot.* **36**, 293–302.

Weissenhorn, I. and Leyval, C. (1995). Root colonization of maize by a Cd-sensitive and a Cd-tolerant *Glomus mosseae* and calcium uptake in sand culture. *Plant Soil* **175**, 233–238.

Weissenhorn, I. and Leyval, C. (1996). Spore germination of arbuscular mycorrhizal fungi in soils differing in heavy metal content and other parameters. *Eur. J. Soil Biol.* **32**, 165–172.

Weissenhorn, I., Leyval, C. and Berthelin, J. (1993). Cd-tolerant arbuscular mycorrhizal (AM) fungi from heavy-metal polluted soils. *Plant Soil* **157**, 247–256.

Weissenhorn, I., Glashoff, A., Leyval, C. and Berthelin, J. (1994). Differential resistance to Cd and Zn of arbuscular mycorrhizal (AM) fungal spores isolated from heavy metal-polluted and unpolluted soils. *Plant Soil* **167**, 189–196.

Weissenhorn, I., Leyval, C. and Berthelin, J. (1995a). Bioavailability of heavy metals and abundance of arbuscular mycorrhiza in a soils polluted by atmospheric deposition from a smelter. *Biol. Fert. Soils* **19**, 22–28.

Weissenhorn, I., Leyval, C., Belgy, G. and Berthelin, J. (1995b). Arbuscular mycorrhizal contribution to heavy metal uptake by maize (*Zea mays* L.) in pot culture with contaminated soil. *Mycorrhiza* **5**, 245–251.

Weissenhorn, I., Mench, M. and Leyval, C. (1995c). Bioavailability of heavy metals and arbuscular mycorrhiza in a sewage-sludge-amended sandy soil. *Soil Biol. Biochem.* **27**, 287–296.

Welsh, R.M., Norvell, W.A., Scgaefer, S.C., Shaff, J.E. and Kochian, L.V. (1993). Induction of iron (III) and copper (II) reduction in peas (*Pisum sativum* L.) roots by Fe and Cu status: does the root-cell plasmalemma Fe(III)-chelate reductase perform a general role in regulating cation uptake? *Planta* **190**, 555–561.

Wilkins, D.A. (1991). The influence of sheathing (ecto-) mycorrhizas of trees on the uptake and toxicity of metals. *Agric. Ecosyt. Environ.* **35**, 245–260.

Wilkinson, D.M. and Dickinson, N.M. (1995). Metal resistance in trees—the role of mycorrhizae. *Oikos* **72**, 298–300.

Wu, L. and Antonovics, J. (1975). Zinc and copper uptake by *Agrostis stoloniferous*, tolerant to both zinc and copper. *New Phytol.* **75**, 231–237.

Yang, W.Q., Goulart, B.L. and Demchak, K. (1996). The effect of aluminium and media on the growth of mycorrhizal and nonmycorrhizal highbush blueberry plantlets. *Plant Soil* **183**, 301–308.

Estimates of the Annual Net Carbon and Water Exchange of Forests: The EUROFLUX Methodology

M. AUBINET, A. GRELLE, A. IBROM, Ü. RANNIK, J. MONCRIEFF,
T. FOKEN, A.S. KOWALSKI, P.H. MARTIN, P. BERBIGIER,
CH. BERNHOFER, R. CLEMENT, J. ELBERS, A. GRANIER,
T. GRÜNWALD, K. MORGENSTERN, K. PILEGAARD, C. REBMANN,
W. SNIJDERS, R. VALENTINI AND T. VESALA

I.	Introduction	114
II.	Theory	116
III.	The Eddy Covariance System	119
	A. Sonic Anemometer	119
	B. Temperature Fluctuation Measurements	120
	C. Infrared Gas Analyser	120
	D. Air Transport System	123
	E. Tower Instrumentation	126
IV.	Additional Measurements	127
V.	Data Acquisition: Computation and Correction	127
	A. General Procedure	127
	B. Half-hourly Means (Co-)variances and Uncorrected Fluxes	130
	C. Intercomparison of Software	134
	D. Correction for Frequency Response Losses	136
VI.	Quality Control	144
	A. Raw Data Analysis	145
	B. Stationarity Test	145
	C. Integral Turbulence Test	146
	D. Energy Balance Closure	147
VII.	Spatial Representativeness of Measured Fluxes	154
VIII.	Summation Procedure	156
IX.	Data Gap Filling	158
	A. Interpolation and Parameterization	158
	B. Neural Networks	158
X.	Corrections to Night-time Data	162
XI.	Error Estimation	164

XII. Conclusions 167
Acknowledgements 168
References 168
Appendix A 173
Appendix B 175

I. INTRODUCTION

The dramatic increase in the concentration of carbon dioxide (CO_2) in the atmosphere since the industrial revolution of the mid-nineteenth century constitutes a chemical change of global proportions. This increase in a radiatively active trace gas is the most obvious consequence of industrialization and is well documented (Houghton *et al.*, 1996). Fossil fuel combustion and deforestation explain most of this long-term increase in atmospheric CO_2 concentration, yet the pathways and processes that determine the dynamics of the global carbon cycle remain largely unknown. For example, the ultimate fate of half of the anthropogenically released CO_2 is uncertain in that this 'missing sink' may be in the terrestrial biosphere, the soil or the ocean, or some combination of all three (Schimel, 1995; Houghton *et al.*, 1998). Such incomplete knowledge as well as uncertainties in the fluxes between parts of the system adds to the uncertainty in estimations of the present state of the global carbon cycle. Inadequate specification of the system also limits predictions about its response to natural and anthropogenic perturbations. Fortunately, long-term studies of gaseous exchanges between the biosphere and the atmosphere can help by filling knowledge gaps and building a stronger predictive capability. It is against this background that the EUROFLUX project was established: this chapter describes the methodology that is common to all partners within the programme.

By definition, long-term studies of the exchange between the biosphere and the atmosphere require direct and continuous measurement of water vapour and CO_2 transfer. Micrometeorological methods such as eddy covariance offer a means to monitor directly the exchange of trace gases between the biosphere and the atmosphere (Moncrieff *et al.* (1997b). In this technique, eddies are sampled for their vertical velocity and concentration of scalar of interest (e.g. H_2O or CO_2). Averaging these recorded fluctuations in gas concentrations over a period of about 30 min yields the net amount of material being transported in the vertical above the surface. Swinbank (1951) proposed and tested the fundamental concepts of eddy covariance but the difficulties associated with instrumentation and data collection were not overcome until some three decades later. A number of systems have now been described in the literature (Lloyd *et al.*, 1984; Businger, 1986; McMillen, 1988; Grelle and Lindroth, 1996; Moncrieff *et al.*, 1997a). This method was used to carry out flux measurement campaigns on increasing time-scales. The first campaigns were limited to 1 or 2 weeks. Reports of such experiments were presented, notably, by Desjardins (1985) on maize, by Verma et al. (1986) on tallgrass prairie, by Valentini et al. (1991) on Mediterranean macchia, and by Verma et al. (1989), Baldocchi and Meyers (1991) and Hollinger et al. (1994) on forests. Campaigns extended on a full growing season were presented later by Grace et al. (1996) on Amazonian forest, by Valentini et al. (1996) on a beech forest, and by Baldocchi et al. (1997) and Jarvis et al. (1997) on different types of boreal forests.

The first annual time-scale CO_2 flux measurement campaigns above forests using the eddy covariance technique were organized in the early 1990s in North America (Wofsy et al., 1993; Black et al., 1996; Goulden et al., 1996a,b; Greco and Baldocchi, 1996) and Europe (Lindroth et al., 1998). In addition, annual time-scale CO_2 flux measurements were performed by teams using other micrometeorological techniques such as the Bowen ratio or aerodynamic techniques (Vermetten et al., 1994; Saigusa et al., 1998). These early results extended our knowledge of the different ecosystem behaviours, but each was limited to a particular ecosystem. However, no attempt at a systematic study across related biomes and across climate scales was made. The conclusions of an International Geosphere-Biosphere Program (IGBP)-sponsored workshop for global change scientists held at La Thuile, Italy, in 1996 highlighted the need for regional networks of flux measurement stations sited across a broad spectrum of ecosystems and climatic environments (Baldocchi et al., 1996).

The EUROFLUX project, sponsored by the Fourth Framework Programme of the European Commission, arose shortly after this workshop to meet this need. A fundamental premise of the EUROFLUX project was to make continuous and long-term (at least three complete annual cycles) of carbon and water exchange between European forests and the atmosphere. EUROFLUX also sought to provide information about the role of the terrestrial biosphere in the climate system. This was attempted for forests of different types and under different climate regimes. The outputs from the project will include not only a better understanding of the functioning of forests but also their role in climate change. Results from EUROFLUX should contribute to mesoscale or general circulation models of the atmosphere by improving surface parameterizations and aggregation schemes in order to scale up to global scale. The important issue of how forest management practices can alter ecosystem water and carbon balances (and the role of forests in the sequestration of carbon) is a further output of the project. The Protocol on Climate negotiated during the third session of the United Nation (UN) Framework Convention on Climate Change (FCCC), signed in Kyoto on 11 December 1997, strengthened the interest in continental-scale measurement networks and the kind of studies described above. The Kyoto Protocol also revealed the immediate policy relevance of EUROFLUX, which thus became much more than a well-orchestrated academic exercise.

The EUROFLUX network includes 14 measuring stations and encompasses a large range of latitudes (from 41.45N to 64.14N), climates (Mediterranean, temperate, arctic) and species (*Quercus, Fagus, Pseudotsuga, Pinus, Picea*). Table 1 provides information on each of the 14 sites. Scaling up measurements from the local scale to the European continent (Martin et al., 1998) and site intercomparisons constitute a central objective of EUROFLUX. This imposes on the partners within the programme the adoption of a series of common methodologies for the measurement of fluxes and for the correction and treatment of data. It is these methodologies, which are used in the EUROFLUX network, that are presented here.

In the first part of the chapter (III to VII), the measurement system and the procedure followed for the computation of the fluxes is described. In the second part (VIII to XI), the procedure of flux summation, including data gap-filling strategy, night flux corrections and error estimation is presented.

II. THEORY

The conservation equation of a scalar is:

$$\frac{\partial \rho_s}{\partial t} + u \frac{\partial \rho_s}{\partial x} + v \frac{\partial \rho_s}{\partial y} + w \frac{\partial \rho_s}{\partial z} = S + D \tag{1}$$

where ρ_s is the scalar density, u, v and w are the wind velocity components, respectively, in the direction of the mean wind (x), the lateral wind (y) and normal to the surface (z). S is the source/sink term and D is molecular diffusion. The lateral gradients and the molecular diffusion will be neglected afterwards. After application of the Reynolds decomposition where: $u = \overline{u} + u'$, $v = \overline{v} + v'$, $w = \overline{w} + w'$, $\rho_s = \overline{\rho}_s + \rho_s'$, where the overbars characterize time averages and the primes indicate fluctuations around the average, integration along z and assumption of no horizontal eddy flux divergence, equation (1) becomes:

$$\underbrace{\int_0^{h_m} S\,dz}_{I} = \underbrace{\overline{w'\rho_s'}}_{II} + \underbrace{\int_0^{h_m} \frac{\overline{\partial \rho_s}}{\partial t} dz}_{III} + \underbrace{\int_0^{h_m} \overline{u}\frac{\partial \rho_s}{\partial x} dz}_{IV} + \underbrace{\int_0^{h_m} \overline{w}\frac{\partial \rho_s}{\partial z} dz}_{V} \tag{2}$$

Term I represents the scalar source/sink term which corresponds to the net ecosystem exchange (N_e) when the scalar is CO_2, and to ecosystem evapotranspiration (E) when the scalar is water vapour. Term II represents the eddy flux at height h_m (the flux measured by eddy covariance systems). It is noted as F_c for carbon dioxide and F_w for water vapour. Under conditions of atmospheric stationarity and horizontal homogeneity, all the other terms on the

Table 1
Site characteristics

Participant	Site	Geographical co-ordinates	Dominant tree species	Age (years)	Canopy height (m)	Mean tree diameter (cm)	Tree density (ha^{-1})	Leaf Area Index (m^2 m^{-2})	Soil type
IT1	Collelongo	41° 52' N, 13° 38' E	*Fagus sylvatica*	100	22	18.1	885	4.5	Calcareous, brown earth (inceptisol)
IT2	Castel Porziano	41° 45' N, 12° 22' E	*Quercus ilex*	50	12.5	16	1500	3.5	Sandy
FR1	Hesse	48° 40' N, 7° 05' E	*Fagus sylvatica*	25–30	13	10	4000	5.5	Luvisol to stagnic luvisol
FR2	Les Landes	44° 42' N, 0° 46' W	*Pinus pinaster*	36	18	26	500	2.5–3.5	Sandy podzol
DK1	Lille Boegeskov	55° 29" 13' N, 11° 38" 45' E	*Fagus sylvatica*	80	25	28	430	4.75	Cambisol
SW1	Norunda	60° 05' N, 17° 28' E	*Picea abies, Pinus sylvestris*	70–120	24	21	678	4–5	Deep sandy till
SW2	Flakkalinden	64° 07' N, 19° 27' E	*Picea abies*	35	8	8.5	2127	2.8	Shallow till
GE1	Bayreuth	50° 09' N, 11° 52' E	*Picea abies*	43	19	23	1000	6.7 ± 1.5	Brown earth (acidic cambisol)
NL1	Loobos	52° 10' N, 5° 44' E	*Pinus sylvestris*	97	15.1	25.4	362	3	Podzollic
UK1	Griffin	56° 37' N, 3° 48' W	*Picea sitchensis*	15	6	7	2500	About 8	Stony podsolized brown earth
GE2	Tharandt	50° 58' N, 13° 38' E	*Picea abies*	106	28	27	652	5	Brown earth
BE1	Vielsalm	50° 18' N, 6° 00' E	*Fagus sylvatica, Pseudotsuga menziesii*	60–90	27–35	34	230	5–5.3	Dystric cambisol
BE2	Braaschaat	51° 18' N, 4° 31' E	*Pinus sylvestris, Quercus robur*	67	22	26.8	542	3	Moderately wet sandy soil
FI1	Hyytiälä	61° 51' N, 24° 17' E	*Pinus sylvestris*	32	12	13	2500	3	Till

More information about the sites may be found at http://www.unitus.it/eflux/euro.html.

right-hand side of equation (2) decay and the measurement by eddy covariance is equivalent to the source/sink term. However, in forest systems (and others), these conditions are not always met and this approximation cannot always be made.

Term III represents the storage of the scalar below the measurement height. Storage of carbon dioxide, S_c, is typically small during the day and on windy nights. However, significant positive values may be observed during poor mixing conditions at night when the CO_2 produced by the ecosystem respiration is accumulating in the ecosystem. On the other hand, a negative peak of S_c is often observed in the morning when the CO_2 accumulated at night is flushed out of the ecosystem or absorbed by ecosystem assimilation (Grace et al., 1996; Goulden et al., 1996a). The daily mean of this term is zero, so ignoring it over the long term is acceptable. However, in the short term, it may be a significant measure of ecosystem response and should be taken into account. The storage of water vapour (S_w) is small at night, the ecosystem transpiration rate being low during this period.

Terms IV and V represent the fluxes by horizontal and vertical advection (V_c and V_w for CO_2 and water vapour respectively). Term IV is significant when horizontal gradients of scalar exist, i.e. in heterogeneous terrain or at night over sloping terrain, when the CO_2 produced by the ecosystem respiration is removed by drainage. The vertical velocity (w) and, consequently, the vertical advection (term V) were found typically to be zero over low crops. However, there is no such evidence above tall vegetation like forest canopies, and Lee (1998) and Baldocchi et al. (unpublished) have shown that this mechanism is not negligible and could even be more important than turbulent transport during calm nights. The advection terms are not measurable with available technology. These terms are probably responsible for the mismatch observed at night between N_e and $F_c + S_c$, as shown in section X below, and a procedure to correct these fluxes empirically is described.

Finally, using the notations introduced previously, equation (2) becomes:

$$N_e = F_c + S_c + V_c \tag{3}$$

F or CO_2 and:

$$E = F_w + S_w + V_w \tag{4}$$

for water vapour.

In the rest of this paper, we will first describe the measurement system (sections III and IV) and the procedures for computation, correction and data-quality analysis (section V) of the eddy fluxes F_c and F_w. The spatial representativeness of the fluxes is analysed in section VII. The procedure followed to obtain annual carbon sequestration is described in section VIII.

It also involves the data gap-filling strategy (section IX), the correction for night flux underestimation (section X) and the estimation of uncertainties (section XI).

III. THE EDDY COVARIANCE SYSTEM

A number of European laboratories collaborated in the early 1990s to produce a common design for an eddy covariance system for the HAPEX-Sahel project (Moncrieff *et al.*, 1997a), and this system formed the basis for that used in EUROFLUX. Other groups in Europe have independently arrived at very similar systems (Grelle and Lindroth, 1996). In essence, the flux systems comprise a three-axis sonic anemometer (Solent 1012R2; Gill Instruments, Lymington, UK), a closed-path infrafred gas analyser (LI-COR 6262; LI-COR, Lincoln, New England, USA) and a suite of analysis software for real-time and post-processing analysis. Gas samples are taken next to the sonic path and ducted down a sample tube to the infrared gas analyser (IRGA).

A. Sonic Anemometer

The sonic anemometer produces the values of the three wind components and the speed of sound at a rate of 20.8 times per second. A set of sound transit times can be obtained 56 times per second. In addition, it has a built-in five-channel analogue to digital (A/D) converter with an input range of 0–5 V and a resolution of 11 bits; this permits simultaneous measurement of analogue sensors and their digitization and integration with the turbulence signals. The analogue channels are sampled at a rate of 10 Hz. In the EUROFLUX system, the analogue channels are used for water vapour and CO_2 concentration output from the IRGA. Additional signals, such as air temperature, air flow through the analyser and analyser temperature, can also be accommodated in this way. Data from the sonic anemometer can be collected as either analogue or digital signals.

To avoid the need to turn the sensing volume of the anemometer into the mean wind direction, the omnidirectional probe head was chosen. The supporting rod and the probe head itself are quite slender and produce relatively little flow distortion (< 5% effect on scalar fluxes; Grelle and Lindroth, 1994). However, a wind tunnel calibration is recommended and at present is applied by only two groups (SW1 and 2, DK1). Before the EUROFLUX project, a sonic anemometer of the same type had been tested under Arctic conditions within the framework of ARKTIS-93 (Peters *et al.*, 1993; Grelle *et al.*, 1994) and proved to work reliably even at temperatures as low as –42°C.

Built-in memory on the circuit board enables the anemometer to store a maximum of 170 s worth of data, depending on the number of analogue inputs sampled. Some of the software in use within EUROFLUX uses the sonic in 'prompted mode' (i.e. data are transferred only by request). In this way, 'quasi on-line' processing of data can easily be carried out in the breaks between transmissions. Other software in use accepts the data stream in real time as it is sent out by the sonic ('unprompted' mode).

In some of the eddy covariance systems (SW1 and 2), the sonic probe head is mounted on a two-axis inclination sensor and/or an adjustable boom to correct for sensor misalignment (Grelle and Lindroth, 1996). In high wind speeds, vibrations of the boom that supports the sonic may introduce a significant bias in flux measurements, but making the boom asymmetrical may reduce these vibrations.

B. Temperature Fluctuation Measurements

Generally, measurement of air temperature fluctuations is done by means of the speed-of-sound output of the sonic anemometer. At wind speeds above about 8–10 m s^{-1}, however, speed-of-sound data become noisier because of mechanical deformation of the probe head (Figure 1). Determination of the sensible heat flux from the sonic measurements is impossible in these cases (Grelle and Lindroth, 1996). Thus, additional fast thermometers are applied at sites with frequent high wind speeds (SW1 and 2). Particularly at high wind speeds and over forest canopies, even relatively robust sensors possess a sufficiently fast frequency response to measure temperature fluctuations without considerable loss; for example, the use of platinum resistance wires was described by Grelle and Lindroth (1996).

C. Infrared Gas Analyser

The concentrations of water vapour and CO_2 are measured by a LI 6262 infrared gas analyser (IRGA). The IRGA is a differential analyser which compares the absorption of infrared energy by water vapour and CO_2 in two different chambers within the optical bench. Use of this instrument has been described in detail by Moncrieff *et al.* (1997a). In EUROFLUX, the analyser is used in absolute mode, i.e. the reference chamber is filled with a gas scrubbed of water and CO_2. For this purpose, two methods may be used: the manufacturer suggests a closed pump-driven air circuit through the reference cell with desiccant and CO_2-absorbing chemicals. Four teams (IT1 and 2, FR1, BE2) use this configuration (Figure 2a). However, it often results in a zero drift of the signals, which is apparently caused by the release of CO_2 by soda lime or ascarite when the reference air becomes too dry, and in some cases by leakage in the reference line. Leakages occur, for example, when the O-ring seals of the scrubber tubes are damaged by the influence of the chemicals. On the other hand, an open reference air stream is not practical for unattended long-term measurements, as the chemicals would need to be changed too often. The

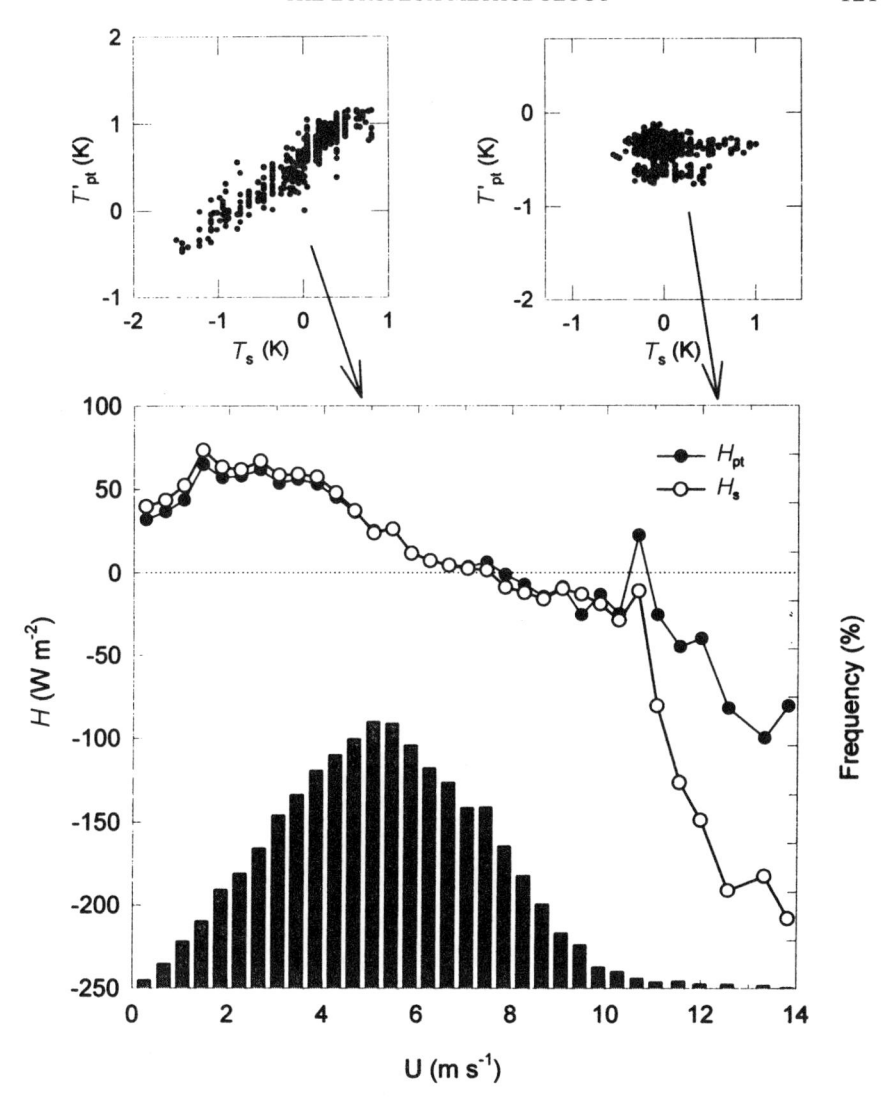

Fig. 1. Sensible heat flux by sonic temperature (H_s) and platinum wire thermometer (H_{pt}), classified by wind velocity. *Upper graphs*: Time series of temperature fluctuations, sonic temperature (T_{son}) versus platinum wire temperature (T_{pt}) at wind speeds of 4.4 and 12.2 m s^{-1}.

second method, which overcomes these problems, consists of flushing the reference chamber with nitrogen gas from a cylinder, maintaining a flow rate of about 20 ml min^{-1}. Ten teams use this configuration (FR2, DK1, UK1, SW1 and 2, NL1, GE1 and 2, BE1, Fl1) (Figure 2b). In systems that have different

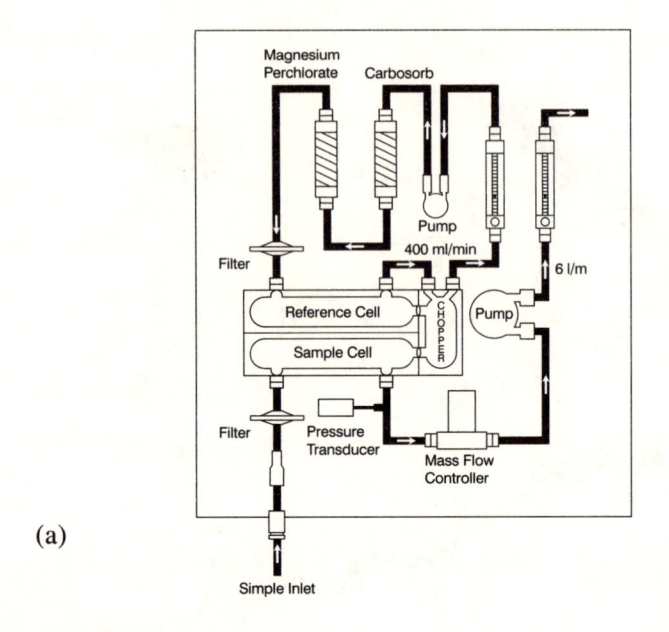

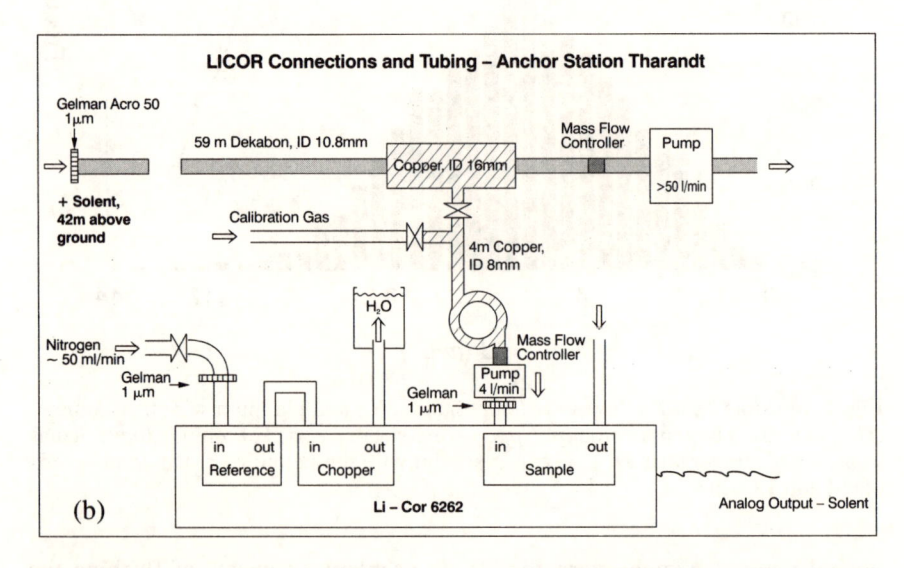

Fig. 2. Two typical flow configurations used in EUROFLUX. (a) System with through flow in the sample circuit and chemicals in the reference circuit (Moncrieff et al., 1997a). (b) System with secondary flow in the sample circuit and nitrogen in the reference.

analysers at several levels on a tower, and when N_2 from one common source has to pass to different gas analysers, the chopper chambers of the gas analysers cannot be flushed by the common N_2 path because these chambers usually leak. In these cases, scrubber chemicals are used (SW1 and 2, GE1).

Calibration of the IRGA is carried out according to the preferences of the teams: some perform it automatically every day while others perform it manually at a lesser frequency (fortnightly). A lower calibration frequency may affect the mean CO_2 concentration measurement, as the LI-COR 6262 is subject to significant zero drift. However, as the slope drift is very low, the impact of the calibration frequency on the fluxes is limited.

The signals for water vapour and CO_2 concentration can be output in a variety of units from the IRGA, either as analogue voltage signals, analogue current signals, or as ASCII for serial communication. However, if a fast response is desired, the raw voltage signals for the CO_2 and water vapour mixing ratio should be used. With this configuration, the 95% time constant of the analyser is 0.1 s (LI-COR, 1991). Additionally, the internal analyser temperature can be obtained as an analogue signal.

D. Air Transport System

A variety of different airflow configurations is used by teams within the project. The technical specifications are given for each site in Table 2. The choice of flow configuration is determined according to a number of criteria:

(1) To minimize frequency loss (i.e. by maintaining flow with high Reynolds number (Re) in the inlet tube (Leuning and Moncrieff, 1990).
(2) To avoid condensation within tubes and analyser.
(3) To avoid pressure fluctuations and air contamination caused by the pump.
(4) To comply with the IRGA's range of operational parameters (e.g. pressure, chamber temperature).
(5) To stabilize and monitor the air flow.
(6) To keep the analyser chamber clean.

Most commonly, the analyser is placed on the tower, some metres away from the sonic. Some groups use long sample tubes with the analyser on the ground surface as this permits easy access to service the gas analyser (DK1, GE2).

To minimize frequency losses, turbulent airflow through the tube is sometimes chosen (GE2, SW1 and 2) (Moore, 1986), but this usually requires powerful pumps with a high power consumption. Laminar flow in the tube is maintained by less powerful pumps and the extra loss in frequency response can be accounted for adequately (Moncrieff et al., 1997a). The data in Table 2 show that the EUROFLUX systems operate between Re 1800 and 6550 (where Re of about 2030 is generally considered as the cut-off for transition to turbulent flow) within the types of sample tube used here. However, several

Table 2
Technical specifications of the sites

Site	Pump type	Circuit	Q (l min^{-1})	Filters	L_t (m)	r_t (10^{-3} m)	L_s (m)	τ_{lr}(real) (s)	τ_{lc} (comp.) (s)	$F_{co\text{-}ss}$ (Hz)	a_{ss} (m^{-1})	Re
IT1	D	TS	12	2A	9.0	2.0	0.35	2.1	0.56	19.14	0.31	4245
IT2	D	TS	10.5	2A	3.0	2.0			0.22	26.07		3712
FR1	R	TS	6.0	A+S	30.0	2.0	0.2	3.9–4.1	3.8	2.48	0.54	2122
FR2	R	TS	8.0	A+S	3.0	2.0	0.2	1.9–2.2	0.28	15.30	0.54	2828
DK1	P	M	20.0		48.0	4.0	0.6	7.5–9.5	7.24	2.10	0.18	3535
SW1	2D	TB	9.9	2A	7.2	2.0	0.1	1.3–1.3	0.55	15.08	1.07	3503
SW2	2D	TB	11.6	2A	6.5	2.0	0.1	1.05	0.42	21.21	1.07	4105
GE1	R	TS	6.0	2A	7.0	3.15	0.2	2.5	0.55	12.9	0.54	2660
NL1	D	M	7.4		7.0	2.0			0.71	8.47		2616
UK1	D	TS	6.0	2A	18.0	3.2	0.15	6–6.5	5.2	0.53	0.71	1885
GE2	LD	M	50.0	2A	59.0	5.4	0.2	10–11.5	6.48	3.46	0.54	6546
BE1	D	TS	6.0	A+S	8.0	2.0	0.2	2.6–3.1	1.01	4.81	0.54	2122
BE2	R	TS	5.5	2A	3.0	2.2	0.48	1.5–2.0	0.48	2.37	0.22	1796
FI1	D	TS	6.3	1A	7.0	2.0	0.15	1.5–1.7	0.84	5.82	0.71	2227

Q, flow rate in tube; L_t, tube length; r_t, tube radius; L_s, separation distance between the sonic anemometer (centre of the measurement volume) and the inlet tube; τ_l, time lag (r, measured; c, computed); $F_{co\text{-}ss}$, cut-off frequency characterizing the sensor separation effect; a_{ss}, slope of the relation between $F_{co\text{-}ss}$ and the mean wind speed (equation (27)). Re, Reynolds number; D, diaphragm; R, rotative; P, piston; 2D, double diaphragm; LD, linear diaphragm; TS, sucked through the analyser; TB, blown through the analyser; M, two pumps; 2A, 2 Acro 50 1 µm; A + S, 1 Acro 50 1 µm + 1 Spiral cap 0.2 µm.

experimenters have been able to produce laminar flows in tubes at higher values of Re (up to 100 000) and it is apparently impossible to find a universal critical value of Re for flows in tubes (Monin and Yaglom, 1971). Consequently it is difficult to characterize, with certainty, the flow regime in each system. Anyway, the problem is not critical provided that adequate corrections for high frequency losses are applied. These corrections are discussed in section V.D.2).

To prevent condensation, in most cases the air is sucked through the tube and the analyser (i.e. the air pump is the last link in the air transport chain). The data are then corrected for the effect of under-pressure within the analyser's sample cell (Figure 2a) (Moncrieff *et al.*, 1997a). A sensor (LI-COR 6262-03) that measures the IRGA chamber pressure at 1 Hz may be used to that end. Another possibility (SW1 and 2) is to blow the air through the analyser, giving high flow rates and a reduced number of corrections. With a suitable air pump and appropriate heating, pump-induced distortions and condensation can be prevented more effectively (Grelle, 1997). A third possibility is to transport the air at a high flow rate close to the measurement point with subsampling at a lower flow rate for H_2O and CO_2 (Figure 2b). This method produces very high flow rates without creating too large an under-pressure in the measuring chamber, but it requires two pumps. It is used in long tube systems (DK1, GE2).

To prevent contamination of the IRGA chamber, filters must be placed upstream of the analyser. Except for sites with very clean air (FI1), a number of filters placed in series is required to reduce the zero drift of the analyser that becomes inevitable after a couple of months. It is therefore recommended to place two filters in series in the pumping circuit. The first filter (ACRO 50 PTFE 1 μm; Gelman, Ann Arbor, Michigan, USA) should be placed at the inlet of the tube and replaced every fortnight (more frequent changes have been necessary in winter at some sites when reduced mixing conditions produces an acceleration in filter contamination (FR1, BE1)). The second filter (ACRO 50 PTFE 1 μm, or Spiral cap 0.2 μm; Gelman) may be placed close to the analyser and does not need to be changed as frequently. The positioning of a filter at the inlet of the tube also prevents contamination of the tube by hygroscopic dirt which could damp the high-frequency water vapour fluctuations and so lead to under-estimation of water vapour flux (Leuning and Judd, 1996). In sites near oceans (FR2), sea salt could damage the IRGA chamber coating. To avoid such problems a coalescent filter (Balston A944) is placed at the input of the IRGA chamber.

The time lag that is introduced in a ducted system is taken into account by software, and the lag varies significantly from site to site according to tube length and flow rate. The exact determination of the time lag is more crucial for systems with long tubes (DK1). The dominant parameter determining time lag is the flow rate through the sample tube (Leuning and Moncrieff, 1990) and the filters. Thus, if some sort of flow regulator is used, the time lag is kept fairly constant; otherwise, a dynamic correction must be applied. Generally, the time lag is computed by finding the maximum in the covariances (McMillen, 1988; Grelle and Lindroth, 1996; Moncrieff *et al.*, 1997a) and is 1–2 s higher than that predicted by flow equations in the tube. The difference between theory and practice arises because the fluid flow equations do not take into account the influence of the filters and the time constant of the analyser. In practice, however, time lag variations are usually small and can be specified in the software within a range of ± 1 s of the true lag. This permits faster computation and a reduction in errors, especially when the fluxes are low and the maximum in the covariance is difficult to detect.

To enable automatic calibration of the CO_2 signal, a calibration gas can be supplied to the gas analyser at predetermined intervals through a remotely controlled solenoid valve. Reference H_2O sources such as the LI-COR dewpoint generator are not suitable for unattended long-term use and high flow rates. Thus, separate measurements of reference air humidity are more appropriate to ensure calibration of the H_2O signal.

E. Tower Instrumentation

Given the aim of operating flux systems for extended periods of time, some protection of the system components from adverse weather is essential. Owing to the large climate variability between the different EUROFLUX sites, the technical requirements for appropriate protection differ. At northern latitudes (SW1 and 2, FI1), the air analysis system is placed in thermally insulated boxes which are ventilated by filtered air and heated to keep the internal temperature constant (Grelle, 1997). In some cases, the sonic probes are heated to prevent rime (SW1 and 2, GE1, FI1). In the Mediterranean region, it is generally sufficient to keep the system in a clean and dry environment.

The signals of air temperature, water vapour concentration, CO_2 concentration, gas analyser temperature, and air flow through the analyser are connected to the sonic's A/D converter. Here, the non-linear signals of the gas analyser offer the fastest frequency response, whereas the linearized outputs offer higher signal resolution when used with the Solent sonic anemometer's built-in A/D converter. The limitation in frequency response is larger for the H_2O signal, whereas in practice the limited resolution affects only the CO_2 signal (Grelle, 1997). Thus, utilization of the linear CO_2 signal and the non-linear H_2O signal is the optimum configuration. Signals from the sonic anemometer are transferred by serial communication according to the RS 422 standard.

The power supply and interface unit originally delivered with the SOLENT sonic does not have electrically insulated signal lines and it has only single-ended inputs using the same reference potential as the sonic probe itself. In particular, this means that, if additional devices with a different reference potential (e.g. the LI-COR gas analyser) are connected to the sonic in the original configuration, differences in ground potential between the mast and the ground can cause currents through the signal lines and thus data errors. To overcome this problem and to improve the lightning protection, the anemometer has been modified by one group (SW1 and 2), which now uses a floating power supply and a modified converter interface with insulated signal lines (Grelle and Lindroth, 1996). Another group (FI1) uses optical fibres to transfer the data from the tower to the measurement building at the bottom of the instrumentation tower. This requires a modification of the Solent output (Haataja and Vesala, 1997).

IV. ADDITIONAL MEASUREMENTS

At each EUROFLUX site, the eddy covariance measurements are supported by a set of meteorological measurements that are also collected every half an hour. The role of these measurements is:

(1) To characterize the meteorological conditions under which fluxes occur. This allows further analysis as the inference of the flux responses to meteorological variables (at a half-hourly scale) or the response of the annual carbon sequestration to climate and the interpretation of the interannual variability of the net carbon storage.
(2) To correct the eddy covariance measurements. In particular, the CO_2 storage estimation (term III in equation (2)) requires half-hourly measurements of the CO_2 concentration below the measurement point; corrections of fluxes for pressure or for high-frequency losses require additional meteorological measurements.
(3) To control the quality of the measurements. The energy balance closure test is performed currently to check the data quality (see section VI.D). It requires, in particular, the net radiation and the energy storage.
(4) To provide a set of prognostic data that can be used to fill in the gaps in the flux data series (see section IX).
(5) To provide a set of input data for Soil Vegetation Atmosphere Transfer (SVAT) models that can be used for calibration as well as for validation.

The list of variables that are measured at all sites is given in Table 3.

V. DATA ACQUISITION: COMPUTATION AND CORRECTION

A. General Procedure

The general procedure followed to collect and process the data is presented in Figure 3.

At intervals of 1/20.8 s, instantaneous measurement of values for the three components of the wind velocity, the speed of sound and the molar fractions of

Table 3
Meteorological variables measured in the EUROFLUX sites

Symbol	Unit	Variable	Instrument	Status
Radiation				
R_g	W m^{-2}	Global radiation	Pyranometer	O
R_n	W m^{-2}	Net radiation	Net radiometer	O
PPFD	μmol m^{-2} s^{-1}	Photosynthetic photon flux density	Photodiode	O
R_{ref}	W m^{-2}	Reflected radiation	Pyranometer	F
R_d	W m^{-2}	Diffuse radiation	Pyranometer + screen	F
APAR		Light interception	Photodiodes	F
Temperature				
T_a	°C	Air temperatures (profile)	Resistance or thermocouple	O
T_{bole}	°C	Bole temperature	Resistance or thermocouple	O
T_s	°C	Soil temperature (profile 5–30 cm)	Resistance or thermocouple	O
T_c	°C	Canopy radiative temperature	Infrared sensor	F
G	W m^{-2}	Soil heat flux density	Heat flux plates	O
Hydrology				
P	mm	Precipitation	Rain gauge	O
RH	%	Relative humidity profile	Resistance, capacitance or psychrometer	O
SWC	% by volume	Soil water content (0–30, 40–70, 80–110)	TDR or theta probe (15 days)	O
SF	mm	Stemflow	Tree collectors (15 days)	F
T_f	mm	Throughfall	Pluviometer (15 days)	F
SNOWD	mm	Snow depth	Sensor (15 days)	F
Miscellaneous				
P_a	kPa	Pressure	Barometer	O

O, obligatory; F, voluntary.

CO_2 and water vapour is collected and stored. Software computes the mean values, variances, covariances and the so-called 'uncorrected fluxes' at this stage, corrections for high-frequency losses are not performed). The software that performs these computations is described in section V; the operations performed and the recommended equations are given in section V.B; the software intercomparison exercise, carried out among EUROFLUX teams, is presented in section V.C; and the difference between different high-pass filtering methods is discussed in section V.D.1.

The amount of raw data collected at these sample rates is about 600 Mb per month and is typically stored on tapes, compact disk or removable drives for post-processing (e.g. further analysis of the time series, data quality analysis or spectral analysis). The supporting meteorological measurements are also averaged over intervals of half an hour. In particular, energy storage in the biomass and air (possibly in the soil) or CO_2 storage in the air is computed at

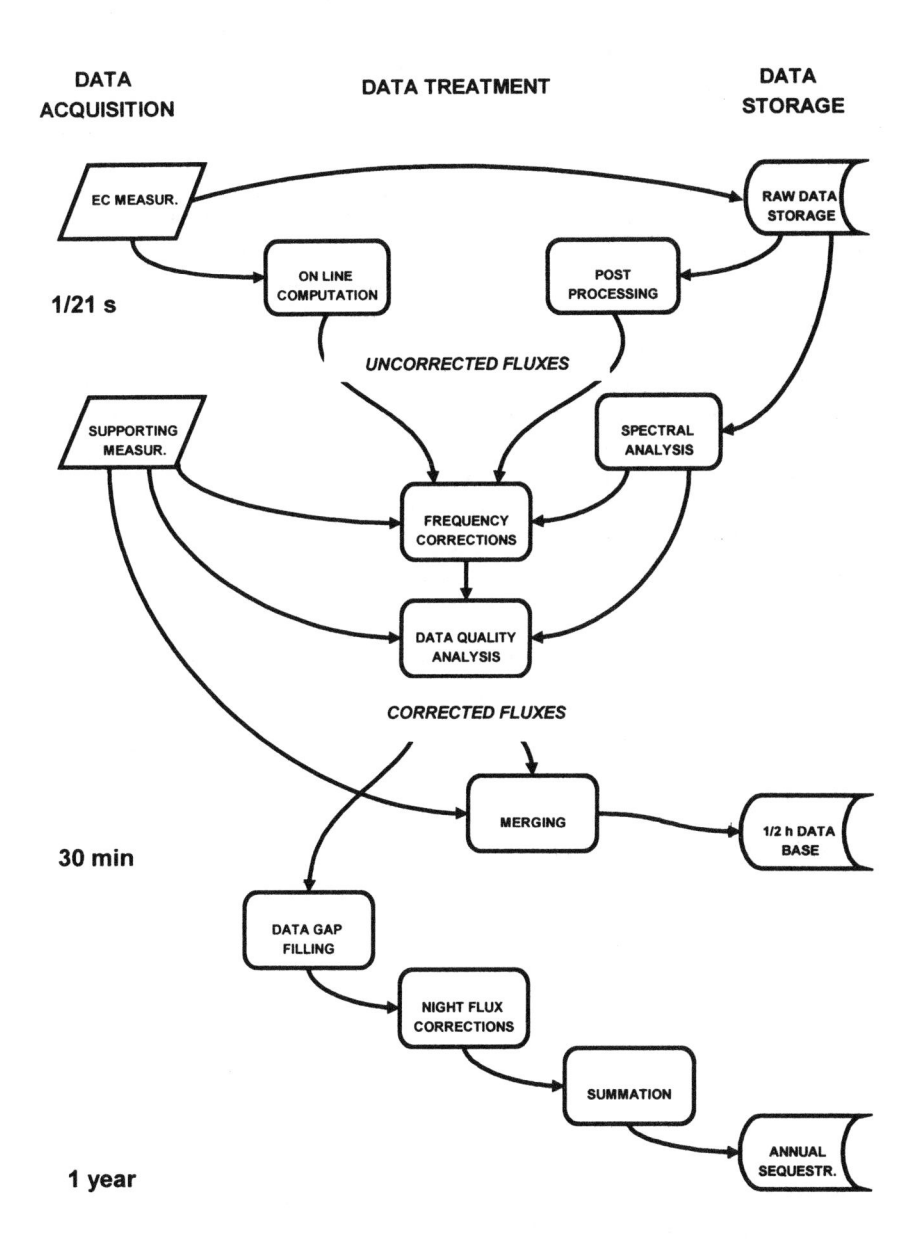

Fig. 3. Schematic of data acquisition, processing and storage in EUROFLUX systems.

this stage. Meanwhile, the corrections of fluxes are applied. They include corrections for latent heat or lateral momentum fluxes (Schotanus *et al.*, 1983) and corrections for high-frequency losses (Moore, 1986; Leuning and Judd, 1996; Moncrieff *et al.*, 1997a). The latter require the cut-off frequency of the system that can be obtained by spectral analysis. The procedure is described in section V.D.2. There is no need for air density corrections on the CO_2 flux (Webb *et al.*, 1980), as will be shown below.

The eddy covariance fluxes are then submitted to a series of tests to eliminate data that fail certain criteria. Different tests (based on statistical analysis, stationarity or similarity criteria, energy balance closure) are presented in section VI. After correction and quality analysis, the eddy covariance and supporting measurements are merged and constitute the input from each group to the EUROFLUX database.

At a yearly time scale, the fluxes are summed to provide an annual carbon sequestration estimation. This sum requires a complete data series, i.e. a procedure to fill in the unavoidable gaps in the flux data. A procedure based on neural networks is described in section IX. The errors that affect the annual sequestration are discussed later: the selective systematic error due to underestimation of the fluxes at night is discussed in section X and the other measurement uncertainties are discussed in section XI.

B. Half-hourly Means, (Co-)variances and Uncorrected Fluxes

1. Computation of the Means and Second Moments

The input data are the instantaneous values of the three components of the velocity $(u, v, w$ (ms^{-1}), of the sound speed (U_{son}(ms^{-1}), and of the CO_2 and H_2O mole fractions (c(µmol mol^{-1}), h (mmol mol^{-1})). In the following part, variables may be designed in a general way by ξ or η, and a scalar concentration by s.

The mean of ξ and the second moment of ξ and η are computed as:

$$\bar{\xi} = \frac{1}{n_s} \sum_{k=1}^{n_s} \xi_k \qquad (5)$$

and

$$\overline{\xi'\eta'} = \frac{1}{n_s} \sum_{k=1}^{n_s} \xi'_k \eta'_k \qquad (6)$$

where n_s is the number of samples. The fluctuations (ξ'_k) around the mean at the step k (ξ_k) are computed as:

$$\xi'_k = \xi_k - \bar{\xi}_k \qquad (7)$$

In autoregressive filtering algorithms, the mean $\overline{\xi}_k$ is computed as (McMillen, 1986, 1988; Baldocchi *et al.*, 1988; Kaimal and Finnigan, 1994):

$$\overline{\xi}_k = e^{-\Delta t/\tau_f}\overline{\xi}_{k-1} + (1 - e^{-\Delta t/\tau_f})\xi_k \quad (8)$$

where Δt is the measurement time interval and τ_f is the running mean time constant. As $\Delta t \ll \tau_f$, equation (8) may also be written (after first-order Taylor expansion):

$$\overline{\xi}_k = \left(1 - \frac{\Delta t}{\tau_f}\right)\overline{\xi}_{k-1} + \frac{\Delta t}{\tau_f}\xi_k \quad (8')$$

It is to be noted that the running mean $\overline{\xi}_k$ cannot be taken for the mean $\overline{\xi}$, which is computed on the whole data set.

In linear detrend algorithms, $\overline{\xi}_k$ is computed by least squares regression as (Gash and Culf, 1996):

$$\overline{\xi}_k = \overline{\xi} + b\left(t_k - \frac{1}{n_s}\sum_{k=1}^{n_s}t_k\right) \quad (9)$$

where t_k is the time at the step k and b is the slope of the linear trend of the sampling. It is computed as:

$$b = \frac{\sum_{k=1}^{n_s}\xi_k t_k - \frac{1}{n_s}\sum_{k=1}^{n_s}\xi_k \sum_{k=1}^{n_s}t_k}{\sum_{k=1}^{n_s}t_k t_k - \frac{1}{n_s}\sum_{k=1}^{n_s}t_k \sum_{k=1}^{n_s}t_k} \quad (10)$$

Non-linear transformations of the variables must be applied before any mean or second moment computation. This is the case with the sonic temperature T_{son} (K) which is deduced from sound velocity (U_{son}) by (Schotanus *et al.*, 1983; Kaimal and Gaynor, 1991):

$$T_{son} = \frac{U_{son}^2}{403} \quad (11)$$

For the computation of $\overline{w'c'}$ or $\overline{w'h'}$, the time series of w'_k must be delayed for synchronization with c'_k or h'_k to take into account the time taken for the air to travel down the sample tube. Recommendations concerning the time lag computation were given in section III.C. Some algorithms in use within EUROFLUX (SW1 and 2) use a constant time lag, whereas others (e.g. EdiSol) estimate it by maximization of the covariances.

2. Co-ordinate rotation

Co-ordinate rotations are applied on the raw means and second moments. The first two-axis rotation aligns u parallel to the mean wind velocity, and nullifies v and w. Its aim is to eliminate errors due to sensor tilt relative to the terrain surface or aerodynamic shadow due to the sensor or tower structure. However, it always suppresses the vertical velocity component and that is not always appropriate as non-zero vertical velocity components may appear above tall vegetation when flux divergence or convergence occurs (Lee, 1998).

The two-axis co-ordinate change on mean velocity components and on covariances, including scalar concentrations, may be expressed in matrix form (McMillen, 1986; Kaimal and Finnigan, 1994; Grelle, 1996):

$$\overline{u_{2,i}} = \sum_j A_{02,i,j} \cdot \overline{u_{0,j}}$$
$$\overline{s'u'_{2,j}} = \sum_j A_{02,i,j} \cdot \overline{s'u'_{0,j}} \tag{12}$$

where the numerical index (n) refers to the number of rotations applied to the means and covariances, $A_{02,i,j}$ are the elements (ith line, jth column) of the two-axis rotation matrix and $u_{n,j}$ is the jth element of the velocity vector. In particular, the mean wind speed is $U = \overline{u_{2,1}}$. The expression of A_{02} in terms of the non-rotated velocity components is established in Appendix A and given in equation (A4).

The two-axis rotation is applied on the variances and covariances following:

$$M_2 = A_{02} M_0 A_{02}^\tau \tag{13}$$

where A_{02}^τ is the transposed A_{02} and M_n are the (co)variance matrices defined as:

$$M_n = \begin{pmatrix} \overline{u'_n u'_n} & \overline{u'_n v'_n} & \overline{u'_n w'_n} \\ \overline{v'_n u'_n} & \overline{v'_n v'_n} & \overline{v'_n w'_n} \\ \overline{w'_n u'_n} & \overline{w'_n v'_n} & \overline{w'_n w'_n} \end{pmatrix} \tag{14}$$

The third rotation is performed around the x-axis in order to nullify the lateral momentum flux density ($\overline{v'w'}$ covariance). Indeed this term is zero over plane surfaces and is likely to be very small over gentle hills. A complete justification on this point is given by Kaimal and Finnigan (1994). McMillen (1988) stresses that this rotation is not well defined in low-speed conditions and recommends its application with care and, in any event, to limit it to 10°. The co-ordinate change for means and covariances including a scalar is expressed by:

$$\overline{u_{3,j}} = \sum_j A_{23,i,j} \cdot \overline{u_{2,j}}$$
$$\overline{s'u'_{3,j}} = \sum_j A_{23,i,j} \cdot \overline{s'u'_{2,j}} \quad (15)$$

The expression of A_{23} in terms of the two-axis rotated velocity components is established in Appendix B (equation (B4)). For the variances and the covariances, the change of co-ordinates is:

$$M_3 = A_{23} M_2 A_{23}^\tau \quad (16)$$

Since computation of the A_{23} matrix elements requires two rotated variances and covariances computations, it is not possible to combine A_{23} and A_{02} to build a unique three-axis rotation matrix. In the following sections, the three components of the three-axis rotated instantaneous velocity will be noted, for simplicity, u, v and w.

3. Conversions and Corrections

The mean air temperature $\overline{T}_a$ (K) is deduced from sonic temperature using (Schotanus *et al.*, 1983; Kaimal and Gaynor, 1991):

$$\overline{T}_a = \frac{\overline{T}_{son}}{1 + 3.210^{-4} h} \quad (17)$$

The fluxes of CO_2 (F_c, μmol m^{-2} s^{-1}) or H_2O (F_w, mmol m^{-2} s^{-1}) are directly deduced from the rotated covariances as:

$$F_s = \frac{P_a}{\Re \overline{T}_{son}} \overline{w's'} \quad (18)$$

where P_a is atmospheric pressure and $\Re$ is the gas constant (8.314 J K^{-1} mol^{-1}).

A paper by Webb *et al.* (1980) pointed out the necessity to correct eddy covariance fluxes in order to remove air density fluctuations. These fluctuations, due either to temperature or to water vapour concentration fluctuations, can be erroneously attributed to fluctuations in scalar concentrations. The correction term was computed for closed path by Leuning and Moncrieff (1990) and Leuning and King (1992). They showed that, if the air was brought in the chamber at a constant temperature, the density corrections due to temperature fluctuations could be removed. Leuning and Judd (1996) estimated the minimum sampling tube length required to reach such constant temperature when the tube is made by a good thermal conductor. Rannik *et al.* (1997) developed a similar computation for poor thermal conductors like Teflon, which is currently used in standard eddy covariance systems. They showed that the minimum tube length to reduce fluctuations to a fraction of 0.01 from the initial temperature is of the order of 1000 times the tube inner diameter, a length reached in most of the experimental set-ups. They concluded that there is no need to apply the density correction due to sensible heat flux. In addition, as humidity fluctuations are automatically corrected by the LI-COR software, there is no need to apply the density corrections due to latent heat flux either.

The latent heat flux is defined as $\lambda M_w F_w/1000$ where M_w is the molar mass of water (0.0180153 kg mol^{-1}) and λ (J kg^{-1}) is the latent heat of vaporization of water, which varies with air temperature:

$$\lambda = (3147.5 - 2.37 T_a) 10^3 \qquad (19)$$

Finally, the sensible heat flux, H, is given as:

$$H = \rho_m C_m \overline{w'T_a'} = \frac{P_a M_d}{\Re \overline{T}_{son}} C_d \left(\overline{w'T'}_{son} - 3.210^{-4} \overline{T}_{son} \overline{w'h'} + \frac{2U\overline{u'w'}}{403} \right) \qquad (20)$$

where M_d is the molar mass of dry air (0.028965 kg mol^{-1}), ρ_m and C_m are the density and specific heat of the moist air (which is practically the same for humid and dry air, i.e. less than 0.5% difference in the meteorological range). The second term in parentheses accounts for the difference between sonic and real temperature (see equation (17)), the third term corrects the sound speed for lateral momentum flux perturbations (Schotanus *et al.*, 1983; Kaimal and Gaynor, 1991). Equation (20) is rigorous only when the speed of sound is measured along a vertical sound path. Thus, in the case of the SOLENT, where the axis is tilted with respect to the vertical, this relation is not strictly valid. However, the departure from this relation is small. When the temperature fluctuations are measured with a fast thermometer (platinum resistance wire), these corrections are not necessary.

C. Intercomparison of Software

Although common software for data acquisition and flux computation was proposed, several teams within EUROFLUX opted to develop their own software in order to retain flexibility and to be able to adapt the software for particular instrumentation configurations (e.g. simultaneous monitoring of several eddy covariance systems). As one of the main objectives of EUROFLUX is to perform comparisons among sites, it was necessary to ensure that the differences in measured flux from site to site was not an artefact of the different algorithms in use. An intercomparison between the available software was therefore carried out.

First, a series of raw data files ('golden files') obtained from a standard eddy covariance system was established. This data set comprises 19 successive 30-min digital files containing raw time-series data from the sonic anemometer, namely the three velocity components, speed of sound (all ms^{-1}), CO_2 (μmol mol^{-1}) and H_2O concentrations (mmol mol^{-1}) sampled at 20.8 Hz. The data cover a range of meteorological conditions, at night and during the day, and were obtained at the BE1 site on 1 June 1997 between 12.00 and 9.00 p.m. (The 'golden files' are available on fsagx.ac.be\ftp\euroflux\serie2.zip). A description of this site is given by Aubinet *et al.* (unpublished).

Each software package was run independently on this data set and the results were compared. When convergent, they were considered as standards (available on fsagx.ac.be\ftp\euroflux\results.xls) and the maximum difference between them as the tolerance. Six software packages were used to generate the standards; three used a running mean algorithm with a 200-s time constant: *EdiSol* (Edinburgh University, UK), *UIA* (University of Antwerpen, Belgium) and *SMEAR Solent* (University of Helsinki, Finland), and the other three used a linear detrending algorithm *UBT* (University of Bayreuth, Germany), *RISOE* (RISOE National Laboratory, Denmark) and *IBK* (Georg August University of Goettingen, Germany). All of them output the means and variances of wind velocity (three components), temperature, water vapour and CO_2 concentrations, the fluxes of momentum (or the friction velocity), sensible heat, water vapour (or latent heat) and CO_2.

All software packages were in very good agreement for mean values: the maximum difference between them was 1% for wind velocity, 0.01% for CO_2 concentration and 0.03% for the water vapour concentration. The agreement for software using either linear detrend or running mean was better than 2% for the latent heat and better than 1% for the other fluxes.

The main sources of disagreement between software were:

(1) Individual errors: some programming errors were detected as a result of this exercise, which fully justifies it. This test is thus recommended for each software designer.
(2) Differences in software initialization for running mean algorithm: the initialization period of these algorithms may exceed the time constant by a factor of 5 or 6. Under these conditions, the first measurement of any series generally gives poor flux agreement. An ideal initialization procedure would be to process the first series of data in the reverse order and initialize with the running mean obtained then. This is impractical, of course, for on-line measurements.

(3) Time lag estimations: software optimizing the time lag without restricting its variation range biased (up to 50% underestimation) the flux values during some periods when the correlation optimum was not clearly defined.
(4) Differences between the hypotheses used for the computation: to avoid such divergence, the use of a standard set of hypotheses and equation is recommended. In EUROFLUX, we used the equations presented in section V.B.1.
(5) Differences between linear detrend and running mean algorithms reached 10% and flux estimates were generally higher for software employing the running mean algorithm. However, the present data series is too short to lead to a definitive conclusion.

D. Correction for Frequency Response Losses

The turbulent flow in the atmospheric boundary layer may be considered as a superposition of eddies of different sizes. At one measurement point they generate velocity and scalar concentration fluctuations of different frequencies. It is common to describe the frequency repartitions of the fluctuations by introducing the co-spectral density. In the case of turbulent fluxes of a scalar, the co-spectral density, C_{ws} is related to the covariance (Stull, 1988; Kaimal and Finnigan, 1994):

$$\overline{w's'} = \int_0^\infty C_{ws}(f)\,df \qquad (21)$$

where f is the cyclic frequency. In practice, the integral range is limited at low frequency by the observation duration and/or the high-pass filtering, and at high frequency by the instrument response. Consequently, the measured turbulent fluxes of a scalar s may be thought of as the integration over frequencies of the co-spectral density multiplied by a transfer function characterizing the measurement process:

$$\overline{w's'}_{meas} = \int_0^\infty TF(f)\,C_{ws}(f)\,df \qquad (22)$$

The fluxes must therefore be corrected to take these effects into account. A correction factor is introduced which is the ratio of the flux free from filtering (equation (21)) and the measured flux (equation (22)):

$$CF = \frac{\int_0^\infty C_{ws}(f)\,df}{\int_0^\infty TF(f)\,C_{ws}(f)\,df} = \frac{\int_0^\infty C_{ws}(f)\,df}{\int_0^\infty TF_{HF}(f)\,TF_{LF}(f)\,C_{ws}(f)\,df} \qquad (23)$$

where the low (TF_{LF}) and high (TF_{HF}) frequency parts of the transfer function are introduced separately.

1. Effect of High-pass Filtering on Fluxes

As shown in section V.B.1, turbulent fluxes of scalars are calculated over a finite averaging time as the average products of the fluctuations of vertical wind speed and scalar concentration, obtained from the original records by linear detrending (LD) or high-pass running mean filtering (RM). The transfer functions for linear detrending and high-pass filtering generally remove variability (also covariance) at low frequencies, passing the high-frequency contribution. The spectral transfer functions for variances and covariances due to LD and RM are (Kristensen, 1998):

$$TF_{LF}^{LD}(f) = 1 - \frac{\sin^2(\pi f \tau_c)}{(\pi f \tau_c)^2} - 3 \frac{\left(\frac{\sin(\pi f \tau_c)}{\pi f \tau_c} - \cos(\pi f \tau_c)\right)^2}{(\pi f \tau_c)^2}$$

$$TF_{LF}^{RM}(f) = \frac{(2\pi f \tau_f)^2}{1 + (2\pi f \tau_f)^2}$$

(24)

where τ_c is the time period of the trend calculation and averaging, and τ_f is the time constant of the high-pass filter. In the case of the RM method, it is assumed that high-pass filtered series are used directly as the fluctuating components in the flux calculation. The high-pass filtering effect of the LD method depends on the time interval, τ_c, but the effect of the RM depends only on the time constant of the filter, τ_f. Because the transfer functions are less than unity at low frequencies, application of LD or RM in flux calculation leads to systematic underestimation of turbulent fluxes.

The filtering effect of the RM with a time constant 200 s is stronger than that of LD (Figure 15 in Kristensen, 1998), where the curve of high-pass filtering for $\tau_f/\tau_c = 60^{-1/2}$ corresponds roughly to a running mean with a time constant of 200 s. In other words, the absolute average flux estimates obtained by RM with a time constant of 200 s, with negligible random uncertainty due to the stochastic nature of turbulence, are always less than the estimates obtained when applying the LD algorithm. This is not necessarily valid for the short-period average fluxes with significant random errors, being calculated from a realization of turbulence in virtually a point in space over a relatively short time.

Due to detrending or high-pass filtering, estimated average fluxes are always biased. For the unstable, but neutral limit of the atmospheric surface layer co-spectra given in Horst (1997), the systematic errors of flux estimates due to LD and RM with time constant 200 s for the height to wind speed ratios from 1 to 10, vary from 0.7 to 6.1% and from 1.5 to 11.4% (without taking into account the underestimation of fluxes at high frequencies) respectively. The corresponding error range for the RM with time constant 1000 s is from 0.2 to 1.8%. The systematic errors are bigger for unstable conditions and smaller for stable conditions. The relatively small errors in short-period fluxes translate to bigger errors in long-term averages. The systematic errors of fluxes can be minimized by choosing a suitable filtering method or averaging time, but there are also other factors which should be accounted for when choosing the filter for calculation of statistics.

To observe negligible systematic errors in fluxes, the RM has to be applied with moderately long time constants, but this leads to systematic overestimation of variances in the periods of non-stationarity of first moments (Shuttleworth, 1988). In addition, it was shown by Rannik and Vesala (1999.) that fluxes are subject to increased random errors during episodes of non-stationarity. Thus, it seems justified to apply the method with a moderate time constant to avoid these problems at the expense of small systematic errors in fluxes, which should be then accounted for. In the case of atmospheric surface layer measurements, reasonable values of the time constant of the RM seem to be between 200 and 1000 s, probably around 500 s, dependent on the measurement height.

Application of the LD method with commonly accepted averaging times does not generally lead to overestimation of variances and increases in random errors, to the advantage of the method. However, the systematic errors in fluxes due to LD are not negligible under certain experimental conditions. For evaluation of systematic errors, co-spectral information is needed. The model co-spectra of turbulence can be used in the estimation of errors, but there is remarkable uncertainty in the low-frequency co-spectral densities under unstable stratification (Kaimal et al., 1972), when the underestimation of fluxes is greatest. Alternatively, the co-spectra could be estimated as site specific from a series of measurements under similar conditions of atmospheric stability to obtain the low-frequency densities with sufficient accuracy.

2. Effects of Low-pass Filtering

Low-pass filtering results from the inability of the system to resolve fluctuations associated with small eddies and induces an underestimation of the measured turbulent flux. The transfer function characterizing the high-frequency losses (TF_{HF}) is 1 at low frequencies, decays to zero at high frequencies and may be characterized by its cut-off frequency (f_{co}), that is, the frequency at which the transfer functions equals $2^{-1/2}$. When the cut-off frequency is known, the transfer function may be combined in equation (23) with the high-pass transfer function (equation (24)) and model co-spectra as defined by Kaimal *et al.* (1972) or Horst (1997) in order to estimate the correction factor. Moncrieff *et al.* (1997a) showed in this way that, for a given eddy covariance system, the correction factor is a function of wind speed (U) and measurement height above the displacement height ($h_m - d$). We show in Figure 4 the relation between the correction factor and the cut-off frequency for five different ($h_m - d$)/U values. The correction factor values are listed in Table 4. The transfer function and its cut off frequency may be estimated either theoretically or experimentally. Both approaches will be described here.

The theoretical description of eddy covariance system transfer functions was first given by Moore (1986) for momentum and sensible heat fluxes. It was later applied to scalar flux measurements by Leuning and Moncrieff (1990), Leuning and King (1992), and Leuning and Judd (1996), who treated in particular the systems that use a closed-path gas analyser. A complete description of the transfer functions of eddy covariance systems with closed-path analysers is given by Moncrieff *et al.* (1997a). Following this theory, the transfer function of an eddy covariance system is considered as the product of individual functions, each describing a particular instrumental effect. The instrument effects that damp the high-frequency fluctuations are: the dynamic frequency response of the sonic anemometer and of the IRGA, the sensor response mismatch, the scalar path averaging, the sensor separation (Moore, 1986) and the attenuation of the concentration fluctuations down the sampling tube, which is typical of closed-path systems (Leuning and Moncrieff, 1990; Leuning and King, 1992; Leuning and Judd, 1996).

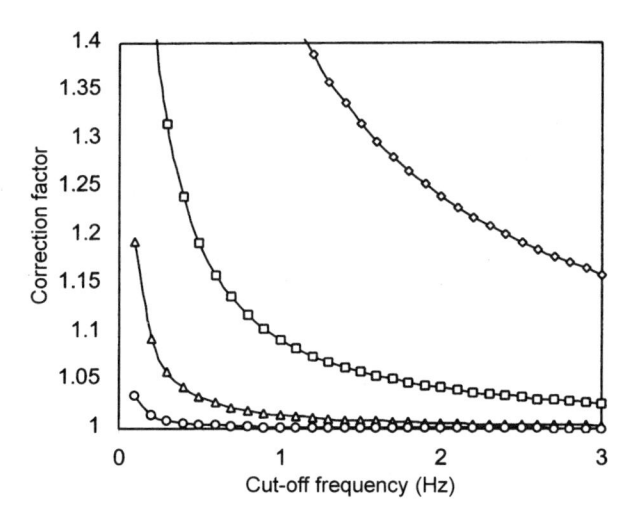

Fig. 4. Evolution of the correction factor according to the cut-off frequency. ◇, ($h_m - d$)/U = 0.2 s; □, ($h_m - d$)/U = 1 s; △, ($h_m - d$)/U = 5 s; ○, ($h_m - d$)/U = 25 s.

Table 4
Correction factor for high-frequency damping in relation to the cut-off frequency (f_{co}) and to the ratio $(h_m - d)/U$

f_{co} (Hz)	$(h_m - d)/U$ (s)			
	0.2	1	5	25
0.1	4.136	1.819	1.194	1.035
0.2	2.742	1.456	1.094	1.015
0.3	2.246	1.316	1.060	1.009
0.4	1.984	1.241	1.043	1.006
0.5	1.819	1.193	1.033	1.005
0.6	1.704	1.160	1.027	1.004
0.7	1.619	1.137	1.022	1.003
0.8	1.552	1.119	1.019	1.003
0.9	1.499	1.105	1.016	1.002
1.0	1.456	1.093	1.014	1.002
1.2	1.388	1.076	1.011	1.001
1.4	1.337	1.064	1.009	1.001
1.6	1.297	1.055	1.008	1.001
1.8	1.266	1.048	1.007	1.001
2.0	1.240	1.043	1.006	1.001
2.2	1.219	1.038	1.005	1.001
2.4	1.201	1.035	1.005	1.001
2.6	1.185	1.031	1.004	1.001
2.8	1.172	1.029	1.004	1.000
3.0	1.160	1.026	1.003	1.000

The first four functions are described in detail by Moore (1986) and will not be restated here. They depend only on the sonic and IRGA characteristics and on the wind speed, and are thus the same for all the EUROFLUX sites. The cut-off frequency of the transfer function that results from the combination of these four effects depends slightly on the wind speed, being 2.26 Hz at 1 ms^{-1} and 3.5 Hz at 5 ms^{-1}. Above 5 ms^{-1}, it remains practically constant. These values correspond to the maximal cut-off frequency that can be achieved in EUROFLUX systems. The resulting error is significant only under high wind speeds or at low measurement height (it exceeds 1% when $U > 2.5$ ms^{-1} at $h_m - d = 4$ m, $U > 4.5$ ms^{-1} at $h_m - d = 8$ m, $U > 7.0$ ms^{-1} at $h_m - d = 12$ m). These effects can be neglected most of the time.

The function describing the sensor separation effect is given by:

$$TF_s(f) = \exp\left\{-9.9\left(\frac{fL_s}{U}\right)1.5\right\} \tag{26}$$

where L_s is the separation distance between the sensors. Its cut-off frequency ($f_{co\text{-}ss}$) is given by:

$$f_{co\text{-}ss} = \frac{0.107U}{L_s} = a_{ss}U \tag{27}$$

It increases linearly with wind speed. The value of the slope of the increase, a_{ss}, is given for each site in Table 2. It varies from 0.2 to 0.9 m^{-1}. The error is thus particularly important at low wind speeds, especially at sites with a L_s, and this distance ought to be kept to a minimum.

The functions describing the attenuation of the fluctuation concentrations down the sampling tube are (Lenshow and Raupach, 1991; Leuning and King, 1992):

$$TF_t(f) = \exp\left\{\frac{-\pi^2 r_t^2 f^2 L_t}{12 D_s U_t}\right\} \tag{28}$$

when the flow in the tubes is laminar, or:

$$TF_t(f) = \exp\left\{\frac{-80 Re^{-1/8} r_t f^2 L_t}{U_t^2}\right\} \tag{29}$$

when it is turbulent. L_t and r_t are the tube length and radius, U_t the air speed in the tube and D_s the molecular diffusivity of the scalar. The argument of the exponentials in equations (28) and (29) are half the values given by Leuning and King (1992) because the latter refer to variance calculations whereas ours refer to covariance calculations. On the other hand, the values given by Moncrieff *et al.* (1997a, p. 610, last two formulae) are too big by a factor of 2. In addition, following a typographic error, the exponent –1/8 was omitted in their last formula.

The corresponding cut-off frequencies ($f_{co\text{-}t}$) are given, respectively, by:

$$f_{co\text{-}t} = 0.649 \sqrt{\frac{U_t D_s}{r_t^2 L_t}} \tag{30}$$

in the laminar case, and:

$$f_{co\text{-}t} = 0.666 Re^{1/16} \frac{U_t}{\sqrt{r_t L_t}} \tag{31}$$

in the turbulent case. For the latter, Leuning and Judd (1996) have proposed an alternative expression derived from the work of Massman (1991):

$$f_{co\text{-}t} = \ln(0.76 Re^{0.039}) \frac{U_t}{\sqrt{r_t L_t}} \tag{32}$$

The tubing cut-off frequencies predicted by theory are given for each site in Table 2. The cut-off frequencies are high under a turbulent regime and the tubing effect appears insignificant. On the other hand, under a laminar regime the cut-off may be reduced to 0.4 or 0.5 Hz and thus it becomes the main influence on frequency loss.

In conclusion, the theoretical approach shows that the transfer function depends only on the wind speed and on the system characteristics. In the case of the closed chamber system used in the EUROFLUX network, the most important characteristics are the tube length and diameter, the mass flow in the tubes and the separation distance between the tube inlet and the sonic. However, it will be shown that the cut-off frequencies predicted by the theory are generally overestimated and consequently that the predicted correction is generally underestimated.

Besides the theoretical approach, an experimental procedure may be developed to estimate the transfer functions and correction factors, as follows:

(1) Select a long time period (3 h at least in order to reduce the uncertainties on the low-frequency part of the co-spectra) with sunny conditions in addition that meets both fetch and stationarity criteria (see section VI.B).
(2) Calculate the co-spectra for heat (C_{wT}), and for the scalar (C_{ws} for CO_2 or water vapour).
(3) Calculate the transfer function as the ratio of the normalized co-spectral densities as:

$$TF_{HF}^{exp}(f) = \frac{N_T C_{ws}(f)}{N_s C_{wT}(f)} \quad (33)$$

where N_T and N_s are normalization factors.

(4) Calculate the correction factor by fitting an exponential regression, i.e.:

$$TF_{HF}(f) = \exp\left\{-0.347\left(\frac{f}{f_{co}}\right)^2\right\}$$

on the experimental transfer function and introducing it in equation (23).

The normalization factors N_T and N_s should be the real covariances (respectively $\overline{w'T'}$ or $\overline{w's'}$) or, if similarity between heat and scalar transport is assumed, the real standard deviation (respectively σ_T and σ_s). However, the measured covariances and standard deviations are lower than the real ones, as affected by the high-frequency attenuation. To avoid this bias, an alternative way of computing the normalization factor is:

$$\frac{N_T}{N_s} = \frac{\int_0^{f_0} C_{wT}(f)df}{\int_0^{f_0} C_{ws}(f)df} \quad (34)$$

where f_0 is a limit frequency, chosen to be low enough for the attenuation be negligible in the integrals in equation (34), but high enough for the number of points used to estimate the integrals to be sufficient and the uncertainty on the normalization factors to be low.

The co-spectra for sensible heat (the denominator of equation (33)) are obtained experimentally and are therefore affected by the dynamic response of the sonic (or the high-frequency (HF) thermometer), the scalar path averaging and the sensor response mismatch. As the corresponding functions are not the same for the sonic or the HF thermometer and the IRGA, these effects are not taken into account correctly by the experimental transfer function. However, we showed above that these effects were small compared with the tubing or with the sensor separation effect, and could be neglected.

It was shown above that the transfer function depends only on the set-up. Consequently, it is not necessary to calculate it every time. An estimation every month in order to follow the wear of the pumps (if the flow is not controlled) is sufficient.

Aubinet *et al.* (unpublished) followed this method and found significantly lower cut-off frequencies than predicted by theory. The comparison between the theoretical and the experimental transfer function they found for their site is given in Figure 5. They found $f_{co} = 0.2$ Hz for the experimental transfer function, whereas the theoretically predicted value was 0.6 Hz. Consequently, the theoretically predicted error was half the observed value. In addition, the experimental cut-off frequencies are always lower for water vapour than for CO_2, which cannot be explained only by the differences in molecular diffusivities in equation (28).

Reasons for these discrepancies may come from uncertainties in the mass flow in systems that do not employ a mass flow controller, or from uncertainties in the real mass flow regime, the Reynolds number not being a sufficient criterion, as discussed in section III.D. The main reason is likely to be that the theory ignores the impact of the filters on the transfer function. This effect is significant, depends on the filter type and number (increasing with the filtration surface) and may be higher for water vapour than for CO_2, the former being absorbed and released by the filters. For these reasons, it is recommended: (1) to use the experimental procedure to obtain the transfer function and its cut-off frequency; and (2) to deduce the correction factor from the cut-off frequency, wind velocity and measurement height, using the results of Figure 4 or Table 4.

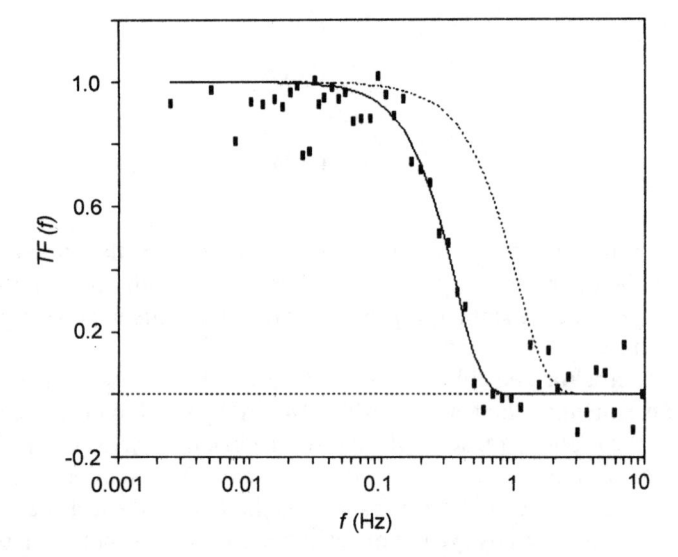

Fig. 5. Comparison between the theoretical (dotted line) and experimental (points and solid line) transfer functions (TF) of the Vielsalm measurement system.

VI. QUALITY CONTROL

To be able to compare the results of different flux measurement stations, the quality of the flux data has to be assessed, because otherwise individual flux patterns cannot be distinguished from site-specific influences of the methodology. However, the quality of flux measurements in the atmospheric boundary layer is difficult to assess, because no standard methods are available to calibrate a given experimental design. There are various sources of errors in flux measurements, ranging from failure to satisfy any of a number of theoretical assumptions to failure of the technical set-up. The resulting errors may be systematic and/or randomly distributed (Moncrieff *et al.*, 1996). Conflicts with the assumptions made in the derivation of the turbulent flux equation arise for certain meteorological conditions and site properties. A more detailed description is given by Moncrieff *et al.* (1997a) and Ibrom *et al.* (unpublished). As these effects cannot be quantified solely from eddy covariance data, a classical error analysis and error propagation will remain incomplete. The alternative chosen here is to investigate empirically whether the fluxes meet certain plausibility criteria.

Four criteria are investigated here. They concern, respectively, the statistical characteristics of the raw instantaneous measurements (section VI.A), the stationarity of the measuring process (section VI.B), boundary layer similarity (section VI.C) and energy balance closure (section VI.D).

A. Raw Data Analysis

The quality control of raw instantaneous data may be performed using a software package (QC) which applies a series of statistical tests as described by Vickers and Mahrt (1997). The FORTRAN package is freely available (http://mist.ats.orst.edu/Software/qc/qc.html) and is intended as a safety net to identify instrument and data-logging errors before data analysis. The user may select which tests are run, limits and thresholds for flagging of data by each test, and local averaging scales. Only the tests that we recommend are described here, including spikes, higher moment statistics, absolute limits and discontinuities. The tests (in italics), described briefly below, warn of suspect data.

Spikes can be caused by random electronic spikes or sonic transducer blockage (during precipitation, for example). An algorithm similar to that described by Højstrup (1993) detects and removes spikes, replacing them with the expected value (based on a local averaging scale and point-to-point auto-correlation) before analysis by the other QC tests. A warning is noted (flagged) when the percentage of replaced data exceeds a threshold. Data are further required to fall within *absolute limits* and normal ranges of *skewness* and *kurtosis*. Discontinuities in the mean and variance of a time series (coherent on a local averaging scale) are detected using the Haar transform (Mahrt, 1991). Finally, each variable is tested for a lower limit in *absolute variance* and a variance ratio for consecutive local windows.

The QC package was run on all 1997 eddy covariance data from BE2 site. Some 13% of all data records were flagged for some problem or another; many appeared to be associated with plumbing problems for LICOR sampling, or with precipitation on the sonic anemometer. Visual inspection confirmed that flagged data were in fact bad (obvious non-physical behaviour in time series, due to instrument, logger or computer problems). No effort was made to evaluate data that were not flagged. For the flagged data, the breakdown by variable is as follows: 10% for u, 16% for v, 13% for w, 19% for T_{son}, 19% for c, and 24% for h. Sixty-nine per cent of the bad data were flagged by more than one QC test; 27% of the flags were due to the absolute limits test. The remaining bad data were flagged by the following individual tests: kurtosis (3%), Haar tests (1%), spikes (< 1%).

Computation of fluxes using bad data generally leads to the addition of random noise. The effects of such noise on long-term Flux estimates have not been quantified. However, it is clearly desirable to minimize problems associated with bad data, and the QC program is an objective means of achieving this goal.

B. Stationarity Test

One hypothesis used to simplify equation (2) is the stationarity of the measuring process. This assumption has to be fulfilled for using the eddy covariance method.

The non-stationarity test has been used since the 1970s by Russian scientists (Gurjanov et al., 1984; see Foken and Wichura, 1996) and is based on equation (6) for the determination of fluxes. The measured time series of about 30-min duration will be divided into $n_s/m = 4...8$ intervals of about 5 min. The covariance of a measured signal ξ and η (similar algorithm for dispersions with $\xi = \eta$) of the interval l with $m = 6000$ (6000 measuring values in 5 min for 20-Hz scanning, i.e. $n_s = 36\,000$ in 30 min) measuring values is:

$$\overline{\xi'\eta'} = \frac{1}{m-1}\left[\sum_{k=1}^{m}\xi_{kl}\eta_{kl} - \frac{1}{m}\left(\sum_{k=1}^{m}\xi_{kl}\right)\left(\sum_{k=1}^{m}\eta_{kl}\right)\right] \qquad (35)$$

For the test, the mean covariance of the n_s/m single intervals is used:

$$\overline{\xi'\eta'} = \frac{1}{n_s/m}\left[\sum_{l=1}^{n_s/m}\overline{\xi'_l\eta'_l}\right] \qquad (36)$$

On the other hand, the value of the covariance for the full period will be determined according to (compare with equation (35)):

$$\overline{\xi'\eta'} = \frac{1}{n_s-1}\left[\sum_{l=1}^{n_s/m}\sum_{k=1}^{m}\xi_{kl}\eta_{kl} - \frac{1}{n_s}\left(\sum_{l=1}^{n_s/m}\sum_{k=1}^{m}\xi_{kl}\right)\left(\sum_{l=1}^{n_s/m}\sum_{k=1}^{m}\eta_{kl}\right)\right] \qquad (37)$$

If there is a difference of less than 30% between the covariances (or dispersions) determined with equations (36) and (37), then the measurement is considered to be stationary. This is an initial criterion that characterizes the quality of the measurements. For practical use, all data with differences < 30% are of high quality and those with differences between 30% and 60% have an acceptable quality (Foken et al., 1997).

C. Integral Turbulence Test

The integral (normalized) turbulence characteristics (flux variance similarity) may be used together with the non-stationarity test for a check on data quality. This parameter has been investigated extensively by Wichura and Foken (1995).

The integral characteristics of the vertical wind are defined by:

$$\frac{\sigma_w}{u_*} = a_1[\varphi_m(\zeta)]^{b_1} \qquad (38)$$

and the integral characteristics of the temperature by:

$$\frac{\sigma_T}{T_*} = a_2[\zeta\varphi_h(\zeta)]^{b_2} \tag{39}$$

where σ_w and σ_T are the vertical velocity and temperature standard deviations, φ_m is the surface layer similarity function, u_* is the friction velocity and ζ is the stability parameter defined as $(h_m - d)L$, where L is the Obukhov length. The empirical coefficients, a_i and b_i, are obtained with the model of Foken et al. (1991) and given in the updated form by Foken et al. (1997) in Table 5. According to Foken (1999a) these parameters can also be used for stable stratification. The data quality is good if the difference between the measured integral characteristics and the calculated value differs by not more than 20–30%. For neutral stratification, this test cannot be used for scalar fluxes.

By definition, the integral characteristics are basic similarity characteristics of the atmospheric turbulence (Obukhov, 1960; Wyngaard et al., 1971). Because of the similarity in the basic equations, there is a close connection to the correlation coefficient. Therefore, they characterize whether or not the turbulence is well developed according to the similarity theory of turbulent fluctuations. It is possible to discover some typical effects of non-homogeneous terrain. First, if there is additional mechanical turbulence caused by obstacles or generated by the measuring device itself, the measured values of integral characteristics are significantly higher than predicted by the model. Second, the measured values of integral characteristics are significantly higher than the model for terrain with an inhomogeneity in surface temperature and moisture conditions, but not for inhomogeneities in surface roughness. This was found by De Bruin et al. (1991) and confirmed by Wichura and Foken (1995).

D. Energy Balance Closure

A requirement that should be met despite any ecological and climatological differences among the flux study sites is the conservation of energy in the systems, according to the first law of thermodynamics. Thus, the closure of the

Table 5
Coefficients of the normalized turbulence characteristics according to Foken et al. (1991, 1997)

	ζ	a_1	b_1	a_2	b_2
σ_w/u_*	$0 > \zeta > -0.032$	1.3	0		
	$-0.032 > \zeta$	2.0	1/8		
σ_u/u_*	$0 > \zeta > -0.032$	2.7	0		
	$-0.032 > \zeta$	4.15	1/8		
σ_T/T_*	$0 > \zeta > -0.0625$			0.5	$-1/2$
	$-0.0625 > \zeta$			1.0	$-1/4$

energy balance is a useful parameter to check the plausibility of data sets obtained at different sites. In this approach, the sum of turbulent heat fluxes is compared with the available energy flux (the net radiative flux density less the storage flux densities in the observed ecosystem, including soil, air and biomass). If the energy terms balance each other, the quality of the flux data is considered to be sufficient. However, there are serious objections to this concept, because the radiation fluxes and heat storage terms are also subject to errors. Moreover, under certain meteorological conditions, processes, such as melting, freezing or heat conductance to cold intercepted rain, which are usually not considered in the budget calculation when using a standard set of meteorological observables, may contribute considerably to the energy balance. Under these conditions the closure cannot be taken as a plausibility criterion for flux observations. On the other hand, the apparent lack of energy balance does not constitute conclusive evidence for erroneous turbulent flux measurement, but might indicate the occurrence of other non-vertical and turbulent fluxes (such as advection and subsidence).

Unclosed energy balance has been published for many study sites covering either grasslands or forests, but a comprehensive review of the problem, taking all possible reasons for an unclosed energy balance into account, has still to be carried out (Foken, 1999b). Despite the likely need for a literature review on that particular topic, it is not within the scope of this chapter as it mainly uses the closure of the energy balance as an empirical plausibility criterion for the question of whether or when the vertical turbulent fluxes represent the total fluxes of a scalar.

Ibrom *et al.* (unpublished) developed a statistical procedure to tackle this problem and were able to derive characteristic relationships between the closure of the energy balance and certain meteorological conditions at their study site. These relationships are expected to be influenced by site properties such as fetch, roughness and mesoscale effects. In this investigation, this hypothesis will be tested by comparing the long-term flux data of six EUROFLUX sites.

Table 6 lists the locations and equipment used at the six selected EUROFLUX sites. Although most of the technical details of eddy flux measurements have been harmonized among the EUROFLUX teams, some important aspects of the experimental set-ups, in particular the net radiometer type, were chosen by each group individually. In a field intercomparison of different net radiometers within the Boreal Ecosystem Atmosphere Study (BOREAS) project, Smith *et al.* (1997) observed deviations of up to 16% from their standard device, due mainly to different calibrations of the sensors. The implication is that flux networks such as EUROFLUX need to intercompare the meteorological sensors in addition to the sensors used for flux determination.

To represent the relative degree of energy balance closure ($\overline{EBC}$) in a data set, the slope of a linear regression, s_1, and the slope of a linear regression

Table 6
Site parameters and sensors used for assessment of energy flux terms at six flux stations

	BE1	FR2	FR1	GE1	GE2	GE3
Location	50° 18' N 6° 00' E	44° 42' N 0° 46' E	48° 40' N 7° 05' E	50° 09' N 11° 52' E	50° 58' N 13° 38' E	51° 46' N 9° 35' E
Elevation	450	60	300	780	380	500
Fetch	300–1500 m	300 to > 1000 m	500 m	200 m	150 to > 500 m	270 to > 500 m
Slope	3%	None	None	None	< 5°	3°
$h_m - d$	11.5 m	12 m	8 m	20 m	20 m	16 m
Net radiation budget	Schenck type 8111	REBS Q7	REBS Q7	Schulze/Däke	Schulze/Däke	Schulze/Däke
Turbulent fluxes	Solent R1024	Solent R1024	Solent R1024	Solent R1024	Solent R1024	USAT-3
	LI-6262	LI-6262	LI-6262	LI-6262	LI-6262	LI-6262
High-pass filter	RM (200 s)	RM (200 s)	RM (200 s)	LD	None	LD
Rate of change of heat storage derived from						
Soil	1 profile in 5 depths + humidity profile	4 profiles in 8 depths + humidity profile	Not derived	6 heat flux plates	Heat flux plates	4 profiles in 5 depths
Canopy air space	2 air temperatures in canopy	11 temperatures in canopy	Sonic temperatures at reference height	Temperature and humidity in canopy air	Temperature and humidity in canopy	6 temperatures and humidities in canopy
Biomass	8 temperatures in tree trunks	54 temperatures at different locations	5 × 4 temperatures in tree trunks	Bole temperatures and radiation temperature	Temperature at reference height	1 air temperature

RM, running mean; LD, linear detrend.

forced through the origin, s_2, are used. This is necessary because the linear relationship between turbulent energy fluxes and available energy usually produces a significant intercept. Consequently, *EBC* cannot be represented by one single parameter. The two estimators describe the sensitivity of turbulent heat fluxes to available energy, s_1, and the relative location of the centre of the data with regard to the 1 : 1 line. Under optimal conditions, both should be equal to 1. The data sets are subdivided into classes according to the parameter of interest to investigate the functional relationship between *EBC* and various turbulence parameters (Ibrom *et al.*, unpublished). The calculated *EBC* estimators are then compared with the average value of the parameters given in Table 7.

Figure 6 and Table 7 present individual relationships between the turbulent energy fluxes and the available energy for six EUROFLUX sites. Only data measured at temperatures higher than 1°C were taken into account in order to exclude, for example, the effects of snow and hoarfrost on the measurements. However, rain events that could also have affected the quality of the measurements have not been filtered out from all of the data sets considered here. All data sets have a very similar shape, where a large proportion of the data (60–75%) is located in the region of ± 100 W/m^{-2}. The standard deviation around the regression line is of the same order of magnitude, ranging from 40 to 84 W/m^{-2}. In most cases, the variability of available energy fluxes explains more than 85% of the variance of the fluxes of turbulent energy (Table 7). *EBC* values for whole data sets range from 0.7 to full closure if expressed by the estimator s_1. In all cases, the slope of a linear regression forced to the origin, s_2,

Table 7
Parameters of linear regression of turbulent energy fluxes (i.e. the sum of latent and sensible heat flux against available energy).

Site	Type	Slope	Intercept (W m^{-2})	R^2	Root mean square error (W m^{-2})
BE1	s_1	0.936	−4.2	0.90	63.0
BE1	s_2	0.926	0.0		63.0
FR2	s_1	0.997	−32.3	0.93	43.6
FR2	s_2	0.892	0.0		49.1
FR1	s_1	0.756	−19.0	0.91	52.8
FR1	s_2	0.713	0.0		54.6
GE1	s_1	0.726	−10.3	0.88	62.4
GE1	s_2	0.704	0.0		62.9
GE2	s_1	0.944	−9.0	0.84	84.1
GE2	s_2	0.922	0.0		84.4
GE3	s_1	0.798	−9.7	0.92	39.4
GE3	s_2	0.773	0.0		40.3

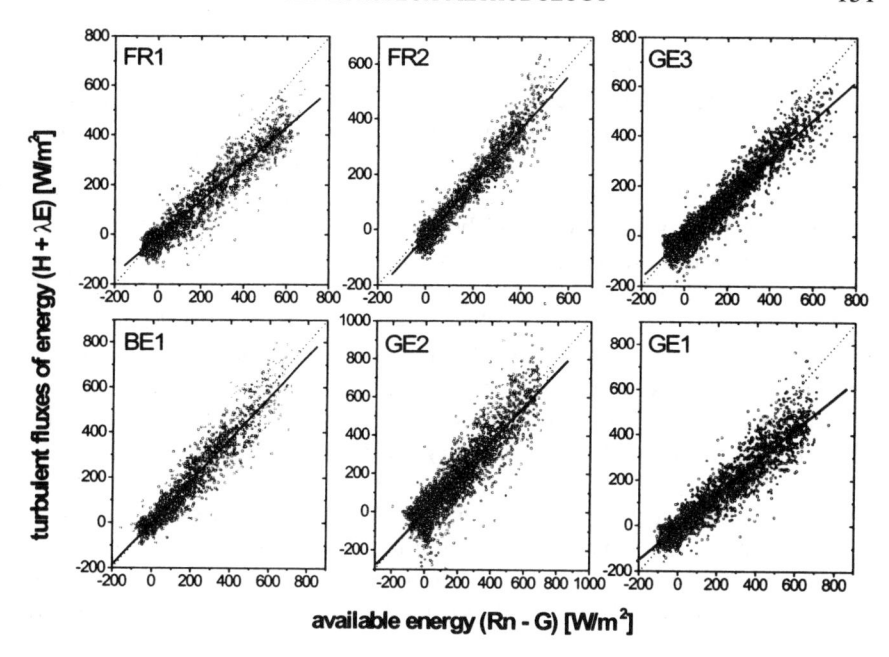

Fig. 6. Scatter diagrams and regression lines of relationship between the turbulent fluxes of energy and available energy fluxes.

is lower than that of the regular regression, s_1, although this effect is not important for the BE1 site. The data set with the highest estimator s_1, FR2, has the lowest negative intercept. Accordingly, the estimator s_2 is about 10% lower than s_1. The estimated relative contribution of the energy storage terms to energy transfer at site FR2 is, on average, twice that for the other sites.

These findings confirm the common observation that the energy balance tends to be more or less unclosed, an indication that non-vertical or non-turbulent fluxes need to be incorporated in the budget calculation, or that there are substantial errors in other energy flux terms which lead to an underestimation of available energy fluxes.

The use of linear regression for the whole data set suppresses the effects of low absolute fluxes. It can be seen at first glance that the relative agreement between turbulently transported energy and available energy is much worse for lower than for higher available energy. Thus it is necessary to go into more detail by using data sets selected by different meteorological conditions. In practice, a relatively large number of parameters could be used to separate the data into subsets and investigate whether they serve as an indicator for the data quality or not, but in most cases the parameters are not independent. The most effective parameters are those connected to turbulence generation and vertical exchange, such as the normalized stability parameter ζ and friction velocity u_*.

Results of the selective statistical analysis are shown in Figures 7 and 8. All data sets show a common pattern. In conditions of unstable stratification, $\overline{EBC}$ is shown to be high, whereas it is low at stable stratification. The values under neutral stratification are less accurate, because many of the energy flux measurements approach zero and become much smaller than the average standard deviation around the regression line. This effect is shown by the larger standard error of the estimate (bars in Figure 7) and some outliers in the near neutral region. By comparing $\overline{EBC}$ values classified into different stratification regimes with the values for the whole data set (Table 7), it turns out that only the highest $\overline{EBC}$ values at unstable stratification fall in the same range as the high values calculated for the complete data set; already starting with neutral stratification, $\overline{EBC}$ becomes very low at stable stratification.

To complete this brief analysis, $\overline{EBC}$ and friction velocity are compared under two different stratification regimes. Again, the data show a common picture. At both stable and unstable conditions, $\overline{EBC}$ grows with increasing u_* over the whole range at unstable conditions and in the range between zero and 0.4 ms⁻¹ at stable stratification. Above the threshold of 0.4 ms⁻¹ the estimates

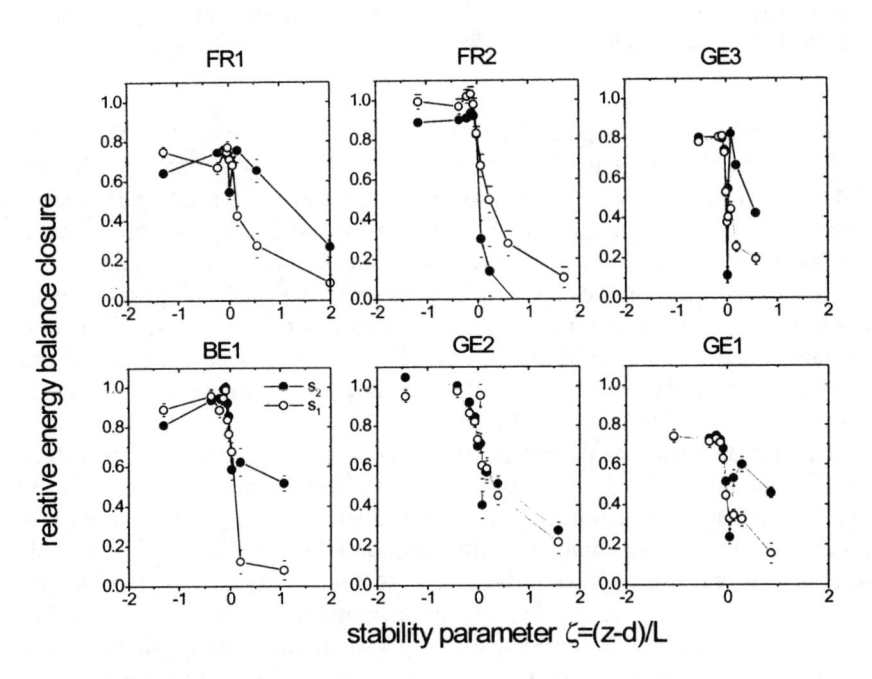

Fig. 7. Functional relationships between the relative energy balance closure and the non-dimensional stability parameter ζ for six EUROFLUX stations.

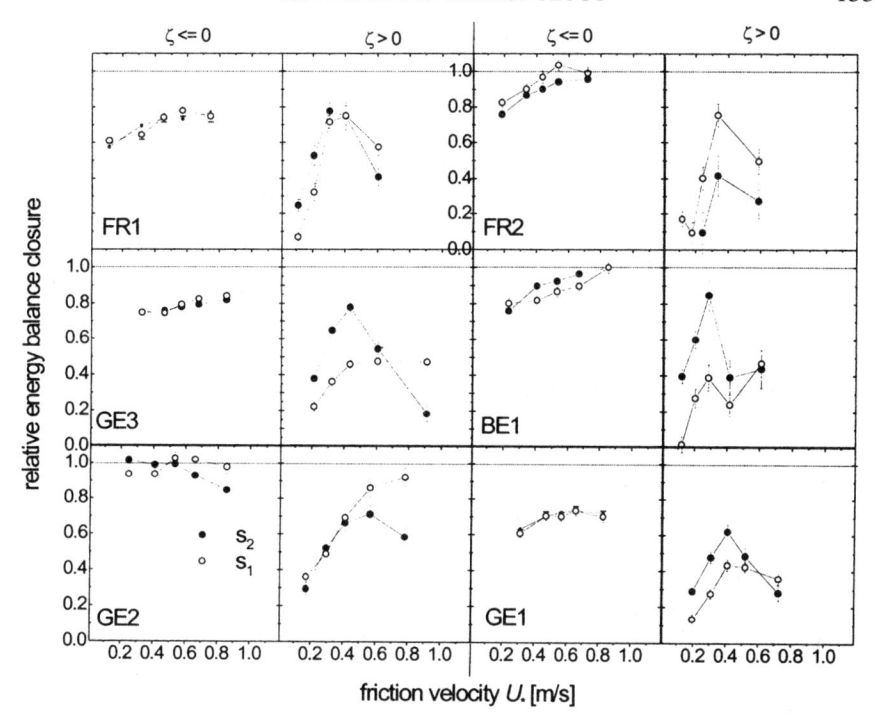

Fig. 8. Functional relationships between the relative energy balance closure and the friction velocity for six EUROFLUX stations.

become less accurate. For near neutral conditions, the energy is very low and small absolute errors become important.

The sensitivity of $\overline{EBC}$ to u_* is much larger at stable than at unstable conditions. This suggests that, at night time, the fluxes could be underestimated owing to a lack of turbulent transport. A similar process is observed for CO_2 fluxes at night (section X). Even if $\overline{EBC}$ appears to be influenced by u_* at stable stratification, this is no evidence for a cause–effect relationship. As expected from its definition, u_* is tightly correlated with the stability parameter ζ. Consequently, classifying the data with u_* will sort the data according to stability as well. Even though distinction cannot be made between different causes, the statistical evidence can be used to derive thresholds of ζ or u_* that are related with a certain degree of $\overline{EBC}$.

The relationships between $\overline{EBC}$ and turbulence parameters indicate that vertical exchange plays an important role in the distribution of energy fluxes into different non-vertical turbulent fluxes, which are consequently lacking if merely eddy covariance data are used. These effects seem to be general, as they have occurred at all sites investigated. The differences of closure with time, due to distinct meteorological conditions, seem to be much more important for the quality of flux data than the averaged differences between the sites.

At the moment we are not able to explain the relatively small differences between the sites, because the intercomparison of the net radiometers has not yet been carried out and estimation of the rate of change of energy storage in the biomass and in the soil is still done differently at the different sites. Thus their relative contribution to the energy balance is more likely to be due to different calculation procedures than to real flux differences at the site. These relationships suggest that an even more detailed analysis will prove to be of benefit.

VII. SPATIAL REPRESENTATIVENESS OF MEASURED FLUXES

To cover the widest climate range and as many forest types as possible, the flux measurement network has to consider sites that are not necessarily ideal for eddy covariance measurements. In particular, heterogeneous forests are studied more and more often. The main question for these sites is to know to what extent the fluxes (and the annual sums) are representative of the real ecosystem flux. To know this, it is necessary to locate the flux sources and to establish the distribution of the frequencies at which they influence the flux measurements.

The footprint analysis allows estimation of the source location. It is based on lagrangian analysis and relates the time-averaged vertical flux of a quantity at the measurement point to its turbulent diffusion from sources located upwind from the measurement point. The extent to which an upwind source located a certain distance from the measurement point contributes to observed flux has been termed the source weight function or flux footprint (Schmid, 1994). In footprint analysis, the relationship between the surface sources (downward flux is equivalent to negative sources at the surface) and the measured flux is studied. This can be achieved by analytical (e.g. Schuepp et al., 1990; Horst and Weil, 1994; Schmid, 1994) and stochastic simulation approaches (Leclerc and Thurtell, 1990).

Frequently, the cross-wind integrated footprint models are used to evaluate the adequacy of the fetch of a homogeneous stand in a certain wind direction and under certain stability conditions. This is done by estimating the fraction of flux contributed by the sources within the homogeneous fetch assuming uniform surface sources. Alternatively, an upwind distance giving a significant fraction of flux can be estimated. Different analytical models differ in their complexity. For example, the simplest analytical model by Schuepp et al. (1990) for the cross-wind integrated flux footprint, assumes constant wind speed in the vertical and does not explicitly include the effects of atmospheric stability. For the evaluation of cross-wind integrated footprint by a stochastic approach, a uniquely defined one-dimensional stochastic model (along-wind turbulent dispersion is neglected) for particle trajectories with parameterization for the atmospheric surface layer (Wilson and Sawford, 1996) can be applied.

Table 8 gives estimates of the horizontal position of the footprint peak and the distance for 80% of cumulative footprints by three different footprint models for two measurement height ($h_m - d$) to roughness length (z_0) ratios. The results corresponding to model HW are obtained by applying the approximate version of the analytical model by Horst and Weil (1994), and in stochastic simulations the uniquely defined one-dimensional model was applied. Distances are normalized by height. Of the three models, the HW and stochastic models are in better agreement, while the simplest model by Schuepp et al. (1990) predicts further distances for peak location as well as for the 80% cumulative footprint value. The difference is greater for low observation levels relative to roughness length. Of the three models, the more realistic cross-wind integrated footprint models—the Horst and Weil (1994) or the stochastic model—are recommended. Of these, the HW model is easier to apply but, in practice, it cannot be solved for very low measurement height : roughness length ratios.

For measurements above forests, such low values apply (often 20 or even less). Under these conditions, the footprint models designed for application inside the surface layer are not strictly applicable. The stochastic models can be parameterized for more complicated flow conditions and also to account for the roughness sublayer effects above the forest. Baldocchi (1997) has studied the footprint characteristics for the levels inside and above the forest, assuming that the sources are located on the forest floor (in the case of surface-layer footprint analysis, the height of sources is assumed to be $d + z_0$, roughly the upper levels of the canopy, and this level is not permeable for flow). The footprints for measurements inside the forest were observed to be much closer than footprints for measurements inside the surface layer (Baldocchi, 1997). For measurements above the forest, the level of sources is not well defined, but this is not likely to introduce great uncertainty owing to the relatively slow horizontal dispersion inside the canopy.

Table 8
Normalized location of footprint peak ($X_{max} h_m^{-1}$) and 80% of cumulative footprint ($X_{80} h_m^{-1}$) for three cross-wind integrated footprint models

Model	$X_{max} h_m^{-1}$		$X80 h_m^{-1}$	
	$h_m z_0^{-1} = 20$	$h_m z_0^{-1} = 100$	$h_m z_0^{-1} = 20$	$h_m z_0^{-1} = 100$
Schuepp et al. (1990)	6.7	11.4	60.3	102.3
Horst and Weil (1994)	4.2	8.5	44.7	83.5
Wilson and Sawford (1996)*	3.8	9.1	35.4	71.8

*Stochastic simulations with one-dimensional trajectory model.

VIII. SUMMATION PROCEDURE

The net ecosystem exchange is the sum of different physiological components: the gross leaf assimilation, A, and the leaf, wood, root and heterotrophic respiration rates, R_l, R_w, R_r, R_h:

$$N_e = -A + R_l + R_w + R_r + R_h \qquad (40)$$

This equation is derived from Ruimy *et al.* (1995) with sign adaptations to fit the meteorological conventions (i.e. positive upwards, negative downwards) According to equation (1), the storage and advection terms are included in N_e. The assimilation and all the respiration terms are positive. The integration of N_e over a given period τ_s gives the net ecosystem production (*NEP*):

$$\text{NEP} = - \int_{t}^{t+\tau_s} N_e \, dt \qquad (41)$$

This sum may be directly deduced from eddy covariance and storage flux measurements. The procedure used for the summation, including data gap filling (section IX), night flux corrections (section X) and estimates of uncertainties (section XI), is detailed below.

Other sums of physiological relevance are the net primary productivity (*NPP*) which quantifies the net CO_2 flux exchanged by the vegetation (soil excluded):

$$\text{NPP} = - \int_{t}^{t+\tau_s} (N_e - R_h) \, dt \qquad (42)$$

and the gross primary production (*GPP*) that quantifies the total mass of synthesised glucides:

$$\text{GPP} = - \int_{t}^{t+\tau_s} (N_e - R_l - R_w - R_r - R_h) \, dt \qquad (43)$$

However, neither of these can be deduced from flux measurements. The *NPP* calculation requires knowledge of the heterotrophic respiration. This is not available directly even when the eddy covariance system is complemented by soil chambers that measure soil CO_2 fluxes. Indeed, this system does not distinguish between heterotrophic and root respiration.

The *GPP* computation would require the knowledge of the total ecosystem respiration. In particular, it would take into account the difference between day and night-time leaf respiration, which is significant (Ruimy et al., 1995; P.G. Jarvis, personal communication, 1998). This difference cannot be deduced from flux measurements.

Alternative sums may be deduced from the flux measurements as the daytime (night-time) *NEP* is obtained by integration of N_e only on day (night) periods. These quantities do not depend on any data treatment but they cannot be directly related to the fluxes of physiological relevance. A sum currently (but erroneously) named gross primary production (*GPP'*) may currently be deduced using the following procedure:

(1) To liken the night-time CO_2 flux to the total ecosystem respiration (R_{tot} = $R_l + R_w + R_r + R_h$).
(2) To fit a regression equation on the respiration to temperature response.
(3) To use this equation to estimate daytime respiration.
(4) To calculate the summed respiration over day and night.
(5) To deduct this sum from the *NEP*.

Caution is required when calculating this sum to avoid several errors. The night-time fluxes must be selected according to a turbulence criterion in order to eliminate the underestimation error (section X). The temperature used as a reference for the respiration response must be chosen appropriately. To avoid extrapolation errors, it is necessary to choose a temperature for which daytime and night-time ranges are close together. To that end, the soil temperature (i.e. at 2-cm depth) is advisable as a reference as its daily cycle is damped and logged compared to air temperature. It is, of course, a gross simplification to choose a unique reference temperature given that the respiring elements are spread in the ecosystem and are submitted to different temperatures.

Again, in order to minimize the extrapolation error, the regression equation must be chosen so as to provide a non-biased residual distribution. In particular, the modified Arrhénius equation is advisable (Lloyd and Taylor, 1994):

$$R_{tot} = R_{10} \exp\left\{308.56\left(\frac{1}{56.02} - \frac{1}{T_s - 227.13}\right)\right\} \quad (44)$$

whereas the classical exponential regression using R_{10} and Q_{10} parameters is not advisable because it does not reflect correctly the dependence on temperature, Q_{10} being shown to decrease with temperature (Lloyd and Taylor, 1994). The parameterization of *R* according to temperature could be completed by a response to the soil water content, which is significant on dry soils (FR2). Finally, *GPP'* obtained with the procedure described above differs from *GPP* as it ignores the difference between daytime and night-time leaf respiration.

IX. DATA GAP FILLING

Gaps in long-term data series are almost unavoidable. They may occur as a result of system breaks for routine maintenance or calibration, inadequacy of the meteorological conditions or complete system failures. In EUROFLUX, the effort put into maintenance of the measuring systems reduces the data gaps to less than 15% of the total data. The data gaps may be filled in different ways: interpolation, parameterization or use of a (non-) linear regression equation. By way of example of one approach, gap filling by neural network techniques is described below.

A. Interpolation and Parameterization

The following procedure, using both interpolation and parameterization, may be followed in order to fill the CO_2 flux gaps:

(1) To gather flux data per several day-periods. The length of the period may vary from site to site: it must be short enough for seasonal trends to be insignificant but long enough to allow a reasonable interpolation. For temperate regions, a 10-day period is a good compromise.

(2) For day-time, when Photosynthetic photon flux density (PPFD) is available, to fill data gaps, using a parameterized response of N_e to *PPFD* (exponential, see for example Aubinet *et al.* (unpublished); or hyperbolic, for example Valentini *et al.* (1996). Night-time, when soil temperature is available, to fill data gaps using a parameterized response of N_e to T_s (equation (44)). The parameterization of the N_e to *PPFD* response must be re-evaluated each month to take seasonal fluctuations into account. If necessary, the parameterizations may consider additional variables, such as the air temperature and saturation deficit for day-time data, the soil water content for both day- and night-time data. SVAT models are not recommended for data gap filling when the measured data obtained from this procedure are used for their own calibration and validation. If PPFD or temperature is not available, the data remain missing.

(3) For each half-hour, to sum on the period the corresponding fluxes (measured and parameterized) and to estimate the possible missing data as the corresponding mean.

B. Neural Networks

The data collected at the EUROFLUX stations can be separated into *prognostic* (such as wind speed, temperature, specific humidity, global radiation and incoming long-wave radiation) and *diagnostic* (such as latent heat flux and CO_2 flux) data. For most modelling applications, complete time series of prognostic data are compulsory, whereas diagnostic data are used for calibration and validation of the model only. In that case, complete time series of the diagnostic data are not needed: short periods with reliable data are generally sufficient. On the other hand, for comparison of diagnostic data between sites, complete time series of those data are a necessity.

As is well accepted by most modellers, the best model to fit a time series is a non-linear regressor (see for example Mihalakakou et al., 1998). The drawback of using it is that there is no information on the system; in other words, it is a complete black box. However, for gap-filling purposes, understanding the underlying physics is not really of interest. For this reason, a non-linear regressor such as a neural network is an excellent tool to fill data gaps. A neural network is especially useful if a full theoretical model cannot be constructed (Gardner and Dorling, 1998). This may be the case if data from other sites are used to fill the gaps, or if relations are described by ill-defined empirical functions, as is the case between global radiation and CO_2 flux. For the examples shown here, the neural network tool of Saxén and Saxén (1995) is used.

1. General Network Architecture

Although it is not possible to prescribe a single configuration of a neural network for all problems, some general considerations can be taken into account. It has been shown (e.g. Huntingford and Cox, 1997) that the architecture of a neural network that will be able to simulate almost any relationship consists of, at least, one hidden layer with four or more nodes. The input nodes have a linear activation function by default; the nodes in the hidden layers should have a logistic or hyperbolic tangent function. The output node may have a linear activation function, which enables the output signal to have any value. It is generally recommended to scale the input signals by using the means and standard deviations of the whole data series. This is to prevent the network from ignoring input nodes with a small range, which may happen if the randomly assigned initial weights are small for the input node with the small range. Also not strictly necessary but recommendable is to remove as much of the expected behaviour of the target variable as possible. This may be done by using as a target value the residue of the original variable minus, for example, the same variable simulated with a simple known physical relation or minus the same variable measured by another instrument. The network can then be used to predict the differences, instead of the actual variable.

Selection of the number of input and hidden nodes is not straightforward. If the number of nodes is too small, the network will fail to converge. On the other hand, too many nodes will result in over-fitting. To decide on the number of hidden nodes, a separate test data set may be fed to the network simultaneously with the training data set. Increasing the number of nodes will increase the degrees of freedom and therefore the goodness-of-fit of the training data set. However, if the goodness-of-fit of the test data set decreases at the same time, this is an indication of over-fitting and the number of nodes should be decreased.

2. Examples

As an example, global radiation is simulated. In Figure 9 the network configuration is depicted. As input variables for the network, the air temperature, relative humidity, wind speed and incoming global radiation at the top of the atmosphere are used. The target variable of the network is the residual of the measured global radiation minus a fixed fraction of the global radiation at the top of the atmosphere. The fraction was chosen such that, on a clear day, the residual would be almost zero. As activation function for the input and output nodes, a linear function is used, while for the hidden nodes a sigmoid varying between 0 and 1 is used.

Here the network was used to simulate the effects of clouds on global radiation. The network was trained on a data series of 60 days. Figure 10 shows the results of using the trained network for a separate data set. Although the results are not excellent (on some days the differences may be as large as 200 W m^{-2}), they show that the network is capable of simulating relatively complicated relationships. Possibly the results may be improved by using a longer data set for training and/or other input variables.

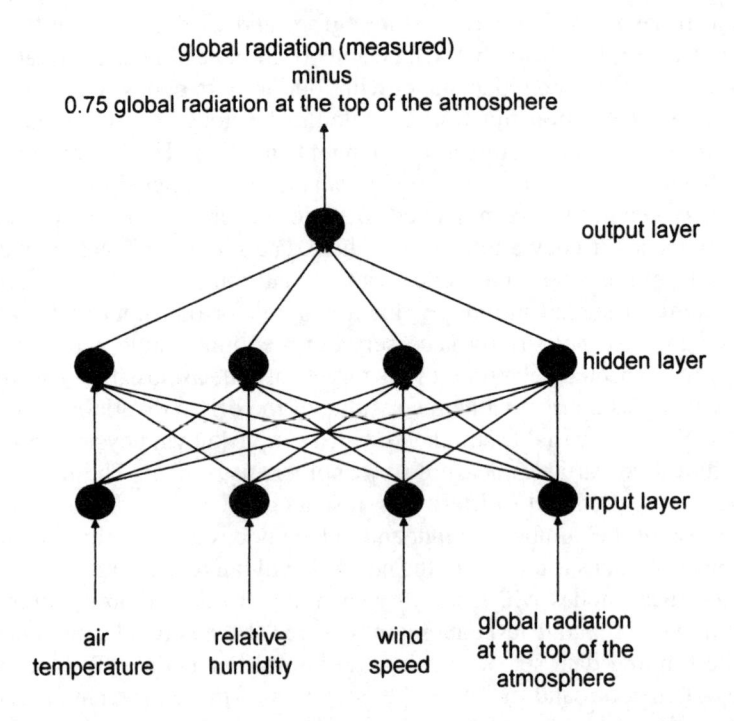

Fig. 9. Example of a neural network configuration with four input signals, one hidden layer with four nodes and one output signal.

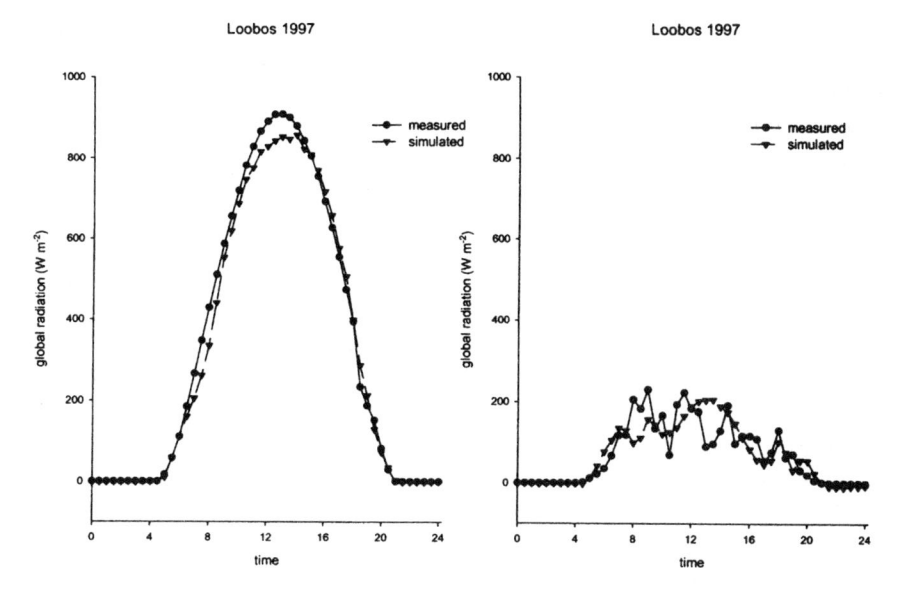

Fig. 10. Simulation of global radiation at a clear day (left) and a cloudy day (right).

In Table 9 the regression results are presented for the training and validation data sets. Here the regression results are also shown filling gaps in data series of the latent heat flux and the CO_2 flux. For the latent heat flux, the target variable was the residual between the measured flux and the flux calculated using the Penman–Monteith equation with a fixed surface resistance of 100 s m^{-1} for dry conditions and of zero for wet conditions. The global radiation, temperature, humidity deficit, soil moisture content at 50 cm depth and precipitation rate were used as driving variables to explain

Table 9
Results of the regression of simulated versus measured data

Variable	Constant	Slope	R^2
Global radiation (c)	25.60 W m^{-2}	0.85	0.85
Global radiation (v)	33.68 W m^{-2}	0.89	0.79
Latent heat flux (c)	7.95 W m^{-2}	0.78	0.68
Latent heat flux (v)	40.61 W m^{-2}	0.60	0.55
Carbon flux (c)	−1.04 μmol m^{-2} s^{-1}	0.79	0.75
Carbon flux (v)	−0.15 μmol m^{-2} s^{-1}	0.90	0.69
Carbon flux as a function of global radiation	−4.47 μmol m^{-2} s^{-1}	0.36	0.44

The (c) indicates calibration results, while (v) stands for validation results.

the residual. One month of data was used for calibration and another month for validation. This period is probably too short. Huntingford and Cox (1997) used a neural network to simulate the stomatal conductance and found that the neural network could explain 74% of the variance in their test data set. The carbon flux was simulated using a neural network and an empirical relationship. The neural network used global radiation, air temperature, net radiation, soil temperature, soil moisture content, CO_2 concentration in the trunk space, sensible heat flux, latent heat flux and friction velocity as input variables, with carbon flux as the target variable. To train the network, 40 days of 30-min data were used. For validation, an extra 10-day period was used. Only daytime values were considered. For the training period, the network explained 75% of the variance, while for the validation data set this decreased to 69%. To compare this with a conventional method of filling data gaps, a non-linear relationship between carbon flux and global radiation was used. The performance of such a relation is much worse in comparison with the network results (see Table 9).

These results show the benefits of using a neural network, especially if there is no need to investigate the physical relations of the processes, as when the prime purpose is to fill gaps in data series. It should be borne in mind that a neural network is suitable only to interpolate. Extrapolation outside the range of the training data may lead to unexpected results. More suggestions on the use of a neural network for atmospheric sciences are given by Gardner and Dorling (1998).

X. CORRECTIONS TO NIGHT-TIME DATA

There is some likelihood that, during stable night-time conditions, CO_2 exchange is underestimated by the eddy covariance measurements. This is supported by the apparent correlation between u_* (used as a measure for turbulent mixing) and CO_2 efflux during the night (Grace et al., 1996; Goulden et al., 1996a). Indeed, night-time efflux resulting from ecosystem respiration is controlled mainly by temperature and soil water content; it should be independent of turbulence if correlations between u_* and temperature are removed. Turbulence may, however, affect soil respiration by 'pressure pumping' at high levels of u_*, but in the range of interest for stable night-time conditions there is no evidence for that. Moreover, the relation between $\overline{EBC}$ and u_* observed under stable conditions at different sites (Figure 8) confirms that the observed underestimation of the CO_2 fluxes at night comes from a lack of turbulence rather than a varying CO_2 source strength. Underestimation of the night-time CO_2 flux is an example of a selective systematic error and as such can be a serious problem, particularly when long-term budgets are estimated by integration of short-term flux measurements (Moncrieff et al., 1996).

Stable nights increase the importance of accounting for storage of CO_2 in the layer below the eddy flux system. CO_2 stored within the forest during the night can be partly assimilated in the morning and partly flushed out. Although this leads to a discrepancy between released CO_2 and measured flux in the short term, it does not affect the long-term budget as long as the morning flush is captured by the measurement. It would appear that storage may not account for the total loss of fluxes, and some of the CO_2 that is released by the soil and the vegetation seems to leave the forest by as yet unquantified routes. Possible mechanisms may be katabatic flow or high-frequency fluctuations that are not detected (Goulden et al., 1996a; Moncrieff et al., 1996), slow diffusion or local convection cells. There is some evidence that some CO_2 does leave the forest vertically, without being measured by the eddy covariance system, when storage of CO_2 above the eddy correlation system exceeds the measured flux (Figure 11).

A way to correct for flux underestimation during stable nights is to replace the measured fluxes by the simulated efflux estimated by a temperature function derived during well-mixed conditions. Two general problems with

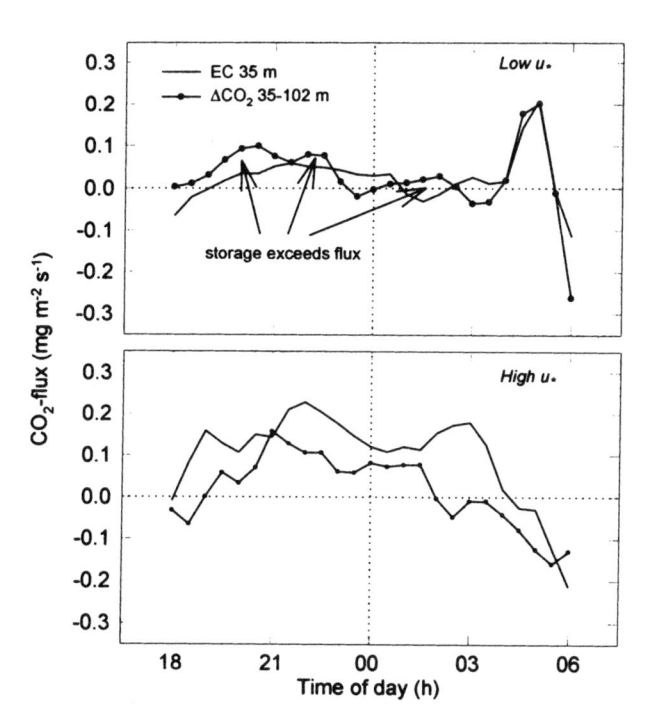

Fig. 11. CO_2 flux measured by eddy correlation and CO_2 storage above the eddy correlation system during a calm and a well-mixed night at Norunda, central Sweden.

this approach are: (1) the sensitivity of the CO_2 budget to the threshold value of u_* that is used to distinguish between stable and well-mixed conditions (Grelle, 1997); and (2) the potential risk of 'double counting' if there is a morning flush of CO_2.

To estimate the effects of night-time stability on night-time fluxes at various sites, nocturnal CO_2 release was plotted against friction velocity for ten EUROFLUX sites (Figure 12). To reduce the scatter, data have been sorted by u_* in intervals of 0.05 m s^{-1}. To eliminate correlations between u_* and temperature, the CO_2 fluxes have been normalized with the simulated flux estimated by a temperature function. For some sites, however, a representative temperature function could not be established with the available data, either because variations in temperature were too small or because there were too few occasions with well-mixed conditions. In these instances, data were confined to a narrow temperature range (1.5 K) and normalized by the 'saturation value' reached at high friction velocities. These two approaches have also been used by Goulden et al. (1996a).

The results for the different EUROFLUX sites are shown in Figure 12. By weighting the relative underestimations with the frequency distribution of the friction velocity, a relative loss of CO_2 efflux can be estimated (Figure 13). Because of the different normalizing procedures used for individual sites, the saturation value of the normalized efflux was taken as the reference for zero loss rather than the 1.0 constant.

This is a rather rough estimation which generally depends on the time period and the number of data points. However, it gives a clue to the importance of storage for the correction of short-term night-time fluxes at the individual sites.

XI. ERROR ESTIMATION

Many of the potential sources of error have been described in preceding sections, for example errors introduced by non-steady-state conditions, advection, complex terrain and instrumentation error. Every flux network operates within the constraints imposed upon it by natural variability in atmospheric and surface properties. Wesely and Hart (1985) estimated that such variability, in combination with instrumentation errors, restricts the accuracy of an individual turbulent flux measurement to between 10% and 20%. If the errors are random, then integrating the individual half-hourly flux estimates over a day will improve this figure, and thus daily integrals are likely to be reasonable. Businger (1986) investigated errors between different eddy covariance systems and found that, in general, systematic errors of the order of 30% or more, and random errors of the same magnitude, existed. The network approach, of course, removes some of this potential source of error by harmonizing the instrumentation and software across sites. At any one site, a careful review of all the components in an eddy covariance system is essential. In one

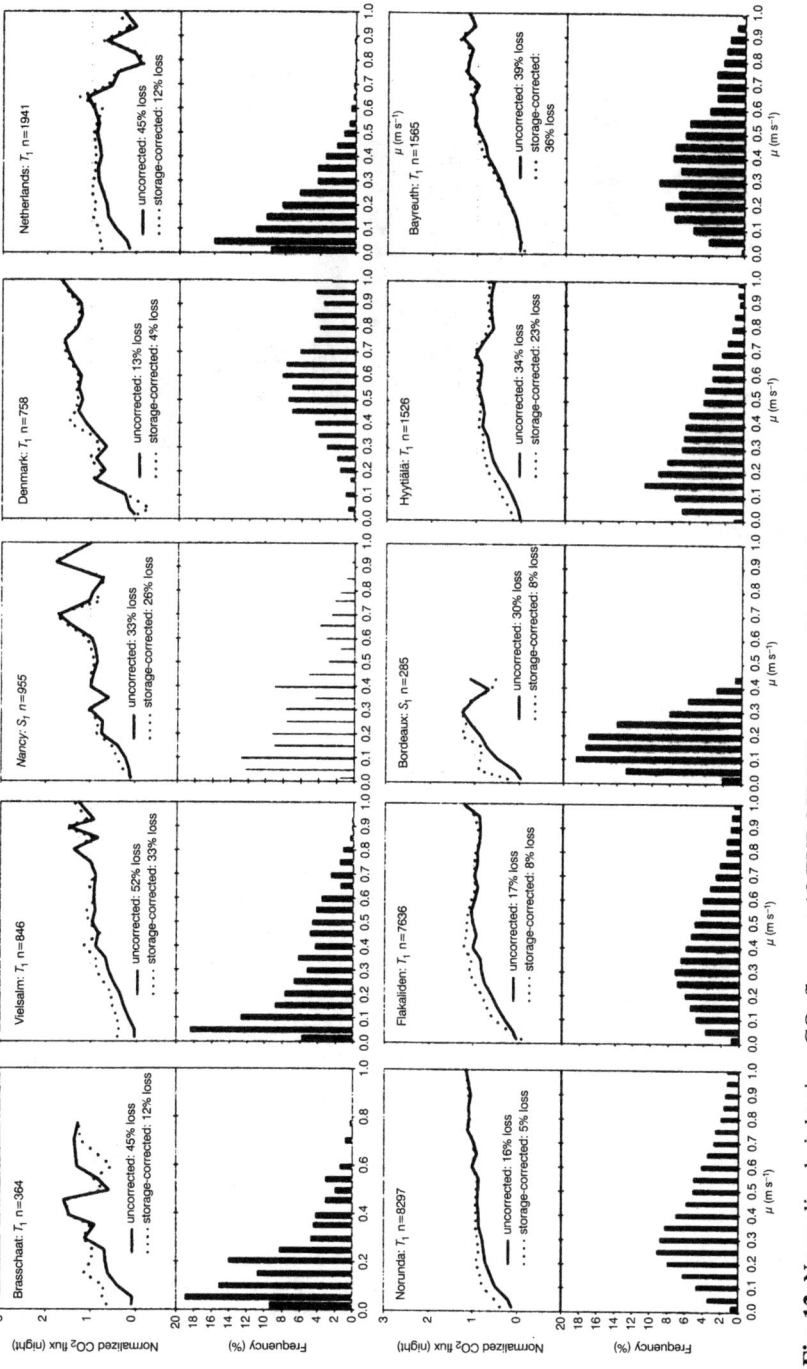

Fig. 12. Normalized night-time CO_2 fluxes at 10 EUROFLUX sites sorted by friction velocity into intervals of 0.05 m s^{-1}. Normalization is done either by a temperature response function (T) or by the saturation value (S). n = number of data points.

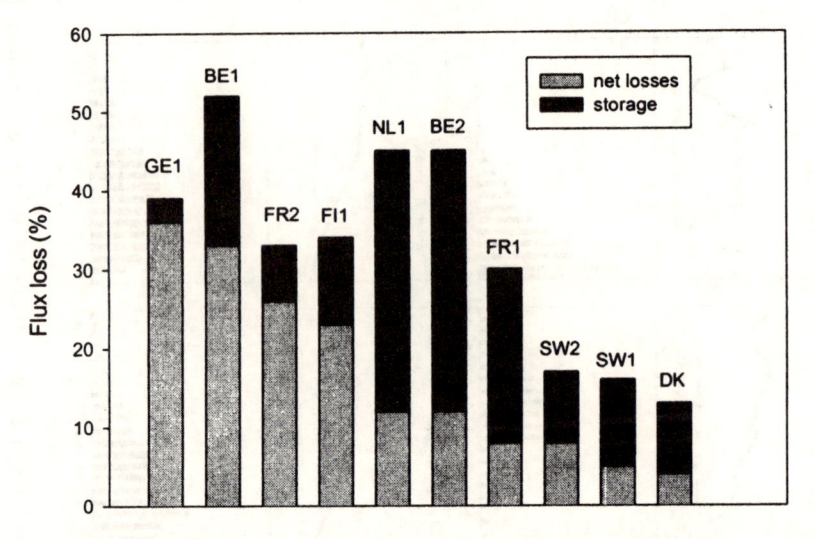

Fig. 13. Relative loss of night-time fluxes for 10 EUROFLUX sites and the storage component.

of the few papers describing errors in long-term flux studies, Goulden *et al.* (1996a) suggested that the long-term precision of their eddy covariance flux measurements was ± 5%, and they estimated that their annual net canopy exchange of carbon should be regarded as being within ± 30 g C m^{-2} year^{-1}.

The EUROFLUX partners have adopted the error terminology defined by Moncrieff *et al.* (1996), i.e. that errors can loosely be either random and/or systematic (the latter being subdivided into full or selective errors) and that a full error analysis is essential for published flux estimates. In addition, flux estimates from EUROFLUX will be published as $X \pm a \pm b$, where X is the mean value of the result, a is the estimate of stochastic (random) uncertainty, and b is an assessment of systematic uncertainty because of possible systematic errors in the system. Many of the quality checks applied within EUROFLUX as part of the accepted methodology will isolate gross errors and inconsistencies, and careful calibration and analysis routines will help to prevent the more insidious selective systematic errors. Errors will remain, however, and must continually be checked for and eliminated.

One method that is useful for predicting the likely impact on integral flux estimates of different types of error follows the procedure outlined in Moncrieff *et al.* (1996). In brief, it is desirable to produce daily plots of CO_2 and H_2O fluxes averaged over a particular time period, such as a week or a month, and then to apply 'errors' of different types and magnitudes to those mean values at different times of the day or night. The advantage of this approach is that it permits the unequivocal nature of the source or sink of carbon, for instance, to be stated with some justification. The net flux over the mean daily cycle is directly proportional to the overall net flux over the measurement period. This approach is valid as long as the basic character of the daily cycle does not vary significantly from day to day.

XII. CONCLUSIONS

To meet the requirement for information about biosphere–atmosphere interactions, several measurement networks have been originated. The EUROFLUX network, aiming at measuring long-term net carbon dioxide and water vapour exchanges between European forests and the atmosphere, was first initiated in 1996. It was followed in 1998 by the Ameriflux (northern and central America) and Medeflu (Mediterranean countries) networks, as well as the tentative global NETFLUX network, which pursue the same goals. Further networks in Japan, Australasia and southern America are planned. At present, in 1998, about 80 stations measuring CO_2 and H_2O fluxes exchanged by different ecosystems with the atmosphere are to be found around the world; the number is growing and there is a call for the stations to constitute a permanent network operating like surface weather station networks (Running *et al.*, unpublished).

The history of surface weather stations teaches us that an essential requirement for such a task is the standardization of the measurement procedure. In the present case, the eddy covariance method was chosen on all sites because it is the only method that allows direct measurement of the fluxes at the ecosystem scale (several hectares), at a half-hourly time-scale and continuously over several years. However, standardization of the eddy covariance method requires not only the choice of a common measurement system, but also the setting up of a procedure for computing, correcting, summing and checking the quality of the measurements.

In EUROFLUX, standardization of the measurement system was achieved from the beginning, as each team agreed to use the same system. It comprises a three-dimensional sonic anemometer coupled with a closed-path infrared gas analyser. Reliability, handling facilities but also commercial availability of the system motivated this choice. In addition to this, a standard measurement procedure had to be achieved. The aim of this chapter was to present this.

Computation of the fluxes from instantaneous measurements is a complex procedure (to give an idea, one half-hourly flux derives from more than 22 400 instantaneous measurements) that includes high-pass filtering, time-lag estimation, mean and (co)-variance computation, co-ordinate rotation, and conversion from covariances to fluxes. In this chapter we have proposed a standard way of achieving these operations and have presented a method for testing the software that performs them.

The fluxes are also subject to full or selective systematic errors due to instrument limitations or non-ideal meteorological conditions. Thus, a correction must be introduced. We have compared different approaches for correcting the fluxes and, finally, have proposed a standard method. In addition, we have proposed a series of tests for assessing the quality of the measurements as well as different methods that allow prediction of the spatial representativeness of the fluxes.

Half-hourly fluxes must be summed in order to give the annual net exchanges of CO_2 and H_2O. We have discussed the relevancy of different variables and have presented procedures for data gap filling. Finally, we described a way of estimating the error.

All the procedures described in this chapter represent the present state-of-the-art in EUROFLUX. They are followed by the 12 teams of the group. We propose them as a standard procedure to be used by all eddy covariance measuring stations.

ACKNOWLEDGEMENTS

This research was supported by European Commission, Programme Environment and Climate 1994–1998, Project EUROFLUX under contract ENV4-CT95-0078. This work is the result of many people's efforts. The authors acknowledge all the scientists as well as all the technicians who, thanks to their continuous work, made the realization of this paper possible (with apologies to people who have been forgotten here): Jean-Marc Bonnefond, Reinhart Ceulemans, Lars Christensen, Michael Courtney, Han Dolman, Patrick Gross, John Hansen, Paul Jarvis, Poul Hummelshoej, Niels-Otto Jensen, Petri Keronen, Fred Kockelbergh, Tapio Lahti, Anders Lindroth, Giorgio Matteucci, Eddy Moors, Toivo Pohja, S.L. Scott, Erkki Siivola, John Tenhunen and Michel Yernaux.

REFERENCES

Aubinet, M., Chermanne, B., Vandenhaute, M., Longdoz, B., Yernaux, M. and Laitat, E. (unpublished). Long term measurements of water vapour and carbon dioxide fluxes above a mixed forest in Ardenne's region.

Baldocchi, D. (1997). Flux footprints within and over forest canopies. *Boundary-Layer Meteorol.* **85**, 273–292.

Baldocchi, D. and Meyers, T.D. (1991). Trace gas exchange above the floor of a deciduous forest. 1. Evaporation and CO_2 efflux. (1991) *J. Geophys. Res.* **96**(D4), 7271–7285.

Baldocchi, D., Hicks., B.B. and Meyers, T.D. (1988). Measuring biosphere–atmosphere exchanges of biologically related gases with micrometeorological methods. *Ecology* **69**, 1331–1340.

Baldocchi, D., Valentini, R., Oechel, W. and Dahlman, R. (1996). Strategies for measuring and modelling carbon dioxide and water vapour fluxes over terrestrial ecosystems. *Global Change Biol.* **2**, 159–168.

Baldocchi, D., Vogel, C.A. and Hall, B. (1997). Seasonal variation of carbon dioxide exchange rates above and below a boreal jack pine forest. *Agric. For. Metereol.* **83**, 135–146.

Baldocchi, D., Wilson, K. and Paw, U.K.T. (unpublished). On measuring net ecosystem carbon exchange in complex terrain over tall vegetation.

Black, T.A., Den Hartog, G., Neumann, H.H., Blanken, P.D., Yang, P.C., Russell, C., Nesic, Z., Lee, X., Chen, S.G., Staebler R. and Novak, M.D. (1996). Annual cycles of water vapour and carbon dioxide fluxes in and above a boreal aspen forest. *Global Change Biol.* **2**, 219–229.

De Bruin, H.A.R., Bink, N.J. and Kroon, L.J.M. (1991). Fluxes in the surface layer under advective conditions. In: *Workshop on Land Surface Evaporation Measurement and Parametrization* (Ed. by T.J. Schmugge and J.C. André), pp. 157–169. Springer, New York.

Businger, J.A. (1986). Evaluation of the accuracy with which dry deposition can be measured with current micrometeorological techniques. *J. Clim. Appl. Meteorol.* **25**, 1100–1124.

Desjardins, R.L. (1985). Carbon dioxide budget of maize. *Agric. For. Meteorol.* **36**, 29–41.

Foken, Th. (1999a). The turbulence experiment FINTUREX at the Neumayer-Station/Antartica. *Ber. des Deutschen Wetterdienstes,* Berlin (in press).

Foken, Th. (1999b). Die scheinbar ungeschlossene Energiebilanz am Erdboden-, eine Herausforderung an die Experimentelle Meteorologie. *Sitzungsberichte der Leibnitz-Sozietaet,* Berlin (in press).

Foken, Th. and Wichura, B. (1996). Tools for quality assessment of surface-based flux measurements. *Agric. For. Meteorol.* **78**, 83–105.

Foken, Th., Skeib, G. and Richter, S.H. (1991). Dependence of the integral turbulence characteristics on the stability of stratification and their use for Doppler–Sodar measurements. *Z. Meteorol.* **41**, 311–315.

Foken, Th., Jegede, O.O., Weisensee, U., Richter, S.H., Handorf, D., Görsdorf, U., Vogel, G., Schubert, U., Kirzel, H.-J. and Thiermann, V. (1997). Results of the LINEX-96/2 experiment. *Deutscher Wetterdienst, Geschäftsbereich Forschung und Entwicklung, Arbeitsergebnisse* **48**, 75 pp.

Gardner, M.W. and Dorling, S.R. (1998). Artificial neural networks (the multilayer perception)—a review of applications in the atmospheric sciences. *Atmos. Environ.* **32**, 2627–2636.

Gash, J.H.C. and Culf, A.D. (1996). Applying linear detrend to eddy correlation data in real time. *Boundary-Layer Meteorol.* **79**, 301–306.

Goulden, M.L., Munger, J.W., Fan, S.-M., Daube, B.C. and Wofsy, S.C. (1996a). Measurements of carbon sequestration by long-term eddy covariance: methods and a critical evaluation of accuracy. *Global Change Biol.* **2**, 159–168.

Goulden, M.L., Munger, J.W., Fan, S.-M., Daube, B.C. and Wofsy, S.C. (1996b). Exchange of carbon dioxide by a deciduous forest: response to interannual climate variability. *Science* **271**, 1576–1578.

Grace, J., Malhi, Y., Lloyd, J., McIntyre, J., Miranda, A.C., Meir, P. and Miranda, H.S. (1996). The use of eddy covariance to infer the net carbon dioxide uptake of Brazilian rain forest. *Global Change Biol.* **2**, 209–217.

Greco, S. and Baldocchi, D. (1996). Seasonal variations of CO_2 and water vapour exchange rates over a temperate deciduous forest. *Global Change Biol.* **2**, 183–197.

Grelle, A. (1996). *SOLCOM: Gill Solent Communication software.* Technical note. Department for production ecology. Faculty of forestry. Swedish University of Agricultural Sciences, Uppsalla.

Grelle, A. (1997). *Long-term Water and Carbon Dioxide Fluxes from a Boreal Forest: Methods and Applications.* PhD thesis. Swedish University of agricultural studies, Uppsalla.
Grelle, A. and Lindroth, A. (1994). Flow distortion by a solent sonic anemometer: wind tunnel calibration and its assessment for flux measurements over forest and field. *J. Atmos. Oceanic Technol.* **11**, 1529–1542.
Grelle, A. and Lindroth, A. (1996). Eddy-correlation system for long term monitoring of fluxes of heat, water vapour, and CO_2. *Global Change Biol.* **2**, 297–307.
Grelle, A., Lohse, H. and Peters, G. (1994). Central ice station. In: *The Expedition ARKTIS-IX/1 of RV "Polarstern" in 1993. Berichte zur Polarforschung* 134.
Gurjanov, A.A., Zubkovskii, S.L. and Fedorov, M.M. (1984). Mnogoknal'naja avtomatizirov-annaja sistema obrabotki signalov no baze EVM. *Geod. Geophys. Veröff.* **RII**(26), 17–20.
Haataja, J. and Vesala, T. (1997). *SMEAR II, Station for Measuring Forest Ecosystem–Atmosphere Relation.* University of Helsinki, Department of Forest Ecology Publications, Helsinki, Finland.
Højstrup, J. (1993). A statistical data screening procedure. *Meas. Sci. Technol.* **48**, 472–492.
Hollinger, D.Y., Kelliher, F.M., Byers, J.N., Hunt, J.E., McSeveny, T.M. and Weir, P.L. (1994). Carbon dioxide exchange between an undisturbed old-growth temperate forest and the atmosphere. *Ecology* **75**(1), 134–150.
Horst, T.W. (1997). A simple formula for attenuation of eddy fluxes measured with first order response scalar sensors. *Boundary-Layer Meteorol.* **82**, 219–233.
Horst, T.W. and Weil, J.C. (1994). How far is far enough? The fetch requirements for micrometeorological measurement of surface fluxes. *J. Atmos. Oceanic Technol.* **11**, 1018–1025.
Houghton, J.J., Meiro Filho, L.G., Callander, B.A., Harris, N., Kattenberg, A. and Maskell, K. (1996). *Climate Change 1995: The Science of Climate Change.* Cambridge University Press, Cambridge.
Houghton, R.A., Davidson, E.A. and Woodwell, G.M. (1998). Missing sinks, feedbacks, and understanding the role of terrestrial ecosystems in the global carbon balance. *Global Biogeochem. Cycles* **12**, 25–34.
Huntingford, C. and Cox, P.M. (1997). Use of statistical and neural network techniques to detect how stomatal conductance responds to changes in the local environment. *Ecol. Model.* **80**, 217–246.
Ibrom, A., Morgenstern, K., Richter, I., Falk, M., Oltchev, A., Constantin, J. and Gravenhorst, G. (unpublished). The energy balance: implications for the quality assessment of eddy covariance measurements above a forest canopy.
Jarvis, P.G., Massheeder, J.M., Hale, S.E., Moncrieff, J.B., Rayment, M. and Scott, S.L. (1997). Seasonal variation of carbon dioxide, water vapor, and energy exchanges of a boreal black spruce forest. *J. Geophys. Res.* **102**(D24), 28953–28966.
Kaimal, J.C. and Finnigan, J.J. (1994). *Atmospheric Boundary Layer Flows: Their Structure and Measurement.* Oxford University Press, Oxford.
Kaimal, J.C. and Gaynor, J.E. (1991). Another look at sonic thermometry. *Boundary-Layer Meteorol.* **56**, 401–410.
Kaimal, J.C., Wyngaard, J.C., Izumi, Y. and Cote, O.R. (1972). Spectral characteristics of surface-layer turbulence. *Q.J.R. Meteorol. Soc.* **98**, 563–589.
Kristensen, L. (1998). *Time Series Analysis. Dealing with Imperfect Data.* Risø National Laboratory, Roskilde, Denmark.
Leclerc, M.Y. and Thurtell, G.W. (1990). Footprint prediction of scalar fluxes using a Markovian analysis. *Boundary-Layer Meteorol.* **52**, 247–258.

Lee, X. (1998). On micrometeorological observations of surface–air exchange over tall vegetation. *Agric. For. Meteorol.* **91**, 39–49.

Lenshow, D.H. and Raupach, M.R. (1991). The attenuation of fluctuations in scalar concentrations through sampling tubes. *J. Geophys. Res.* **96**, 5259–5268.

Leuning, R. and Judd, M.J. (1996). The relative merits of open- and closed-path analysers for measurements of eddy fluxes. *Global Change Biol.* **2**, 241–254.

Leuning, R. and King, K.M. (1992). Comparison of eddy-covariance measurements of CO_2 fluxes by open-and-closed-path CO_2 analysers. *Boundary-Layer Meteorol.* **59**, 297–311.

Leuning, R. and Moncrieff, J. (1990). Eddy-covariance CO_2 measurements using open- and closed-path CO_2 analysers: corrections for analyser water vapor sensitivity and damping of fluctuations in air sampling tubes. *Boundary-Layer Meteorol.* **53**, 63–76.

LI-COR (1991) *LI-6262 CO_2 H_2O Analyser Instruction Manual.* LI-COR, Lincoln, Nebraska.

Lindroth, A., Grelle, A. and Moren, A.-S. (1998). Long-term measurements of boreal forest carbon balance reveal large temperature sensitivity. *Global Change Biol.* **4**, 443–450.

Lloyd, C.R., Shuttleworth, W.J., Gash, J.H.C. and Turner, M. (1984). A microprocessor system for eddy-correlation. *Agric. For. Meteorol.* **33**, 67–80.

Lloyd, J. and Taylor, J.A. (1994). On the temperature dependence of soil respiration. *Funct. Ecol.* **8**, 315–323.

McMillen, R.T. (1986). *A BASIC Program for Eddy Correlation in Non Simple Terrain.* NOAA Technical Memorandum, ERL ART-147. NOAA, Silver Spring, Maryland.

McMillen, R.T. (1988). An eddy correlation technique with extended applicability to non-simple terrain. *Boundary-Layer Meteorol.* **43**, 231–245.

Mahrt, L. (1991). Eddy asymmetry in the sheared heated boundary layer. *J. Atmos. Sci.* **4**, 153–157.

Martin, P.H., Valentini, R., Jacques, M., Fabbri, K., Galati, D., Quaratino, R. *et al.* (1998). A new estimate of the carbon sink strength of EU forests integrating flux measurements, field surveys, and space observations: 0.17–0.35 Gt(C). *Ambio* **27**, 582–584.

Massman, W.J. (1991). The attenuation of concentration fluctuations in turbulent-flow through a tube. *J. Geophys. Res.* **96**, 5269–5273.

Mihalakakou, G., Santamouris, M. and Asimakopoulos, D. (1998). Modelling ambient air temperature time series using neural networks. *J. Geophys. Res.* **103**, 19509–19517.

Moncrieff, J.B., Mahli, Y. and Leuning, R. (1996). The propagation of errors in long-term measurements of land–atmosphere fluxes of carbon and water. *Global Change Biol.* **2**, 231–240.

Moncrieff, J.B., Massheder, J.M., de Bruin, H., Elbers, J., Friborg, T., Heusinkveld, B., Kabat, P., Scott, S., Soegaard, H. and Verhoef, A. (1997a). A system to measure surface fluxes of momentum, sensible heat, water vapour and carbon dioxide. *J. Hydrol.* **188–189**, 589–611.

Moncrieff, J.B., Valentini, R., Greco, S., Seufert, G. and Ciccioli, P. (1997b). Trace gas exchange over terrestrial ecosystems: methods and perspectives in micrometeorology. *J. Exp. Bot.* **48**, 1133–1142.

Monin, A.S. and Yaglom, A.M. (1971). *Statistical Fluid Mechanics, Mechanics of Turbulence.* MIT Press, Cambridge, Massachusetts.

Moore, C.J. (1986). Frequency response corrections for eddy correlation systems. *Boundary-Layer Meteorol.* **37**, 17–35.

Obukhov, A.M. (1960). O strukture temperaturnogo polja i polja skorostej v uslovijach konvekcii. *Izv. AN SSSR. ser. geofiz.* 1392–1396.

Peters, G., Claussen, M., Lohse, H., Grelle, A., Kornblüh, L. and Fischer, B. (1993). Ice floe station. In: *ARKTIS 1993. Ber. Nr. 11 Ber. aus dem ZMK. A: Meteorology*. Hamburg University, Department of Meteorology, Hamburg.

Rannik, Ü. and Vesala, T. (1999). Autoregressive filtering versus linear detrending in estimation of fluxes by the eddy covariance method. *Boundary-Layer Meteorol.* **91**, 259–280.

Rannik, Ü., Vesala, T. and Keskinen, R. (1997). On the damping of temperature fluctuations in a circular tube relevant to the eddy covariance measurement technique. *J. Geophys. Res.* **102**, 12789–12794.

Ruimy, A., Jarvis, P.G., Baldocchi, D. and Saugier, B. (1995). CO_2 fluxes over plant canopies and solar radiation: a review. *Adv. Ecol. Res.* **26**, 1–69.

Running, S.W., Baldocchi, D., Cohen, W., Gower, S.T., Turner, D., Bakwin, P. and Hibbard, K. (unpublished). A global terrestrial monitoring network scaling tower fluxes with ecosystem modeling and EOS satellite data.

Saigusa, N., Oikawa, T. and Liu, S. (1998). Seasonal variations of the exchanges of CO_2 and H_2O between a grassland and the atmosphere: an experimental study. *Agric. For. Meteorol.* **89**, 131–139.

Saxén, B. and Saxén, H. (1995). *NNDT—A Neural Network Development Tool. Version 1.2*. Åbo Akademi University, Åbo, Finland.

Schimel, S.D. (1995). Terrestrial ecosystems and the global carbon cycle. *Global Change Biol.* **1**, 77–91.

Schmid, H.P. (1994). Source areas for scalars and scalar fluxes. *Boundary-Layer Meteorol.* **67**, 293–318.

Schotanus, P., Nieuwstadt, F.T.M. and de Bruin, H.A.R. (1983). Temperature measurement with a sonic anemometer and its application to heat and moisture flux. *Boundary-Layer Meteorol.* **26**, 81–93.

Schuepp, P.H., Leclerc, M.Y., MacPherson, J.I. and Desjardins, R.L. (1990). Footprint prediction of scalar fluxes from analytical solutions of the diffusion equation. *Boundary-Layer Meteorol.* **50**, 355–373.

Shuttleworth, W.J. (1988). Corrections for the effect of background concentration change and sensor drift in real-time eddy correlation systems. *Boundary-Layer Meteorol.* **42**, 167–180.

Smith, E.A., Hodges, G.B., Bacrania, M., Cooper, H.J., Owens, M.A., Chappell, R. and Kincannon, W. (1997). *BOREAS Net Radiometer Engineering Study: Final Report*. Grant NAG5-4447. NASA.

Stull, R.B. (1988). *An Introduction to Boundary Layer Meteorology*. Kluwer Academic, Dordrecht.

Swinbank, W.C. (1951). The measurement of vertical transfer of heat and water vapour by eddies in the lower atmosphere. *J. Meteorol.* **8**, 135–145.

Valentini, R., Scarascia Mugnozza, G.E., De Angelis, P. and Bimbi, R. (1991). An experimental test of the eddy correlation technique over a Mediterranean macchia canopy. *Plant Cell Environ.* **14**, 987–994.

Valentini, R., De Angelis, P., Matteucci, G., Monaco, S., Dore, S. and Scarascia Mugnozza, G.E. (1996). Seasonal net carbon dioxide exchange of a beech forest with the atmosphere. *Global Change Biol.* **2**, 199–207.

Verma, S.B., Baldocchi, D.D., Anderson, D.E., Matt, D.R. and Clement, R.J. (1986). Eddy fluxes of CO_2, water vapor and sensible heat over a deciduous forest. *Boundary-Layer Meteorol.* **36**, 71–91.

Verma, S.B., Kim, J. and Clement, R. (1989). Carbon dioxide, water vapor and sensible heat fluxes over a tallgrass prairie. *Boundary-Layer Meteorol.* **46**, 53–67.

Vermetten, A.W.M., Ganzenveld, L., Jeuken, A., Hofschreuder, P. and Mohren, G.M.J. (1994). CO_2 uptake by a stand of Douglas fir: flux measurements compared with model calculation. *Agric. For. Meteorol.* **72**, 57–80.

Vickers, D. and Mahrt, L. (1997). Quality control and flux sampling problems for tower and aircraft data. *J. Atmos. Oceanic Technol.* **14**, 512–526.

Webb, E.K., Pearman, G.I. and Leuning, R. (1980). Correction of flux measurements for density effects due to heat and water vapour transfer. *Q. J. R. Meteorol. Soc.* **106**, 85–100.

Wesely, M.L. and Hart, R.L. (1985). Variability of short term eddy-correlation estimates of mass exchange. In: *The Forest–Atmosphere Interaction* (Ed. by B.A. Hutchison and B.B. Hicks), pp. 591–612. D. Reidel, Dordrecht.

Wichura, B. and Foken. Th. (1995). *Anwendung integraler Turbulenzcharakteristiken zur Bestimmung von Beimengungen in der Bodenschicht der Atmosphäre.* DWD, Abteilung Forschung, Arbeitsergebnisse, No. 29 Deutscher wetterdierst, Offenbuch.

Wilson, J.D. and Sawford, B.L. (1996). Review of lagrangian stochastic models for trajectories in the turbulent atmosphere. *Boundary-Layer Meteorol.* **78**, 191–210.

Wofsy, S.C., Goulden, M.L., Munger, J.W., Fan, S.-M., Bakwin, P.S., Daube, B.C., Bassow, S.L. and Bazzaz, F.A. (1993). Net exchange of CO_2 in a mid-latitude forest. *Science* **260**, 1314–1317.

Wyngaard, J.C., Coté, O.R. and Izumi, Y. (1971). Local free convection, similarity and the budgets of shear stress and heat flux. *J. Atmos. Sci.* **28**, 1171–1182.

APPENDIX A

Computation of the Two-axis Rotation Matrix Elements

Two rotations are successively applied around the z and y axes. The first aligns u into the x direction in the x–z plane and nullifies v, according to:

$$u_{1,i} = \sum_j A_{01,i,j} \cdot u_{0,j} \tag{A1a}$$

with

$$A_{01} = \begin{pmatrix} \cos\theta & \sin\theta & 0 \\ -\sin\theta & \cos\theta & 0 \\ 0 & 0 & 1 \end{pmatrix} \tag{A1b}$$

and where the rotation angle, its cosine and sine are:

$$\theta = \tan^{-1}\left(\frac{\bar{v}_0}{\bar{u}_0}\right), \quad \cos\theta = \frac{\bar{u}_0}{\sqrt{\bar{u}_0^2 + \bar{v}_0^2}}, \quad \sin\theta = \frac{\bar{v}_0}{\sqrt{\bar{u}_0^2 + \bar{v}_0^2}} \tag{A1c}$$

The second rotation forces u to point along the mean wind direction. It nullifies w according to:

$$u_{2,j} = \sum_j A_{12,i,j} \cdot u_{i,j} \tag{A2a}$$

with

$$A_{12} = \begin{pmatrix} \cos\phi & 0 & \sin\phi \\ 0 & 1 & 0 \\ -\sin\phi & 0 & \cos\phi \end{pmatrix} \tag{A2b}$$

and where the rotation angle is:

$$\phi = \tan^{-1}\left(\frac{\overline{w}_1}{\overline{u}_1}\right) \tag{A2c}$$

That is, in terms of the non-rotated components of the velocity:

$$\phi = \tan^{-1}\left(\frac{\overline{w}_0}{\sqrt{\overline{u}_0^2 + \overline{v}_0^2}}\right), \quad \cos\phi = \frac{\sqrt{\overline{u}_0^2 + \overline{v}_0^2}}{\sqrt{\overline{u}_0^2 + \overline{v}_0^2 + \overline{w}_0^2}}, \quad \sin\phi = \frac{\overline{w}_0}{\sqrt{\overline{u}_0^2 + \overline{v}_0^2 + \overline{w}_0^2}} \tag{A2d}$$

The two-rotation matrix is obtained by multiplication of the single rotation matrices:

$$A_{02} = A_{12} \cdot A_{01} \tag{A3}$$

Finally, by introducing in equation (A3) the expressions (A1b) and (A2b) of the single-axis rotation matrix, in which the sine and cosine functions are expressed according to (A1c) and (A2d), we obtain:

$$A_{02} = \begin{pmatrix} \dfrac{\overline{u}_0}{\sqrt{\overline{u}_0^2 + \overline{v}_0^2 + \overline{w}_0^2}} & \dfrac{\overline{v}_0}{\sqrt{\overline{u}_0^2 + \overline{v}_0^2 + \overline{w}_0^2}} & \dfrac{\overline{w}_0}{\sqrt{\overline{u}_0^2 + \overline{v}_0^2 + \overline{w}_0^2}} \\ -\dfrac{\overline{v}_0}{\sqrt{\overline{u}_0^2 + \overline{v}_0^2}} & \dfrac{\overline{u}_0}{\sqrt{\overline{u}_0^2 + \overline{v}_0^2}} & 0 \\ -\dfrac{\overline{u}_0\overline{w}_0}{\sqrt{\overline{u}_0^2 + \overline{v}_0^2 + \overline{w}_0^2}\sqrt{\overline{u}_0^2 + \overline{v}_0^2}} & -\dfrac{\overline{v}_0\overline{w}_0}{\sqrt{\overline{u}_0^2 + \overline{v}_0^2 + \overline{w}_0^2}\sqrt{\overline{u}_0^2 + \overline{v}_0^2}} & \dfrac{\sqrt{\overline{u}_0^2 + \overline{u}_0^2}}{\sqrt{\overline{u}_0^2 + \overline{v}_0^2 + \overline{w}_0^2}} \end{pmatrix} \tag{A4}$$

APPENDIX B

Computation of the Third Axis Rotation

The third rotation is performed around the x axis:

$$\begin{pmatrix} u_3 \\ v_3 \\ w_3 \end{pmatrix} = \begin{pmatrix} 1 & 0 & 0 \\ 0 & \cos\psi & \sin\psi \\ 0 & -\sin\psi & \cos\psi \end{pmatrix} \cdot \begin{pmatrix} u_2 \\ v_2 \\ w_2 \end{pmatrix} \tag{B1}$$

and nullifies the $v'w'$ covariances. Kaimal and Finnigan (1994) showed that the angle that fulfilled this condition is:

$$\psi = \frac{1}{2}\tan^{-1}(2Y), \text{ with } Y = \frac{\overline{v_2 w_2}}{\overline{v_2^2} - \overline{w_2^2}} = \frac{\overline{v'_2 w'_2}}{\overline{v'^2_2} - \overline{w'^2_2}} \tag{B2}$$

Consequently, the sine and cosine functions are:

$$\cos\psi = \left(\frac{1+(1+4Y^2)^{-1/2}}{2}\right)^{1/2}, \quad \sin\psi = \left(\frac{1-(1+4Y^2)^{-1/2}}{2}\right)^{1/2} \tag{B3}$$

and the matrix A_{23}:

$$A_{23} = \begin{pmatrix} 1 & 0 & 0 \\ 0 & \left(\dfrac{1+(1+4Y^2)^{-1/2}}{2}\right)^{1/2} & \left(\dfrac{1-(1+4Y^2)^{-1/2}}{2}\right)^{1/2} \\ 0 & -\left(\dfrac{1-(1+4Y^2)^{-1/2}}{2}\right)^{1/2} & \left(\dfrac{1+(1+4Y^2)^{-1/2}}{2}\right)^{1/2} \end{pmatrix} \tag{B4}$$

The Cost of Living: Field Metabolic Rates of Small Mammals

J.R. SPEAKMAN

I. Summary 178
II. Introduction 179
    A. The Importance of Energy in Living Systems 179
    B. Limitations on Animal Energy Expenditure 181
    C. The Extrinsic Limitation Hypothesis 189
    D. The Intrinsic Limitation Hypothesis 192
    E. Experimental Studies of the Limitation Hypotheses 197
    F. The Central Limitation Hypothesis and Links between
       FMR and RMR 200
    G. Interspecific Reviews of the Link between DEE and RMR 202
    H. Summary and Aims 204
III. Methods 205
    A. Measuring Energy Expenditure by Indirect Calorimetry 205
    B. Basal and Resting Energy Expenditure 206
    C. Time and Energy Budget Estimates of Daily Energy Expenditure . . 209
    D. Direct Measurements of Free-living Energy Expenditure 212
    E. Summary and Data Inclusion Criteria for the Present Review 217
IV. Results 218
    A. Overview of the Database 218
    B. Factors Influencing Daily Energy Expenditure 221
V. Discussion 242
    A. Links Between FMR and RMR 242
    B. Sustainable Metabolic Scope 255
Acknowledgements 264
References 264
Appendix A 283
Appendix B 289
Appendix C 294

I. SUMMARY

Energy is a universal currency that is required for all biological functions. It has been frequently suggested that the rates at which animals can expend energy are limited in some manner. These limits may be extrinsic, set by the availability of energy resources in the environment, combined with the inability of animals to harvest these resources effectively. Alternatively the limits may be set intrinsically by aspects of the animal's physiology. Some evidence from field manipulations of birds and insects suggests that the limits are more likely to be intrinsic than extrinsic. Intrinsic limits may be set centrally (e.g. by capacities of the alimentary tract or the respiratory system's ability to provide oxygen) or peripherally at the sites where energy is utilized (such as the muscles during exercise, brown adipose tissue during thermoregulation, or mammary tissue during lactation).

In the late 1980s it was widely believed that the intrinsic limitations were centrally mediated. In part this belief originated in observations that free-living energy demands of animals appeared to be linked to the level of their basal metabolic rates. This link was believed to arise because the major tissues contributing to basal metabolism are those that limit sustainable metabolism, namely the alimentary tract and associated structures such as the liver and kidneys. It was suggested that the physiological aspects of this linkage impose a limit on expenditure at around either 4× or 7× basal metabolism. Laboratory manipulations of lactating mice have confirmed a link between sustainable expenditure and basal metabolism but simultaneously cast severe doubt on the notion of a central limit, and it is currently more widely believed that limits on sustainable expenditure are set peripherally.

In this review I have summarized 185 measurements of the cost of living (daily energy expenditure or field metabolic rate) made on 73 species of small mammals (weighing less than 4 kg), using the doubly labelled water method. Field metabolic rate (FMR) was dependent on body mass, ambient temperature, and latitude of the study site. I confirmed that the effect of body mass was not a statistical artefact of using species as the sampling unit, by calculating the phylogenetically independent contrasts of mass and FMR. Diet affected the FMR in the total sample of 185 measurements, but not in a sample where each species was entered only once. Phylogeny had no significant effect in either sample. A predictive equation was derived which allows estimation of energy demands of free-living small mammals from knowledge of their body mass, combined with the ambient temperature and latitude of the study site, which has a mean error rate of 21%. This was five times more precise than predictions using body mass alone (mean error rate 124%). Estimates of resting metabolic rates in the thermoneutral zone (RMR) were available for 60 of the 73 species. Resting metabolism was also dependent on body mass and latitude of the site where the subjects were collected. The effect of mass remained when the phylogenetically independent contrasts of mass and RMR were calculated. There was a strong relationship between FMR and RMR. However, this relationship was much weaker when the shared effect of mass was removed. There was a significant relationship between the residuals of the phylogenetic contrast relationships of FMR and RMR to body mass. When the shared variance due to temperature and latitude was removed, the relationship was further weakened, and dependent only on two outlying data points.

Evidence supporting a link between FMR and RMR is very weak. The ratio of FMR to RMR averaged 3.4. The maximum ratio was 7.3 and the minimum was 1.6. Twenty per cent of values exceeded the postulated limit at 4× RMR. Most small mammals are working at well below their supposed physiological limits. In this sample there was no link between the ratio of FMR to RMR and body mass, and no evidence that the ratio was lower in the Chiroptera than in other orders.

The absence of these associations does not support the trade-off explanation for the observed discrepancy between habitual levels of expenditure and the supposed limits. Expenditures by semelparous marsupial mice were greater than the average ratio, supporting the trade-off hypothesis. There was a significant link between the ratio and diet. Small mammals exploiting more energy-abundant resources had greater ratios. These latter data support the extrinsic limitations hypothesis.

Overall these data provide no support for the notions of intrinsic physiological limits acting at either 4× or 7× RMR. Such limitations may be illusory. Independent evaluations of limits on performance from ultra-distance runners and migrating birds suggest that animals may be physiologically capable of expending up to between 9× and 15× RMR daily. They routinely do not do this because of limits imposed by the extrinsic supplies of energy and possibly trade-offs with future fecundity. In small mammals the former may be most important, but in other groups (for example, birds) trade-offs may dominate.

II. INTRODUCTION

A. The Importance of Energy in Living Systems

Living things are the most complex and organized structures that we are currently aware of. Chemically the elemental composition of living objects is not particularly noteworthy. It is the complexity and organization of these elements that sets the living and non-living apart. However, there is a potential downside to being so complex and organized: the second law of thermodynamics. This is the inexorable trend towards increased entropy in a closed system. Fortunately, living things are not closed systems. They are consequently able to maintain a high level of complexity and organization for a sufficient time to enable them to reproduce, and thereby propagate the genetic information they are carrying into the future gene pool. They do this by taking materials from outside their own systems and using these materials to shore up and repair the perpetually decaying fabric of their organization. Eventually, however, once the propagation of genes has been completed, the rates of repair decline, but the rates of decay do not (Barnett and King, 1995). It has been calculated, for example, that the DNA in every cell is hit by oxidative damaging agents approximately 10^4 times each day (Ames et al., 1995). Our capacity to repair this damage is considerably greater in our midlife than in later life (Barnett and King, 1995). Physiological attrition of our entire system consequently increases (Meyer et al., 1991) until eventually the organization becomes reduced to such an extent that the unspectacular amalgamation of elements becomes just that. It is no longer sufficiently organized to be classed as an animate object. Life and death are convenient points of reference along a continuum of complexity and organization. Life is the quasi-steady state in which the rate of repair matches the rate of degradation. Death is the unsteady state that results when it does not.

Automobiles are also fairly complex structures. They are not quite so organized and complex as living things, but at a lower level they illustrate the need of any organized system to be continually maintained. Left alone, without any maintenance, automobiles gradually decay, until eventually they fail to function as vehicles any more. Maintenance of complexity in a car, or a living system, requires two distinctly different activities to be performed continuously, and two distinct types of material are involved. Anyone who has had a car repaired at a garage will be familiar with these two separate activities when they have received a bill separated into the costs for parts and labour.

Parts and labour are the two things that all organized systems need: materials (parts) to repair damaged components of the existing system, and the energy (labour) to perform the work necessary to remove the damaged pieces and replace them with undamaged bits. Living systems require nutrients in the form of macronutrients, such as protein and fat, or micronutrients, such as vitamins and trace elements. It is important, however, to distinguish between the elements of nutrition that concern replacement of the essential building blocks of the system (parts) and the nutrients whose role is to supply the energy necessary to integrate the building blocks into the system as it is repaired (labour). The requirement for energy to sustain the complexity and organization of life is a distinct requirement from the needs for specific essential building blocks. Energy is often called a nutrient, but I think this usage obscures the very real difference between energy and other essential aspects of the diet. I think this confusion arises in part because animals can mobilize parts of their body that have been taken in as essential building blocks and use them to provide energy, if they require to do so. Normally, however, this is only a desperate response to a shortage of energy intake in the diet. It is a bit like selling the seats in your car to pay for some petrol—a measure that would work if needed, but not something you would (could) do every day. By using this analogy I do not mean to imply that animals do not store energy in their bodies for use at some later time, as most animals do store energy in their bodies as lipids, and withdraw these reserves to cover shortfalls in the external food supply. In the above analogy this is equivalent to the petrol in the fuel tank, not the seats.

Energy is required by animals not only to sustain the complexity of their systems, it is also required to allow them to perform work and also to maintain homoeostasis. Animals perform work whenever they do anything at that involves movement. During movement animals transform chemical energy into kinetic energy, potential energy and heat. Ultimately the kinetic and potential energy also becomes heat. This requirement for energy includes all the movement going on inside the animal: pumping of the heart, breathing in and out, contracting the alimentary tract, as well as all the more visible and audible aspects of movement such as running, jumping, fighting, copulating, singing and flying. All movements made by living things, no matter how trivial, require energy to be performed.

Animal systems also generally exist a long way from natural equilibria. Animals maintain non-equilibrium ionic potentials in their cells, for example, and their body temperatures are perturbed from the levels they would reach if the system were left to reach equilibrium. To maintain these non-equilibrium states, animals must also use energy, for example to operate ionic pumps in cells to sustain the ionic gradients (Swaminathan *et al.,* 1982, 1989; Poehlman *et al.,* 1993). The ubiquitous demands for energy to fuel all biological processes means that energy is a fundamentally important currency of life (Kleiber, 1961; Bartholomew, 1982; Brafield and Llewellyn, 1982; Blaxter, 1989; McNeil Alexander, 1999).

B. Limitations on Animal Energy Expenditure

On 16 June 1999, during the Tsiklitiria International Track Meeting in Athens, Greece, the 100-m sprint for men was won by the American Maurice Greene in a time of 9.79 s. This was the fastest time that a human being had ever covered the distance entirely under their own steam (and without the assistance of drugs—the Canadian Ben Johnson has also clocked this time but his run is not formally recognized as a record). Several animal species are also capable of making very fast sprints of short duration. Cheetah *(Acinonyx jubatus)*, for example, can sprint at speeds approaching 75 km h^{-1} for short periods. Measurements of energy expenditure during sprints by cheetah suggest it might be as high as 50 times the resting metabolic rate (RMR), measured inside the thermoneutral zone (Taylor, 1974). Rheas *(Rhea americana)* also expend energy at over 35× RMR when sprinting (Bundle *et al.,* 1999).

In the Athens games the 400-m sprint race was run in 44.7 s. This was also a remarkable running feat, but it was not a world record. The fastest that an athlete has ever run 400 m was in Seville, Spain, in 1999 when Michael Johnson covered the distance in 43.18 s. Although it covers four times the distance, Johnson's record is 4.02 s slower than four times the 100-m record time set by Greene. The 400-m athlete cannot sustain the same power output as a 100-m athlete even though the race still lasts for well under 1 min. This phenomenon is further exemplified by the fact that, despite the potential hindrance of baton changes, the 4 × 100-m relay (37.4 s) is run almost 6 s faster than a single person can run 400-m. (It is run faster than four times the 100-m race because athletes have running starts at three of the four sections). Figure 1 shows a plot of the speeds averaged during current (1999) world records for men running different distances, against the duration of the races. This figure demonstrates that there is a progressive decline in the speeds achieved by world record-breaking athletes as the distance and duration over which they run increases.

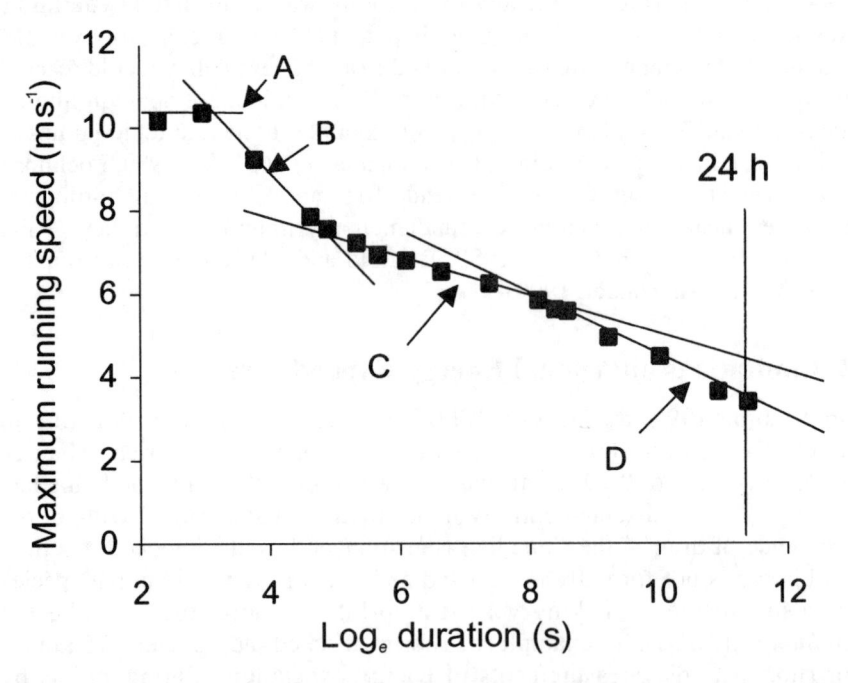

Fig. 1. Maximum running speed (m s^{-1}) attained by men in competitive athletics events as a function of $\log_e$ duration (s) of the event. There is a progressive decrease in the achieved maximum speeds. The pattern of decline follows four distinct phases, which appear to reflect different physiological limitations on the capacity to expend energy.

It may seem obvious to state that a marathon runner cannot run the standard marathon distance of 26 miles 385 yards at the same pace as a 100-m runner. As we can be reasonably confident, however, that the people in all these races, independent of the distance, are running as fast as they possibly can, the implication of this observation is that there are limitations on the physiological ability of humans to expend energy. In the context of running athletes, these limitations appear to be related to the availability of metabolites to support the running activity (Farfel, 1960; McGillvery, 1971) and the capacity of tissues to utilize these metabolites and transform them into useful power output.

The shape of the curve relating maximum running speed to distance indicates that there is a substrate supporting very intense short duration activities (such as 100 m sprinting) which in a fit adult human, exercising at maximum capacity, lasts for approximately 20 s (Figure 1, line A). This substance is creatine phosphate. When the muscles contract and utilize adenosine triphosphate (ATP), the concentration of adenosine diphospate (ADP) in the cells rises. The ADP reacts with creatine phosphate generating creatine and reforming ATP, which is available to support further contraction. In 20 s of peak activity the supply of creatine phosphate is exhausted and a different pathway must then supply the ATP (Goldstein, 1977). It is well established, in greyhound racing circles, that feeding the dogs a diet rich in meat enhances their running performance. It has recently been shown that this performance change is not due to increased protein in the diet (Hill *et al.*, submitted). Greyhound races typically last about 30 s, thus supply of creatine phosphate is probably critically important for performance. The improvement in running times when dogs are fed meat may thus be dependent on a change in creatine phosphate levels in muscles that accompanies meat feeding. Food-supplement manufacturers have not been slow to recognize the potentially pivotal importance played by the supply of creatine phosphate in sprinting performance, and it is now possible to buy pure creatine as a dietary supplement for human athletes to 'ensure Olympian performance'. Whether orally administered creatine leads to raised muscle levels of creatine phosphate, however, remains to be demonstrated.

Once creatine phosphate has been exhausted, the switch to a second pathway, which can support only lower rates of ATP generation, leads to a different gradient relating speed against duration. This gradient appears to remain the same for exercise up to about 5 min (Figure 1, line B). During this phase of activity ADP concentration increases because it is no longer phosphorylated to ATP by creatine phosphate. The increase in ADP stimulates the production of pyruvate from glucose, which, via the tricarboxylic acid (TCA) cycle, produces ATP by oxidative phosphorylation. However, the enzymes catalysing conversion of glucose to pyruvate exceed the capacity of mitochondrial enzymes to catalyse its oxidation. The excess pyruvate is converted to lactate, by lactate dehydrogenase, recycling nicotinamide–adenine dinucleotide (NAD) from

NADH and simultaneously converting some ADP to ATP. The production of lactate therefore allows a greater rate of production of ATP than by oxidative phosphorylation alone. Measurements of lactate accumulation, however, suggest that this route can be sustained for only about 3–4 min. Thereafter the production of ATP is supported entirely by oxidative phosphorylation. This corresponds to the third phase in the relationship of running speed to duration (Figure 1, line C).

Readers who have competed in marathon events may be surprised that the curve relating record speeds to duration gives no indication of a decline in performance at around 1.75 h. For many amateur marathon runners this represents a time when substrate availability appears to decline, and the runner appears to hit a 'wall' where sustaining further activity at the same pace is extremely difficult. The absence of this phenomenon in the world record data may, however, only suggest that individuals capable of performing at world record levels do not hit this limit until much later. Measurements of substrate utilization suggest that there is a progressive shift, from using predominantly glycogen, to predominantly fat, to fuel oxidation as duration of exercise increases (Edwards *et al.*, 1934). The 'wall' probably reflects exhaustion of glycogen reserves and difficulty in mobilizing fat at the same high rate as glycogen. Many marathon athletes pre-load their systems with carbohydrate to boost their glycogen stores before performing to prevent hitting this substrate limitation before the end of the race.

Competitive human running events, which are commonly competed for in international games, last a maximum of 2.5 h. The pattern of speeds for these records establishes that physiological barriers limit energy expenditure over these time periods. Extrapolation of Figure 1 (line C) suggests that if a competitive event existed that involved running continuously for 24 h, and if no further substrate changes occurred, humans could probably sustain a running speed of about 4.0 m s^{-1} and the race would cover a distance of about 346 km. This would be equivalent to eight sequential marathons in a single day.

The furthest distance ever run to date (1999) in 24 h was 295.03 km. A Greek called Yiannis Kouros in Canberra, Australia, achieved this on 1–2 March 1997. The discrepancy between the predicted and observed distances may indicate that further changes in substrate utilization occur after 2.5 h (possibly a delayed 'wall' effect; Figure 1, line D), or it may be that this event is less keenly competed for than the events of shorter duration. We do not know for certain how much energy Yiannis Kouros expended during his 24-h record-breaking run. However, we can make an estimate because there have been many studies of the energy costs of human locomotion as a function of running speed. Given the distance he ran, and the time it took him, we know that Yiannis ran at an average speed of 3.4 m s^{-1}. Using the relationship between running speed and oxygen consumption (Margaria *et al.*, 1963), a person sustaining this speed for 24 h would consume a total of about 5040 litres of oxygen. Assuming a respiratory quotient of 0.85 leads to a predicted energy expenditure of 101 MJ (see the section on Methods below and Appendix A for details of the methodology for measuring oxygen consumption and its conversion to energy expenditure). The human resting metabolic rate at thermoneutral is about 7.1 MJ per 24 h (Schmidt Nielsen, 1975) and thus Yiannis was probably expending energy at a rate of about $101.2/7.1 = 14.2 \times$ RMR. This is probably the maximum level of energy expenditure that could be sustained over a 24-h period.

Numerous other animal species emulate this level of performance. Several species of migrating birds, for example, may fly continuously for protracted periods when they cross areas of open ocean. Barnacle geese (*Branta leucopsis*) migrating from the islands of Svalbad well above the Arctic Circle to spend the winter months in Scotland are a well studied example (Butler *et al.*, 1998). Satellite tracking studies have shown that these birds take off from Svalbad and fly in several stages down the Norwegian coast, occasionally flying continuously for up to 14 h, until they reach their winter habitat in south-west Scotland (55°N) (Butler *et al.*, 1998). Because flight is energetically expensive, with the energy costs normally ranging between 13 and 18 times resting metabolic rate (Tucker, 1966; Thomas, 1975; Masman and Klaassen, 1987; Rothe *et al.*, 1987; Butler and Bishop, 1998; Winter and von Helversen, 1998), these animals are probably expending energy at this level for the majority of the duration of the trip. Measurements of heart rate in migrating barnacle geese (Butler *et al.*, 1998) suggest that their heart rates are actually substantially lower than those recorded in the same species trained to fly behind a truck (Butler and Woakes, 1980). This might suggest that flight during migration is much cheaper than that reported in other studies. However, there are several alternative explanations for the low heart rate, and at present it is not possible to extrapolate the energy demands of migrating geese from the heart rates of geese flow in different situations (Butler *et al.*, 1998). Indeed, by making assumptions about oxygen extraction efficiency and stroke volume, and combining these with the observed heart rate, Butler *et al.* (1998) estimated that the oxygen consumption during flight was 302 ml min^{-1} at the start and 215 ml min^{-1} at the end of the flight. These values are equivalent to approximately 21 and 18 times the resting oxygen consumption at the same stages respectively. Since some birds flew for between 60% and 80% of the time on migration, this would imply sustained expenditure of between 13 and 15 times resting metabolic rate over the entire 2.5–4-day trip, very similar to the levels expended by a human running for an entire day.

Although mammals weighing around 100 kg (such as humans) and birds can perhaps sustain maximum energy expenditures of around 12–15× RMR, it would appear that smaller terrestrial mammals do not have the same capabilities. Taylor (1981) summarized the data across all species of the maximum reported levels of oxygen consumption, and found that maximum oxygen consumption (V_{O_2} max) scaled with body mass with the relationship:

$$V_{O_2} \text{ max (ml O}_2 \text{ min}^{-1}) = 0.43 M_b (g)^{0.81}$$

whereas the standard or resting metabolic rate scales with the relationship:

$$V_{O_2} \text{ (SMR) (ml O}_2 \text{ min}^{-1}) = 0.063 M_b (g)^{0.75}$$

(after Kleiber, 1975) where M_b is body mass. Because the exponents of these relationships are slightly different, the curves describing these trends diverge slightly as mass increases. This means the V_{O_2} max as a multiple of resting metabolic rate also increases as animals gets bigger. This effect can be illustrated by considering predicted levels of oxygen consumption for a 70-kg and for a 70-g animal. For the 70-kg animal the ratio is 13.3× (close to the value predicted for a human based on running speeds). However, the predicted ratio for a 70-g animal is only 8.8×. This lower ratio for small terrestrial mammals is reinforced by the review of McMillan and Hinds (1992), who compared the maximum oxygen consumption of rodents during exercise with the resting oxygen consumption. Their results are plotted in Figure 2. In this sample of animals the maximum oxygen consumption induced by exercise followed a curve that was parallel to the resting metabolic rate curve, leading to a fixed ratio between the two of about 7×. This lends support to the idea that, for terrestrial small mammals, the maximum daily energy expenditure may be more constrained at the upper margin than the expenditure of either larger mammals or birds (and presumably bats).

History does not record what Yiannis Kouros did on 3 March 1997, the day after his record-breaking 24-h run. It is a pretty safe bet, however, that he did not immediately set out on a repeat of his performance. Similarly, migrating birds cannot migrate over expanses of open ocean endlessly. They must prepare for the trips by depositing energy stores which they utilize during the flight (Marsh, 1983; Biebach, 1998; Butler et al., 1998; Jenni and Jenni-Eierman, 1998), and perhaps must recuperate once they arrive at their destination. The performances of 24-h record-breaking distance runners and transoceanic migrants are consequently not sustainable feats, since the energy expenditure over the periods when the heavy exercise occurs cannot be sustained indefinitely.

There has been considerable interest in the levels of the maximum average daily rates of energy expenditure that animals can maintain for indefinite periods. In practice this means the maximum expenditure that animals can engage in over protracted periods, whilst still performing all the behaviours essential for longer-term survival (for example, sleeping sufficiently to avoid prolonged sleep deprivation). Moreover they must intake

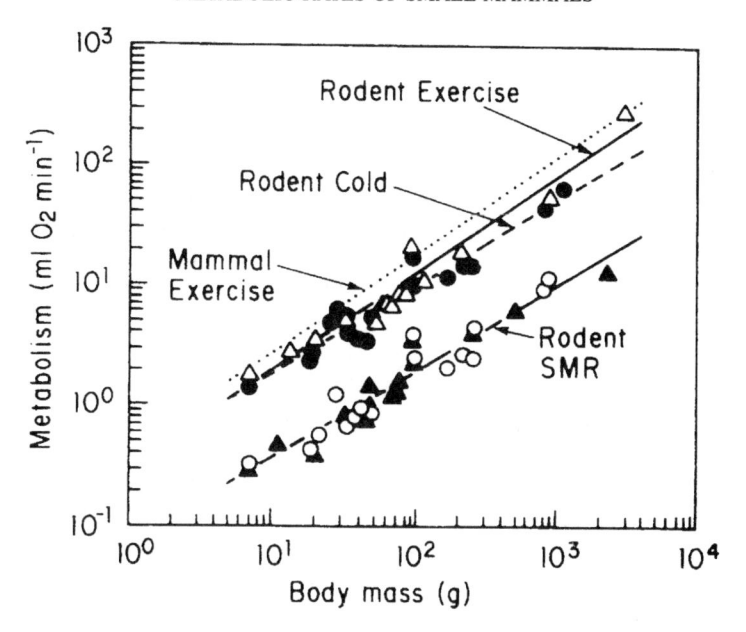

Fig. 2. Maximum oxygen consumption of small rodents in relation to resting (standard) metabolic rate (SMR). Mammals stressed by cold temperatures (and mixtures of helium and oxygen); △, mammals stressed by exercise. Reversed symbols (○, ▲) reflect the corresponding standard metabolic rate. The levels for exercise exceed those for cold exposure and indicate a maximum expenditure of around 7–8× RMR. From McMillan and Hinds (1992).

sufficient food that the energy expenditure is covered completely, i.e. the animal is not relying on depletion of stored energy reserves to support the high rates of expenditure. When animals perform at this level, it is generally called the maximum sustainable metabolic rate (Peterson *et al.*, 1990; Hammond and Diamond, 1997).

The reasons why ecologists are interested in this maximum sustainable level of energy expenditure is that it forms an interesting upper boundary on the sum total of activities in which an animal can engage. Because all activities require energy, the summed requirements of the activities that an animal performs must fit within the envelope of the total sustainable energy requirements (apart from interesting multiple uses of resources such as using the heat generated from activity to pay the costs of thermoregulation; e.g. Paladino and King, 1984; Webster and Weathers, 1990; Zebra and Walsberg, 1992). If we knew what the sustainable limits were, this would provide a powerful tool for predicting the limitations on animal performance.

For example, imagine that we know a given animal species is capable of expending a maximum sustainable energy expenditure of 40 kJ each day. If the animal is an endotherm, we also know that its energy requirements increase in relation to decreases in ambient temperature. If each drop in temperature by 1°C below the thermoneutral zone increased daily energy demands by 1 kJ, and the basal requirement was 10 kJ, then we would know that the species in question could not survive for protracted periods in areas with ambient temperatures more than 30°C below its lower critical temperature. Imagine the same animal requires 5 kJ per day to raise an offspring. In warm regions the spare capacity to expend energy beyond that committed to thermoregulation might allow animals to raise up to four offspring, but in cold conditions the animals may be able to raise only two, and in some regions there may be no spare capacity for reproduction at all, although survival might be possible. The notion of a sustainable maximum energy expenditure therefore provides an attractive framework for understanding many aspects of animal ecology, such as geographical distributions and breeding ranges (Root, 1988; Bozinovic and Rosenmann, 1989). Most of life history theory is founded on the assumption that energy resources are limited and animals must therefore trade-off their use of these resources (Gadgil and Bossert, 1970; Pianka and Parker, 1975; Stearns, 1976; 1983; 1993; Calow, 1979; Townsend and Calow, 1981).

Part of the problem in sustaining their energy expenditures at 12–15× basal metabolic rate (BMR) for transoceanic migrant birds, and perhaps also for ultra-long distance runners, is the fact that while they are engaged in their high rates of energy expenditure they are unable to feed, and perhaps also to sleep properly. This is particularly the case for the birds, because over the open ocean, at the high altitudes at which they migrate, there is no food available. Runners also need to stop to consume adequate amounts of food and to sleep. Consequently, in both examples, the high rates of expenditure over 24 h cannot be sustained indefinitely, because over the long term the expenditure cannot be balanced by sufficient energy intake.

Ultra-long distance runners compete in events that normally last for up to 6 days, although there are some exceptional events that last for over 20 days and cover up to 3100 km (for example, the Sri Cinnmoy event in the USA). The relationship between speed and distance for these events follows a different slope, indicating further limits on the performance capabilities as the duration of events increases (Figure 1, line D). This is mostly related to a decrease in the proportion of the period that athletes are able to remain actively running, as opposed to a decline in the running speed. The 6-day world distance record is also held by Yiannis Kouross. Between 2 and 8 July 1984 he covered 1022 km in New York. Note this is an average of 170 km per day, which is only 58% of the 24-h record distance set by the same man. Assuming that he was running for 60% of each day and had an expenditure during this time of 14.2× RMR (above) and that during the rest of the time he expended energy at only 2× RMR, over this more prolonged period his expenditure was probably down to around 9× RMR. For humans this is probably the maximum possible metabolic rate that can be sustained over a protracted (possibly indefinite) period.

Some birds do fly almost continuously and feed when they are flying, for example several species of swifts (Apodidae), terns (Sternidae) and the albatrosses (Diomedeidae). It is even suggested that swifts sleep and copulate on the wing (Lack, 1954) and from fledging to their first breeding attempts, 2 years later, they may never land, covering approximately 500 000 km in continuous flight. It is interesting that all these bird species have extreme adaptations of their wing morphology (Greenwalt, 1962; Norberg, 1990). The wings have very high aspect ratios, which means they are very long and thin. Theoretical aerodynamic modelling suggests that this form of wing shape minimizes the energy costs of flying (Pennycuick, 1969; 1989; Rayner, 1979; Norberg, 1990), but has some disadvantages; for example, manoeuvrability is reduced and it is difficult to fly slowly (Norberg, 1986, 1990), hence the birds have difficulty taking off from flat surfaces.

Direct measurements of the energy costs of flight, in these animals, have been measured. They confirm the predictions of the aerodynamic models as the costs are between about 2× and 5× BMR (e.g. Utter and LeFebvre, 1973; Hails, 1979; Bryant and Westerterp, 1982; Flint and Nagy, 1984; Westerterp and Bryant, 1984; Costa and Prince, 1987; Bevan *et al.*, 1999a; Adams *et al.*, 1986), compared with 13–18× BMR for most other birds (Tucker 1966; Masman and Klaasen, 1987; Rothe *et al.*, 1987) and bats (Speakman and Racey, 1991; Winter, 1998; Winter and von Helversen, 1998; Winter *et al.*, 1998). There are no continuously flying birds with wing shapes that would require continuous energy expenditure at the much higher levels. This observation suggests that there is some other limiting factor on the rates at which birds can expend energy, apart from the physiological aspects of substrate utilization by actively metabolizing tissues. If another limit existed, for example at around 5× BMR, this would constrain the types of animals that fly continuously to be only those with flight morphologies permitting such cheap flight.

There are two different types of hypothesis that aim to explain the nature of the limitations on sustainable daily energy expenditure. These might be termed the extrinsic and intrinsic hypotheses.

C. The Extrinsic Limitation Hypothesis

The extrinsic limitation hypothesis suggests that the dominant limitation on the expenditure of most wild animals is the availability of their food supply from the environment. Animals must forage for energy, and must therefore expend energy in its acquisition. If food is widely scattered, and the energy spent finding it is significant, the net energy return when foraging may be relatively small. Animals may be limited in their total energy demands, therefore, by the duration of time they can devote to searching for and ingesting food (Weiner,

1987). In richer habitats, where the net gain when foraging is higher, or where the time available for foraging is greater, the maximum sustainable energy demands would be predicted to be higher if this hypothesis were correct.

The extrinsic limitation hypothesis is intuitively attractive, and formed a fundamental assumption underpinning much of the development of optimal foraging ideas in the late 1960s and 1970s. Optimal foraging theory, in its classical form, was based on the premise that animals should be selected to maximize their net intake of energy during the periods that they spend foraging (Emlen, 1966; McArthur and Pianka, 1966; Schoener, 1971; Charnov, 1976; Krebs, 1978). This would likely be important to animals if either energy supply in the environment were limiting or, alternatively, if there were a selective advantage in minimizing the time spent foraging to meet their energy demands (Schoener, 1971). The former of these fundamental assumptions of the theory is equivalent to the extrinsic limitation hypothesis.

There are some data that are consistent with the extrinsic limitation hypothesis. For example, it is widely observed that clutch sizes of small birds get larger as one moves from the tropics to the temperate and arctic regions (Skutch, 1949). One interpretation of this trend is that energy supply is lower in the tropics, and increases as one moves further north during the summer, because day length is longer, thereby increasing the time available for foraging. Longer periods of input of solar radiation may also lead to greater environmental productivity in temperate and arctic regions when compared with tropical areas during the summertime. The greater amounts of energy available to animals in temperate and arctic regions may therefore allow them to raise more offspring, the energy costs of which rise in relation to the number being raised (e.g. Bryant and Westerterp, 1983; Kenagy et al., 1989).

There can be no doubt that, as summer turns into winter in the temperate and arctic regions, the extrinsic supply of energy becomes a potent limitation on the energy expenditure of many small endothermic animals. There are three fundamentally different ways that animals can respond to the decline in food supply during the winter. The first response is literally to flee the problem by migrating away from these regions to more equable climates. To migrate from temperate and arctic regions to tropical or subtropical areas, animals need to travel long distances. Typically migrations need to be at least 1500 to 2500 km to avoid successfully the extrinsic limitations that winter imposes. These distances pose strict limitations on the types of animals that can use this strategy. Studies of short-tailed field voles (*Microtus agrestis*) in captivity suggest that the absolute maximum distance that voles can travel during the night-time, when they are active, is about 15 km (Redman P. and J.R. Speakman, unpublished data). If voles were to migrate 2500 km it would take them over 300 days to travel this distance and return for the breeding season the following year (Speakman and Rowland, 1999; see also similar arguments in McNeil Alexander, 1998, 1999). In contrast, small birds can fly at speeds of

15 m s⁻¹ for protracted periods, and with tail winds may achieve ground speeds of 25 m s⁻¹. At these speeds they could cover approximately 300–500 km each day (Hedenstrom and Alerstam, 1998) and migrate 2500 km in less than a fortnight. Migration over land is a feasible strategy only for small animals that can fly. It is used by approximately 30% of temperate and arctic birds (Baker, 1978) and several species of bats (Davis and Hitchcock, 1965; Strelkov, 1969) but is unknown, for example, in rodents.

The second strategy for coping with the decline in the extrinsic supply of energy is to flee physiologically, by using hibernation. Hibernation involves regulating body temperature at a considerably reduced level (Hock, 1951) and suppressing metabolic rates (Geiser, 1988; Heldmaier *et al.,* 1993; Geiser and Ruf, 1995) so that energy demands are massively reduced to match more closely the reduced extrinsic availability of food supply (Kenagy, 1987). Even this may be insufficient and the animals usually also store considerable deposits of fat to supplement the available extrinsic energy supply during this period (Krzanowski, 1961; W.W. Baker *et al.,* 1968; Kunz *et al.,* 1998b; Speakman and Rowland, 1999). Hibernation is a strategy adopted by many small mammals to survive the winter months.

The third approach to surviving the winter is to remain endothermic but to reduce energy requirements by behavioural or morphological adaptations rather than physiological modifications. For example, in several species there is a breakdown in territoriality and animals become sociable and gregarious, allowing them to conserve energy by huddling together in shared nests (Vogt and Lynch, 1982; Karasov, 1983). Many species build well insulated nests during winter (Casey, 1981); they may also increase the insulation of their pelage to reduce heat loss (Hart and Heroux, 1953; Chappell, 1980); and often they also reduce their body masses to reduce total energy requirements (Iverson and Turner, 1974; Heldmaier, 1989). In some species, such as shrews, this may include reductions not only in body mass but also skeletal remodelling and shrinkage (Pucek, 1970; Pasanen, 1971; Merritt, 1986).

Even after adopting these different approaches, many animals also find it necessary to supplement the winter energy supply by caching food in autumn so that they can use it over the winter (Lyman, 1954). In addition, many small endothermic animals need to feed all day in winter to meet their energy requirements (Gibb, 1957) and often suffer their greatest seasonal mortality rate when the weather conditions deteriorate and they are unable to meet energy requirements from the available extrinsic energy supply (Chitty, 1952; Berry, 1968). The behavioural, morphological and physiological responses of temperate and arctic animals, in combination with the fact that starvation in winter is a common cause of death, strongly suggest that extrinsic supply of energy is the dominant factor limiting energy expenditure in many (most?) small endothermic mammals in the temperate and arctic regions during the winter months.

In summer, and in tropical regions, however, the limitations are less apparent and there is considerable evidence to suggest that if limitations do apply they are unlikely to be extrinsic. For example, many food supplementation studies have been performed in summer in temperate and arctic regions. The responses of animals to this supplementation do not indicate that energy supply before supplementation was limiting. For example, if food supply was limiting we might expect that animals would respond to supplementation by increases in reproductive output, increased home range areas and increased activity. Normally, however, reproductive output remains unchanged by supplementation, and home ranges and activity commonly decline rather than increase (e.g. Akbar and Gorman, 1993a,b; Cucco and Malacarne, 1997; Monadjem and Perrin, 1998). This suggests that energy demands are limited by some other factor, and the animals use the supplemented energy to meet these demands more rapidly than they would do if feeding from the unmanipulated environment.

Further evidence that animals are capable of harvesting more energy from the environment than they routinely do, in summer, comes from elegant manipulations where animals are experimentally tricked into gathering more food from the environment than they routinely collect. Masman *et al.*, (1989), for example, manipulated kestrels (*Falco tinnunculus*), rearing young by waiting until the parents had delivered a prey item to the offspring, and stealing the food item from the chicks via a trapdoor cunningly placed in the back of the nest. This activity meant the chicks did not become satiated, and kept begging their parents for more and more food. In these conditions the adult kestrels continued foraging far longer than they normally would, and managed to harvest about three times more food from the environment than they would have otherwise. These data strongly suggest that, at least during the summer, there is far more energy out there than animals are capable of using (but also see Wiehn and Korpimaki, 1997). Limitations on energy demands would appear to be more intrinsically set than extrinsically determined.

D. The Intrinsic Limitation Hypotheses

If energy expenditure is not limited by the external supply of energy, it must be limited by some intrinsic aspect of the animal itself. The potential limiting processes can be appreciated by considering the whole process from energy ingestion to final energy utilization (Figure 3). Apart from a few compounds such as nectar, the foods that animals eat are generally complex macromolecules, which must be digested in the alimentary tract and broken down into more simple compounds which are absorbed across the gut lining. Further processing of these absorbed compounds occurs in the liver, and waste products of the digestive process are eliminated by the kidneys. The first potential limitation may be the process of ingestion. It has been often

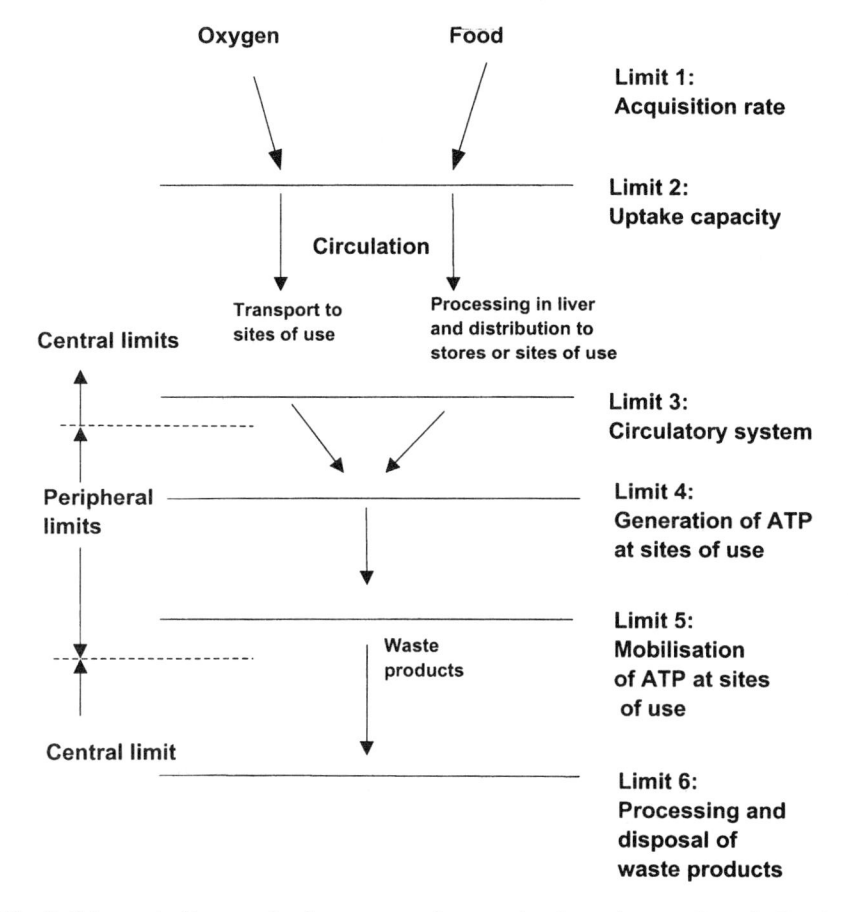

```
                Oxygen              Food
                   \                 /
                    \               /              Limit 1:
                     \             /               Acquisition rate
                      ↓           ↓
                  ─────────────────────────        Limit 2:
                      │           │                Uptake capacity
                      │           │
                      │ Circulation
                      │           │
                      ↓           ↓
                 Transport to  Processing in liver
Central limits   sites of use  and distribution to
                               stores or sites of use
   ↑
   │          ─────────────────────────            Limit 3:
   ┆ ─────────────                                 Circulatory system
   ↕              \           /
   ┆               \         /
   │                ↘       ↙
Peripheral       ─────────────────────────         Limit 4:
limits                   │                         Generation of ATP
                         │                         at sites of use
                         ↓
   │
   ┆             ─────────────────────────         Limit 5:
   ↕ ─────────────    Waste                        Mobilisation
   ┆                  products                     of ATP at sites
   │                                               of use
   ↓
Central limit            ↓
                 ─────────────────────────         Limit 6:
                                                   Processing and
                                                   disposal of
                                                   waste products
```

Fig. 3. Schematic diagram for the process of energy intake and expenditure indicating the six systems where limits on sustainable energy expenditure might apply.

suggested that this may impose a limit; for example, McNab (1980, p.106) stated that the '...rate of energy expenditure may be limited by the rate of acquisition', and it is frequently observed that animals reach an asymptote in their food intake as a function of prey density. Early studies inferred that the asymptote was a consequence of animals reaching a capacity for intake linked to handling times (Holling, 1959). However, studies in the 1980s showed that the asymptote actually occurs long before animals are spending all their time handling prey, and the asymptote reflects a complex interaction of prey density and diet choice (Sutherland, 1982). Nevertheless the process of ingestion does impose a theoretical potential limit on expenditure. A parallel, potentially limiting, process may be the capability to acquire oxygen at a sufficient rate to oxidize the ingested foodstuffs (Pasquis *et al.*, 1970). Once the food has been ingested, the second potential limiting process is the rate at which it can be digested by the alimentary tract and processed by the liver.

Processed food may be utilized for respiration immediately, or it may be stored for later use. In general only a small proportion of the food processed by the liver is utilized at this site. The primary end-products of digestion—amino acids, triglycerides, fatty acids and simple sugars—must all be transported around the body to storage depots or to sites of utilization, such as muscles and other organs. The distribution of the processed products of digestion from the alimentary system to storage depots and utilization sites is performed by the circulatory system. This may represent another limiting process on the total energy expenditure of animals.

The chemical energy in substrates derived from ingestion is released during systematic oxidation of the substances and the ultimate formation of water and CO_2. The released energy is trapped by conversion of ADP and phosphate to ATP. The generation of ATP may depend on the availability of substrates (Wang, 1978; Weber, 1992) and activities of enzymes that control the TCA cycle and the cytochrome system. Alternatively the system may be limited by the capability of the circulatory system to transport oxygen to the cells in sufficient quantity (Karas *et al.*, 1987). ATP is the energetic currency of cellular reactions, and ultimately conversion of ATP back to ADP and phosphate underpins almost all the energy-consuming processes that animals engage in. A notable exception is the generation of heat in brown adipose tissue, where the coupling of energy release to generation of ATP is deliberately disrupted to generate heat directly. The efficiency of ATP generation may act as another limitation on sustainable energy expenditure (Wang, 1978), as might the capacity of different processes to utilize ATP once it has been formed.

There are six basic processes, therefore, that might serve as intrinsic limits on sustainable energy demand. The first is the process of ingestion of the food and acquisition of the oxygen to oxidize it. The second process is digestion of the food and generation of metabolic substrates. The third is the process of distribution of metabolic substrates and oxygen around the body. Fourth is the efficiency of conversion of metabolic substrates to ATP during oxidative metabolism, and fifth is the capability of tissues to utilize ATP at major sites of energy expenditure, such as the major organs and muscles. Finally, a limit may exist in the capacity to dispose of waste products generated by the whole process.

Previous treatments of the potential areas where intrinsic limitations might occur have generally recognized only two processes or systems that might serve as limitations on sustained energy expenditure: the process of energy assimilation and the process of energy utilization. Definitions of these have been rather informal, and energy assimilation has been taken to include not only the processes of absorption by the gut, transformation by the liver and excretion by the kidneys, but also the distributional and storage processes involving the heart and circulatory systems. Energy utilization has been taken to include all the processes involved in conversion of substrates to ATP, and the subsequent utilization of ATP to generate work and heat. Separating the complex processes in Figure 3 into two broad categories certainly simplifies the conceptual understanding of where limits might apply in this system. However, as is often the case, such simplicity, although conceptually appealing, may obscure the reality of where intrinsic limitations apply.

The idea that capacity for energy assimilation (in its broadest sense) imposes a limit on sustainable energy expenditure has been termed the 'central limitation hypothesis' (Gross *et al.,* 1985; Karasov and Diamond, 1985; Weiner, 1987, 1989, 1992; Peterson *et al.,* 1990). It is termed 'central' because the process is common to and independent of all the various methods by which energy might be expended. Central limitation imposes the same limits on sustained energy expenditure whether the animal is stressed by decreases in ambient temperature, increases in physical activity, reproduction or combinations of stressors. A direct prediction of the central limits model, therefore, is uniformity in the observed maximal rates of sustained energy expenditure. In contrast, the idea that energy expenditure is limited at the sites of energy utilization has been generally called the 'peripheral limits hypothesis'. In contrast to the central limitation idea, it predicts that limitations imposed by different processes need not necessarily be the same. On the face of it, therefore, a suitable test of the contrasting hypotheses might be to push a group of animals to their limits of sustainable energy expenditure in several different ways, and to compare the sustained limits.

1. Comparisons of Sustainable Energy Expenditure under Different Stressing Factors

Several attempts have been made to compare the sustainable metabolic rates of animals under different stressors to establish whether their maximum sustainable capacities are equivalent. McMillan and Hinds (1992), for example, compiled data on maximum oxygen consumption of rodents not only during exercise but also during cold exposure (Figure 2). They found that rodents routinely did not expend energy at the same high levels during cold exposure as they managed to achieve during exercise. However, these were not sustainable rates, because, as we have discovered, animals generally are unable to exercise continuously. However, they may be able to thermoregulate continuously, and this difference might bring the sustainable rates of expenditure much more closely in line.

Direct measurements of maximal 24-h energy expenditures, under different stressors, in small mammals are more sparse than measurements of short-duration maximal rates, and researchers have relied on using lactation as an alternative to exercise, presumably because of difficulties in getting animals to exercise for prolonged periods (but see also Perrigo, 1987). Comparative data are presented in Table 1. In these cases the maximum for each species is expressed as a multiple of the resting metabolic rate measured at thermoneutral. These are generally termed sustained metabolic scopes (Peterson *et al.,* 1990).

Across these five comparisons, the maximum sustainable scope during lactation exceeds the maximum during cold exposure (paired t = 8.16, 4d.f., P = 0.0012). This would appear to support the peripheral limitation hypothesis, because the central limitation hypothesis would predict these maxima to be equal. There are, however, several problems with these comparisons.

First, food intake has been very widely employed as a measure of sustainable energy expenditure. This is because over protracted periods, for animals in mass balance, the ingested food is mostly oxidized. However, there will be a discrepancy to true levels of expenditure because some of the food will not be assimilated. Because assimilation rates are generally high, however, food intake does provide a convenient approximate measure of long-term expenditure. This assumption, however, of the approximate equivalence of intake and expenditure, breaks down when considering lactation as a stressor. During lactation animals take in energy but not all of their intake is

Table 1.

Sustainable daily energy expenditure as a multiple of resting metabolic rate (Sustainable metabolic scope) for small mammals engaged in different activities presumed to be maximizing daily energy expenditure.

Species	Maximum sustainable metabolic scope		
	Lactation	Cold	References
Acomys caharinus	6.9	4.0	Koteja *et al.* (1994)
Peromyscus maniculatus	4.35	5.0	Koteja (1996a,b)
	7.7		Stebbins (1977)
	6.2		Millar (1979)
Mus musculus	7.2	4.7	Hammond and Diamond (1992), Konarzewski and Diamond (1994)
	6.0		Speakman and McQueenie (1996)
Microtus agrestis	4.3	2.7	M.S. Johnson and J.R. Speakman (unpublished data), Migula (1969), McDevitt and Speakman (1994a)
Phodopus sungorus	3.4	1.8	Weiner (1987)
	–	3.1*	

In all cases the maximum daily expenditure is calculated from the food intake and the resting metabolic rate (RMR) is quantified from the oxygen consumption. Maximum sustainable metabolic scope is the maximum expenditure/RMR.
*Refers to animals after a period of thermal acclimation to low temperatures.

oxidized and appears as energy expenditure. This is because a substantial proportion of energy ingested is converted into milk and re-exported. Inevitably, then, the energy expenditure during lactation will be lower than the peak energy intake estimated from food intake. Few studies have quantified the extent to which these two measures differ. Scantlebury *et al.* (submitted) measured the food intake, milk production and energy expenditure of lactating dogs (miniature schnauzers and labradors). They found that the peak energy demands in lactation, expressed from food intake, averaged 5.0× and 6.5× RMR for the two breeds respectively. However, actual energy expenditures were considerably lower at only 2.4× and 2.9× RMR respectively. Clearly, then, the measurements of lactational food intake are not closely linked to the true levels of energy expenditure. This means that, although the levels of food intake are generally higher during lactation than during prolonged cold exposure (Table 1), the levels of actual expenditure may not be.

The central limitation model predicts that expenditure will be limited by central processing capacity. The key question is whether maximal food intake is a valid measure of central processing capacity. Although animals may eat more food in lactation, and must absorb this across the gut walls, there may also be a single system which controls the distribution of this energy for expenditure. Hence, although the levels of food intake may differ, if the levels of expenditure do not then we cannot rule out the possibility that there is a central control mechanism involved (see Figure 3 for different levels of control).

The second problem is that all these comparisons involve animals measured under laboratory conditions. In this situation it is possible to question whether the animal comprehends the basis of the experiment, and complies by expending energy at its maximal possible rate. We do not fully understand the motivations of animals when they are kept in captivity, and therefore this may undermine attempts to measure maximal rates of performance. Finally, even if we were to measure accurately the levels of expenditure in both situations and they proved to be the same (or at least not significantly different), this would not necessarily support the central processing model because the expenditures might be limited peripherally, yet by chance have equal values.

E. Experimental Studies of the Limitation Hypotheses

Perhaps the best set of experimental studies that have been performed to examine the nature of limitations on the sustained energy expenditure of small mammals are those by Kim Hammond and Jared Diamond (and colleagues) using the lactating Swiss Webster mouse as a model system. Swiss Webster mice normally eat about 4 g of food each day. During lactation, however, their food intake increases enormously to around 18–20 g per day, almost equalling their own body mass. If the mice are given five pups to raise, they eat only about 11 g, but for litters of 8 and 11 offspring the food intake is higher. Food intake, however, does not rise in direct proportion to the number of offspring, so the intake for litters of 8 and 11 is actually fixed at the 18–20-g level. In consequence, there is less food available to support the production of milk and therefore there is an inverse relationship between the size of the litter and the size of the individual offspring in that litter. Bigger litters contain smaller offspring, presumably because they are getting less milk per pup. This trade-off in litter and offspring size is not a specific feature of Swiss Webster mice and is observed in other strains of mice and many other species of small rodent during lactation (see examples in Stearns, 1993). Findings from the study of these mice are therefore likely to be more generally applicable.

Because food intake in these mice reached an asymptote as litter size increased, Hammond and colleagues reasoned that this might represent a central limit on the system, ultimately controlling the production of milk. To test this idea they attempted to manipulate the system in different ways to see whether they could force the mice to increase their food intake. First, they added pups to the litters so that they were abnormally large (Hammond *et al.*, 1992). The mice responded by producing even smaller individuals, and did not increase their food intake. Second, the mice were placed in cages where the female had to climb up a wire funnel to get to the food. This was not designed to make her work harder for the food, but rather prevented the young pups from gaining access to it. Consequently, when the pups came to the day 18 of lactation, at which point they normally start to eat solids and thus demands on the female start to decline, they were unable to, and instead of declining the demands on the females continued to increase until 24 days postpartum (Hammond *et al.*, 1992). Nevertheless, the females continued to eat only 18–20 g of food, strongly suggesting that there was a central limit operating in the system which was involved in food intake. These observations are also consistent with earlier work performed by Perrigo (1987), who forced mice to work for their food. When mice were forced to work to get food, they had to expend more energy in its acquisition. They could respond in theory to this increased demand by increasing their food intake. However, they did not do this. Instead they kept eating the same amount of food. This meant they had less to allocate to the process of lactation and in response to this reduction in energy available for lactation, they actually killed some of their offspring, to reduce the litter size to a level compatible with the resources that were being allocated.

Taken together, the data from Hammond, Diamond and colleagues, and Perrigo, strongly suggest that there is a central limit controlling the maximum sustainable energy expenditure. Measurements of the resting metabolic rate of the Swiss Webster mice indicated that when they were working at this maximum limit they were ingesting energy at about 7.2× RMR. This appeared to be the maximum rate at which they could take in resources. Hammond *et al.* (1994) performed yet another manipulation to test this idea of centrally mediated sustainable limit during lactation. In the same laboratory they had previously shown (Konarzewski and Diamond, 1994) that during cold exposure the mice also increase their food intake to meet the increased thermoregulatory requirements. How would mice respond when faced with these cold demands during lactation? If the central limit idea was correct, the mice would face exactly the same dilemma that had been faced by the mice studied by Perrigo (1987). They would have fewer resources available to support lactation and consequently they would have only two options: kill some of the offspring to sustain their weaning size (as Perrigo's mice had done), or reduce the average size of the offspring (as Hammond and Diamond's mice had previously done during manipulations).

Against all the flow of the data, the response of the mice was completely unexpected. They increased their food intake to cover the extra demands of being in the cold. Instead of eating 18–20 g they upped their intake to about 23 g each day, which is 10–25% higher than it had been. This is really a remarkable result. Remember that previous manipulations had all failed to get the mice to eat more food. In Perrigo's experiment the mice had actually been prepared to kill their babies rather than eat more. Yet, in the cold, the mice demonstrated that they were capable of eating more food. The impact of this experiment was enormous because it forced a whole rethink of what was actually going on in the model system. Hammond *et al.* (1996) reasoned that all the previous manipulations had involved not only trying to get the mouse to eat more, but also attempting to get her to produce more milk (although this was not the case in the Perrigo (1987) experiment). By giving her more offspring to raise, or delaying weaning, the manipulations aimed to increase milk demands and thereby increase food consumption. Perhaps, then, the limit was not actually at the central level of processing the food, but at the peripheral level of milk production. Mice had been unable to produce more milk, and therefore had responded by producing smaller offspring, and there was no point in increasing food intake because the extra energy could not be channelled into raised milk production.

Hammond *et al.* (1996) performed a further experiment to test this idea. This involved surgically removing mammary tissue. They reasoned that, if the milk production capacity of the tissue was already at a maximal level, removing mammary tissue would necessarily reduce milk production. The mice would be unable to respond by cranking up production in the remaining tissue, and there would be several knock-on consequences: litters would grow less and the mother would eat less food in response to the lower demand. However, if the limit acted centrally, the mice would increase milk production in the remaining tissue, milk production would continue to sustain the offspring who would have unaffected growth, and food intake would also be unaffected. The experiment was slightly more complex than this, as it involved manipulations of the ratio of offspring to teat number, but the bottom line was that the results supported the peripheral limitation model rather than the central limitation model. The mammary tissue remaining after surgery appeared unable to respond to replace the production of tissue that had been removed.

Similar work has been performed more recently by Rogowitz and McClure (1995) and Rogowitz (1996, 1998), who have demonstrated directly that milk production appears to be the limiting stage in lactation of a different species, the cotton rat (*Sigmodon hispidus*). These data further support the suggestion that sustainable energy demands are set at the peripheral rather than the central level.

F. The Central Limitation Hypothesis and Links between FMR and RMR

One of the major ideas that had led to the suggestion that animals were controlled by a central limit was the observed linkage between resting metabolic rate and sustainable metabolic rate (King, 1974; Drent and Daan, 1980; Weiner, 1987, 1989, 1992). The idea that this link is connected with the central limitation theory stems from observations that different tissues *in vitro* respire at different rates. In particular, fat tissue and muscle tissue have relative low metabolic rates but organs such as the alimentary tract, the liver and kidneys have very high rates, as does neuronal tissue and the brain (e.g. Field *et al.*, 1939; Krebs, 1950; Martin and Fuhrman, 1955; Scott and Evans, 1992). This information was combined with data showing that when animals have increased energy requirements (such as following cold exposure or during lactation) they respond to the increased demands by growing these tissues. Kennedy *et al.* (1958), for example, documented growth of the liver, and Jolicoeur *et al.* (1980) reported changes in the size of the pancreas during lactation, and similar changes in the alimentary tract were reported by Fell *et al.* (1963) and by Cripps and Williams (1975). This phenotypic flexibility in organ size in relation to demand is a very general phenomenon, for which there are many demonstrations (e.g. Myrcha, 1964; Drozdz, 1968; Gebcznska and Gebcznski, 1971; Gross *et al.*, 1985; Green and Millar, 1987; Bozinovic *et al.*, 1990; Hammond, 1993; Nagy and Negus, 1993; Konarzewski and Diamond, 1994; Derting and Noakes, 1995; Derting and Austin, 1998).

If animals grow their alimentary tract and processing machinery (notably the liver) when they have high energy demands, it follows that in parallel with greater levels of food intake there will be an increase in resting metabolic rate (Szarski, 1983; Else and Hulbert, 1985; Karasov and Diamond, 1985; Weiner, 1987, 1992), because these organs have high metabolic rates. Indeed, several studies have demonstrated that interspecific and within-species (interstrain) variation in resting metabolic rate is linked to differences in the relative organ sizes. Animals with higher metabolic rates generally also have larger hearts, kidneys, livers and alimentary tracts (e.g. Daan et al., 1989, 1990a; Konarzewski and Diamond, 1995). Given these studies, it was suggested that a limit on sustainable metabolic rate exists at around 7× RMR, which reflects the link between RMR and the central processing machinery (Weiner, 1987, 1992; Daan et al., 1990a; Peterson et al., 1990). Maximum achievable metabolic rates were therefore viewed as limited by the sizes (or activities) of the central processing system at around seven times the prevailing resting rate of metabolism.

These linkages can be illustrated by considering a study from my own group, on lactation performance in a different strain of mice to that studied by Hammond et al.: the MF1 strain (Speakman and McQueenie, 1996). When these mice are virgins they consume about 4.5 g of food daily. Their RMRs at rest and in thermoneutral average 1.0 ml O_2 min^{-1} and their energy expenditure for RMR is thus about one-third of the daily energy expenditure (DEE) expressed as food intake. The ratio of DEE to RMR is thus about 3.1. During lactation, the food intake increases dramatically. During the last 6 days of lactation the mice eat, on average, 23 g each day. The precise amount depends on the litter size, but for litters above 10 offspring the females all eat 23 g per day. If RMR had remained at the level reported before mating, the mice would be taking in food at about 12 times their resting expenditure. This is well beyond the suggested alimentary limit at around 7× RMR. The animals achieve this because during pregnancy and lactation they grow their guts and their livers to accommodate the increased food intake. A lactating MF1 mouse actually grows its small intestine by about 12 cm compared with the length before breeding, and the gut wall thickens and becomes more folded. In addition the liver increases from a wet mass of around 1.0 g to over 2.4 g. Because of these changes in the organ structures, the lactating MF1 mouse also has an increased resting metabolic rate, averaging 2.1 ml O_2 min^{-1}. This brings the ratio of food intake to resting metabolic rate down to about 5.9× RMR, which is within the supposed limit of around 7× RMR, supporting the general model.

On the other hand, there are several studies that do not support the model of linkages between food intake, RMR and variation in the alimentary tract. For example, we have studied the process of thermal acclimation in the short-tailed field vole (*Microtus agrestis*). During cold exposure small mammals experience increases in their energy demands and food intakes, at the same time as having changes in their resting metabolic rates (Adolph, 1950; Adolph and Lawrow, 1951; Hart, 1953a,b; Krog et al., 1954; Depocas et al., 1957; Chaffee and Roberts, 1971; Rosenmann et al., 1975; Feist and Rosenmann, 1976; Klaus et al., 1988; Heldmaier, 1989). One interpretation of the pattern of change is that the energy demands increase above the sustainable capacity of the alimentary tract, and that in response the tract grows, leading to an increase in RMR. Detailed studies, however, in *M. agrestis* reveal that this is not the case (McDevitt and Speakman, 1994a). Before cold exposure the voles have a food intake that averages about 1.3 times their basal rate of expenditure. During chronic cold exposure (5°C) the animals respond by increasing their food intake immediately to about twice the non-breeding level, but this still averages only 2.8× RMR. Exposure to colder temperatures direct from the warm is fatal (McDevitt and Speakman, 1994b), indicating that the voles reach a critical limit in their capacities at around this temperature. These seem unlikely, however, to be mediated centrally, given the low multiple of RMR involved. Over the following 15 days the RMR slowly increases, following, rather than preceding or occurring in parallel with, the change in food intake. This change in RMR appears to be linked most closely to variation in the brown adipose tissue, rather than alterations in gut length and mass or the liver mass (McDevitt and Speakman, 1994a,b). These interactions suggest that during cold exposure these rodents are peripherally rather than centrally limited.

G. Interspecific Reviews of the Link Between DEE and RMR

Several previous reviews have addressed the nature of the linkage between RMR and DEE, summarizing data collected across species (Drent and Daan, 1980; Bryant, 1990; Peterson *et al.*, 1990; Bryant and Tatner, 1991; Koteja, 1991; Daan, 1990b; Degen and Kam, 1995; Ricklefs *et al.*, 1996; Hammond and Diamond, 1997). These reviews stemmed from the suggestions in the 1970s (King, 1974) that DEE and RMR might somehow be linked. The study by Drent and Daan (1980) explicitly suggested that there was a limit on DEE of small birds set at around 4× RMR. Subsequently, as more data accumulated, the 4× RMR limit was breached on several occasions. Bryant and Tatner (1991) and Bryant (1990) reviewed the bird literature again, and concluded there was no evidence favouring a rigid 4.0× RMR limit as it was breached in over 10% of studies. Peterson *et al.* (1990) reviewed data from both mammals and birds, and concluded that the limit sits at a higher level of around 7× BMR. This was more consistent with the data reviewed by Bryant and Tatner (1991), and would also prove to be close to the limits observed later in laboratory experiments on mice pushed to their lactation limits (see Hammond papers reviewed above).

To an extent, RMR and DEE will be correlated because both are dependent on body mass. Koteja (1991) examined the links between sustainable energy expenditures and resting metabolic rates in birds and mammals, and removed the covariable effects of body mass to establish how close the links where in the absence of this shared variation. He found that in both birds and mammals there was a link even after the shared variation due to mass had been removed. Subdividing these groups, however, revealed some groups where there was no significant effect, for example in marsupials and in passerines and seabirds.

Whether this lack of significance in some groups was a real effect, however, was difficult to assess because the subdivided groups had much lower sample sizes than the pooled groups (e.g. nine in each of the passerine and marsupial groups). Degen and Kam (1995) suggested that the ratio of sustainable metabolic rate to RMR was not constant with body size, and that larger animals consistently have lower ratios than smaller animals. This suggestion recalls earlier studies which had pointed out that the scaling effects of mass on daily energy demands differ from the scaling effects on BMR (Bennett and Harvey, 1987; Koteja, 1987; Nagy, 1987) and thus 'FMR/BMR scopes are not constants but depend on body size' (Koteja, 1991).

Ricklefs et al. (1996) addressed the problem of correlated variation due to both variables being related to body mass, as had been done by Koteja (1991), and confirmed that the relationships remain when this shared variation is removed. However, they also took the analysis a stage further, by eliminating the problem that data collected across different species are not independent because of their shared phylogenetic history. This revealed that there was still a relationship evident in mammals, but that the relationship in birds was perhaps a consequence of a phylogenetic artefact. The reasons for the differences between birds and mammals, however, remained obscure. Finally, Hammond and Diamond (1997) reviewed a diverse group of data that comprised some field, some laboratory and some data compiled for humans to reiterate the suggestion that there is a limit imposed on sustainable metabolic rate at around 7× RMR.

The possibility consistently sought in these reviews that there is a link between the resting rate of energy expenditure and the sustainable rate of expenditure (SusMR) is attractive because it has wide implications. In particular, such a link would provide a useful framework for understanding the variations in RMR that are observed between different animal species. One theory for the evolution of high sustained levels of resting energy expenditure in endotherms is the possibility that high resting rates potentiate high maximal rates of metabolism—the so-called aerobic capacity model for the evolution of endothermy (Bennett and Ruben, 1979; Dawson et al., 1979; Taigen, 1983; Bozinovic, 1992; Chappell and Bachman, 1995; Hayes and Garland, 1995; Ruben, 1995). If a link is established between RMR and SusMR, it could also be argued that high rates of RMR potentiate high sustainable metabolic rates. Such high rates may then allow endotherms to pursue life histories that are unavailable to exotherms. It has been argued, for example, that a certain level of RMR is necessary to sustain the mammalian pattern of reproduction (Thompson and Nicoll, 1986; Nicoll and Thompson, 1987; Thompson, 1992).

Within mammals there is also enormous variability between species in their RMR. The supposed link between RMR and SusMR provides one mechanism for understanding this variability. A high RMR may enable a high SusMR, which could be necessary to support high rates of reproductive output.

Supporting this viewpoint, several studies have suggested that there are links between interspecific variations in RMR (or BMR) and life history parameters such as litter size and intrinsic rates of population increase (Henneman, 1983; Padley, 1985; Haim, 1987; McNab, 1987a,b; Koteja and Weiner, 1993). However, other studies have disputed such claims (Hayssen, 1984; Harvey et al., 1991), and attempts to find links at the intraspecific level have been similarly disappointing (Derting and McClure, 1989; Hayes et al., 1992a). The wider implications of the possible link between RMR and sustainable metabolic rate are consequently still quite confused.

H. Summary and Aims

The overall picture that emerges from this analysis is that intrinsic limitations on sustainable energy demands are probably set in the peripheral systems (Figure 3). The energetically demanding absorption system (gut and liver) as well as the elimination system (kidneys) respond to match these requirements, thus generating a link between RMR and FMR. In this paper I will address two particular questions concerning sustainable energy requirements. First, what are the dominant factors that influence the levels of energy expenditure of free-living small mammals? Second, what is the nature of the linkage between RMR and SusMR?

Measurements of energy expenditure have been made on a very wide range of animals including insects, reptiles, birds, and both small and large mammals. The scope of this study has been restricted to small endothermic mammals weighing less than 4 kg. The reasons for this restricted data set are 2-fold. Smaller animals tend to have relatively high metabolic rates (per gram of metabolizing tissue; e.g. Pearson, 1947; Lasiewski, 1963; Chai et al., 1998) and this is particularly so for small endotherms, which regulate their body temperatures by generating heat internally. If limits on rates of energy expenditure are important, this is likely to be the case for this group of animals more than for any other. The second reason is that there are some methodological complexities involved in comparing the daily energy expenditures of large and small animals, which will be explored further in the Methods section. These methodological difficulties mean that it is desirable to compare energy demands for a restricted size class of animals either exceeding or lower than a threshold of around 4 kg. Most data collected to date are for animals below this threshold, and consequently it was decided to review the data from below the threshold rather than those from above it.

The aim of this review is 3-fold. First, I hope to produce a quantitative description of the factors that influence energy demands, which might subsequently be used as a predictive model for scientists involved with modelling energy flows in populations and communities. Second, such a predictive model will be beneficial to other researchers examining aspects of mammalian ecophysiology, by providing a comparative reference point to the levels of demand that might be expected in different circumstances. Significant deviations from these expectations may then indicate that given study animals are performing interesting things with their energy budgets that are worthy of further investigation at different levels (e.g. behavioural, physiological, biochemical and molecular studies). Finally, I intend to investigate more closely the idea of a link between sustainable field metabolic rate and resting metabolic rate. In the Methods section I propose to review the methods available for measurement of energy expenditure, in particular the method of indirect calorimetry and the field techniques of time and energy budgeting and doubly labelled water.

III. METHODS

This section reviews the alternative approaches available for the measurement of daily energy expenditure (DEE) by animals. This review sets the scene for later comparisons of daily energy demands to resting and basal energy expenditure, as well as highlighting the methodological complexities of measuring daily energy demands. This section, therefore, also explains the inclusion criteria of estimates of DEE for the present study.

A. Measuring Energy Expenditure by Indirect Calorimetry

The standard method for quantifying the energy requirements of animals is indirect calorimetry. Indirect calorimetry is based on the fact that animals consume oxygen from the air to oxidize organic compounds, thus releasing the chemical energy stored in the bonds of those chemicals for use by the animal. The exact relationship between consumed oxygen and energy expenditure depends on the organic substrate being utilized. When animals oxidize carbohydrate, the amount of energy released per millilitre of oxygen consumed is about 20.9 kJ. When an animal metabolizes fat, the energy released per millilitres of oxygen consumed is lower, at around 19.66 kJ. Exact conversion values vary slightly depending on the substrate being utilized, but on average the difference between the lowest and highest conversion values is about 6%. Different substrates result in the generation of different amounts of the primary end-product of oxidation, carbon dioxide. The ratio of O_2 consumption to CO_2 production (called the respiratory quotient, or RQ) gives an indication of the substrate utilization. If the RQ is known, the conversion of O_2 consumption to energy expenditure generally involves relatively minor errors (< 0.5%; Gessaman and Nagy, 1988). However, when RQ is unknown an error of varying magnitude can be introduced to the conversion from O_2 consumption.

In theory, measurements of CO_2 production alone can also be used to estimate energy expenditure. Measurements of O_2 consumption, however, are preferred. This is because there is much greater variability in the amount of energy released per millilitre of CO_2 produced, than per millilitre of O_2 consumed. For example, when an animal mobilizes carbohydrate (RQ ≃ 1.0), the energy equivalent of 1 ml CO_2 produced is 20.9 kJ. However, when fats are being mobilized, the energy equivalent of 1 ml CO_2 is 28.1 kJ. The difference between the lowest and highest conversion values for CO_2 is therefore about 34%, compared with a 6% difference for O_2 consumption. Because the error in conversion from CO_2 production to energy (in the absence of information about RQ) is potentially much greater than the conversion from O_2 consumption to energy, measurements of O_2 consumption rather than CO_2 production dominate the literature.

Measurements made by indirect calorimetry generally involve confining the subject animal inside a chamber through which a flow is passing. Oxygen consumption is evaluated by measuring the flow rate and the O_2 content of gasses entering and exiting the chamber. A full discussion of the theory behind this method is presented in Appendix A. The major advantage of using indirect calorimetry to estimate the metabolic rates of animals is this accuracy and precision. The major disadvantage is that the animal needs to be confined in a chamber for the measurement to be made.

B. Basal and Resting Energy Expenditure

It was recognized very early in the study of animal metabolism that many factors affect metabolic rate. Lavoisier and Sequin, for example, at the end of the eighteenth century had already established that O_2 consumption depended on the size of the subjects, whether they were active or not, and whether or not they had recently eaten a meal. In comparing the energy requirements of different species, therefore, it was desirable to define a set of conditions that would be equivalent across all animals. Kleiber (1932, 1961) and Brody (1945) were instrumental in establishing a set of standard conditions for the measurement of animal metabolism that would allow broad comparisons across species. In particular, it was thought desirable to remove the effects of any environmental factors that might increase metabolism to produce a comparable estimate across species of the lowest level of metabolic rate. These conditions, as enumerated by Kleiber (1961), were that the animal should be, 'Mature animals in the post-absorptive state and measured in a range of metabolically indifferent environmental temperatures at rest, or at least without abnormal activity' (Kleiber, 1961, p. 204). Because this was considered to be a minimal estimate of metabolism, measurements conforming to these criteria were defined by Kleiber (1932, 1961) as 'basal metabolic rate' (BMR). A similar definition was used by Brody (1945) for minimal metabolism, which differed from the Kleiber definition only in that the animals were not required to be post-absorptive. This was called 'resting metabolic rate' (RMR). Although it was originally defined as a measurement almost identical to BMR, but lacking the requirement that the animals be post-absorptive, subsequently RMR has been used to describe the metabolism of any animal that is inactive, whether it is inside the thermoneutral zone or not. This has led to some confusion, therefore, about how restrictive the term RMR actually is. To avoid any ambiguity in this chapter, I have used two terms: RMRt, to reflect resting metabolism without any control of temperature, and RMR, to reflect the use of RMR in the original sense defined by Brody (1945) as measured inside the thermoneutral zone.

As more and more animals have been measured, it has become increasingly obvious that the criteria for measurement of basal metabolism are most useful for the small set of animals that Kleiber had measurements for at the time the criteria were established. All of these animals are strict thermoregulators and maintain their body temperatures within very tight limits. A major factor, however, that influences the metabolic rate of an animal is its body temperature. Because Kleiber did not incorporate this effect into the definition, it is not a requirement that an animal be maintaining its body temperature at euthermic levels. Most animals have a circadian rhythm in body temperature and in parallel maintain a rhythm of metabolic rate (Aschoff and Phol, 1971). Yet time of day is also not part of the Kleiber criteria.

Kleiber (1961) did recognize that when the animals in his group were reproducing, or growing, their demands increased and thus the measurement of basal metabolic rate was compromised. However, an important factor he did not consider was the effect of seasonal changes on BMR. This is probably because the domesticated animals in his sample were routinely maintained in constant environments throughout the year and experienced minimal seasonal fluctuations in day length and ambient temperature. In contrast, wild animals experience large shifts in ambient temperatures and photoperiod if they live in temperate or arctic regions, and large fluctuations in rainfall if they inhabit the tropics. These large seasonal changes in environmental conditions also result in large seasonal differences in BMR as animals become seasonally acclimatized. Season of measurement is also not a factor that Kleiber required to be controlled in his definition of BMR.

Several problems also arise because in some animals the criteria are mutually exclusive. In soricid shrews, the response to food deprivation is generally to increase activity to seek out food (Hanski, 1985). The requirements, therefore, that the animals be post-absorptive and inactive are therefore almost impossible to attain in these species (McDevitt and Andrews, 1997), although some researchers claim that all that is required is patience (McNab, 1997). There are, however, some much clearer incompatibilities in the criteria than evidenced by shrews. In many temperate zone insectivorous bats, starving the animals to ensure that they are post-absorptive often forces them to abandon temperature regulation compared with animals that are fed as normal (Kurta, 1991). The animals that are fed and regulating their body temperatures at euthermic levels yield a measurement that is closest to the spirit of measuring a broadly comparable minimal metabolic rate. However, this measure would not conform to the Kleiber criteria because the animals might not be post-absorptive. On the other hand, the animals starved overnight would definitely be post-absorptive and therefore would meet the Kleiber criteria although the low metabolic rate would occur only because the animals had abandoned thermoregulation at euthermic levels. In ruminants there is the problem that the rumen contents are also actively metabolizing. This will add to the observed heat production. Should this be included in the measurement of basal metabolism, or should the rumen be emptied before the measurement is made?

It has been suggested that these are abnormal exceptions and that the Kleiber criteria form a broad basis for comparison across the vast majority of species (McNab, 1997). However, this is an optimistic view of basal metabolism as a unifying measurement. There are several other factors that influence the measurement of BMR, which are universal across all animals but not controlled in the Kleiber definition. These are the duration for which an animal is measured and the time period over which the minimal metabolic rate is defined. Many small rodents, for example, exhibit a stress response to handling and have high metabolic rates immediately after they have been placed into a respirometry chamber. Hayes *et al.* (1992b) found in wood mice (*Apodemus sylvaticus*) that extending the length of time a measurement was made from 1 to 3 h could reduce the minimal metabolic rate by 10%. The minimum duration for which animals should be measured does not feature in the Kleiber definition.

Given the arbitrary nature of the Kleiber criteria, there is a clear need to apply some common sense in deciding whether a measurement is acceptable as a measurement of BMR or not. Gallivan (1992), for example, has suggested that all previous estimates of Cetacea could not be included as BMR measurements because previous measurements in the animals (e.g. Kasting *et al.*, 1989; Innes and Lavigne, 1991) were not completely inactive, or were inactive but the measurements were very short. (This is further complicated by the fact that Kleiber did not specify complete inactivity in his definition— '...*inactive or nearly so*'). As has been pointed out, rejecting these measures is probably not justified on this basis, because making Cetacea completely inactive for prolonged periods is probably unachievable (Speakman *et al.*, 1993). Sometimes, getting close to the Kleiber criteria is as close as we are ever going to get, so rejecting such data does not really advance our understanding very much. Including them does not necessarily imply a complete relaxation of the Kleiber criteria to include anything (McNab, 1997).

Despite these apparent methodological problems with measurement of BMR, estimates of it have proliferated in the literature since its first definition.

Ignoring the fact that some (much?) of the variability between measurements may reflect methodological differences between studies, many attempts have been made to understand the observed interspecific variability that is evident in this trait. Some authors have suggested that environmental factors such as latitude (Ellis, 1984), climate (McNab and Morrison, 1963; Hulbert and Dawson, 1974) and diet (McNab, 1980, 1983, 1986a, 1986b, 1987a, 1988) are the most important factors. Others have emphasized the linkage of variation in BMR to variation in the morphology of the animals (Daan *et al.*, 1989, 1991; Konarzewski and Diamond, 1995), but others have suggested that the variation is mostly linked to differences in phylogeny (Hayssen and Lacy, 1985; Bennett and Harvey, 1987; Elgar and Harvey, 1987; Harvey and Elgar, 1987; Harvey and Pagel, 1991). This argument, however, has been confused because of some misclassifications of diet and because some measurements of BMR probably did not conform to the Kleiber criteria as they were measured below thermoneutral in the species included in reviews favouring phylogenetic effects (McNab, 1987b). In this context an important but seldom referenced set of data are those collected by Stephenson and Racey (Stephenson and Racey, 1993a, 1993b, 1994, 1995; Racey and Stephenson, 1996) concerning the resting energy requirements of the Tenrecidae. This group of insectivorans, isolated in Madagascar, shows remarkable convergence in many metabolic traits with rodents and insectivores occupying equivalent niches in mainland Africa.

C. Time and Energy Budget Estimates of Daily Energy Expenditure

Most animals living in their natural environment expend energy above the basal requirement almost all the time. By observing an animal over a 24-h period it would be possible to classify its behaviour into several distinct classes. The time and energy budget method for estimating the DEE suggests that an estimate of total daily demands can be derived by multiplying the time spent in each activity by the energy costs of that activity (McNab, 1963; Schartz and Zimmerman 1977; Weathers and Nagy, 1980; Weathers *et al.*, 1984; Buttemer *et al.*, 1986; Goldstein, 1988). For example, imagine an animal that spends its time either flying or resting. The daily time budget might consist of 5 h of flight and 19 h of rest. If the cost of flight was 10 W and the cost of rest was 1 W (1 J s^{-1}), the accumulated cost would be:

$$(5 \times 3.6 \times 10) + (19 \times 3.6 \times 1) = 248.4 \text{ kJ}$$

In most circumstances determining a simple behavioural time budget for an animal is not difficult. The problem arises when one attempts to convert the simple time budget into an energy budget. This is because assigning a cost to a particular activity is complicated by the fact that the costs are not constant for a given behaviour but depend on several other factors. Take for example the energy cost of resting (RMRt). The cost of resting depends on a wide range of external factors that relate to the thermoregulatory demands being placed on the animal: ambient temperature, wind speed, solar radiative input, whether it is raining or not, the presence of an insulating nest and the time of day. In addition, even under a set of completely fixed external conditions, the cost of resting also depends on several attributes of the individual animal: its body mass, body condition, whether it is digesting food or not, what the composition of that food was, whether it is growing or not, what ambient conditions it had previously encountered and whether it is reproducing or not.

It might be relatively simple, therefore, to establish that an animal has spent 12 h each day at rest. However, assigning a single energy cost to that behaviour would be impossible, because the factors that can potentially affect resting energy expenditure would be unlikely to remain constant over such a long period of time. To construct an accurate time and energy budget, one would need to subdivide the time spent at rest into the time spent resting in each of the potential situations that might affect the energy expenditure. Only in this way could a realistic energy cost be assigned to the time budget.

The second problem is that, even if it was possible to construct a detailed time budget, such that the time spent in each situation that might affect energy demands was quantified, one would need to know the actual energy demands associated with each unique combination of factors. For a single behaviour (like rest) there are at least five external factors, some of which are continuously variable traits (like ambient temperature), and at least five internal factors (again, several of which are continuously variable) that can affect energy demands. The potential numbers of combinations of these factors, for even a single behaviour, means that establishing energy costs for all potential conditions would take an incredible amount of time to establish. Once one adds the fact that animals routinely perform a complete repertoire of behaviours, the task facing the constructor of an accurate time and energy budget is truly gargantuan.

One development that has provided some hope that time and energy budgeting might be feasible is the use of copper taxidermy mounts to quantify the complex thermal environment of an animal. The theory behind the use of these models is elegant. There are a potentially diverse range of factors that influence thermoregulatory requirements. Quantifying the effects of such a large array of factors on energy expenditure would be time consuming and complex. However, if the different factors could be reduced to a single dimension, one would need only to quantify the metabolic response of the animal to that single dimension and record in the field the effects of the diverse environmental factors on this single trait. The single dimension is called the standard operative temperature. Standard operative temperatures are measured using a heated copper models of the animal in question covered by skins from dead specimens. Imagine that such a model has been constructed. If it was placed into an incubator at a known air temperature, the electrical energy required to heat the model would be related to the air temperature and the setting of the thermostat inside the model. If the thermostat was set at the body temperature routinely maintained by the animal, the electrical energy required to heat the model would reflect the thermal load placed on an animal at any given temperature. It would be possible to place a live animal into the incubator as well and to measure its metabolic rate in relation to incubator temperature by indirect calorimetry.

The clever part is subsequently to take the heated model out into the environment. Here the model is exposed to a multitude of factors that affect its thermal load: air temperature, ground temperature, sunshine intensity, wind speed, rain, etc. By measuring the electrical energy required to heat the animal in any given circumstance, one can estimate the equivalent incubator temperature for any environmental situation. Complex multidimensional changes in the environmental conditions can all be reduced to their effect on the electrical energy required to heat the model. The multidimensional nature of heat loading is thus reduced to a single dimension (the standard operative temperature). The energy expenditure of a real animal can be estimated knowing the relationship between real animal metabolic rate and incubator temperature. This approach greatly simplifies the construction of time and energy budgets, and its development owes much to the work of George Bakken of the University of Indiana (Bakken and Gates, 1975; Bakken, 1976, 1980, 1992). The method has been criticized recently because of inconsistencies in the responses shown by different models (Walsberg and Wolf, 1996). However, at least part of the problem is in the method used to express error (Larochelle, 1998) and lack of standardization of the models, which can remove much of the intermodel variability.

Although the use of heated models may provide a useful method of condensing the environmental variability into a single dimension that is equivalent to thermoregulatory demands, there is still the problem of quantifying the energy demands connected with activity. Added to this complexity is the fact that many studies have suggested that the costs associated with thermoregulation and activity are not simply additive. This is because animals appear able to utilize heat generated by activity to supply their thermoregulatory requirements (Paladino and King, 1984; Webster and Weathers, 1990; Zebra and Walsberg, 1992; Chai *et al.,* 1998). Consequently there is a saving to the animal when compared with the predicted additive costs. Although thermal substitution is probably the best known, and best quantified example of lack of additivity in energy budgetting, there are other less obvious cases, for example the utilization by bats of the same muscles to generate the forces necessary for them to fly, and the respiratory burst necessary for them to echolocate. Consequently, although the costs of echolocation vocalizations for a stationary bat are very high (Speakman et al., 1989), when a bat is flying these costs disappear (Speakman and Racey, 1991).

Even taking the magnitude of the task of quantifying the energy costs of different behaviours into account, and evaluating whether these costs are additive to the thermoregulatory requirement, there are further logistical obstacles that impair the ability to measure energy expenditure in some circumstances. These are limits imposed by the difficulties of mimicking accurately the situation for which one requires an energy cost, inside a respirometry chamber. It seems probable that there are many situations in which animals routinely find themselves for which we are unlikely ever to derive a realistic energy cost estimate for indirect calorimetry, for example the energy cost of a bird flying in rain. Finally, assuming everything could be measured, each term of a time and energy budget is measured with an error. Travis (1982) has shown that these errors in individual components of the budget can accumulate to produce very wide confidence limits in the eventual estimates of daily energy demands.

D. Direct Measurements of Free-living Energy Expenditure

1. Heart Rate Method

When an animal consumes oxygen it must transport that oxygen from outside its body to the sites where it is being consumed. This involves the ventilatory system to move O_2 from the atmosphere into the blood, and the circulatory system to transport the oxygenated blood to the sites of its utilization. In the ventilatory system, the delivery of O_2 into the blood is a function of the ventilation rate and the tidal volume, as well as the partial pressure of O_2 in the air, relative to the loading characteristics of the respiratory pigment(s) involved. In the circulatory system O_2 delivery to tissues is governed by heart rate, stroke volume and the extent to which O_2 is off-loaded from the blood at the tissues where it is being utilized. Animals may respond to changes in demand by varying different parameters of the supply lines. For example, during activity an increase in O_2 demand at the muscles could be met by an increase in ventilation rate and tidal volume to drive more O_2 into the blood, combined with an increase in heart rate and O_2 extraction efficiency to deliver it to the sites of utilization. If the dominant mode of response involved a single parameter, such as ventilation rate or heart rate, then monitoring of this parameter would provide a method of continuously measuring the animal's metabolism without the need to restrict it in a respirometer. Ventilation rate and heart rate provide the most convenient parameters in this respect, because the electrical signals involved in muscle contraction are relatively easily picked up by appropriately located electrodes, and the resultant electromyographic or electrocardiographic signal can be monitored to measure ventilation and heart rates.

The potential of this approach is enormous because it provides not only a direct field estimate of metabolic rate over protracted periods, enabling an estimate of sustainable metabolism to be derived, but also enables subdivision of the total costs into its components. There are, however, several methodological complexities that need to be surmounted when applying the method. The first problem is that externally mounted electrodes (even if the animal will tolerate their presence) seldom produce signals of sufficient clarity for use. Ideally, then, the whole package, consisting of the electrodes and a unit to transmit or store the information, should be mounted internally. The current minimum size of such packages means that this method is restricted to animals weighing more than 100 g. Moreover, the method is consequently much more invasive than the alternative approaches detailed below. In at least some areas of the world performing such invasive work on wild animals might involve legislative problems.

A second potential problem is that several studies have highlighted the individual nature of the relationships between heart rate and O_2 consumption (Morhardt and Morhardt, 1971; Bevan *et al.*, 1994; Boyd *et al.*, 1995). To generate precise estimates of O_2 consumption, therefore, it is necessary to generate a relationship for each individual animal involved in the experiment. This requires a further period of holding the study animal to establish such a relationship, which may be inappropriate if the animal is performing behaviours in the field where a period spent in the laboratory would be disruptive, such as caring for offspring. Finally, the link between heart rate and O_2 consumption may vary according to the source of energy demands placed on the animal. In response to low-level activity, for example, an animal may cover the increased O_2 requirements completely by changing heart rate, yet during cold exposure increases in O_2 extraction efficiency and stroke volume may make additional contributions. Even during exercise there may be a limit in the manner of response by heart rate alone. At low levels increased O_2 demands appear to be met by changes in heart rate alone, but at high levels changes in O_2 extraction efficiency become important (Grubb, 1982; Jones *et al.*, 1989; Butler, 1991) and stoke volume remains unaffected.

Perhaps because of these potential problems and technical difficulties, relatively few studies have utilized techniques based on this theoretical outline. For example, Weatherly *et al.* (1982) used the opercular movements of fish (= ventilation rates) to evaluate their metabolic rates, but no studies have attempted to link ventilation rates to O_2 consumption in free-living mammals and birds. Probably the most development has occurred in the use of heart rate to estimate metabolism. This has been used extensively to evaluate the metabolic rates of fish (Priede and Tytler, 1977; Priede, 1983; Armstrong, 1986; Lucas and Armstrong, 1991) and birds (Gessaman, 1980; Stephenson *et al.*, 1986; Bevan *et al.*, 1994, 1995a,b,c, 1997), but has been applied only infrequently to free-living mammals.

Although I have detailed above some of the problems of using heart rate to monitor O_2 consumption, it is important to note that several validation studies have been performed using heart rate monitoring at the same time as the doubly labelled water method (detailed below, and on which most of this paper is based), and comparing both of these to standard indirect calorimetry (Nolet et al., 1992; Bevan et al., 1994, 1995c; Boyd et al., 1995; Hawkins et al., submitted). In all these studies, the predicted DEE by heart rate was at least as accurate as that derived by the doubly labelled water method, and in many cases individual estimates were better matched to the estimates by indirect calorimetry. This is perhaps in part because the use of heart rate involves prediction of O_2 consumption, while the doubly labelled water method involves prediction of CO_2 production, and conversion of the latter to energy expenditure is potentially less accurate. The improved accuracy when using the heart rate approach occurs in addition to the wealth of data furnished on temporal partitioning of the costs. There is no doubt that, where it has proved possible to apply this method, the insights generated into the components of field metabolism have been spectacular (Bevan et al., 1995a; Butler et al., 1998).

2. Isotope Elimination Methods

Although there are several methods based on the elimination rates of various isotopes (e.g. Baker et al., 1968; Chew, 1971; Baker and Dunaway, 1975; McLean and Speakman, 1995; Peters et al., 1995) which appear to provide reliable estimates of food intake for free-living animals, none of these methods, with the possible exception of ^{22}Na elimination (e.g. Green 1978; Green and Dunsmore, 1978a,b; Green and Eberhard, 1979; Green et al., 1984; Tedman and Green, 1987; Gales, 1989), has been applied to sufficient animals to provide a large enough database for comparative analysis. The only method that has been used sufficiently for this purpose is the doubly labelled water (DLW) technique. The method was invented by Lifson and colleagues in the 1950s (Lifson et al., 1955) and a complete history of the development of the method can be found in Speakman (1997a, 1998).

3. Theory of Using Doubly Labelled Water

I have elaborated in detail elsewhere the theoretical basis of the technique (Speakman, 1997a). Other treatments can be found in Lifson and McClintock (1966), Nagy (1980), Speakman and Racey (1988), Tatner and Bryant, (1989) and Bryant (1989). Briefly, the method depends on the fact that isotopes of oxygen in body water are in complete and rapid exchange equilibrium with the oxygen in respiratory carbon dioxide. This means that if an isotopic label of oxygen is introduced into body water it will be eliminated from the body primarily by the flow of water and CO_2 leaving the body. In contrast, an isotopic label of hydrogen will leave the body primarily only as water. If both isotopic labels are introduced at the same time (hence doubly labelled) the difference in their respective elimination will reflect the CO_2 production and thus indirectly the energy expenditure (see above). It is perhaps important to reiterate at this point that conversion of CO_2 production to energy expenditure is less accurate than the conversion using O_2 consumption. The problem of converting DLW derived estimates of CO_2 production to energy expenditure, in the absence of a known RQ, may therefore limit its accuracy.

After an isotopic label has been injected into an animal, the time course of its enrichment follows a complex path reflecting several different processes (Figure 4). Before injection the isotopic enrichment is at some background level. For heavy oxygen (^{18}O) the background level of the isotope in most living systems is around 2000 ppm and for heavy hydrogen (^{2}H) it is around 150 ppm. After injection, the enrichment in the blood rises very rapidly as the pool of isotope at the injection site diffuses into the bloodstream. Over time, isotopes

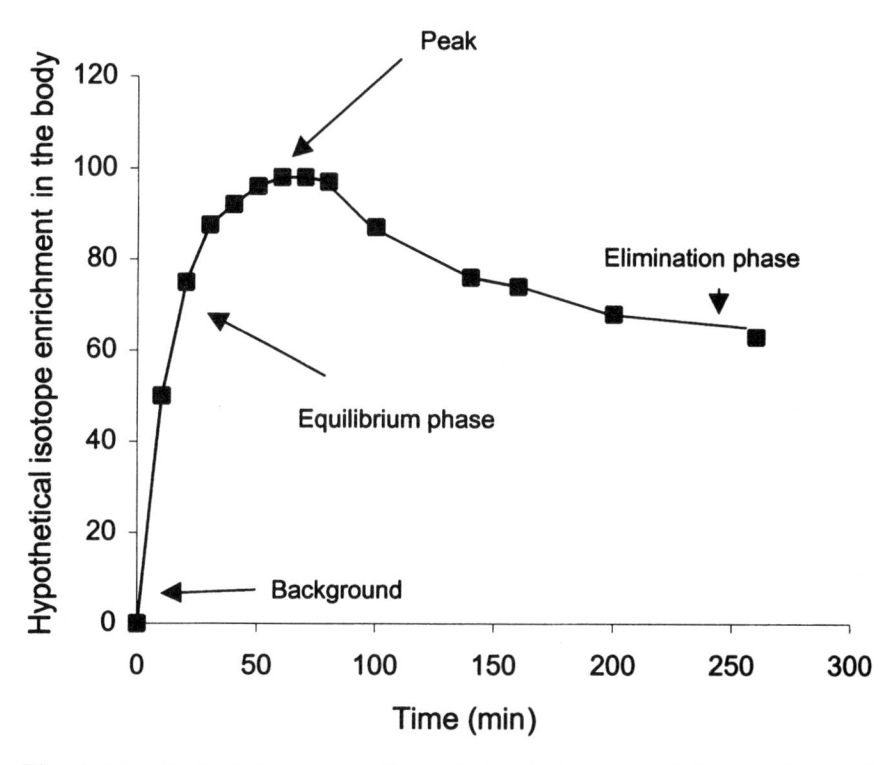

Fig. 4. Hypothetical time course for variation in isotope enrichment of a small mammal injected with labelled water. The animal is injected at time 0.

will be eliminated from the system in water and (for oxygen) expired CO_2. At some point a dynamic equilibrium will occur where the rate at which isotopes flood into the system from the injection site exactly equals the rate of elimination. The result is a stable isotope enrichment in the blood which is generally called the plateau phase. The enrichment may home in on the plateau from above or below depending on the method of isotope administration. The plateau may also last a variable time depending on the dynamics of the system. Eventually, however, when all the isotope has flooded into the system the enrichment starts to decline because only the elimination process dominates it. The elimination follows a negative exponential. At first lots of isotope is removed because its enrichment in the body water and eliminated products is high. However, as the enrichment declines, the amount lost in each volume of water and CO_2 also declines, and the rate of decline becomes progressively slower and slower, until the enrichment reaches the background level again.

If the difference between the isotope enrichment in the body and the background enrichment is converted to logarithms, the curved exponential becomes linear, and the gradient of this linear relationship allows us to characterize the rates of isotope elimination from the system. The gradient is generally called k_o or k_d for the oxygen and hydrogen isotope respectively. One other thing is needed to convert these elimination rates into actual flows of materials carrying the isotopes, and that is the dilution volumes in which the isotopes are distributed (generally called N_o and N_d respectively for oxygen and hydrogen). In general, the flow of material carrying oxygen is approximated as $k_o \cdot N_o$ and the flow of material carrying hydrogen is approximated as $k_d \cdot N_d$.

There has been considerable debate in the literature, however, about the ideal manner in which k_o, k_d, N_o and N_d should be combined to estimate the CO_2 production of an animal. One of the alternatives concerns the use of both N_d and N_o in the equation, or whether N_o should be used alone. These have been respectively termed the two-pool and single-pool methods. The difference between these models in the estimated CO_2 production depends on several factors but the most important is the ratio of k_o to k_d. When this ratio approaches unity, the models produce widely divergent estimates of CO_2 production. For most animals, however, the difference is likely to be between 5% and 25%. The choice of pool model is consequently not trivial. Validation studies in recent years are strongly pointing towards both methods being appropriate, but in animals of different body sizes (reviewed in Speakman, 1997a). Moreover, there are good theoretical grounds for expecting this to be the case (Speakman, 1987).

The exact point at which the different models become appropriate has not yet been determined precisely. However, it appears that for animals weighing in excess of 4–5 kg the two-pool model is likely to be most appropriate, whereas smaller animals should be measured using the single-pool model. Unfortunately most studies of animals, independent of their size, have utilized the single-pool model. This generally leads to an overestimate of the metabolic energy expenditure of larger animals. This would not be too serious a problem if it were not for the fact that it is generally impossible to recalculate the CO_2 production estimates using the two-pool model because the data necessary to make these recalculations are generally not quoted in the papers. This is a potentially important problem, which might compromise reviews of energy expenditure that utilize data from animals drawn from across the boundary, which delimits single-pool and two-pool determinations. Unfortunately this encompasses almost all the reviews of DEE based on the DLW technique that have been published to date.

E. Summary and Data Inclusion Criteria for the Present Review

In conclusion, the measurement of energy expenditure by free-living animals is difficult. The time and energy budget method provides an estimate of expenditure. However, in general, even if the time and energy budget methods employed are thorough, and include detailed quantification of the effects of thermal and activity regimes on energy demands, combined with detailed time budgets, the complexity of interactions in energy demands makes the resultant estimates prone to considerable error (Travis, 1982).

In contrast to the time and energy budget approach, the heart rate telemetery and isotope elimination methods provide more direct estimates of daily energy demands which take into account the wide diversity of potential interactions in the individual elements that go to make up the total expenditure. Heart rate estimates have been applied to too few animals yet to make a substantive review possible. For the doubly labelled water isotope elimination method, validation studies suggest that the resultant estimates have a mean individual error of around 12–15%. Individual estimates, therefore, may be unreliable. However, group mean estimates of CO_2 production are generally very accurate (1–3% in error). What one sacrifices with the isotope approaches is a detailed breakdown of the component costs, which contribute towards the total cost. In theory, at least, this is provided by the time and energy budget method, but the reliability of the subdivision of the budget is open to considerable speculation, given the poor conformation of the totals to simultaneous DLW estimates. Only the heart rate method may allow a detailed component breakdown as well as a reliable overall estimate to be derived.

Although several isotope methods are available, only one (the DLW method) has been used on sufficient species to provide a comparative database for investigation. Previous reviews of this database have generally covered the complete body mass range. However, the general use of the single pool DLW model for studies of larger animals makes inclusion of larger animals suspect.

In view of these facts, the present study concerns a review of all DLW measurements made on free-living mammals weighing up to 4 kg. The database searched for included information on all studies published between 1970, when the first measurements were published for a free-living wild mammal (Mullen, 1970), and June 1998. Many studies involve groups of animals measured in different conditions, for example in different seasons or at different stages of their annual cycles. In all cases I have taken the mean energy expenditures for homogeneous groups of animals and have not pooled the data across all conditions for any particular species. This is because it is apparent that energy demands fluctuate with many different factors and pooling data together masks these effects. However, this does raise problems of independence in the data, since some species provide several data points to the analysis. I have treated this problem differently and explicitly at different points throughout the paper. In addition to estimates of energy expenditure, the papers were scoured for information on the following for each group that provided an energy expenditure estimate: body mass of the animals, ambient temperature, latitude of the study site, altitude of the study site, diet of the group, sex and reproductive status of the animals.

IV. RESULTS

A. Overview of the Database

I found a total of 69 papers containing data on small mammal field metabolic rates measured using the doubly labelled water technique that had been published between 1970 and 1998. I added to these studies an unpublished study from my own research group. The published and unpublished studies together included a total of 184 measurements of homogeneous groups of adult mammals in different situations (e.g. seasonal, altitudinal or reproductive classes) on a total of 74 different species (Appendix B). I excluded from the review estimates that had been made on immature or juvenile mammals that were still growing. Each 'measurement' includes between 1 and 20 individual estimates of energy expenditure across a group of individual animals (median 8 per measurement). The total number of individuals contributing to the database is uncertain. In some studies the same individuals contributed repeated measurements, but, while this is mentioned, the authors did not quantify it precisely. Moreover, it is possible (probable) that in some cases the same individual measurements contributed to multiple papers by the same authors. In the database I have treated these as independent measurements, which they may not be. Thus, for example, Berteaux and colleagues published three papers in 1996 and 1997 on the meadow vole (*Microtus pennsylvanicus*). These papers address different biologicial aspects of the species and each includes a different sample size, different mean values for body mass and metabolic rate. These probably represent different subsets of a single data set. The extent of overlap between papers, however, is impossible to ascertain from the published information. The same is true of the two papers by Salsbury and Armitage (1994, 1995) on yellow-bellied marmots (*Marmota flaviventris*) which also include different means and sample sizes but probably refer to subsets of the same group of animals. Despite these problems, the total number of individuals contributing to the 184 measurements is probably in excess of 1000.

There are currently 20 recognized orders of extant mammals (Novacek, 1992). Of these orders, however, five contain no representatives that weigh less than 4 kg (the cetaceans, sirenians, proboscidens, perissodactyls and tubulidentates). Of the remaining 15 orders, data were available for representatives of eight of them. At present we are still lacking any information on the daily energy requirements of small (< 4 kg) representatives of the monotremes, dermopterans, tupiids, artiodactyls, pholidotids, macroscelids, hyracoids and primates. I am aware that measurements for small representatives of at least three of these groups were in progress in late 1998, and early 1999.

The presently available data set was dominated by measurements made on rodents, which accounted for almost half the total ($n = 89$ measurements on 32 species). The second largest order represented in the data was the marsupials with a total of 68 measurements on 24 species. Five other orders contributed smaller numbers (Chiroptera, $n = 11$ measures of nine species; Insectivora, $n = 6$ measures of five species; Lagomorpha, $n = 2$ measurements of a single species: Edentata, $n = 3$ measures of a single species; and Carnivora, $n = 5$ measurements of three species). To an extent this bias reflects the large diversity of small rodents. However, this is not the entire reason. The bats, for example, comprise almost 1000 species, all of which weigh less than 4 kg (Altringham, 1996), yet we have data on daily energy expenditures for only nine of them (less than 1% of the total). The lagomorphs are similarly grossly under-represented in the available data. Despite rodents being well represented, there are still some large groups of rodents for which we have no data on free-living energy demands (e.g. the hystricomorphs).

The latitudinal distribution of available measurements (Figure 5) shows a large peak around 30–40°N, with 71 of the 184 measurements in this 10° latitude band. This is in part a consequence of the intensive research efforts of three researchers (Ken Nagy at the University of California at Los Angeles in the USA, Allan Degen at Side Boquer in Israel, and Brian Green at CSIRO in Australia). Between them, these three research workers with their collaborators and students have made 127 (nearly 70%) of the 184 published measurements. Since these researchers all live around 35°N or S and have studied their local faunas extensively, this accounts for the large number of measurements made at this latitude. As two of these workers (Nagy and Degen) have specialized in desert ecology, the data at these latitudes pertains mostly to desert living mammals, in particular desert-living rodents. We have comparatively few data for mammals that inhabit the tropical and temperate forests, and almost no data ($n = 5$ measurements for two species) for the arctic.

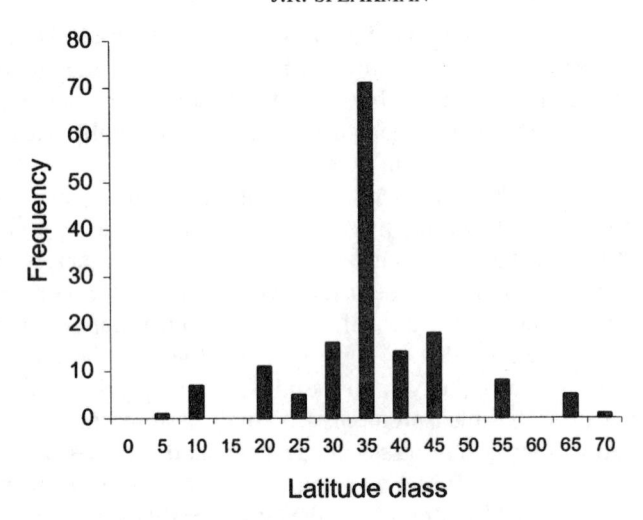

Fig. 5. Latitudinal distribution of measurements of field metabolic rate included in the database. There was a very strong bias around 35°N and S.

Geographically the data originate mostly from the North America ($n = 86$), Eurasia ($n = 22$) and Australia ($n = 67$). Relatively few data are available for South America ($n = 3$) and all these refer to measurements made north of the Amazon. Similarly, sparse data are available for Africa ($n = 6$ including Madagascar). This geographical bias may reflect, at least partly, the high financial costs of applying the DLW methodology. We still have no measurements of the energy demands of small mammals living in mainland Asia (including all of Russia and China) and the Indian subcontinent.

Temporally the pace at which measurements are being added to this database is increasing. In the 1970s data were accumulating at about two measurements each year. This doubled during the 1980s. The first serious attempt to summarize the available information was made by Nagy (1987), who included 19 measurements from 13 species of small mammal in a more wide-ranging review, which also took into account field energy expenditures of larger mammals and birds. Koteja (1991) included 24 species of mammal weighing less than 4 kg in his review of field energy demands, and Karasov (1992) included 17 species of mammal in his review of links between FMR and RMR. The pace of data collection, however, doubled again in the 1990s, so that currently about eight measurements per year are being added to the database. More recent summaries have therefore been able to include larger samples. Hence, Nagy (1994) included 34 measurements, Degen and Kam (1995) included 38 measurements, and Hammond and Diamond (1997) included 35 measurements of small mammals (although in this latter paper many of the 'field' measurements referred to captive animals and included measurements based on food intake rather than DLW). The current review is the most comprehensive data set yet accumulated for this group of animals. Undoubtedly, however, it is not complete and there are probably more data out there in sources that are inaccessible to current bibliographic searching methods (notably abstracts, conference proceedings and books). I apologize to those authors whose work I have neglected, and would be glad to receive information to improve the database and rectify the omissions.

Given this overview it is apparent that the data on field energy demands of small mammals are very sparse. I estimate that there are about 2500 species of mammal weighing less than 4 kg and consequently we currently have measurements for about 3% of them. Even if the pace at which data accumulates continues to double in each of the next two decades, by the year 2020 we will still have measurements for fewer than 20% of the total. Such a pace of increase, however, seems unlikely to be sustained with trends for science funding increasingly being dominated by molecular work at the expense of whole-animal physiology. The patterns of data accumulation are currently very biased in favour of rodents, and in particular North American and Eurasian representatives of this group. These biases should be borne in mind when evaluating the data. As information continues to accumulate for different groups of mammals living in more diverse habitats and geographical regions, our views about the habitual levels of energy demand and the relative importance of factors influencing them may change, certainly in minor ways and perhaps radically. Because of the inherent limitations in the available information it is probably the case that this review can serve only to summarize the extent of our current ignorance rather than document the degree of our insight.

B. Factors Influencing Daily Energy Expenditure

1. Body Mass

Including all the data pooled across all animals in all conditions, ignoring any potential problems of pseudoreplication and lack of independence of the data, the dominant factor influencing the field metabolic rates of small mammals was body mass. There was a linear relationship between $\log_e$ body mass and $\log_e$ field metabolic rate (Figure 6) which explained 85.9% of the variation in field metabolism. The least squares fit equation was:

$$\log_e \text{FMR (kJ day}^{-1}) = 2.022 + 0.627 \log_e \text{ body mass (g)} \qquad (1)$$

($F = 1111.8$, 1,182 d.f., $P < 0.001$). The gradient of this relationship was significantly shallower than the expected gradient for basal metabolism based on the Kleiber relationship (0.75; $t = -6.55$, $P < 0.001$), and was also significantly shallower than the value 0.67 ($t = 2.29$, $P < 0.05$) which is expected from the surface law. Inspection of these data (Figure 6), however, reveals a potential bias because the largest mammal represented is the three-toed sloth (*Bradypus tridactylus*; Nagy and Montgomery, 1980), which has a particularly low

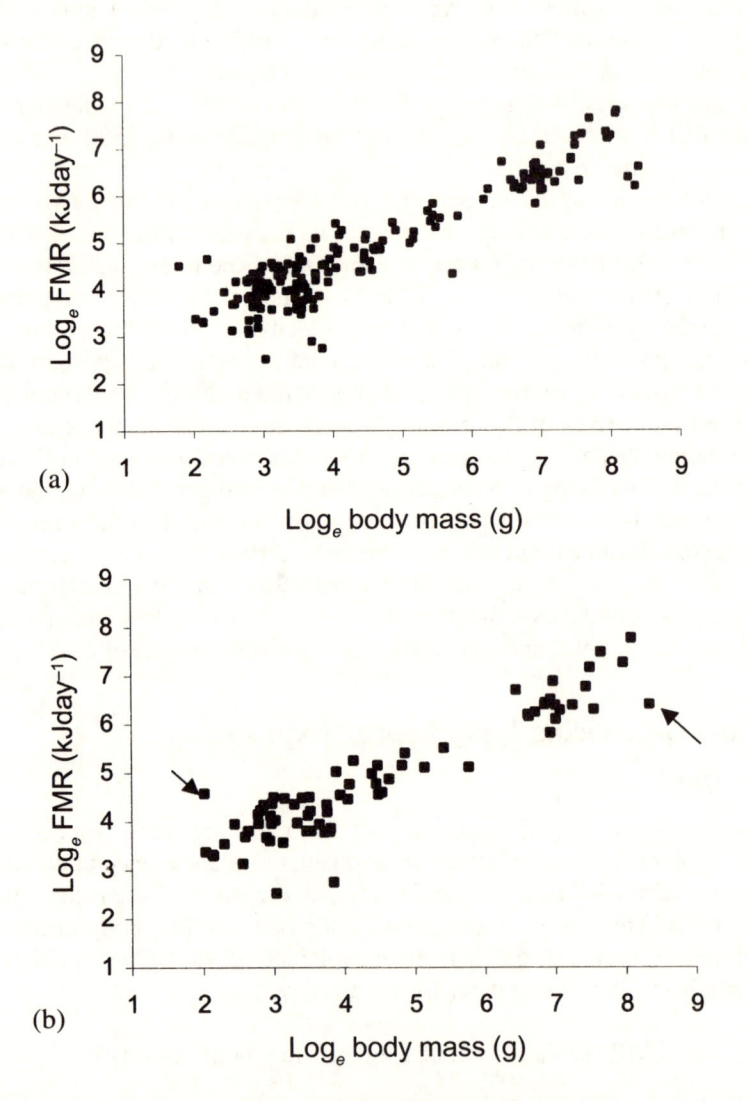

Fig. 6. Logged field metabolic rate (FMR) plotted against body mass for (a) all 184 data points collected across 74 species and (b) data averaged across all measurements for each species to produce single species points ($n = 74$). Data representing the three-toed sloth and the common shrew are indicated by arrows.

metabolic rate for its body mass, and among the smallest mammals represented is the common shrew (*Sorex areneus*; Poppitt et al., 1994), which has a very high metabolic rate for its mass. Because of the large leverage on gradients exerted by data at the extremes, this unfortunate coincidence will tend to make the observed gradient including all the data shallower. Excluding these two sets of measurements ($n = 5$ measurements in total) yields the following equation:

$$\log_e \text{FMR (kJ day}^{-1}) = 1.878 + 0.659 \log_e \text{body mass (g)} \qquad (2)$$

($r^2 = 88.0\%$; $F = 1302.3$, 1,179 d.f., $P < 0.0001$). The gradient of this relationship is almost identical to that anticipated by the surface law.

The validity of using least-squares regression to fit the gradient fitted to these data is in some doubt because the body masses of the animals are not measured without error (Riska, 1981; Rayner, 1985; LaBarbera, 1989). The reduced major axis (RMA) fit gradient fitted to the uncensored data set (including shrew and sloth) was 0.676, which did not differ significantly from 0.66 but was still significantly different to the Kleiber value of 0.75. In the censored data set (excluding shrew and sloth data), the RMA gradient was 0.702, which was marginally significantly greater than the surface law prediction ($t = 1.98$, $P < 0.05 > 0.01$) but substantially lower than the 0.75 prediction.

Using the censored data set the least-squares fitted curve predicts energy expenditures for average 10, 100, 1000 and 4000-g small mammals of 29.9, 136.8, 625.9 and 1564 kJ per day. The RMA curve predicts field metabolic rates of 27.8, 135.5, 630.2 and 1577.8 kJ per day^{-1} respectively. The logged relationships tend to minimize the perceived extent of differences to these mean values. The 95% predictive intervals for these mean predictions at any given body mass spans an approximate 4-fold range. This can be illustrated by considering the 95% predictive interval at each of the above mean masses. At 10 g the range is 14.0–63.7, at 100 g it is 64.4–290.5, at 1000 g it is 293.2–1336.2 and at 4000 g it is 727.3–3363 kJ per day. It is apparent from these ranges that the top of the 95% predictive interval is approximately equal to the bottom of the interval for a body mass 10 times greater. It is literally true, therefore, that there are some small mammals that have absolute metabolic rates equal to those of other mammals 10 times their own body mass, and this is true in the data set that excludes the most extreme examples of the sloth and the shrew. These two species make the most remarkable comparison. The shrew weighs only 9 g and has a metabolic rate of about 100 kJ per day (Poppitt et al., 1994), while the sloth weighs over 400 times more but expends energy only five times faster (489 kJ per day; Nagy and Montgomery, 1980).

Although body mass explains almost 90% of the variability in metabolic rates of small mammals included in the data set, these ranges clarify that the predictive value of the relationship between mass and field metabolic rate for any particular species is minimal. This is apart from the inherent biases in the selection of species that has already been highlighted. Using the predictive equation, the expected values for all the available measurements can be generated, and the difference between these values and the actual values calculated. Given the nature of least-squares regression, these deviations sum to zero. However, ignoring the signs of the differences provides an estimate of the average error that would occur if the predictive equation were utilized to predict field metabolic rates. On average, predicted metabolic rates would be in error by 124% using this approach. This is an optimistic estimate because the same data used to derive the gradient are also used to evaluate the errors.

In the preceding analysis I included each 'measurement' as an independent datum. This follows the procedure adopted by Nagy (1987) in his review of field metabolic rates. Such an approach has been strongly criticized because it includes multiple measurements for some species (up to 16 in the case of the kangaroo rat *Dipodomys merriami*). The multiple appearance of data for some species might be regarded as pseudoreplication. For example, Green (1997, p. 153) states: 'The regression analyses (by Nagy, 1987) are flawed due to the multiple representation of some species'. In response to this type of criticism, Nagy (1994) reanalysed and updated his analysis using only a single datum for each species. However, this is open to a different criticism because the condensing of data obscures much of the variability between species which can be attributed to the different environmental conditions during their measurement: some mammals are measured in winter, for example, and others in summer. Hence Green (1997, p. 153) states: 'Another review by Nagy...does not use multiple data for any species. However, the mean FMR value derived for some species is derived from different seasons and cohorts'. This is a clear case of being damned if you do and damned if you don't. Whatever approach you take with these data, somebody will take offence at it.

I have taken two approaches to the problem. The first is to emulate the approach taken by Nagy (1994) and to condense the data for each species into a single datum reflecting the average body mass and average field metabolic rate across all the 'measurements' available for that species. As pointed out by Green (1997), these species data are not directly equivalent because some represent averages across several seasons and conditions, whereas others represent measurements for only single seasons or particular groups (e.g. males). Pooling the data in this manner yields a total of 74 'species' data points. In this more restricted sample there was still a very dominant effect of body mass on metabolic rate (Figure 6b). Including all the data the least-squares fit regression:

$$\log_e \text{FMR} = 2.062 + 0.621 \log_e \text{body mass (g)} \qquad (3)$$

explained 86.4% of the variation in FMR ($F = 450.9$, 1,72 d.f., $P < 0.001$). Excluding the data for the shrew and sloth (above), which exert high leverage on the gradient, results in the equation:

$$\log_e \text{FMR} = 1.929 + 0.650 \log_e \text{body mass (g)} \qquad (4)$$

($r^2 = 88.8\%$; $F = 523.5$, 1,70 d.f., $P < 0.001$). These equations, their r^2 and probability values are almost identical to those derived for the entire data set. In practice, therefore, although inflation of degrees of freedom is a valid critique of the analysis which includes multiple data for each species, the problems of lack of independence of the data and pseudo-replication appear to have only minor effects on the fitted equations for the effect of body mass on FMR.

There is, however, a further problem with using species averages as independent data in the relationship between FMR and body mass, which has not been considered previously in the majority of allometric summaries of FMR in mammals published to date (Nagy, 1987, 1994; Koteja, 1991; Bryant, 1997; Green, 1997; Hammond and Diamond, 1997; Speakman, 1997b). This problem is the potential lack of independence between species because of their shared evolutionary history (Felsenstein, 1985; Harvey and Pagel, 1991; Garland *et al.*, 1992). Two closely related species may be both large and have high field metabolic rates. In the preceding analyses, these two species would contribute two independent data to the analyses. However, the body masses and metabolic rates of these species might not reflect an independent effect of mass on metabolism, but might occur because of the shared evolutionary history of the two species which has resulted in both of them having the same character traits for metabolism and body mass.

For example, in the database there are several closely related pairs of species. The two spiny mice, *Acomys russatus* and *A. caharinus*, and the two shrew tenrecs, *Microgale talazaci* and *M. dobsoni*, are examples of such pairs. Within each pair the component species have shared the vast majority of their evolutionary history and have diverged from each other only recently. In both cases the species within the pairs have very similar values for body mass and field metabolic rate. For *A. russatus* and *A. caharinus* the body masses are 38.3 and 45 g respectively and the metabolic rates are 51.8 and 47.6 kJ day^{-1}. However, for *M. dobsoni* and *M. talazaci* the values for mass are 42.6 and 42.8 g respectively and the estimated metabolic rates are 77.1 and 66.5 kJ day^{-1}. The masses of the mice and shrew tenrecs are also similar, but the metabolic rates of the shrew tenrecs are much higher (by over 40%). The problem is whether these are really four independent data, or whether the sample size is inflated by the inclusion of such recently diverged species as independent points. Most researchers agree that these closely related species should not be regarded as truly independent data because of this shared evolutionary history (Felsenstein, 1985).

Several methods have been derived in the past decade to overcome this problem. The most popular of these methods is the calculation of phylogenetically independent contrasts for the respective character traits and seeking relationships between these phylogenetically independent data (e.g. Harvey and Pagel, 1988, 1991; Garland et al., 1992). As far as I am aware, the only study that has performed such an analysis on the context of the field metabolic rates of small mammals is the study of Ricklefs et al. (1996) which involved 32 species (see also analysis by Nagy et al., 1999). Unfortunately, this previous analysis is slightly compromised by two mistakes in the phylogenetic tree (p. 1059) used to derive the independent contrasts. First, the bat *Macrotus californicus* is misspelt (*Microtus californicus*) and, perhaps as a result of this misspelling, is misclassified among the microtine rodents (genus *Microtus*). The node connecting this bat to the microtines (node z5) exerts a large influence in the regression of the residuals (Ricklefs et al., 1996. Figure 3) and probably has further knock-on effects in the phylogeny. A second potential problem is the geomyids in their phylogeny are linked more closely to the sciurids than to the muroids, yet several authorities suggest the geomyids should be placed among the muroidomopha (e.g. Eisenberg, 1981). The fact that removing the effect of phylogeny in their analysis had no impact might thus in part reflect the erroneous phylogeny used to test the hypothesis, since the contrasts methods depend on a 'known and correct phylogeny' (Harvey and Pagel, 1991).

These misclassifications may, however, be less important than it may first appear, if only because the phylogeny of the mammals as a whole has been a matter of considerable debate. Over the past 15 years disputes have occurred over the positioning of several major groups. For example, there has been a protracted debate over the Chiroptera, with many authorities claiming that this is a monophyletic group (Simmons and Geisler, 1998; Baker et al., 1991), but others claiming that the megachiropterans should be placed among the primates (Pettigrew et al., 1989, 1995). Even among those favouring monophyly of the Chiroptera as a whole, there has been some recent debate over whether the microchiroptera is a paraphyletic grouping (Hutcheon et al., 1998). Most recently the placing of the Chiroptera among the Archonta is being questioned, following separation of the bats from other Archonta on the basis of sequencing of the Cytochrome oxidase II genes (Adkins and Honeycutt, 1991). Another example is the dispute over the placing of the hystricognath rodent, the guinea-pig (*Cavia porcellus*), with some authorities claiming this species should be removed from the Rodentia and perhaps placed among the lagomorphs (Graur et al., 1991; Novacek, 1992). Several other major disputes concern groups that contain only members weighing more than 4 kg and do not directly concern us here (for example, the association of hyraxes with the elephants and sirenians; Novacek, 1992). Nevertheless it is clear that our understanding of the phylogenetic relationships of the mammals is not fully resolved. Since applying the phylogenetic independent-contrasts method assumes that there is a complete known phylogeny to work from, the incomplete and disputed nature of the mammalian phylogeny is a stumbling block to applying this method. If there is no effect of phylogeny, this may reflect only the fact that the phylogeny being used is wrong. At present there is no way out of this caveat.

To apply the phylogenetic contrasts method to the data on FMR and body mass, I constructed a phylogeny for the 74 species for which data were available. Since even the nature of connections in the mammalian phylogeny is disputed, I assumed the branch lengths for each taxonomic level were equal (following Ricklefs et al., 1996) and did not attempt to construct a phylogeny with branch lengths directly proportional to evolutionary time. The phylogeny is shown in Appendix C with each of the 73 nodes coded. The phylogeny is based on the following sources: Eisenberg (1981), Altringham (1996), Simmons (1998), Simmons and Geisler (1998), Novacek (1992, 1994), Ellerman (1941), Luckett and Hartenberger (1985), Brownwell (1983), Koopman (1994), Hall (1981), Sarich (1985), Ricklefs et al. (1996). Higher level nodes, particularly the placement of the glires (rodents and lagomorphs) between the edentates and the bats/insectivores/carnivores (node 3), follows Novacek (1992, Figure 1). Placement of the geomydis with the muroids rather than the sciurids (node r11) follows Eisenberg (1981). I calculated the phylogenetically independent contrasts for FMR and body mass for each of the nodes using the procedures outlined by Harvey and Pagel (1991). I used the square root of the product of branch length and variance to standardize the contrasts (Felsenstein, 1985; Harvey and Pagel, 1991). The sign of the contrast was ignored and the variance in the rates of change along the branches was calculated for each of the major groups separately, to account for any differences in variance between the major groups (see Martins and Garland, 1991).

There was significant heterogeneity in the variances along the branch lengths of the different groups. In particular the variance for the Chiropetra and Insectivora was only about one-tenth that of the other groups. This highlighted the importance of calculating the variance in change rates for each group separately, and incorporating these differences into the standardization. There was a significant positive relationship between the $\log_e$ of the standardized phylogenetically independent contrast of field metabolic rate and the $\log_e$ of the standardized independent contrast for body mass (Figure 7). The figure shows several outlying points at the lower end of the plot. These were the contrast between the two *Microgale* species (node i1), between *Peromyscus crinitus* and the other two *Peromyscus* species (node r26), between the two *Mus* species (node r17), and between *Hemibeldideus lemuroides* and the two *Pseudochirus* species (node m15). In all these four cases the situation was the same. The branch length was short and the contrast in mass was very small, but there was a larger difference in the field metabolic rates. This effect might be expected to occur occasionally under the Brownian motion model of evolution on which the contrast calculation is founded, since early in the diversification of the traits there may be periods when by chance

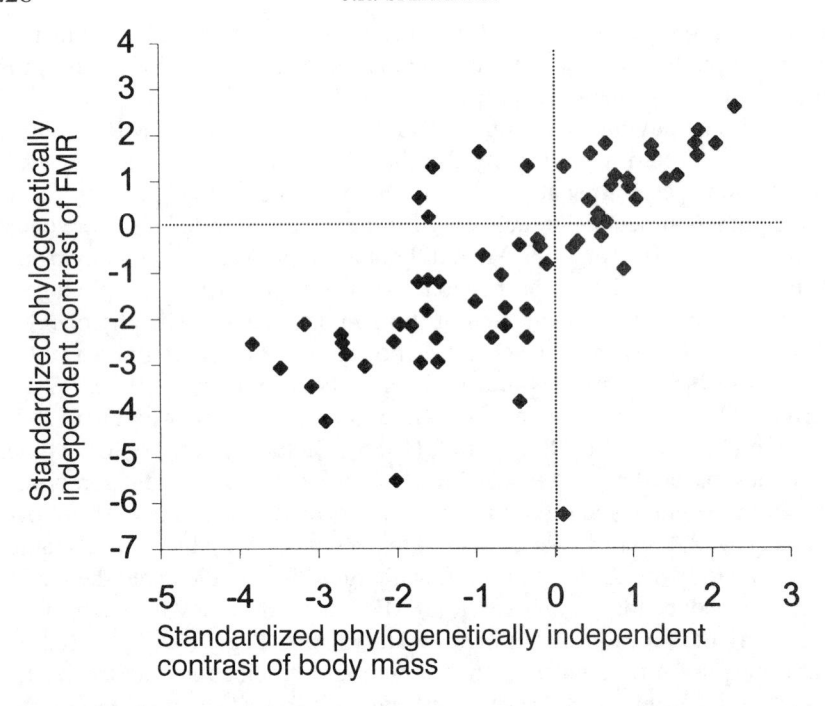

Fig. 7. Standardized phylogenetically independent contrasts of field metabolic rate (FMR) plotted against the standardized phylogenetically independent contrasts for body mass.

one of the traits drifts back to the starting value while the other trait does not. Perhaps a certain degree of divergence is necessary before the contrasts are sufficient to detect associations between the traits. There was no indication of any problem with the nodes around *Ammospermophilus* and *Thomomys bottae* which Ricklefs *et al.* (1996) eliminated as outliers in a similar analysis. This is perhaps because I associated the geomyids with the muroids, rather than the sciurids (see above).

Even including these four outliers, there was a significant relationship between body mass and field metabolic rate. In the least-squares fit regression, body mass contrast explained 52.4% of the variation in the field metabolic rate contrast. The equation was:

$$\log_e \text{FMR contrast} = -0.07 + 0.588 \log_e \text{body mass contrast} \qquad (5)$$

($F = 78.01$, 1,71 d.f., $P < 0.001$). Excluding the four data where mass change was negligible at the node resulted in an improved r^2 and a different equation:

$$\log_e \text{FMR contrast} = -0.09 + 0.76 \log_e \text{body mass contrast} \quad (6)$$

($F = 125.6$, 1,68 d.f., $P < 0.001$; $r^2 = 65.3\%$). Because the explained variation was much lower than the regressions involving the tip data, the variation around these fitted gradients was much greater. Including all the data, the gradient did not differ from the expectation based on the surface law ($t = 1.08$, $P > 0.05$) but did marginally reach significance at the 5% level for difference to the expectation from Kleiber ($t = 2.42$, $P = 0.045$). The gradient fitted excluding the outliers did not differ from either prediction. The significant relationship in the phylogenetically independent contrasts of FMR and mass indicates that the link between field metabolic rate and body mass is not a phylogenetic artefact of using species as the sampling unit.

Although in one sense repeated measurements for a given species (and even species points) are not independent, in all the cases included in the database the measurements are separated because they differ with respect to some presumed variable which is likely to affect energy expenditure. In a real sense each individual datum represents an independent set of conditions where species is one of only a number of multivariate predictors for FMR. The problem of multiple representations of each species is only a problem therefore to the extent that the data are not fully balanced with respect to all the predictor variables. Compressing the data so that each species appears once (and calculating the phylogenetically independent contrasts) removes the independence problem but introduces a second problem, as highlighted by Green (1997), that the data are not equivalent. A third problem, however, is that all the within-species variability, reflecting the effects of variables apart from body mass, is completely lost. My second approach to the problem is to leave all the individual measurements included in the analysis as 'independent' data (without phylogenetic correction), and attempt to explain the resultant residual variability in FMR as a function of these other factors.

2. Ambient Temperature

Ambient temperature measurements were available, or inferred from location and time (Oliver and Fairchild, 1984) for 160 of the 184 FMR measurements. Body mass and temperature were unrelated predictor variables ($r^2 = 0.5\%$). There was a strong negative relationship between the residual FMR (from the least-squares regression on mass) and the ambient temperature (T_A) (Figure 8). The least-squares fitted regression:

$$\log_e \text{residual FMR} = 0.429 - 0.0258\, T_A\, (°C) \quad (7)$$

explained 32.6% of the residual variation ($F = 76.2$, 1,158 d.f., $P < 0.001$). Examination of the plot reveals three data at very low temperatures, which

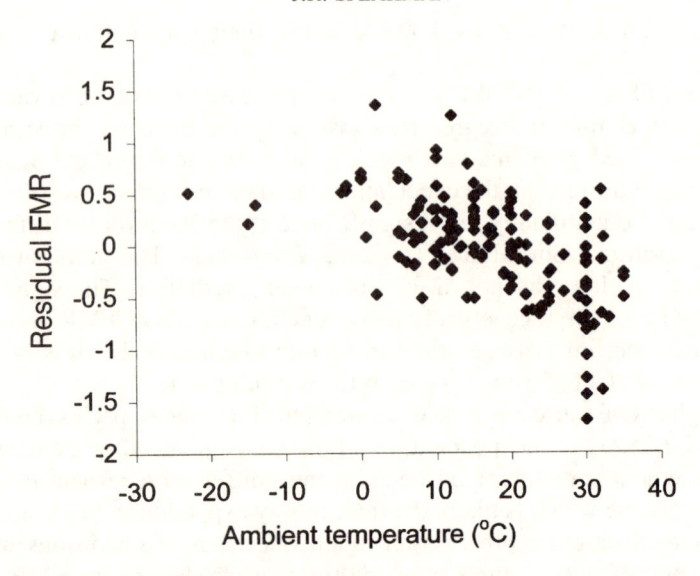

Fig. 8. Residual field metabolic rate (FMR) after taking account of the effect of body mass plotted against the ambient temperature (°C) at the study site. All data were included ($n = 162$ measurements across 60 species).

make the temperature data negatively skewed. These data could have a strong leverage on the regression and hence I removed them to explore their effect on the relationship between residual FMR and temperature. The effect of these three points, however, was minor, resulting in an elevated intercept and gradient but unaltered r^2 and F value (r^2 excluding the data = 34.5% and the equation was $y = 0.506 - 0.0297x$; $F = 81.6$, $P < 0.001$).

Including all the data and entering both temperature and body mass as independent predictors resulted in the following equation:

$$\log_e \text{FMR (kJ day}^{-1}) = 2.382 + 0.644 \log_e \text{ body mass (g)} - 0.0261\, T_A\, (°C) \quad (8)$$

($F = 695.4$, 2,157 d.f., $P < 0.001$, $r^2 = 0.899$). Excluding shrew and sloth data results in a slightly different equation where the effect of mass is more pronounced and the effect of temperature slightly diminished, and the r^2 improved:

$$\log_e \text{FMR (kJ day}^{-1}) = 2.22 + 0.670 \log_e \text{ body mass (g)} - 0.0236\, T_A\, (°C) \quad (9)$$

($F = 790.8$, 2,152 d.f., $P < 0.001$; $r^2 = 0.912$).

Although the effect of ambient temperature on residual FMR in small mammals is very marked, understanding the reasons for this relationship is less clear. On one hand it may seem intuitively obvious that endothermic animals require greater heat production to sustain their high body temperatures as ambient temperature declines. All other factors being equal, one would expect decreases in ambient temperature to require increased energy expenditure. Indeed this effect has been demonstrated many times by indirect calorimetric studies of many species of small mammal in the laboratory. In the field, however, all other factors are not equal, and animals respond the changes in their immediate ambient temperature in a variety of different ways to ameliorate the magnitude of the temperature effects. For example, during winter when it gets colder, animals may respond to the decreased temperature by building better insulated nests (Casey, 1981) and reducing their aggressive behaviours to allow huddling together as a mechanism to conserve heat loss (Contreras, 1984; Karasov, 1983). Temperatures within nests of huddling animals are consequently substantially higher than the reported ambient temperature, and the summed thermal environment experienced by the animals may differ substantially from that expected from ambient temperature alone. A rather simplistic interpretation that this observation represents the expected thermoregulatory effect of decreased ambient temperature on metabolism is consequently not warranted. This is particularly the case because attempts to establish such an effect in other data on field metabolic rate, pertaining to the endothermic birds, have failed to demonstrate the same relationship (Bryant, 1997).

For two species there were multiple measurements covering a wide range of ambient temperature conditions. These were the pocket mouse (*Perognathus formosus*) (Mullen, 1970; Mullen and Chew, 1973) and the kangaroo rat (*Dipodomys merriam*) (Mullen, 1970; Nagy and Gruchacz, 1994). Pooling data across studies in both these cases, there were strong effects of the temperature of the measurement site at the time of the measurement and the FMR (kJ day^{-1}) (Figure 9). The effect of temperature across species was consequently evident in at least two species where sufficient repeated measurements were available to examine the effect.

The extent of the effect of temperature on FMR can be compared for these two species to the effects of ambient temperature on RMRt, since *Dipodomys merriami* and a closely related species of *Perognathus* (*Perognathus californicus*) have been studied in the laboratory using standard indirect calorimetry. The effects of temperature on RMRt in both cases were greater than the observed effects on FMR (Hart, 1971). In both species RMRt increased approximately 4-fold between the lower critical temperature and 0°C, compared with the 2-fold increase observed in FMR (Figure 9). This suggests that in both cases the mammals effected mechanisms when exposed to the cold that ameliorated the effects of the cold exposure. In both species, however, these mechanisms were insufficient to remove the effect of temperature on FMR completely.

A: Pocket mouse

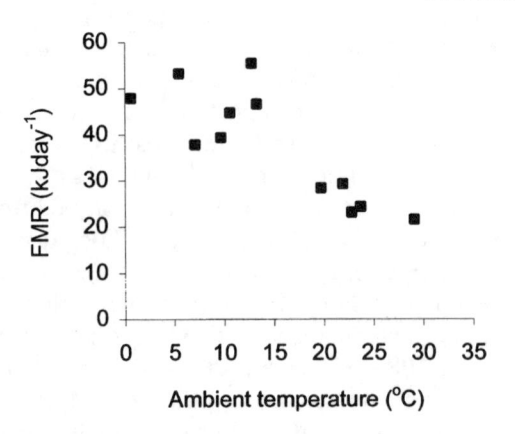

B: Kangaroo rat

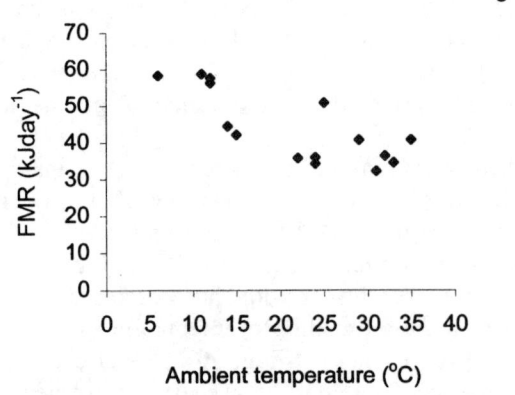

Fig. 9. Field metabolic rate (FMR) as a function of ambient temperature for two well studied species where sufficient measurements had been made across a range of environmental conditions. Each point represents the mean of 8–10 determinations at that temperature.

3. Season

For many species data were available for more than one season. Generally only two seasons of data were available (summer and winter), although in some species data were available almost monthly (e.g. pocket mouse and kangaroo rat; see above). I examined these seasonal effects in three ways. First, I compared the ratio of the summer FMR to the winter FMR. Since body mass may differ between summer and winter mammals, and indeed this has been suggested to be a mechanism used by mammals to ameliorate the impact of reduced winter temperatures, I also calculated the residual logged FMRs in summer and winter, and took the difference between these values. Finally, on average one would expect winter temperatures to be lower than summer temperatures, and thus on average the metabolic rate in winter should exceed that observed in summer if the temperature impact on residual FMR is a consequence of summed seasonal effects across several species. Accordingly, I made a third comparison using the residuals to the multiple regression predictions of logged FMR using both mass and temperature as predictors. This was feasible only where temperature data for winter and summer were also available in addition to the FMR and mass data (Table 2).

In total, data were available for 16 species that had been measured at single sites in both winter and summer (including the kangaroo rat and pocket mouse; see above). Several species had also been measured in spring and autumn but there were too few for any formal analysis. There was no significant difference in the mean body mass across species between summer and winter (paired $t = 0.56$, 15 d.f., $P > 0.58$). In four species the difference between summer and winter masses was less than 10%. In five species the winter mass was greater than 10% higher than the summer mass, and in the remaining seven species the reverse was the case, with the summer mass being more than 10% greater than the mass in winter. In all cases the winter was 10–20°C cooler than in the summer at the same site. Given this large temperature difference and the significant effects of temperature on FMR across all the pooled da'a (Figure 8), one might *a priori* expect that, in spite of the confounding effects of body mass variation between seasons, FMR would on average be greater in winter than in summer. However, there was also no significant difference between the mean FMR in summer and in winter (paired $t = -0.67$, 15 d.f., $P = 0.51$). In some species the metabolic rates were almost identical, despite there being large differences in body mass; for example, for the pocket gopher *Thommys bottae* the summer mass was 10% lower than the winter mass, but the respective values for summer and winter FMR were 126.6 and 127.7 kJ day^{-1}. In direct contrast, in the ground squirrel (*Ammospermophilus leucurus*) the body masses in summer and winter were almost identical (97.5 and 96.1 g respectively) yet the FMR differed by over 50%, with the summer FMR averaging 130.6 kJ per day but the winter FMR averaging only 82.6 kJ per day. In total, in seven of the species summer and winter FMR differed by less than 10%. In five species the summer rate exceeded the winter rate by more than 10%, and in four species the winter FMR exceeded the summer FMR by more than 10%.

When body mass effects were removed from the calculated FMR there was also no significant difference between summer and winter measurements for the same species at single sites (paired $t = -0.5$, 15 d.f., $P = 0.62$). This suggests that, on average, mass changes between summer and winter did not

Table 2

Seasonal comparison of body mass, field metabolic rate (FMR), residual field metabolic rate accounting for body mass (rFMR) and the residual FMR accounting for both body mass and ambient temperature (rFMRt)

Species	Season	Mass	FMR	rFMR	rFMRt
Ammospermophilus leucurus	W	96.1	82.6	−0.47	0.303
	S	97.5	130.6	−0.02	−0.380
Clethrionomys rutilus	W	13.6	64.9	0.497	0.269
	S	15.9	66.1	0.418	−0.375
Thomomys bottae	W	108.2	127.7	−0.116	0.024
	S	99.4	126.6	−0.07	−0.217
Psamomys obesus	W	175.7	184.5	−0.049	−0.034
	S	165.6	146.3	−0.244	0.092
Vulpes cana	W	1016	640	0.102	0.418
	S	874	568	0.076	1.065
Sorex areneus	W	6.0	90.3	1.33	1.18
	S	9.0	104.8	1.23	0.03
Vulpes velox	W	1990	1488	0.527	0.69
	S	2220	2079	0.794	−0.31
Perognathus formosus	W	16.7	47.9	0.08	−0.439
	S	16.2	23.1	−0.63	0.214
Lepus californicus	W	1800	1175	0.343	0.883
	S	1800	1416	0.530	−0.237
Spermophilus saturatus	W	224	232	0.028	0.109
	S	256	248	0.011	−0.108
Antechinus swainsonii	W	43.0	130	0.53	0.057
	S	40	74	−0.07	−0.167
Antechinus stuartii	W	22.8	96.2	0.58	0.117
	S	30	160.2	0.91	0.719
Phascogale calura	W	36	43.0	−0.509	0.151
	S	35	70	−0.004	−0.423
Dipodomys merriami	W	33.6	58.2	−0.162	−0.008
	S	32.9	40.7	−0.506	−0.106
Lagochestes hirsuitus	W	1700	856	0.062	−0.054
	S	1453	661	−0.097	0.009
Macrotis lagotis	W	1208	626	−0.03	0.125
	S	1132	455	−0.314	

Measurements refer to species measured at single sites in winter (W) and summer (S).
For references refer to Appendix B.

underpin differences in FMR. For example, if animals decreased their mass to offset the effect of reduced temperature, one would anticipate an effect of season would be obscured in the data for FMR alone, but revealed in the data where body mass effects were removed.

In contrast, when the effects of both body mass and temperature were removed, there was a very clear difference between the data for summer and winter. In all the mammals the residual metabolic value for the summer was higher than that for the winter. The difference was highly significant (paired $t = -5.5$, 15 d.f., $P < 0.0001$). This suggests that during winter the mammals had lower than anticipated metabolic rates, taking their body masses and the ambient temperature into account. The mean difference across all the mammals amounted to 0.38 log units (SE 0.094, minimum 0.11, maximum 0.67) and was independent of mean body mass (averaged across both seasons, $r^2 = 0.086$). This is a large effect on the metabolic rate. On average a difference of 0.38 log units indicates that the metabolism of mammals in winter was 46% lower than the equivalent summer value (once body mass and ambient temperature had been accounted for).

This difference indicates that during winter small mammals generally activate thermal acclimatization mechanisms which reduce the impact of the lower winter temperatures. These mechanisms are not generally based on reductions in body mass between summer and winter. Rather they reflect other factors that reduce expenditure relative to what would be expected from the reduced temperature and thus increased thermoregulatory requirement. These mechanisms have been known to exist in small mammals since at least the 1950s (Hart and Heroux, 1953). The actual mechanisms probably vary between species but include an increase in social gregariousness to enable huddling behaviour (e.g. Karasov, 1983), an increase in nest-building activity to increase external insulation available when mammals are resting (e.g. Hayward, 1965; Chappell, 1980; Casey, 1981), an increase in surface insulation to retard heat loss when the animals are outside the nest, and efficient use of favourable microclimates when foraging (Hayward, 1965; Peterson and Batzli, 1975), including the subnivean space (Chappell, 1980). The present analysis shows that these mechanisms are an important aspect of field energy expenditure budgeting in wild small mammals.

4. Latitude

There was a significant effect of latitude on the residual FMR (accounting for body mass) (Figure 10). The least-squares fit regression:

$$\text{Residual FMR} = -0.6596 + 0.0182 \text{ latitude (°N or °S)} \quad (10)$$

explained 22% of the variation in the residual FMR ($F = 42.86$, 1,169 d.f., $P < 0.001$). This effect was accounted for mostly because there was a strong [figure: scatter plot of Residual FMR vs Latitude (°N or S)]

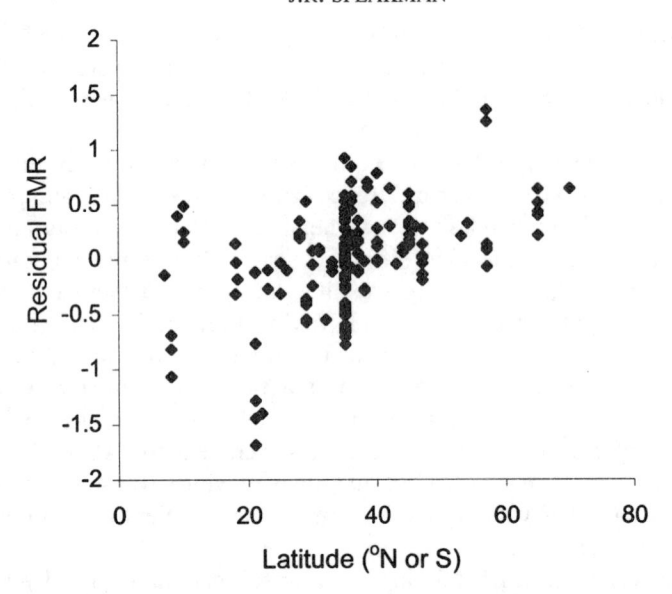

Fig. 10. Residual field metabolic rate (FMR) after accounting for the effect of body mass as a function of latitude of the study site. All data were included ($n = 165$ measurements on 70 species).

relationship ($r^2 = 40.3\%$) between the latitude of the study site and the ambient temperature (Figure 11a) described by the relationship:

$$\text{Temperature (°C)} = 37.5 - 0.573 \text{ latitude (°N or °S)} \qquad (11)$$

However, temperature did not explain all of the latitude effect on FMR since there was also a weak but significant effect of latitude on the residual FMR when both mass and temperature effects were taken into account (Figure 11b). The least-squares fit regression:

$$\text{Residual FMR} = -0.185 + 0.0051 \text{ latitude (°N or °S)} \qquad (12)$$

explained 4.1% of the variation in residual FMR ($F = 4.2$, $P = 0.043$). When body mass, ambient temperature and latitude were all entered as independent predictors of FMR, all three variables emerged as significant predictors. The best fit equation was:

$$\log_e \text{FMR (kJ day}^{-1}) = 1.896 + 0.686 \text{ body mass (g)} - 0.0199\, T_A \text{ (°C)} + 0.0091 \text{ latitude (°N or °S)} \qquad (13)$$

which explained 90.3% of the variation in FMR ($F = 482.8$, 3,156 d.f., $P < 0.001$).

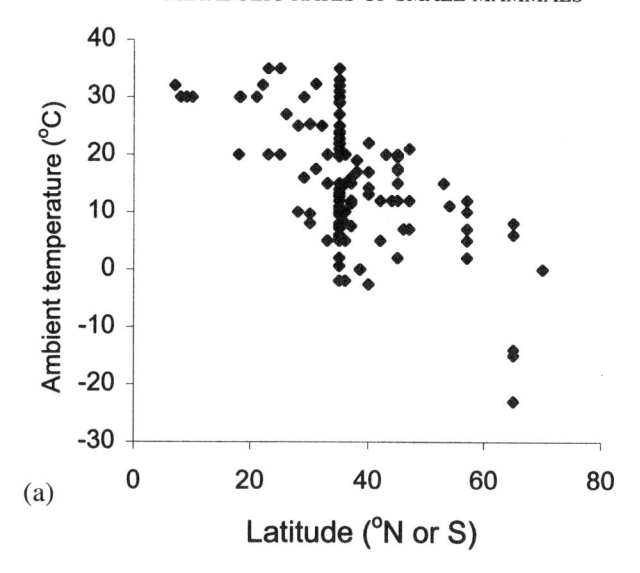

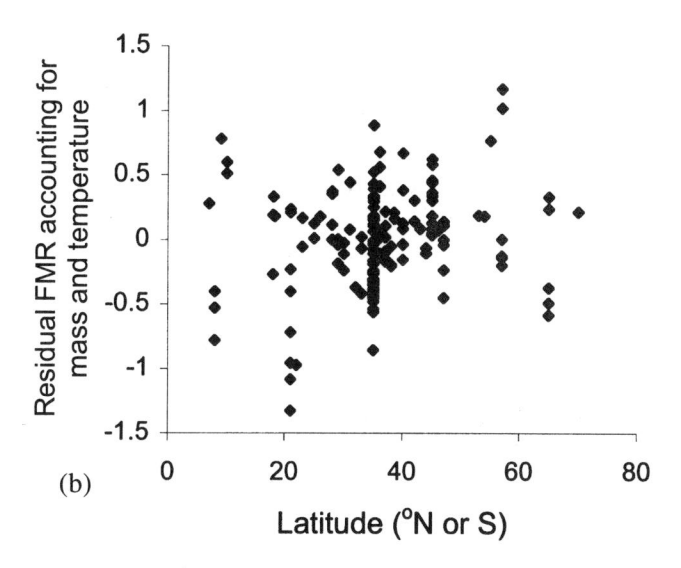

Fig. 11. (a) Relationship between ambient temperature and latitude for all the study sites and measurements. (b) Residual field metabolic rate (FMR) after accounting for the effects of both body mass and temperature plotted against latitude of the study site (all measurements included). The majority of the effect of latitude (Figure 10) was because of the effect of temperature on FMR (Figure 8) and the covariation of temperature and latitude (a), although there was a slight independent effect of latitude that was significant.

Together with the previous section on seasonal effects, this analysis suggests that the dominant cause of the temperature effect in the data set was the consequences of differences in ambient temperatures between study sites across the globe, rather than seasonal temperature effects at given sites. The reason why field metabolic rates were also independently affected by changes in latitude remains uncertain. However, the magnitude of the latitude effect is quite large. If one moves from the equator to 60°N, for example, the predicted latitude effect would amount to 0.546 log units. Over the same latitude change, the ambient temperature would be expected to fall by on average 34°C (equation (11)) and thus the temperature effect over the same range would amount to 0.676 log units.

To assess the predictive usefulness of equation (13), I adopted the same procedures used to assess the predictive power of the allometric equation using body mass alone (equation (1)). This previous analysis had suggested that, although mass explained 85.8% of the variation in FMR, on average predictions differed from actual values by 124%, making the predictive usefulness of the equation minimal. In contrast, when I used the equation employing body mass, ambient temperature and latitude as predictors to derive an estimate of the logged FMR, the mean deviation of the prediction to the actual data averaged only 21.3% (SD 15.4%, minimum 0, maximum 191%). Although the maximum deviation (for the Namibian golden mole *Eremitalpa namibensis*) was large, this prediction was a clear outlier. For the remainder, only six data were more discrepant than 50% from the actual values, and all these deviated by less than 60%. The deviant data included two bat species (*Macrotus californicus* and *Plecotus auritus*) which had significantly lower metabolism than predicted, and two data for the common shrew (*Sorex areneus*) which were much higher than predicted.

On average equation (13) provided an assessment of the expected energy demands of a free-living small mammal that is about five times more accurate (21% average error of prediction compared with 124% average error) than a prediction based on body mass alone (equation (1)). Given the imprecision involved in conversion of CO_2 production to energy expenditure when RQ is unknown, this may be as good as it is possible to get with the current methodology. Although this may at first sight appear an impressive improvement, it is important to be aware of the limitations of this prediction. Most importantly, the error assessment is conservative because the same data used to derive the model were also used to test it (this was also the case for the mass). Even using the same data to test the model, some poorly sampled groups (the Insectivora and Chiroptera) produced discrepant values, and in one case this was enormous. The prediction is consequently most likely to yield useful estimates of field metabolic rate when applied to predictions for rodents and marsupials, which provided the bulk of data for its construction. Nevertheless, in the absence of real data pertaining to a particular species, this approach is probably the best predictive method we have for evaluating the field energy expenditure of a small mammal. As such, this might prove useful in the production of models of bioenergetic flows in ecosystem studies and may also provide a predictive benchmark against which future estimates of metabolism can be judged.

5. Altitude

Ambient temperature varies with altitude as well as latitude, but altitude is also accompanied by changes in barometric pressure and therefore the partial pressure of oxygen, and by changes in other climatic variables such as wind as well as altered productivity. This might be expected to have interesting effects on the energetics of animals living at different altitudes. To date, however, only one study has explicitly examined the effects of altitude on field metabolic rates (Hayes, 1989a,b). In this previous study of Deer mice (*Peromyscus maniculatus*) it was found that the animals living at high altitudes (3800 m) were slightly heavier (18.4 versus 17.6 g) but had a considerably raised FMR compared with mice from lower altitudes (1230 m). At high altitude they had an FMR of 64.8 kJ per day compared with only 48.6 kJ per day at low altitude. This difference in FMR is much greater than would be anticipated by the slight difference in body mass. Indeed, this is confirmed by the fact that the residual FMR accounting for body mass effects derived in the present comparative framework was also much greater at the high altitude site (residual FMR at high altitude was 0.375, and at low altitude 0.013).

A question remains, however, over why the FMR was so greatly increased at high altitude. Dawson and Hulbert (1970), Kinnear and Shield (1975), and Hayes (1989a,b) suggested two potential factors that could be important. First, behaviour might differ between low- and high-altitude sites, combined with differences in the costs of foraging in the two different habitats. Alternatively, the difference might reflect the much lower ambient temperatures reported at the higher altitude. The present comparative analysis allows an assessment of this problem in the framework of the analysis of the temperature effects detailed above. If the effect of altitude were only a consequence of temperature differences between the two sites, we would not anticipate a large difference in the residual FMR for these two sites, once the effects of mass and temperature had been taken into account. Alternatively, if differences in behaviour were solely, or additionally, important, we might expect a much greater difference in the residuals for these two sites. The residual FMR accounting for body mass and temperature effects was 0.166 and 0.155, indicating that the 'altitude' effect on FMR can be fully accounted for by the differences in temperature between the two sites.

With the exception of the studies by Salsbury and Armitage (1995) of high-altitude marmots, and by Green and Crowley (1989) of *Antechinus* living under snow, the altitudes of study sites were not detailed in any of the other papers reviewed, suggesting they were mostly at lower elevations. Further discussion of altitude effects is therefore not possible in the present context.

6. Diet

Mammals seldom feed exclusively on single prey types. Classifying dietary habits of mammals is therefore complicated because there are multidimensional spectra along which diets are selected. For example, many mammals feed on insects combined with other prey such as fruits. Yet other mammals are exclusively insectivorous. The situation is further complicated by the fact that mammals change their diets seasonally and different populations of the same species may feed on radically different diets in different parts of their ranges; for example, the woodmouse *Apodemus sylvaticus* feeds on grains in woodlands (Gorman and Akbar, 1993) but on sand dunes it takes mostly insects (Zubaid and Gorman, 1991). Populations feeding on these very different diets can be found only tens of kilometres apart (Corp *et al.*, 1997a). With this in mind, only very broad dietary classification was possible for the species under study. I reviewed the literature to classify the dietary habits of the mammals that had been studied by DLW techniques. Occasionally this information was present in the papers in which FMR measurements were detailed (e.g. Nagy and Gruchacz, 1994) and the dietary data refer to the particular study population. However, more often than not the information had to be gleaned from general texts concerning feeding behaviour. I classified the diets and foraging strategies of the mammals into one of seven different classes:

(1) Leaves of trees: arboreal folivory
(2) Grass: grazing
(3) Seeds/grains or nuts: granivory
(4) Fungus:
(5) Insects (either alone or combined with other prey but with insects dominating): insectivory
(6) Other vertebrates: carnivory
(7) Exudates/nectar/fruit: nectarivory/frugivory

These seven dietary categories are ranked in the approximate order of the energy content of the food source, combined with the ease of its digestibility. Hence the least energy-rich food was foliage. It is well established that foliage is a poor energy source because the energy is trapped among indigestible components, and the leaves additionally contain many toxic secondary compounds designed to prevent feeding behaviour and retard digestive efficiency. At the other end of the scale are fruits and nectar, which are specifically designed to be energy rich (but protein poor) sources of food. One might argue over the locations of individual classes in this hierarchy. However, overall, I think most researchers would agree (give or take individual classes) that this order reflects the ranked availability of energy in dietary foodstuffs exploited by small mammals.

There was a very strong association between the residual FMR (accounting for the effects of body mass, ambient temperature and latitude) and the dietary class (Figure 12) ($F = 6.83$, 6,150 d.f., $P < 0.001$, analysis of variance (ANOVA)). Approximately 21% of the residual variation in FMR (from equation (13)) was associated with differences in dietary class between the different groups. The most striking pattern in the dietary data (including all the individual data points) was the positive link between the subjective assessment of energy availability from the diet and the field metabolic rate. Since most mammals (70 of 74) had constant dietary assignations, in part this result may reflect the multiple representation of the same species within the data set. I therefore calculated the average residual FMR for each species, and also the average dietary assignation, rounding this to the nearest integer where necessary. When each species appeared only once in the data there was no evidence of a significant link between FMR and diet ($F = 1.65$, 6,57 d.f., $P = 0.15$, ANOVA). The absence of an effect of diet in these data contrasts with the suggestions of McNab (1980, 1983, 1986a) that animal energy expenditure (notably BMR) is strongly influenced by the diet. It has been suggested that the link between BMR and diet occurs because animals feeding on foods with high availability of energy can afford to be extravagant in their use of energy, whereas those feeding on low-quality food must be frugal—hence a link between energy availability from food and BMR. It has often been inferred that a link between FMR and diet underpins this link between BMR and diet. In the current data set there was no evidence to support this suggestion.

The effects of diet on metabolism suggested by McNab (1980, 1983, 1986a) have been criticized because of the failure to take into account the phylogenetic lack of independence of the different species as sampling units (Bennett and Harvey, 1987; Harvey et al., 1991).

7. Phylogeny

To assess the impact of taxonomy on the residual FMR (after accounting for the effects of body mass, temperature and latitude), I coded each species according to the order of its classification: 1, Rodentia; 2, Marsupialia; 3, Chiroptera; 4, Carnivora; 5, Insectivora; 6, Edentata; and 7, Lagomorpha. There was no significant effect of phylogeny on the residual FMR ($F = 1.83$, 6,57 d.f., $P = 0.11$, ANOVA).

The patterns of variation in FMR described here appear to deviate significantly from the patterns of variation that have been suggested previously for BMR, in particular the absence of an effect of diet (even in the data uncorrected for

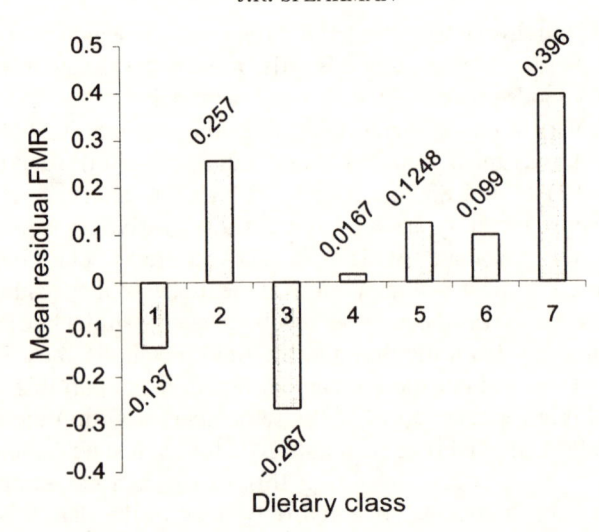

Fig. 12. Mean residual field metabolic rate (FMR) after accounting for the effects of body mass, ambient temperature and latitude for each of seven dietary classes. All data were included. Values above the bars represent the actual mean values. The order 1–7 is arranged such that the better energy sources are given higher numbers; hence 1 is foliage and 7 is fruit. See text for full details of classification.

phylogenetic independence). This raises the question of the extent to which BMR and FMR may be linked. This topic forms the major part of the following discussion.

V. DISCUSSION

A. Links Between FMR and RMR

Rather than concern myself with the relation between FMR and BMR, which would inevitably lead to debate over the quality of the estimates of BMR (see introduction), I will discuss links between FMR and RMR. I have used here the definition of RMR from Brody (1945), which includes all the restrictions for BMR except the animals do not need to be post-absorptive. All the RMR estimates therefore include animals measured at rest, with minimal or no activity and in the thermoneutral zone. I was able to find estimates of RMR for 62 of the 74 species included in the analysis. In 34 of these cases the RMR estimates came from the same studies as the FMR estimates (either the same paper or another paper by the same authors from animals studied in the same area). These estimates probably refer to the same population of animals studied at the same time of year. For the remainder I used published estimates, which inevitably do not come from the same sites or populations.

RMR is not a fixed species-specific trait, but varies with many factors including season and the population under study (e.g. Corp *et al.*, 1997b). Where measurements were made for populations that were not also the populations measured by DLW, it would be inappropriate to apply the same RMR measurement across all the multiple measurements of FMR. The question arises, however, of which measurement the RMR should be applied to, when values were not generated from the same study. To overcome this problem I decided that independent of whether the data had been gathered in the same study or not, I would use only a single mean for each species (pooling data from the multiple situations where this was necessary). I therefore compared the mean estimated RMR for each species with the other traits averaged for the particular species in question.

There was a strong relationship between the measured RMR and body mass (Figure 13). The least-squares fitted regression:

$$\log_e \text{RMR (kJ day}^{-1}) = 0.835 + 0.642 \log_e \text{body mass (g)} \quad (14)$$

explained 90.2% of the variation in RMR ($F = 554.02$, 1,60 d.f., $P < 0.001$). The gradient of this relationship was not significantly different to the expectation from the surface law (0.66), but was much lower than the gradient derived for BMR by Kleiber (0.75; $t = 4.0$, $P < 0.01$). As with the measurements of FMR,

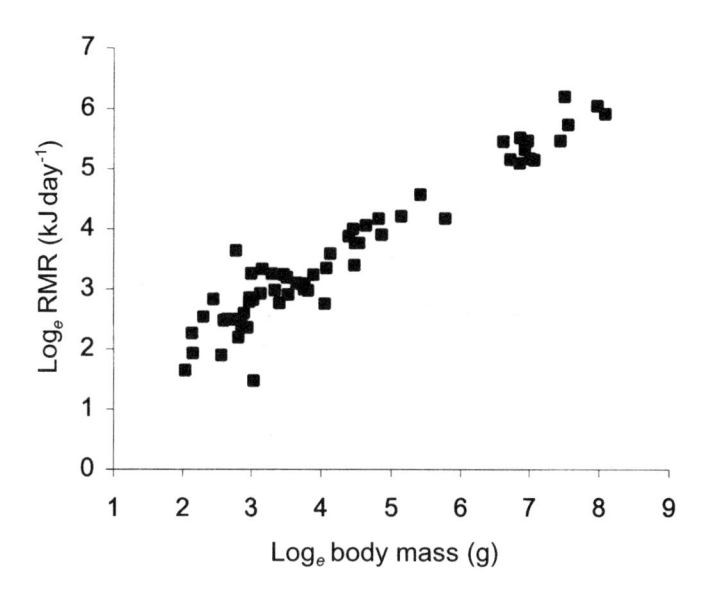

Fig. 13. $\log_e$ resting metabolic rate (RMR) in the thermoneutral zone as a function of body mass. Each point refers to a different species.

however, the gradient of this relationship was possibly affected by the presence of the three-toed sloth as the largest representative, and the common shrew among the smallest. Omitting these data gave the following equation:

$$\log_e \text{RMR (kJ day}^{-1}) = 0.7056 + 0.668 \log_e \text{ body mass (g)} \quad (15)$$

which explained 92.3% of the variation in RMR. This revised gradient was not significantly different to the expectation from the surface law (0.66) but was still significantly lower than the expectation based on the Kleiber curve (0.75; $t = 3.28$, $P < 0.01$). Use of RMA regression produced increased gradients, as would be expected. Using all the data, the RMA regression gradient was 0.675, and using the truncated set excluding sloth and shrew data gave 0.695 as the gradient.

There was a significant negative relationship between the residuals of the relationship between RMR and body mass and the latitude of the sites where the animals were collected (Figure 14). The least-squares fitted regression explained 7.1% of the variation in the residual RMR ($F = 4.56$, 1,57 d.f., $P = 0.04$). Ambient temperature of the sites, averaged across the FMR measurements made at those sites, was not significantly related to the RMR. However, this probably reflects the heterogeneous nature of the data for FMR, with some sites including measurements made across all seasons, but others reflecting

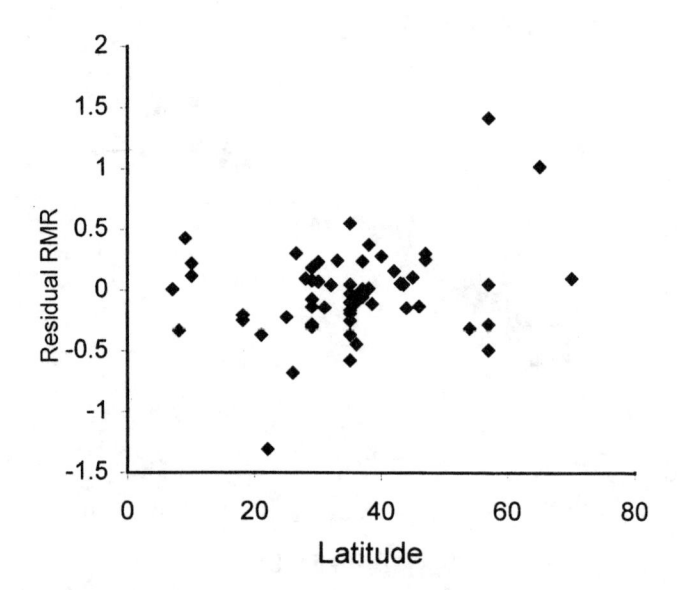

Fig. 14. Residual resting metabolic rate (RMR) (after accounting for body mass) plotted against latitude of the study site. There was a weak but significant relationship.

only summer or winter measurements. When both mass and latitude were entered as predictors, the following predictive equation was obtained ($F = 283.07, 2.56$ d.f., $P < 0.001$; $r^2 = 91.0\%$):

$$\log_e \text{RMR (kJ day}^{-1}) = 0.483 + 0.656 \log_e \text{mass (g)} \\ + 0.00789 \text{ latitude (°N or °S)} \quad (16)$$

The effect of latitude on RMR has been reported previously in seabirds (Ellis, 1984). It has been suggested that RMR is increased at higher latitudes because of the greater thermal demands placed on organisms living at these latitudes. Mammals living in colder conditions must have the necessary machinary to generate heat, and it is suggested that the costs of maintaining this machinary contribute to the RMR. In the woodmouse (*Apodemus sylvaticus*) the noradrenaline-induced metabolic rate, which reflects thermogenic capacity of the brown adipose tissue, is linked to the resting metabolic rate, independent of the body mass effects on both traits (Speakman, 1995), supporting this view. Moreover, it has long been established that thermal acclimation in the cold leads to an increased in the RMR (e.g. Adolph, 1950; Adolph and Lawrow, 1951; Hart, 1953a,b; Depocas *et al.*, 1957; Chaffee and Roberts, 1971; Heldmaier, 1989; Krog *et al.*, 1954; McDevitt and Speakman, 1994a), presumably also linked to altered thermogenic capacity (Pasanen, 1971; Tarkkonen, 1971; Lynch, 1973; Feist and Rosenmann, 1976; Wunder *et al.*, 1977; Bucowieki *et al.*, 1982; Himms-Hagen, 1986; Klaus *et al.*, 1988).

The link between RMR and body mass reported here may again be a consequence of a phylogenetic artefact of using species as independent data. Several previous studies have considered this problem with respect to RMR in various groups (e.g. birds: Reynolds and Lee, 1997), including mammals (Harvey and Pagel, 1991; Ricklefs *et al.*, 1996). In all these cases the effect of mass has been shown not to be a phylogenetic artefact. Previous reviews concerning mammals, however, have not included data collected over such a limited size range as in the present study. I therefore examined whether the mass effect in the current data set was affected by using the phylogenetically independent contrasts.

Using the phylogeny detailed in Appendix C, the independent contrasts were calculated for RMR. As with FMR I calculated the variance in the RMR with branch length independently for each group, and then standardized the contrasts using the square root of the product of branch length and the variance for each group. As with FMR, the variance for the bats and insectivores was an order of magnitude lower than that recorded for marsupials, but for RMR the rodents had an intermediate value between these extremes. There was a significant positive relationship between the logs of the standardized phylogenetically independent contrasts for RMR and body mass (Figure 15). The least squares fitted regression:

$$\log_e \text{RMR contrast} = 0.259 + 0.571 \log_e \text{body mass contrast} \quad (17)$$

explained 53.3% of the variation in the $\log_e$ RMR contrast ($F = 66.3$, 1,58 d.f., $P < 0.001$). The gradient of this curve was not significantly different to the surface law expectation ($t = 1.41$, $P > 0.05$), but was significantly lower than the Kleiber expectation ($t = 2.55$, $P < 0.05$). As with FMR, the outliers at the lower end of the plot reflect situations where mass (or metabolism) hardly changed over a short branch, linking closely related groups. (The exact number of outliers with low mass change was different for RMR because of missing data.) Removing these three outliers improved the r^2 value (to 59.6%) and lowered the gradient of the relationship (0.546), but did not alter the significance of the comparisons to the surface law and Kleiber expectations.

Once the effects of mass and latitude had been removed, there was no evidence in this sample of a link between RMR and diet (classes detailed above: $F = 1.42$, $P > 0.05$, ANOVA) or between RMR and phylogeny (classes also detailed above: $F = 1.95$, $P > 0.05$, ANOVA). This absence of any further effects on RMR was unexpected in the light of previous demonstrations that RMR may be linked to diet (McNab, 1980, 1983, 1986a) and the established effects of phylogeny on RMR, for example the lower RMR of

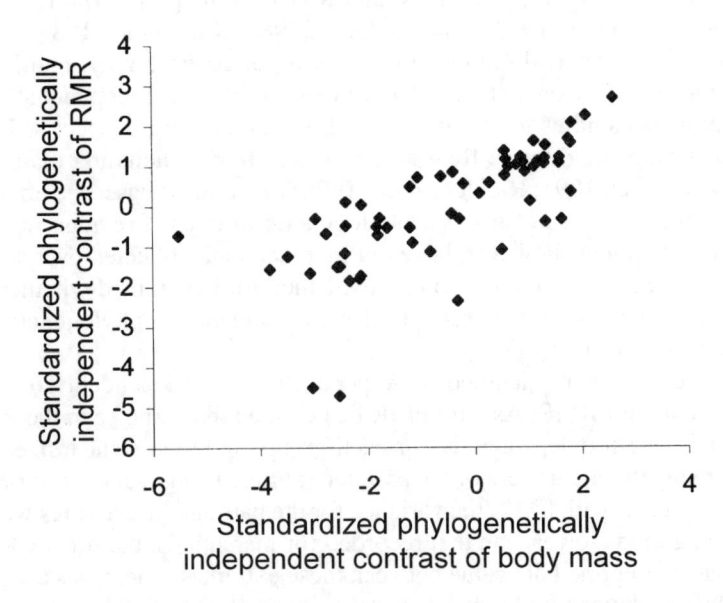

Fig. 15. Standardized phylogenetically independent contrast of resting metabolic rate (RMR) plotted against the standardized phylogenetically independent contrast of body mass.

marsupials (Dawson and Hulbert, 1970; McNab, 1978). The absence of an effect of diet in the present data seems unlikely to be because of insufficient variation in the diets of the mammals included in the database. Although diet may be linked to RMR when a wider comparison is used, in small mammals the evidence appears at best equivocal, even before any assessment of phylogenetic independence of the data is considered (see Bennett and Harvey, 1987). The absence of an effect of phylogeny in the data was probably because several orders (edenates, carnivora and lagomorphs) were represented by only single species. If these single representatives were removed there was an effect of phylogeny in the remaining four orders ($F = 2.95$, 1,55 d.f., $P = 0.04$, ANOVA). However, the pattern of these effects was not exactly what might have been predicted. The bats had lower than average RMRs for their masses and sites of origin, and the insectivores had higher RMRs, but the RMRs of the marsupials were in line with expectations and did not differ from the values for rodents.

Although previous studies of marsupials have clearly indicated that they have reduced resting metabolic rates compared with eutherians, this effect appears to be dependent on the sample of mammals measured (Lovegrove, 1996). In particular, the small Dasyurid marsupials tend to have higher rates than the expected from their masses (Green, 1997), and this group comprised a major proportion of the marsupial species included in the present study.

The pattern of variation in RMR in these data paralleled very closely in several respects the variation in FMR. The gradients of the mass effect were almost identical and changed in similar ways when the data were either truncated or a different model was used to estimate the regression parameters. Moreover, there was a significant positive effect of latitude on RMR, which was similar to the same effect reported in FMR. In addition, there were no profound effects of either diet or phylogeny on the two traits.

Given the large amount of the total variation in both traits that was dependent on differences in body mass between species, and ambient temperature, it was not surprising that there was a strong relationship between FMR and RMR (Figure 16), using the data for averages across FMR and RMR for each species. The least-squares fitted regression:

$$\log_e \text{FMR (kJ day}^{-1}) = 1.472 + 0.911 \log_e \text{RMR (kJ day}^{-1}) \quad (18)$$

explained 91.0% of the variation in FMR ($F = 603.9$, 1,60 d.f., $P < 0.001$). The gradient of this relationship was slightly, but significantly, lower than 1.0 ($t = 2.4$, $P < 0.05$). The reduced major axis gradient was 0.959, and did not differ significantly from a value of 1.0. A gradient of exactly 1.0 would imply a fixed ratio between the RMR and FMR across body mass. In agreement with the gradient being almost equal to 1.0, examination of the actual ratios of FMR

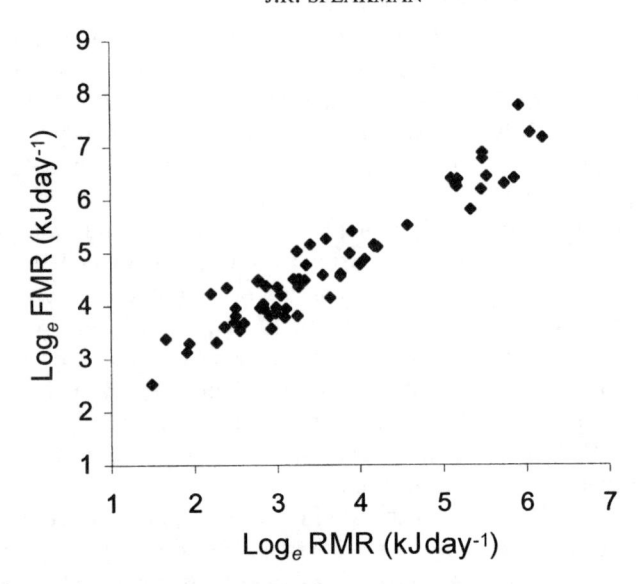

Fig. 16. $\log_e$ field metabolic rate (FMR) plotted against $\log_e$ resting metabolic rate (RMR). Data were averaged across all measurements for each species. Each point represents a different species ($n = 60$).

to RMR, as a function of body mass, reveals no such effect of mass on the ratio (Figure 17) ($F = 0.45$, $P > 0.9$; $r^2 = 0.01$), particularly because one of the largest mammals in the sample (yellow-bellied marmot; Salsbury and Armitage, 1994) also had one of the highest ratios.

Across all measurements the distribution of ratios was positively skewed (Figure 18). The modal class was centred on a ratio of 2.5 (range 2.25–2.75). The median was 3.1 and the mean was 3.4 (SD 1.35, SE 0.171, $n = 62$ species points). The lowest ratio was 1.62 (greater glider, *Petauoides volans*; Foley *et al.*, 1990) and the highest was 7.63 (fat-tailed dunnart, *Sminthropsis crassicaudatus*; Nagy *et al.*, 1988). Although this appears to be a wide range of ratios between FMR and RMR, the range needs to be examined in the context of the total variation in FMR and RMR. Across all body masses there was a greater than 200-fold range in the FMR and a greater than 100-fold range in RMR. Consequently, if RMR and FMR were not closely linked, the variation in the FMR : RMR ratio would be enormous. The likely range in the absence of any link between RMR and FMR can be simulated by randomizing locations of one of the traits (FMR) and recalculating the ratios using this randomized variable where the link of RMR to FMR is broken. When this procedure is performed, the average ratio FMR : RMR turns out to be 11.7 (SD 24.4, median 3.9) and the range stretches from 0.12 to 155.5. In combination with the plot in Figure 16, this illustrates that RMR and FMR are strongly linked in this sample of small mammals.

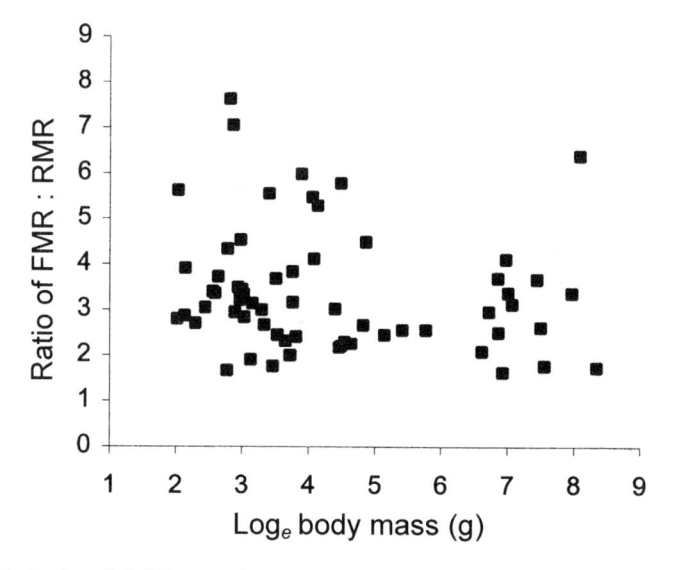

Fig. 17. Ratio of field to resting metabolic rate (FMR : RMR) plotted against $\log_e$ body mass. Each point represents a different species ($n = 60$).

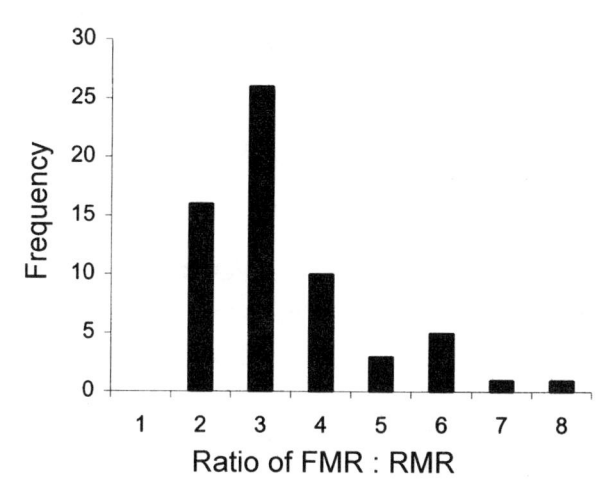

Fig. 18. Frequency distribution of sustained metabolic scopes for the species data ($n = 60$).

There are two reasons why this linkage may reflect only an artefact of the manner in which the analysis has been performed. The first concerns an interpretation of the nature of RMR. There are several ways of considering RMR (Hammond and Konarzewski, 1996). One interpretation is that the metabolic processes constituting RMR are undertaken perpetually. Thus, when an mammal does things other than rest, it layers on top of the RMR a series of additional metabolic events. RMR, however, is sustained underneath these other processes—the so-called 'partitioned pathways model' (Ricklefs et al., 1996). A second interpretation is that the processes of RMR are simply speeded up when an animal performs activity—the so-called 'shared pathways model' (Ricklefs et al., 1996). A third alternative interpretation is that RMR exists only when the animal is at rest, and that when it performs some other activity the metabolic processes constituting RMR do not persist—a 'replacement pathways model'. Since establishing exactly what metabolic processes contribute to RMR has proved elusive, direct testing of these ideas is difficult and, by definition, the alternatives cannot be separated simply by measuring energy expenditure, because the source of this expenditure cannot be partitioned. If the shared pathways and replacement pathways models are correct, it would be appropriate to establish a link between FMR and RMR simply by correlating them together. However, if the partitioned pathways model is correct, there is a potential problem because RMR constitutes a major component of the total FMR. Even if the additional activity and thermoregulation were constant, there would still be a link between RMR and FMR because FMR would equal RMR plus a constant. In effect, one would be correlating RMR against itself. For most wild mammals, however, RMR is a state seldom reached. Consequently, although in the shared pathways and alternative pathways models RMR may also potentially contribute to FMR for the short periods that animals spend in the RMR state, more normally it would not.

To evaluate whether RMR is correlated to FMR, even if the partitioned model is correct, it is necessary to examine whether RMR correlates not with FMR alone but with FMR–RMR (Ricklefs et al., 1996; Speakman, 1997b). There was a significant correlation between RMR and FMR–RMR in the present data set (Figure 19). The least-squares fitted regression:

$$\log_e \text{FMR–RMR (kJ day}^{-1}) = 1.222 + 0.863 \log_e \text{RMR (kJ day}^{-1}) \quad (19)$$

explained 80.0% of the variation in RMR–FMR. Thus, RMR appears to be linked to FMR even if the partitioned model is correct, and the link is not an artefact of RMR being included within the FMR measurement.

A second cause of the link may be the shared variation in both FMR and RMR that is explained by other factors. I have already shown that both FMR and RMR are closely correlated not only with body mass, but also with temperature and latitude (for FMR) and latitude (RMR) (equations (13) and (16)). This shared variation might precipitate the relationships between RMR and FMR. I examined this possibility in three stages: first, by removing the shared effects of body mass using equations (1) and (14) (as has been

METABOLIC RATES OF SMALL MAMMALS 251

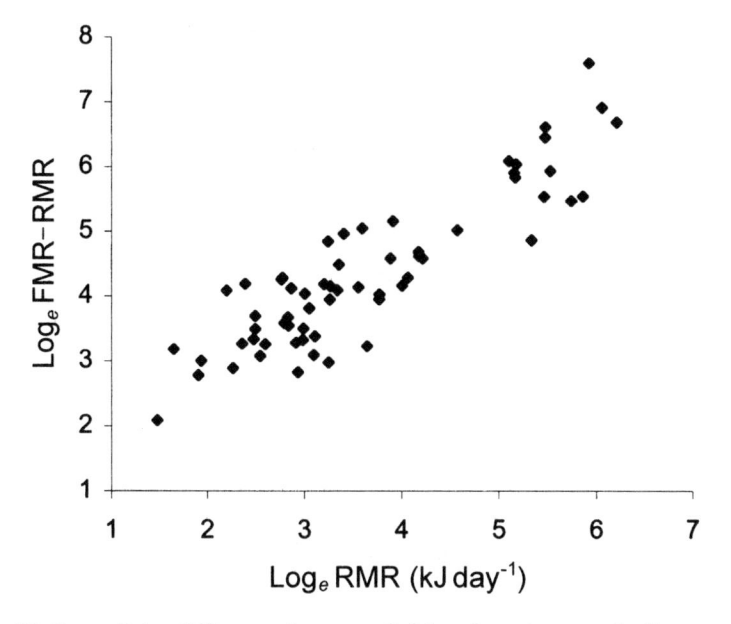

Fig. 19. Log of the difference between field and resting metabolic rates (FMR – RMR) plotted against $\log_e$ resting metabolic rate.

performed in several previous analyses; e.g. Koteja, 1991; Ricklefs *et al.*, 1996; Speakman, 1997b); second, by correlating the residuals derived from the phylogenetically independent comparisons, equations (5) and (17) (as performed by Ricklefs *et al.*, 1996); and third, by removing the effects of body mass, temperature and latitude on both traits using equations (13) and (16) (which, as far as I am aware, has not been performed previously). In all three cases I examined whether there was still a relationship between the residual RMR and the residual FMR.

When the effects of body mass on both traits were removed, there was a significant relationship between residual FMR and residual RMR (Figure 20a) which explained 35.6% of the variation in residual FMR. The least-squares fitted regression equation was:

$$\text{Residual FMR} = 0.0184 + 0.66 \text{ residual RMR} \qquad (20)$$

($F = 33.6$, 1,60 d.f., $P < 0.001$). Examination of this plot reveals two very clear outliers at either end of the plot (common shrew at the upper end and Namibian desert golden mole at the lower end) which could exert undue leverage on the relationship. Removing these outliers reduced the explained variability to 12.2%, but the relationship still remained highly significant

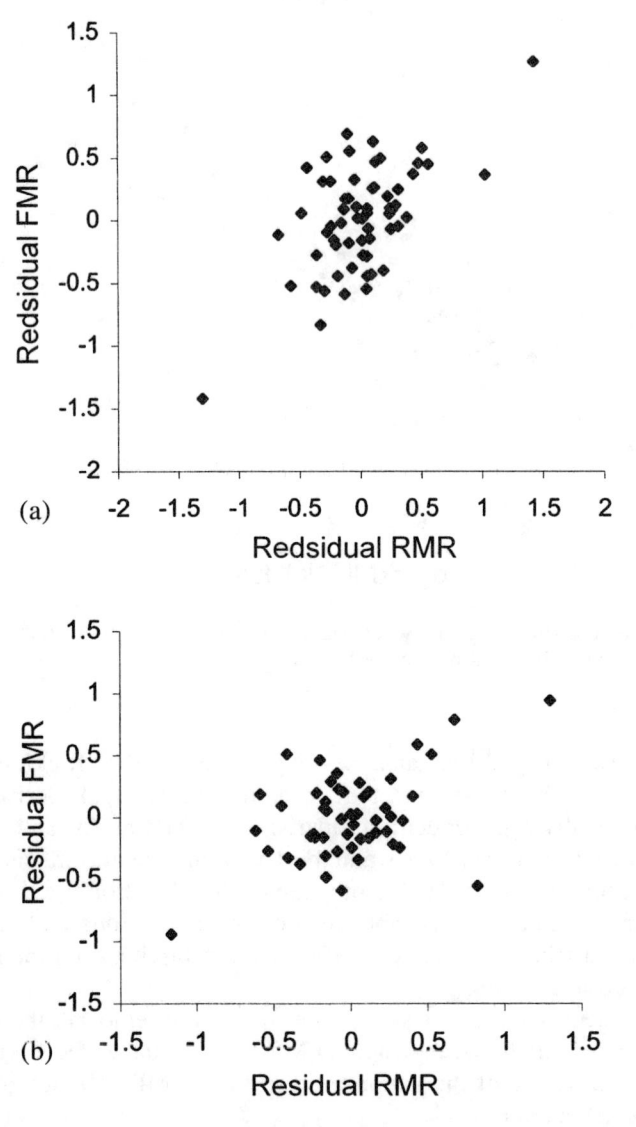

Fig. 20. Residual field metabolic rate (FMR) plotted against residual resting metabolic rate (RMR). (a) Both residuals accounting for body mass only; (b) both residuals accounting for the effects of body mass, temperature and latitude.

($F = 7.9$, $P = 0.007$), indicating that the link between the traits was not only due to these two outliers.

I evaluated residuals to relationships between the logged standardized phylogenetically independent contrasts of FMR and RMR to contrasts of body mass (Figures 7 and 15). There was a significant positive relationship between the residual FMR contrast and the residual RMR contrast (Figure 21). The least-squares fitted regression explained 13.3% of the variation in the residual FMR contrast ($F = 8.91$, 1,57 d.f., $P = 0.004$). Removing the outliers reduced the r^2 value to 0.088 but the relationship remained significant ($P = 0.025$).

Ricklefs *et al.* (1996) found that, while there was a strong correlation between the uncorrected residual RMR and residual FMR in birds, this relationship disappeared when the interrelationships of the residual phylogenetically independent contrasts were considered. This was not the case in their analyses of the mammal data, although this could have been because of the errors (pointed out above) in their mammalian phylogeny. The above analysis confirms that the effect of the residuals of phylogenetic contrasts of the small mammals is not a phylogenetic artefact. This does raise the question, however, of why there were contrasting results for the two major groups. Close examination suggests that the absence of a relationship in the residual contrasts of the birds may reflect errors in the derived values of the contrasts. There are, for

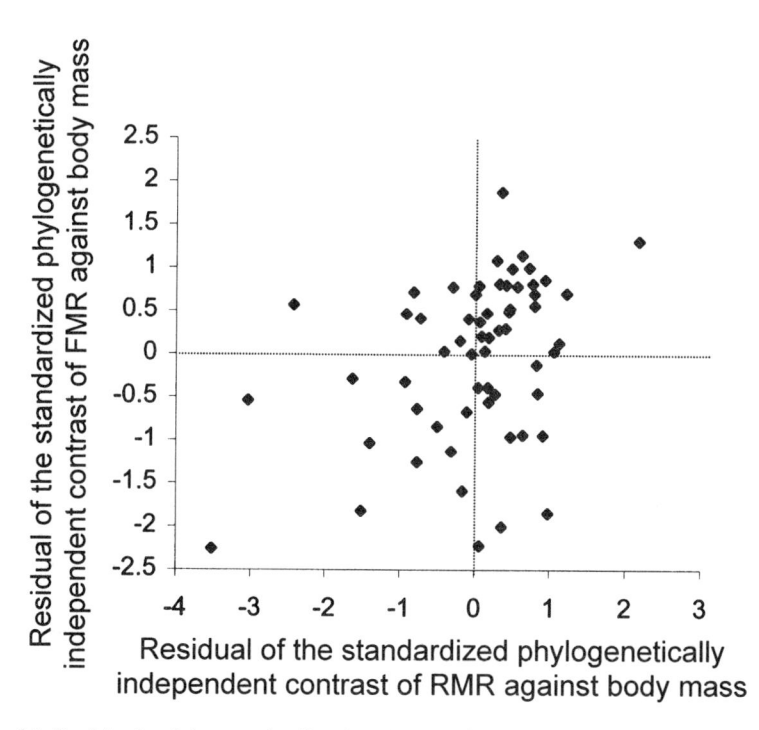

Fig. 21. Residuals of the standardized phylogenetically independent contrasts of field metabolic rate (FMR) and resting metabolic rate (RMR) on body mass plotted against each other.

example, some wide discrepancies between the plots showing the relationships between the FMR and RMR contrasts and the mass contrasts (Ricklefs et al., 1996, Figure 2, p. 1061), and the plots showing the interrelationships of the residuals (Figure 3, p. 1063). Note, in particular, that in both the plots in Figure 2 of contrasts for FMR and FMR–RMR against the mass contrasts the node L represents the lowest residual value, while in the plots of residuals (Figure 3, p. 1063) it is the highest residual. Several other anomalous positions can be identified (e.g. for nodes V, H and C). Together these errors beg the question of whether the absence of a trend in the residuals of the phylogenetically independent contrasts for the birds is a reliable result.

When the effects of both temperature and latitude were removed, as well as body mass, there was still a significant relationship between residual FMR and residual RMR (Figure 20b). However, the significance of the relationship was much reduced ($r^2 = 0.205$) compared with that found when only the effects of body mass were removed. In addition, there were again two outlying points (shrew and mole) which exerted a large influence on the regression. In this instance, removing these outlying data completely removed the significance of the relationhship ($r^2 = 0.028$; $F = 1.4$, 1,50 d.f., $P = 0.242$). This suggests that the existence of a link between FMR and RMR is almost entirely a consequence of the shared variability in the two traits that is explained by the effects of body mass, latitude and temperature. No previously published review has completely accounted for the shared variation, and, as shown above, accounting only for the shared variability due to body mass still leaves a significant relationship, because both residual FMR and RMR depend closely on ambient temperature and latitude.

A question, however, hangs over the validity of removing the outlying data points from the relationship presented in Figure 20. Such outlying data undoubtedly have a disproportionately high leverage in regression analyses, which might be used to justify their removal on statistical grounds. However, another way of looking at this data set is that the total data set represents only 3% of the extant mammal fauna. Both residual FMR and residual RMR were normally distributed, and the data for these two species are exceptional only in the sense that they refer to animals that are apart from the mainstream group which comprise this data set. These are, however, real animals, and there is no reason to doubt the accuracy of the measurements presented for either of them. Although they may be exceptional in the context of this data set, in comparison to the remaining 97% of mammals for which we do not have data they may be very representative. Perhaps these species point to the real underlying relationship between FMR and RMR, which is not exposed when they are omitted. This might occur, for example, simply because the variability in the residual traits is insufficient to detect the trend when they are excluded.

At present, the existence of a link between RMR and FMR appears to be very tenuous and depends only on a few outlying data which may be representative of a wider trend or may not. There are several other species of small mammal that have both high and low residual RMRs. For example, the mustelids have high RMR for their body size (Brown and Lasiewski, 1972; Casey and Casey, 1979) and it would be extremely instructive to know whether they also have high residual FMRs. At the other end of the scale there are several groups with low residual RMR, in particular the tenrecs from Madagascar, for which it would be interesting to establish whether they have similarly low residual FMRs.

A further reason for an artefactual linkage between FMR and RMR is the problem of phylogenetic independence considered above in the context of the effects of body mass on FMR and RMR. Although the correlation of residual FMR and residual RMR accounting for mass is not due to the lack of independence of species as sampling units, it is still possible that a linkage between residual FMR and residual RMR accounting for both mass and temperature/latitude reflects such an effect. In the present context, however, it would be unlikely for the relationship between residual RMR and residual FMR to be a phylogenetic artefact. This is because we already know that the significance of the relationship hinges on data for only two species (one with a high and one with low residual). Removing the effects of both mass and latitude/temperature in the phylogenetically independent contrasts is not possible because of problems of assigning realistic values for the environmental conditions at each of the ancestral nodes of the phylogeny.

B. Sustainable Metabolic Scope

The present estimated ratios of FMR to RMR were all substantially lower than the suggested absolute maximum physiologically possible limit of around 9–12× RMR. The highest ratio in the current data set (7.63) was approximately equal to the maximum sustainable scope of 7× RMR proposed by Hammond and Diamond (1997). However, the overwhelming impression from the observed ratios was that they were all considerably lower than the proposed limits, whether these are set at an absolute maximum of 9–12× RMR (derived in the introduction of the present paper), a limit of around 7× RMR (Peterson et al., 1990; Hammond and Diamond, 1997) or a limit of around 4× RMR (King, 1974; Drent and Daan, 1980). Of the 62 estimates of the ratio, only 12 measurements (19%) exceeded the suggested 4× RMR limit. The average ratio was substantially lower (3.4× RMR), and almost 35% of the ratios were lower than 2.0. These data raise two questions. (1) Why do mammals expend so little energy above their resting requirements? (2) Do the data support the notion of a sustainable metabolic limit?

There are several potential reasons why mammals may routinely expend energy at levels below the putative limits. The first is that by averaging FMR data across all seasons one inevitably pools measurements made in the lowest part of the year with data collected in the highest part of the year. Moreover, the data may also be biased because they routinely refer to mammals that are not working at their maximum capacity because it is easier to study mammals using the DLW methods when they are at phases of their annual cycle that are not the energetically most stressful. There is some support for this suggestion, because the period when small mammals are under most energetic stress is during the period of late lactation, yet relatively few of the measurements refer to this phase of the cycle. Perhaps mammals routinely expend energy at levels below their physiological capabilities for most of the year, but sustain resting rates that are appropriate for the period of late lactation (or some other stressful phase) when they would be expending energy at around the supposed physiological limits of either 4× or 7× RMR.

If this were the case, we might expect that by selecting from those species where multiple measurements at different phases of the annual cycle had been made, the maximum ratios for each species would cluster around a mean value much closer to the supposed limits. To test this I performed this selection on the 56 species for which multiple measurements were available. Although the mean ratio across this selected sample was greater than that across all the measurements, it was raised only slightly (from 3.4 to 3.6, and there were still many values where the ratio was lower than 3.0). These data did not cluster around either of the putative maxima. However, this might be because selecting the maxima from the measurements that had been made does not necessarily also include the very highest field metabolic rates because the highest phase of the cycle might not have been sampled.

To test this idea further I selected only those measurements that included mammals in late lactation. This was a much smaller sample of only 15 species, yet the ratio for this supposedly most stressful energetic phase of the annual cycle averaged only 3.4× RMR (SD 1.2, $n = 15$) and still included mammals that were working at less than 2.5× RMR. This is also an optimistic assessment of the FMR : RMR ratios because the RMR estimates in most of these circumstances do not refer to lactating mammals. As RMR generally increases in lactation, comparing lactating FMR with non-lactating RMR values will overestimate the derived ratio. These two analyses strongly suggest that the low levels of energy expenditure were not a consequence of selecting periods of the annual cycle where the mammals were under the least energetic stress. Moreover, RMR is extremely flexible. It seems improbable that mammals would need to sustain very high basal demands throughout the entire cycle in preparation for only one phase. Indeed, we already know that mammals radically increase their RMR during late lactation (e.g. Speakman and McQueenie, 1996) in response to the high energy demands during this period, and that there is no obvious linkage between pre-breeding RMR and any reproductive parameters (Derting and McClure, 1989; Hayes et al., 1992a), probably because of this flexibility. If mammals were not under maximal stress they could probably reduce their RMR to sustain a working ratio of around 4× or 7× RMR. Clearly, they do not do this routinely.

Why do the mammals not work harder and closer to their putative limits? There are two alternative explanations. The first is that mammals can work harder than they routinely do, up to the supposed limits, but by so doing they would pay a penalty in their life histories. There are several potential penalties that the animals might encounter if they were to increase their energy expenditure. For example, one well established idea concerning the nature of ageing is the free-radical damage hypothesis, first proposed by Harman (1956). This hypothesis suggests that animals age and ultimately die because their bodies are under constant attack by free radicals generated during oxidative phosphorylation.

Mammals could expend more energy but they choose not to because of the potentially negative life history effects that such expenditure might entail. I have suggested elsewhere that this can be envisaged as a curved benefit line that is associated with changes in energy expenditure (Speakman, 1997b). If small mammals expend energy in the field at a level equal to their RMR, they would derive no fitness benefits because they would be unable to do anything apart from rest all day. As expenditure increases, fitness benefits increase because the animal can perform behaviours in addition to resting that enhance its prospects for survival and reproduction. However, increased expenditure brings negative effects as well (in terms of reduced life expectancy), and ultimately these negative effects start to offset the benefits so that the fitness curve reaches a peak at levels well below the physiological maxima at which they are capable of working (Figure 22).

If this model were correct, we might expect that species which have long lives would have greater benefits to derive from staying alive. These animals might therefore have fitness curves that peaked at lower levels than species that are short lived and would have little to lose by expending energy routinely at levels close to their potential maxima. Possible examples include the Dasyurid mice (e.g. *Antechinus*, *Phascogale* and *Sminthopsis*), the males of which have a semelparous reproductive strategy. Males have a period of frenetic reproductive activity followed by death. During this period there would appear to be little benefit to be derived from minimizing expenditure to minimize free-radical damage. Unfortunately estimated energy demands for the two *Antechinus* species in the data set and the wambenger (*Phascogale calura*) do not include measurements made of males during their frantic mating period. However, Nagy and Lee (unpublished results, cited in Nagy *et al.*, 1988) found individual 'breeding season' expenditures of 5–10× RMR for the two *Antechinus* species (*stuartii* and *swainsonii*), and the maximum ratio observed in the present review (7.63× RMR) refers to a sample of predominantly male *Sminthopsis crassicaudata* measured in spring (Nagy *et al.*, 1988). Together, these data strongly suggest that these small mammals are

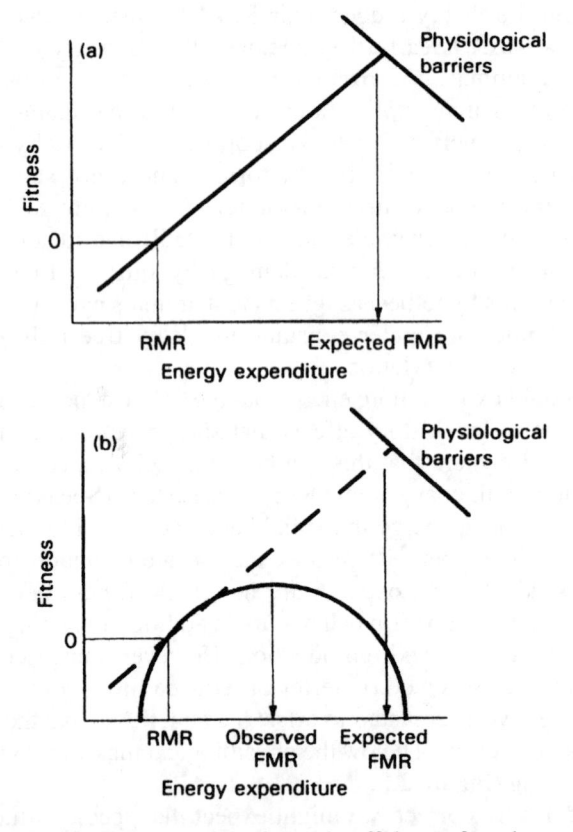

Fig. 22. Schematic diagram to explain the trade-off theory for why most measures of field metabolic rate (FMR) and resting metabolic rate (RMR) fall below the supposed limit at 7× RMR. In (a) there is no trade-off with fitness. Fitness increases with increasing energy expenditure until the physiological barrier is reached. The resultant FMR equals the expected FMR from the physiological limits hypothesis. In (b) the relationship between fitness and energy expenditure is curved. Consequently the observed FMR is substantially below the expectation from the physiological model (after Speakman, 1997b).

working much closer to their supposed limits at a time when there are immediate benefits associated with such expenditure but no long-term consequences. By implication, the remaining animals may be more prudent in their use of energy because of the trade-offs between current expenditure and future life expectancy.

There is considerable circumstantial evidence to support the idea that increased energy demands may be linked negatively with future survival. For example, in birds there have been many brood manipulation studies in which the number of eggs that birds raise is manipulated by adding to or subtracting from those laid naturally by the female. Where investigators have examined it, there is a link between the energy expenditure of females and their brood size (e.g. Bryant and Westerterp, 1983; Dijkstra et al., 1990), although this has been established in relatively few species. Manipulating a brood size upwards therefore probably experimentally increases energy expenditure of the birds. In 36% of studies reviewed by Stearns (1993) there were negative impacts of brood enlargement on future survival of the parent birds. The remainder did not show any negative effects, but in no case was future survival improved by brood enlargement. Other studies have found the opposite effect of brood reduction (e.g. Daan et al., 1990c). The link is circumstantial, however, because generally effects on energy expenditure are not measured directly. A more direct link of expenditure to survival was established by Bryant (1990), who observed that house martins (*Delichon urbica*) that died during the 12 months following determination of their energy demands by DLW techniques, had had greater energy demands than those that survived. However, even this link is still correlational and may reflect other covariable factors.

Several experimental manipulations have been made of animals that tend to support the idea that the level of habitual energy demands is governed by a trade-off in their life histories. Schmidt-Hempel and Wolf (1988) observed that bees that foraged for longer had shorter lifespans. To demonstrate the causal nature of this linkage, Wolf and Schmidt-Hempel (1989) forced bees to carry extra weights, so increasing their energy demands during flight, and found that manipulated bees lived shorter lives than unmanipulated bees. Dijkstra et al. (1990) forced kestrels that were feeding their young to work harder and used DLW techniques to confirm that energy demands had been increased (Deerenberg et al., 1995); they found a negative effect on subsequent survival over the winter (Daan et al., 1996). Priede (1977) and Lucas and Priede (1992) manipulated fish by altering the rations they were offered and found that fish fed high rations were more active, leading to greater energy expenditure (inferred), and died faster.

These data indicate that the study animals were capable of working harder in their respective environments, and thus were unlikely to be limited in their expenditure by extrinsic factors, and that they also paid a penalty in subsequent survival for this extra work. As yet, however, no studies have been performed that involve manipulation of free-living energy demands of small mammals, and the mechanisms underpinning the reported trade-offs remain obscure. One area where considerable work on small mammal longevity has been performed is dietary restriction studies. It has been frequently observed that when small mammals are fed food rations below their habitual intake they experience an increase in survival and hence longevity (reviewed in Weindruch and Walford, 1988; Yu, 1994). Because food intake is reduced in dietary restriction, over the long term this must equate reduced energy expenditure. However, the situation is complicated by the fact that dietary-restricted animals also sustain reduced body masses. In theory, oxidative damage is a phenomenon related to metabolic intensity (metabolism per gram of tissue) rather than whole-animal energy expenditure and, when expressed in this manner, there is no evidence that dietary-restricted animals have reduced metabolic intensity, and some evidence that it may even be increased (Baer et al., 1998). At present, therefore, there is no indication that the increased longevity invariably observed in dietary restriction is mediated via reduced energy expenditure and reduced metabolic intensity, leading to reduced oxidative damage.

To test the idea that small mammals restrict their energy expenditure below the putative intrinsic limitations because of a trade-off with future survival, I made two further analyses. Data for longevity are sparse and generally refer to single individuals maintained in zoos, which may not be representative of larger samples. On a more general level, however, there is a positive correlation between body size and longevity (Peters, 1983; Calder, 1984). On average, we might expect that if animals that live longer have greater benefits to derive from saving themselves there would be a negative relationship between the FMR : RMR ratio and body mass. Although other authors have reported such a linkage (Degen and Kam, 1995), in the present data set there was no evidence to support this prediction (Figure 17: whether bats were included or excluded from the sample; see below). Second, apart from the effects of body mass, some orders of mammals have exceptional longevity. The bats, for example, routinely live five times longer than might be anticipated from their body mass. Because bats have such long lives, we might anticipate they would be prudent in their use of energy above RMR. Refuting this hypothesis there was no significant effect of order on the average ratio ($F = 1.18$, $P = 0.26$, ANOVA:—a similar result either excluding or including the groups represented by single species). Indeed, the bats averaged the highest ratio from the seven orders represented in the data with an average FMR : RMR ratio of 3.97 (SD 1.55, $n = 8$). This high ratio is opposite to that predicted from the trade-off model presented above. One explanation for the lack of such an effect is that longevity is linked to lifetime or annual expenditure, which is only poorly reflected in point measurements. This could be particularly the case for bats, many species of which spend winter in hibernation. However, only four of the nine species involved in the sample of bats used here were species from the temperate zone, which hibernate, and there was no difference in the ratio of FMR : RMR between the species that do, and those that do not, hibernate. Apart from the high energy demands in semelparous male marsupial mice, the present data do not provide any evidence in support of the trade-off model.

Why do small mammals not routinely work at 4×, 7× or even 9–12× RMR? An alternative answer to the trade-off solution to this question is that they may be physiologically capable of working at any level up to the uppermost limit of

9–12× RMR, but they are kept from so doing because of the limited supply of energy from the environment (the extrinsic limitation hypothesis).

If this alternative hypothesis is correct, we might anticipate that mammals feeding on rich and abundant food sources would have higher ratios than the those feeding on poor food sources. I tested this idea using the dietary categories listed above. The FMR : RMR ratio was significantly associated with the diet (Figure 23; $F = 2.85$, $P < 0.05$). The pattern of this effect, however, shows that foods containing more and available energy were associated with higher metabolic rates. Using a dummy variable to code for the dietary energy content there was a significant positive linkage between dietary energy content and the FMR : RMR ratio ($r^2 = 9.3\%$, $P = 0.016$). The highest ratios of FMR : RMR are reported for mammals exploiting the most energy dense and available resources (fruit and nectar). In contrast, mammals that exploit the worst energy source (foliage) have the lowest ratios. Phylogenetic order is not a significant factor ($F = 0.65$, $P > 0.05$) and hence the effect of diet on the ratio is probably not a phylogenetic artefact. A plausible reason why the ratio is dependent on diet, with higher ratios linked to higher density foods, is that different foods provide different supplies of energy, and thus mammals exploiting different food supplies become extrinsically limited in their potential field energy expenditure at different levels (McNab, 1980, 1983, 1986a).

Of the two ideas explaining the lower than maximal field metabolic rates, these data lean more closely to a suggestion that small mammals may be extrinsically limited in their energy budgets (see also Koskela *et al.*, 1998).

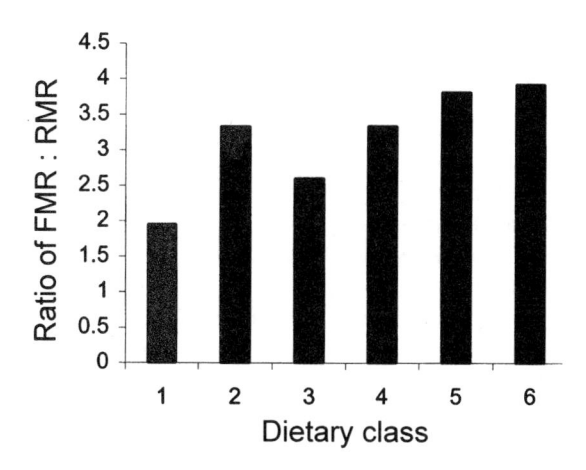

Fig. 23. Mean ratios of field to resting metabolic rate (FMR : RMR) in relation to dietary habits. The order of the dietary habits is related to the availability of energy from the diet (1 = low, 6 = high; see text for details of dietary classes).

This is not very strong evidence, as the effect of diet only barely managed to reach significance. However, dietary assignations are very crude estimates of food availability, and the fact any relationship at all is found is quite surprising. More refined estimates of energy availability should refine the test of the hypothesis. I think this interpretation is unexpected in the light of manipulative experiments performed on birds and insects which suggest that those animals are working under some form of life-history trade-off. If small mammals are extrinsically limited, this discovery leads naturally to some consideration that the supposed intrinsic limits of 4× RMR (Drent and Daan, 1980) and 7× RMR (Hammond and Diamond, 1997) are perhaps illusory. The supposed limit of 4× RMR is based on identical observations to those summarized here, but based on a much smaller database that was available 20 years ago. In that database (for birds) it was noticed that the maximum values of FMR did not exceed 4× RMR, and this was suggested as a physiological upper limit (Drent and Daan, 1980). There is no other reason to suppose that a 4× RMR limit should apply to mammal energy expenditures apart from the fact that the first measures using DLW methods did not exceed this value. Indeed, the fact that as more data have accumulated the 4× RMR limit has been routinely breached in small mammals and in birds (e.g. Bryant, 1990; Bryant and Tatner, 1991) indicates that it is not a physiological threshold. The 7× limit was derived under very similar circumstances. Hammond and Diamond (1997) and Peterson *et al.* (1990), for example, accumulated the available data from both laboratory and field studies, and selected from these the values considered to be maxima. The highest of these was around 6–7× RMR, and therefore they increased the supposed limit to 7× RMR. However, there is little evidence to suggest that this is an actual physiological limitation. Summarizing data in this manner becomes a self-fulfilling prophecy. If one selects the maximum observed FMR : RMR ratio across a sufficiently large data set, and suggests that this is a limit, inevitably most following observations will fall below this exceptional value (i.e. will not breach the limit and thus support its existence). However, data suggestive of higher energy expenditure ratios will be called into question because they are 'physiologically' impossible, which may make them more difficult to publish.

The only appropriate physiological limits are those derived independently of the FMR and RMR data. I have suggested here that continuously exercising large mammals, bats, birds and humans would expend energy at around 9–12× RMR, and that for smaller terrestrial mammals the limit might be lower at around 8–9× RMR. This is currently an extrapolation from measurements of active metabolic rate and it requires verification. However, this represents an independent estimate of a potential physiological limitation on energy expenditure. The closest alternative derivations are estimates of uptake capacity across the brush border of the small intestine (Karasov and Diamond, 1985; Hammond *et al.*, 1994) and maximal rates of food intake rate (Kirkwood,

1983). These latter measurements provide an upper boundary for energy uptake to support expenditure. However, the plasticity of the gut is such that any instantaneous measurements of this capacity may reflect only the demands under which the tract is currently being placed. These are not strictly intrinsic limits but reflections of the existing level of demand.

The current data suggest that RMR is probably constrained by a number of different factors. The most important of these are the overall body mass and the ambient temperatures to which it is exposed. This latter effect is presumably related to changes in brown adipose tissue capacity (e.g. Pasanen, 1971; Tarkkonen, 1971; Feist and Rosenmann, 1976; Wunder *et al.*, 1977; Bucowiecki *et al.*, 1982; Klaus *et al.*, 1988; Speakman, 1995) and overall tissue thermogenic capacity (see, for example, the suggested links between RMR and the levels of UCP-2 in muscle (Boss *et al.*, 1998a; Bouchard *et al.*, 1997) and the increased expression of UCP-2 during cold exposure (Boss *et al.*, 1998b)), but could also reflect resting activities of other organs such as the heart, kidneys and liver (Daan *et al.*, 1989, 1990a), which may vary with the ambient temperature to which the mammal is exposed. The RMR is set primarily by these two parameters, which in the current data set explain 92% of the variability. There may be other factors influencing RMR, which would become evident in wider databases (such as perhaps diet; McNab, 1980, 1983, 1986a) but the current data suggest these are probably of minor importance for the sample of small mammals examined here.

Field metabolic rate depends on a number of factors but is strongly influenced by the demands exerted by the same two factors that dominate RMR: body mass and ambient temperature. This is probably not for exactly the same reasons. Nevertheless, larger body mass and lower temperatures both demand greater field expenditure of energy. Because the direction of these effects on FMR and RMR is the same, there is an apparent linkage between the two, with FMR being about two to three times greater than RMR. Once the effects of body mass and temperature have been removed from both traits, the linkage is virtually eliminated and depends solely on a few outlying data, the significance of which remains to be established. The actual field expenditure, however, depends not only on these demands but also on the supply of energy from the environment. This extrinsic supply of energy depends on the density of the food, its composition and ease of digestibility, and probably also the time available for the animals to collect it. Hence small mammals exploiting rich and abundant foods have raised FMRs because the extrinsic supply of energy allows them to sustain this higher expenditure, whereas those exploiting poor resources have suppressed FMRs. The consequence of this extrinsic limitation of FMR is a link between the ratio of FMR to RMR and the diet of the mammals in question, with low ratios linked to poor-quality diets. At present, there is no evidence for small mammals that favours the hypothesis that field energy expenditures are constrained by their implications for future survival, apart from the high-energy expenditures of some semelparous breeding marsupial mice.

ACKNOWLEDGEMENTS

This review was written as part of a study into links between sustained metabolic rates and reproductive performance funded by the Natural Environmental Research Council (NERC) (grant GR3/9710). I am grateful to David Raffaelli for inviting me to make this contribution *to Advances in Ecological Research*. The original text was greatly improved by the perceptive comments of David Bryant, Sally Ward, Ela Krol and Maria Johnson. My understanding of limits to sustainable metabolic rates has benefited enormously from discussions with David Bryant, Kim Hammond, Marek Konarzewiski and Don Thomas. Professor Moskin, University of Novosibirsk, Russia, kindly supplied the reference to early Russian work on limits to athletic performance.

REFERENCES

Adams, N.J., Brown, C.R. and Nagy, K.A. (1986). Energy expenditure of free-ranging wandering albatrosses *Diomedea exulans*. *Physiol. Zool.* **59**, 583–591.
Adkins, R.M. and Honeycutt, R.L. (1991). Molecular phylogeny of the superorder Archonta. *Proc. Natl. Acad. Sci. USA* **88**, 10317–10321.
Adolph, E.F. (1950). Oxygen consumptions of hyperthermic rats and acclimatization to cold. *Am. J. Physiol.* **161**, 359–373.
Adolph, E.F. and Lawrow, J.W. (1951). Acclimatization to cold air: hypothermia and heat production in the golden hamster. *Am. J. Physiol.* **166**, 62–74.
Akbar, Z. and Gorman, M.L. (1993a). The effect of supplementary feeding upon the demography of a population of woodmice (*Apodemus sylvaticus*) living on a system of maritime sand dunes. *J. Zool. (Lond.)* **230**, 609–617.
Akbar, Z. and Gorman, M.L. (1993b). The effect of supplementary feeding upon the sizes of home ranges of woodmice (*Apodemus sylvaticus*) living on a system of maritime sand dunes. *J. Zool. (Lond.)* **231**, 233–237.
Altringham, J. (1996). *Bats: Biology and Behaviour*. Oxford University Press, Oxford.
Ames, B.N., Shigenaga, M.K. and Hagen, T.M. (1995). Mitochondrial decay in ageing. *Biochim. Biophys. Acta* **1271**, 165–170.
Armstrong, J.D. (1986). Heart rate as an indicator of activity, metabolic rate, food intake and digestion in pike, *Esox lucius*. *J. Fish. Biol.* **29**, 207–221.
Aschoff, J. and Phol, H. (1971). Der Ruheumsatz von Vogeln als funktion der Tageszeit und der Körpergrosse. *J. für ornithol.* **3**, 38–47.
Bartholomew, G.A. (1982). Energy metabolism. In: *Animal Physiology—Principles and Adaptations* (Ed. by M.S. Gordon, G.A. Bartholomew, A.D. Grinnell, C.B. Jorgensen and F.N. White), pp. 46–93, MacMillan, New York.
Baer, D.J., Lane, M.A., Rumpler, W.V., Ingram, D. and Roth, G. (1998). Bioenergetics and aging in monkeys: is less more? *Proc. Comp. Nut. Soc.* **2**, 8–12.
Baker, C.E. and Dunaway, P.B. (1975). Elimination of ^{137}Cs and ^{59}Fe and its relationship to metabolic rates of wild small rodents. *J. Exp. Zool.* **192**, 223–236

Baker, C.E., Dunaway, P.B. and Auerbach, S.I. (1968). *Measurement of Metabolism in Cotton Rats by Retention of Cesium-134.* Oak Ridge National Laboratory Publication (ORNL–TM–2069).
Baker, R.J., Novacek, M.J. and Simmons, N.B. (1991). On the monophyly of bats. *Syst. Zool.* **40**, 216–231.
Baker, R.R. (1978). *The Evolutionary Ecology of Animal Migration.* Hodder and Stoughton, London.
Baker, W.W., Marshall, S.G. and Baker, V.B. (1968) Autumn fat deposition in the evening bat (*Nycticeius humeralis*). *J. Mammal.* **49**: 314–317.
Bakken, G.S. (1976). A heat-transfer analysis of animals: unifying concepts and the application of metabolism chamber data to field ecology. *J. Theor. Biol.* **60**, 337–384.
Bakken, G.S. (1980). Use of standard operative temperature in the study of the thermal energetics of birds. *Physiol. Zool.* **53**, 108–119.
Bakken, G.S. (1992). Measurement and application of operative temperature and standard operative temperatures in ecology. *Am. Zool.* **32**, 194–216.
Bakken, G.S. and Gates, D.M. (1975). Heat transfer analysis of animals: some implications for field ecology, physiology amd evolution. In: *Perspectives of Biophysical Ecology* (Ed. by D.M. Gates and R.B. Schmerl), pp. 255–290. Springer, Berlin.
Barnett, Y.A. and King, C.M. (1995). Investigation of antioxidant status, DNA repair capacity and mutation as a function of age in humans. *Mut. Res.* **338**, 115–128.
Bell, G.P., Bartholomew, G.A. and Nagy, K.A. (1986). The roles of energetics, water economy, foraging behaviour and geothermal refugia in the distribution of the bat *Macrotus californicus*. *J. Comp. Phys. B.* **156**, 441–450.
Bennett, A.F. and Ruben, J.A. (1979). Endothermy and activity in vertebrates. *Science* **206**, 649–654.
Bennett, P.M. and Harvey, P.H. (1987). Active and resting metabolism in birds: allometry, phylogeny and ecology. *J. Zool. (Lond.)* **213**, 327–363.
Berry, R.J. (1968). The ecology of an island population of the house mouse. *J. Anim. Ecol.* **37**, 445–470.
Berteaux, D., Masseboeuf, F., Bonzom, J.-M., Bergeron, J.-M., Thomas, D.W. and Lapierre, H. (1996a). Effect of carrying a radiocollar on expenditure of energy by meadow voles. *J. Mamm.* **77**, 359–363.
Berteaux, D., Thomas, D.W., Bergeron, J.M. and Lapierre, H. (1996b). Repeatability of daily field metabolic rate in female meadow voles (*Microtus pennsylvanicus*). *Func. Ecol.* **10**, 751–759.
Bevan, R.M., Woakes, A.J., Butler, P.J. and Boyd, I.L. (1994). The use of heart rate to measure oxygen consumption of free-ranging black-browed albatrosses, *Diomedea melanophrys*. *J. Exp. Biol.* **193**, 119–137.
Bevan, R.M., Butler, P.J., Woakes, A.J. and Prince, P.A. (1995a). The energy expenditure of free-ranging black-browed albatrosses. *Phil. Trans. R. Soc.* **350**, 119–131.
Bevan, R.M., Woakes, A.J., Butler, P.J. and Croxhall, J.P. (1995b). The heart rate and oxygen consumption of exercising gentoo penguins. *Physiol. Zool.* **68**, 855–877.
Bevan, R.M., Speakman, J.R. and Butler, P.J. (1995c). Daily energy expenditure of tufted ducks: a comparison of indirect calorimetry, doubly-labelled water and heart rate. *Func. Ecol.* **9**, 40–47.
Bevan, R.M., Boyd, I.L., Butler, P.J., Reid, K., Woakes, A.J. and Croxall, J.P. (1997). Heart rates and abdominal temperatures of free-ranging South Georgian shags, *Phalacrocorax georgianus*. *J. Exp. Biol.* **200**, 661–675.
Biebach, H. (1998). Phenotypic organ flexibility in garden warblers, *Sylvia borin*, during long-distance migration. *J. Avian Biol.* **29**, 529–535.

Blaxter, K. (1989) *Energy Metabolism in Animals and Man*. Cambridge University Press, Cambridge.
Boss, O., Muzzin, P. and Giacobino, J.P. (1998a). The uncoupling proteins, a review. *Eur. J. Endocrinol.* **139**, 1–9.
Boss, O., Samec, S., Dulloo, A., Seydoux, J., Muzzin, P. and Giacobino, J.P. (1998b). Tissue dependent upreguation of rat uncoupling protein-2 expression in response to fasting or cold. *FEBS Lett.* **412**, 111–114.
Bouchard, C., Perusse, L., Chagnon, Y.C., Warden, C. and Riquier, D. (1997). Linkage between markers in the vicinity of the uncoupling protein 2 gene and resting metabolic rate in humans. *Hum. Mol. Genet.* **6**, 1887–1889.
Boyd, I.L., Woakes, A.J., Butler, P.J. and Williams, T.M. (1995). Validation of heart rate and doubly-labelled water as measures of metabolic rate during swimming in California sea lions. *Funct. Ecol.* **9**, 151–160.
Bozinovic, F. (1992). Scaling of basal and maximal metabolic rate in rodents and the aerobic capacity model for the evolution of endothermy. *Physiol. Zool.* **65**, 921–932.
Bozinovic, F. and Rosenman, M. (1989). Maximum metabolic rates of rodents—physiological and ecological consequences on distribution limits. *Funct. Ecol.* **3**, 173–181.
Bozinovic, F., Novoa, F.F. and Veluso, C. (1990). Seasonal changes in energy expenditure and digestive tract of *Abrothrix andinus* (Cricetidae) in the Andes range. *Physiol. Zool.* **63**, 1216–1231.
Bradshaw, S.D., Morris, K.D., Dickman, C.R., Ithers, P.C. and Murphy, D. (1994). Field metabolism and turnover in the golden bandicoot (*Isoodon auratus*) and other small mammals from Barrow Island, Western Australia. *Aust. J. Zool.* **42**, 29–41.
Brafield, A.E. and Llewellyn, M.J. (1982). *Animal Energetics*. Blackie, Glasgow.
Brody, S. (1945). *Bioenergetics and Growth*. Hafner, New York.
Brown, J.H. and Lasiewski, R.C. (1972). Metabolism of weasels: the cost of being long and thin. *Ecology* **53**, 939–943.
Brownwell, E. (1983). DNA/DNA hybridisation studies of muriod rodents. Symmetry and roles of molecular evolution. *Evolution* **37**, 1034–1051.
Bryant, D.M. (1989). Determination of respiration rates of free-living animals by double-labelling technique. In: *Toward a More Exact Ecology. BES Symposium* (Ed. by P.J. Grubb and J.G. Whitaker), pp. 85–109. Blackwells, London.
Bryant, D.M. (1990). Constraints on energy expenditure by birds. *Acta XX Congr. Int. Ornithol.* **20**, 1989–2001.
Bryant, D.M. (1997). Energy expenditure in wild birds. *Proc. Nutr. Soc.* **56**, 1025–1039.
Bryant, D.M. and Tatner, P. (1991). Intraspecies variation in avian energy expenditure: correlates and contraints. *Ibis* **133**, 236–246.
Bryant, D.M. and Westerterp, K.A. (1982). Evidence for individual differences in foraging efficiency amongst breeding birds: a study of house martins *Delichon urbica*, using the doubly labelled water technique. *Ibis* **124**, 187–192.
Bryant, D.M. and Westerterp, K.R. (1983). Time and energy limits to brood size in house martins (*Delichon urbica*). *J. Anim. Ecol.* **52**, 905–925.
Bucowiecki, L., Collett, A.J., Follea, N., Guay, G. and Jahjah, L. (1982). Brown adipose tissue hyperplasia: a fundamental mechanism of adaptation to cold and hyperphagia. *Am. J. Physiol.* **242**, E353–359.
Bundle, M.W., Hoppeler, H., Vock, R., Tester, J.M. and Weyand, P.G. (1999). High metabolic rates in running birds. *Nature* **397**, 31–32.
Butler, P.J. (1991). Exercise in birds. *J. Exp. Biol.* **160**, 233–262.
Butler, P.J. and Bishop, C.M. (1998). Flight. In: *Sturkie's Avian Physiology* (Ed. by G.C. Whittow). Academic Press, London.
Butler, P.J. and Woakes, A.J. (1980). Heart rate, respiratory frequency, and wing beat frequency during free-range flights of barnacle geese. *J. Exp. Biol.* **85**, 213–226.

Butler, P.J., Woakes, A.J. and Bishop, C.M. (1998). Behaviour and physiology of Svalbard barnacle geese *Branta leucopsis* during their autumn migration. *J. Avian Biol.* **29**, 536–545.

Buttemer, W.A., Hayworth, A.M., Weathers, W.W. and Nagy, K.A. (1986) Time–budget estimates of avian energy-expenditure—physiological and meteorological considerations. *Physiol. Zool.* **59**, 131–149.

Calder, W.A. (1984). *Size, Function and Life History.* Dover Publications, New York.

Calow, P. (1979). The cost of reproduction—a physiological approach. *Biol. Rev.* **54**, 23–40.

Casey, T.M. (1981). Nest insulation: energy savings to brown lemmings using a winter nest. *Oecologia.* **50**, 199–204.

Casey, T.M. and Casey, K.K. (1979). Thermoregulation of arctic weasels. *Physiol. Zool.* **52**, 153–164.

Chaffee, R.R. and Roberts, J.C. (1971). Temperature acclimation in birds and mammals. *Ann. Rev. Physiol.* **33**, 155–202.

Chai, P., Chang, A.C. and Dudley, R. (1998). Flight thermogenesis and energy conservation in hovering hummingbirds. *J. Exp. Biol.* **201**, 963–968.

Charnov, E.L. (1976). Optimal foraging: the attack strategy of the mantid. *Am. Nat.* **110**, 141–151.

Chappell, M.A. (1980). Thermal energetics and thermoregulatory costs of small arctic mammals. *J. Mammal.* **61**, 278–291.

Chappell, M.A. and Bachman, G.C. (1995). Aerobic performance in Belding's ground squirrel (*Spermophilus beldingi*): variance, ontogeny and the aerobic capacity model of endothermy. *Physiol. Zool.* **68**, 421–442.

Chevalier, C.D. (1989). Field energetics and water balance of desert dwelling ringtail cats *Bassariscus astutus* (Carnivora: procyonidae). *Am. Zool.* **29**, 8A.

Chew, R.M. (1971). The excretion of ^{65}Zn and ^{54}Mn as indices of energy metabolism of *Peromyscus polionotus*. *J. Mammal.* **52**, 337–350.

Chitty, D. (1952). Mortality among voles (*Microtus agrestis*) at Lake Vyrnwy, Montgomeryshire in 1936–1939. *Phil. Trans. R. Soc.* **236B**, 505–552.

Coburn, D.K. and Geiser, F. (1999) Field metabolic rates of the blossom bat *Syconycteris australis*. *J. Comp. Physiol.* **169**, 133–138.

Contreras, L.C. (1984). Bioenergetics of huddling: test of the psycho-social hypothesis. *J. Mammal.* **65**, 256–262.

Corp, N., Gorman, M.L. and Speakman, J.R. (1997a). Apparent absorption efficiencies and gut morphology of wood mice *Apodemus sylvaticus* from two distinct populations with different diets. *Physiol. Zool.* **70**, 610–614.

Corp, N., Gorman, M.L. and Speakman, J.R. (1997b). Seasonal variation in the resting metabolic rates of male wood mice *Apodemus sylvaticus* from two contrasting habitats 15 km apart. *J. Comp. Physiol.* **167**, 229–239.

Corp, N., Gorman, M.L. and Speakman, J.R. (1999) Daily energy expenditure of the wood mouse (*Apodemus sylvaticus*) from two distinct habitats. *Funct. Ecol.* (in press).

Costa, D. P. and Prince, P.A. (1987). Foraging energetics of grey-headed albatrosses *Diomedea chrysostoma* at Bird Island, South Georgia. *Ibis* **128**, 149–158.

Covell, D.F., Miller, D.S. and Karasov, W.H. (1996). Cost of locomotion and daily energy expenditure by free-living swift foxes (*Vulpes velox*): a seasonal comparison. *Can. J. Zool.* **74**, 283–290.

Cripps, A.W. and Williams, V.J. (1975). The effect of pregnancy and lactation on food intake, gastrointestinal anatomy and the absorptive capacity of the small intestine in the albino rat. *B. J. Nutr.* **33**, 17–32.

Cucco, M. and Malacarne, G. (1997). The effect of supplemental food on the time budget and body condition in the black redstart *Phoenicurus ochruros*. *Ardea* **85**, 211–221.

Daan, S., Masman, D., Strijkstra, A. and Verhulst, S. (1989). Intraspecific allometry of basal metabolic rate: relations with body size, temperature, composition and circadian phase in the kestrel (*Falco tinnunculus*). *J. Biol. Rhyth.* **4**, 267–283.

Daan, S., Masman, D. and Groenewold, A. (1990a). Avian basal metabolic rates: their association with body composition and energy expenditure in nature. *Am. J. Physiol.* **259**, R333–340.

Daan, S., Masman, D., Strijkstra, A.M. and Kenagy, G.J. (1990b). Daily energy turnover during reproduction in birds and mammals: its relationship to basal metabolic rate. *Acta XX Cong. Int. Ornithol.* **IV**, 1976–1987.

Daan, S., Dijkstra, C. and Tinbergen, J.M. (1990c). Family planning in the kestrel (*Falco tinnunculus*): the ultimate control of covariation of laying date and clutch size. *Behaviour* **114**, 83–114.

Daan, S., Deerenberg, C. and Dijkstra, C. (1996). Increased daily work precipitates natural death in the kestrel. *J. Anim. Ecol.* **65**, 539–544.

Dawson, T.J. and Hulbert, J. (1970). Standard metabolism, body temperature and surface areas of Australian marsupials. *Am. J. Physiol.* **218**, 1233–1238.

Dawson, T.J., Grant, T.R. and Fanning, D. (1979). Standard metabolism of monotremes and the evolution of endothermy. *J. Zool. (Lond.)* **27**, 511–515.

Davis, W.H. and Hitchcock, H.B. (1965). Biology and migration of the bat, *Myotis lucifugus*, in New England. *J. Mammal.* **46**, 296–313.

Deerenberg, C., Pen, I., Dijkstra, C., Arkies, B.J., Visser, G.H. and Daan, S. (1995). Parental energy expenditure in relation to manipulated brood size in the Eurasian kestrel *Falco tinnunculus. Zool. Anal. Complex Syst.* **99**, 39–48.

Degen, A.A. (1993). Energy requirements of the fat sand rat (*Psammomys obesus*) when consuming the saltbush, *Atriplex halimus*: a review. *J. Basic Clin. Physiol Pharmacol.* **4**, 13–28.

Degen, A.A. and Kam, M. (1995). Scaling of field metabolic rate to basal metabolic rate ratio in homeotherms. *Ecoscience* **2**, 48–54.

Degen, A.A., Kam, M., Hazan, A. and Nagy, K.A. (1986). Energy expenditure and water flux in three sympatric desert rodents. *J. Anim. Ecol.* **55**, 421–429.

Degen, A.A., Hazan, A., Kam, M. and Nagy, K.A. (1991). Seasonal water influx and energy expenditure of free-living fat sand rats. *J. Mamm.* **72**, 652–657.

Degen, A.A., Pinshow, B. and Kam, M. (1992). Field metabolic rates and water influxes of two sympatric Gerbillidae *Gerbillus allenbyi* and *Gerbillus pyramidium. Oecologia* **90**, 586–590.

Depocas, F. and Hart, J.S. (1957). Use of the Pauling oxygen analyser for measurement of oxygen consumption of animals in open circuit and in short lag, closed circuit apparatus. *J. Appl. Physiol.* **10**, 388–392.

Depocas, F., Hart, J.S. and Heroux, O. (1957). Energy metabolism of the white rat after acclimation to warm and cold environments. *J. Appl. Physiol.* **10**, 393–397.

Derting, T.L. and Austin, M.W. (1998). Changes in gut capacity with lactation and cold exposure in a species with low rates of energy use, the pine vole (*Microtus pinetorum*). *Physiol. Zool.* **71**, 611–624.

Derting, T.L. and McClure, P.A. (1989). Intraspecific variation in metabolic rate and its relationship with productivity in the cotton rat, *Sigmodon hispidus. J. Mammal.* **70**, 520–531.

Derting, T.L. and Noakes, E.B. (1995). Seasonal changes in gut capacity in the white footed mouse (*Peromyscus leucopus*) and the meadow vole (*Microtus pennsylvanicus*). *Can. J. Zool.* **73**, 243–252.

Dijkstra, C., Bult, A., Bijlsma, S., Daan, S., Meijer, T. and Zijlstra, M. (1990). Brood size manipulation in the kestrel (*Falco tinnunculus*): effects on offspring and parent survival. *J. Anim. Ecol.* **59**, 269–296.

Drent, R. and Daan, S. (1980). The prudent parent: energetics adjustments in avian breeding. *Ardea* **68**, 225–252.
Drozdz, A. (1968). Digestibility and assimilation of natural foods in small rodents. *Acta Theriol.* **13**, 367–389.
Edwards, Margaria, and Dill, (1934). *Am. J. Physiol.* **108**, 203.
Eisenberg, J.F. (1981). *The Mammalian Radiations.* Athlone Press, London.
Elgar, M.A. and Harvey, P.H. (1987). Basal metabolic rates in mammals: allometry, phylogeny and ecology. *Funct. Ecol.* **1**, 25–36.
Ellerman, J.R. (1941). The families and genera of living rodents. Vols II and III. British Museum (Natural history), London.
Ellis, H.I. (1984). Energetics of free-ranging seabirds. In: *Seabird energetics.* (Ed. by Whittow, G.C. and Rahn, H), New York. pp. 203–233. Plenum Press.
Else, P.L. and Hulbert, A.J. (1985). An allometric comparison of the mitochondria of mammalian and reptilian tissues: the implications for the evolution of endothermy. *J. Comp. Physiol.* **156**, 3–11.
Emlen, J.M. (1966). The role of time and energy in food preference. *Am. Nat.* **100**, 611–617.
Farfel, V.S. (1960). *Physiology of Sport.* National Publications Institute, Moscow (in Russian).
Feist, D.D. and Rosenmann, M. (1976). Norepinephrine thermogenesis in seasonally acclimatized and cold acclimated red-backed voles in Alaska. *Can J. Physiol. Pharmacol.* **54**, 146–153.
Fell, B.F., Smith, K.A. and Campbell, R.M. (1963). Hypertrophic and hyperplastic changes in the alimentary canal of the lactating rat. *J. Pathol. Bacteriol.* **85**, 179–188.
Felsenstein, J. (1985). Phylogenies and the comparative method. *Am. Nat.* **125**, 1–15.
Field, J., Belding, H.S. and Martin, A.W. (1939). An analysis of the relation between basal metabolism and summated tissue respiration in the rat. *J. Cell. Comp. Physiol.* **14**, 143–157.
Flint, E.N. and Nagy, K.A. (1984). Flight energetics of free-living sooty terns *Auk* **101**, 288–294.
Foley, W.J., Kehl, J.C., Nagy, K.A., Kaplan, I.R. and Borsboom, A.C. (1990). Energy and water metabolism in free-living greater gliders. *Aust. J. Zool.* **38**, 1–9.
Gadgil, M. and Bossert, W.H. (1970). Life historical consequences of natural selection. *Am. Nat* **104**, 1–24.
Gales, R. (1989). Validation of the use of the tritiated water, doubly labeled water, and sodium-22 for estimating food, energy, and water intake in little penguins *Eudyptula minor. Physiol. Zool.* **62**, 147–169.
Gallivan, G.J. (1992). What are the metabolic rates of cetaceans? *Physiol. Zool.* **65**, 1285–1287.
Garland, T., Harvey, P.H. and Ives, A.R. (1992). Procedures for the analysis of comparative data using phylogenetically independent contrasts. *Syst. Biol.* **41**, 18–32.
Gebczynska, Z. and Gebczynski, M. (1971). Length and weight of the alimentary tract of the root vole. *Acta Theriol.* **16**, 359–369.
Geffen, E., Degen, A.A., Kam, M., Reuven, H. and Nagy, K.A. (1992). Daily energy expenditure and water flux of free-living Blanford's foxes (*Vulpes cana*), a small desert carnivore. *J. Anim. Ecol.* **61**, 611–617.
Geiser, F. (1988). Reduction of metabolism during hibernation and daily torpor in mammals and birds: temperature effect or physiological inhibition? *J. Comp. Physiol.* **158B**, 25–38.
Geiser, F. and Ruf, T. (1995). Hibernation versus daily torpor in mammals and birds: physiological variables and classification of torpor patterns. *Physiol. Zool.* **68**, 935–967.

Gessaman, J.A. (1980). An evaluation of heart rate as an indirect measure of daily energy metabolism of the American kestrel. *Comp. Biochem. Physiol.* **65**, 273–289.

Gessaman, J.A. (1987). Energetics. In: *Raptor Management Techniques Manual.* (Ed. by B.A. Giron, K.W. Pendleton, K.W. Millsap, D.M. Cline, and Bird), pp. 289–320. National Wildlife Federation, Washington, DC.

Gessaman, J.A. and Nagy, K.A. (1988). Energy metabolism—errors in gas exchange conversion factors. *Physiol. Zool.* **61**, 507–513.

Gettinger, R.D. (1984). Energy and water metabolism of free-ranging pocket gophers *Thomomys bottae*. *Ecology* **65**, 740–751.

Gibb, J.E. (1957). Food requirements and other observations of captive tits. *Bird Study* **4**, 207–215.

Goldstein, D.L. (1988). Estimates of daily energy expenditure in birds: the time–energy budget as an integrator of laboratory and field studies. *Am. Zool.* **28**, 829–844.

Goldstein, L. (1977). *Introduction to Comparative Physiology.* Holt Reinhart Winston, New York.

Gorman, M.L. and Akbar, Z. (1993). A comparative study of the ecology of woodmice *Apodemus sylvaticus* in two contrasting habitats: deciduous woodland and maritime sand dunes. *J. Zool. (Lond.)* **229**, 385–396.

Graur, D., Hide, W.A. and Li, W.H. (1991). Is the guinea pig a rodent? *Nature* **381**, 649–652.

Green, B. (1978). Estimation of food consumption in the Dingo, *Canis familiaris dingo*, by means of ^{22}Na turnover. *Ecology* **59**, 207–210.

Green, B. (1989). Water and energy turnover in free-living Macropodids. In: '*Kangaroos, Wallabies* and *Rat-kangaroos* (Ed. by G. Grogg, P. Jarman and I. Hume, pp. 223–229. Surry-Beatty, New South Wales.

Green, B. (1997). Field energetics and water fluxes in marsupials. In: *Marsupial Biology, Recent Research, New Perspectives (*Ed. by N.R. Saunders and L.A. Hinds), pp. 143–162. UNSW Press, Sydney.

Green, B. and Dunsmore, J.D. (1978a). Turnover of tritiated water and ^{22}sodium in captive rabbits (*Oryctolagus cuniculus*). *J. Mammal.* **59**, 12–17.

Green, B. and Dunsmore, J.D. (1978b). Turnover of sodium and water by free-living rabbits, *Oryctolagus cuniculus*. *Aust. Wildl. Res.* **5**, 93–99.

Green, B. and Eberhard, I. (1979). Energy requirements and sodium and water turnover in two captive marsupial carnivores: the Tasmanian devil, *Sarcophilus harrisii*, and the native cat, *Dasyurus viverrinus*. *Aust. J. Zool.* **27**, 1-8.

Green, B. and Eberhard, I. (1983). Water and sodium intake, and estimated food consumption in free-living eastern quolls, *Dasyurus viverrinus*. *Aust. J. Zool.* **31**, 871–880.

Green, B. and Rowe-Rowe, D.T. (1987). Water and energy metabolism in free-living multi-mammate mice (*Praeomys natalensis*) during summer. *S. Afr. J. Sci.* **22**, 14–17.

Green, B., Anderson, J. and Whateley, T. (1984). Water and sodium turnover and estimated food-consumption in free-living lions (*Panthera leo*) and spotted hyenas (*Crocuta crocuta*). *J. Mammal.* **65**, 593–599.

Green, B., King, D. and Bradley, A. (1989). Water and energy metabolism and estimated food consumption rates of free-living wambengers, *Phascogale calura* (Marsupialia: Daysuridae). *Aust. Wildl. Res.* **16**, 501–508.

Green, B., Newgrain, K., Catlin, P. and Turner, G. (1991). Patterns of prey consumption and energy use in a small carnivorous marsupial *Antechinus stuartii*. *Aust. J. Zool.* **39**, 539–548.

Green, D.A. and Millar, J.S. (1987). Changes in gut dimensions and capacity of *Peromyscus maniculatus* relative to diet quality and energy needs. *Can. J. Zool.* **65**, 2159–2162.

Green, K. and Crowley, H. (1989). Energetics and behaviour of active subnivean insectivores *Antechinus swainsonii* and *Antechinus stuartii* (Marsupialia: Dasyuridae) in the Snowy Mountains (New South Wales, Australia). *Aust. Wildl. Res.* **16**, 509–516.

Greenwalt, C.H. (1962). Dimensional relationships for flying animals. *Smithsonian Miscellaneous Collections* **144**, 1–46.

Grenot, C., Pascal, M., Buscarlet, L., Francaz, J.M. and Sellami, M. (1984). Water and energy balance in the water vole (*Arvicola terrestris* Sherman) in the laboratory and the field (Haut-Doubs, France). *Comp. Biochem. Physiol.* **78A**, 185–196.

Gross, J.E., Weng, Z. and Wunder, B.A. (1985). Effects of food quality and energy needs: changes in gut morphology and capacity in *Microtus ochrogaster*. *J. mammal.* **66**, 661–667.

Grubb, B.R. (1982). Cardiac output and stroke volume in exercising ducks and pigeons. *J. Appl. Physiol.* **53**, R207–211.

Hails, C.J. (1979). A comparison of flight energetics of hirundines and other birds. *Comp. Biochem. Physiol.* **63A**, 581–585.

Haim, A. (1987). Metabolism and thermoregulation in rodents—are these adapted to habitat and food quality. *South Afr. J. Sci.* **83**, 639–642.

Haim, A. and Izhaki, A. (1993). The ecological significance of resting metabolic rate and non-shivering thermogenesis in rodents. *J. Thermal Biol.* **18**, 71–81.

Hall, E.R. (1981). *The Mammals of North America*. Vol. II, 2nd edn. Wiley, New York.

Hammond, K.A. (1993). Seasonal changes in gut size of the wild prairie vole (*Microtus ochrogaster*). *Can. J. Zool.* **71**, 820–827.

Hammond, K.A. and Diamond, J. (1992). An experimental test for a ceiling on sustained metabolic rate in lactating mice. *Physiol. Zool.* **65**, 952–977.

Hammond, K.A. and Diamond, J. (1997). Maximal sustained energy budgets in humans and animals. *Nature* **386**, 457–462.

Hammond, K.A. and Konarzewski, M. (1996). The trade-off between maintenance and activity. In: *Adaptations to Cold. Tenth International Hibernation Symposium* (Ed. by F. Geiser, A. Hulbert and S.C. Nicol), pp. 153–158. University of New England Press, Armidale, Australia.

Hammond, K.A., Konarzewski, M., Torres, R. and Diamond, J. (1994). Metabolic ceilings under a combination of peak energy demands. *Physiol. Zool.* **68**, 1479–1506.

Hammond, K.A., Kent Lloyd, K.C. and Diamond, J. (1996). Is mammary output capacity limiting to lactational performance in mice? *J. Exp. Biol.* **199**, 337–349.

Hanski, I. (1985). What does a shrew do in an energy crisis? In: *Behavioural Ecology: Symposium of the British Ecological Society* (Ed. by R.H. Smith and R.M. Sibley), pp. 247–252. Blackwells, Oxford.

Harman, D. (1956). Aging: a theory based on free-radical and radiation chemistry. *J. Gerontol.* **11**, 298–200.

Hart, J.S. (1953a). The relation between thermal history and cold resistance in certain species of rodents. *Can. J. Zool.* **31**, 80–97.

Hart, J.S. (1953b). Energy metabolism of the white footed mouse *Peromyscus leucopus noveboracensis* after acclimation at various environmental temperatures. *Can. J. Zool.* **31**, 99–105.

Hart, J.S. (1971). Thermoregulation of rodents. In: *The Comparative Physiology of Thermoregulation* (Ed. by G.C. Whittow), pp. 1–149. Academic Press, London.

Hart, J.S. and Heroux, O. (1953). A comparison of some seasonal and temperature induced changes in *Peromyscus*: cold resistance, metabolism and pelage insulation. *Can. J. Zool.* **31**, 528–534.

Harvey, P.H. and Elgar, M.A. (1987). In defence of the comparative method. *Funct. Ecol.* **1**, 160–161.

Harvey, P.H. and Pagel, M.D. (1991). *The Comparative Method in Evolutionary Biology.* Oxford Series in Ecology and Evolution. Oxford University Press, Oxford.

Harvey P.H., Pagel, M.D. and Rees, J.A. (1991). Mammalian metabolism and life histories. *Am. Nat.* **137**, 556–566.

Hayes, J.P. (1989a). Altitudinal and seasonal effects on aerobic metabolism of deer mice. *J. Comp. Physiol.* **159**, 453–459.

Hayes, J.P. (1989b). Field and maximal metabolic rates of deer mice (*Peromyscus maniculatus*) at low and high altitudes. *Physiol. Zool.* **62**, 732–744.

Hayes, J.P. and Garland, T. (1995). The evolution of endothermy: testing the aerobic capacity model. *Evolution* **49**, 836–847.

Hayes, J.P., Garland, T. and Dohm, M.R. (1992a). Individual variation in metabolism and reproduction of *Mus*: are energetics and life history linked? *Funct. Ecol.* **6**, 5–14.

Hayes, J.P., Speakman, J.R. and Racey, P.A. (1992b). Sampling bias in respirometry. *Physiol. Zool.* **65**, 604–619.

Hayssen, V. (1984). Basal metabolic rate and intrinsic rate of increase: an empirical and theoretical re-examination. *Oecologia* **64**, 419–421.

Hayssen, V. and Lacy, R.C. (1985). Basal metabolic rates in mammals: taxonomic differences in the allometry of BMR and body mass. *Comp. Biochem. Physiol.* **81A**, 741–754.

Hayward, J.S. (1965). Microclimate temperature and its adaptive significance in six geographic races of *Peromyscus. Can. J. Zool.* **43**, 341–350.

Hedenstrom, A. and Alerstam, T. (1998). How fast can birds migrate? *J. Avian Biol.* **29**, 424–433.

Heldmaier, G. (1989). Seasonal acclimation of energy requirements in mammals: functional significance of body weight control, hypothermia, torpor and hibernation. In: *Energy Transformations in Cells and Organisms* (Ed. by W. Weiser and E. Gnaiger), pp. 130–139. Springer, Berlin.

Heldmaier, G., Steiger, R. and Ruf, T. (1993). Suppression of metabolic rate in hibernation. In: *Life in the Cold: Ecological, Physiological and Molecular Mechanisms* (Ed. by C. Carey, G.L. Florant, B.A. Wunder and B. Horwitz), pp. 545–548, Westview Press, Boulder, Colorado.

Henneman, W.W. (1983). Relation between body mass, metabolic rate and intrinsic rate of natural increase in mammals. *Oecologia* **56**, 104–108.

Hill, R.C., Lewis, D.D., Scott, K.C., Omori, M., Jackson, M., Sundstrom, D., Speakman, J.R., Doyle, C.A., and Butterwick, R.F. (submitted). Increased dietary protein slows racing greyhounds.

Hill, R.W. (1972). Determination of oxygen consumption by use of the paramagnetic oxygen analyser. *J. Appl. Physiol.* **33**, 261–263.

Himms-Hagen, J. (1986). Brown adipose tissue and cold acclimation. In: *Brown Adipose Tissue* (Ed. by P. Trayhurn and D.G. Nicholls), pp. 214–228. Arnold, London.

Hock, R.J. (1951). The metabolic rates and body temperatures of bats. *Biol. Bull.* **101**, 289–299.

Holleman, D.F., White, R.G. and Feist, D.D. (1982). Seasonal energy and water metabolism in free-living Alaskan voles. *J. Mammal.* **63**, 293–296.

Holling, C.J. (1959). Functional response of invertebrate predators to prey. *Mem. Ent. Soc. Can.* **48**, 1–86.

Horvath, S.M., Folk, G.E., Craig, F.N. and Fleischmann, W. (1948). Survival time of various warm blooded animals in extreme cold. *Science* **107**, 171–172.

Hulbert, A.J. and Dawson, T.J. (1974). Standard metabolism and body temperature of parameloid marsupials from different environments. *Comp. Biochem. Physiol.* **47A**, 583–590.

Hume, I. (1989). Marsupial metabolism and nutrient requirements. In: *Kangaroos, Wallabies and Rat-kangaroos* (Ed. by G. Grogg, P. Jarman and I. Hume), pp. 1–26, Surry-Beatty, New South Wales.

Hutcheon, J.M., Kirsch, A.W. and Pettigrew, J.D. (1998). Base-compositional biases and the bat problem. III. The question of microchiropteran monophyly. *Phil. Trans. R. Soc.* **353**, 607–617.

Innes, S. and Lavigne, D.M. (1991). Do cetaceans really have elevated metabolic rates? *Physiol. Zool.* **64**, 1130–1134.

Iverson, S.L. and Turner, B.N. (1974). Winter weight dynamics in *Microtus pennsylvanicus*. *Ecology* **55**, 1030–1041.

Jenni, L. and Jenni-Eierman, S. (1998). Fuel supply and metabolic constraints in migrating birds. *J. Avian Biol.* **29**, 521–528.

Jolicoeur, L., Asselin, J. and Morisset, J. (1980). Trophic effects of gestation and lactation on rat pancreas. *Biomed. Res.* **1**, 482–488.

Jones, J.H., Longworth, K.E., Lindholm, A., Conley, K.E., Karas, R.H., Kayar, S.R. and Taylor, C.M. (1989). Oxygen transport during exercise in large mammals. I. Adaptive variation in oxygen demand. *J. Appl. Physiol.* **67**, 862–870.

Kaczmarski, F. (1961) Bioenergetics of pregnancy and lactation in the bank vole. *Acta Theriol.* **11**, 409–417.

Karas, R.H., Taylor, C.R., Rosler, K. and Hoppeler, H. (1987). Adaptive variation in the mammalian respiratory system in relation to energetic demand. V. Limits to oxygen transport by the circulation. *Resp. Physiol.* **69**, 65–79.

Karasov, W.H. (1981). Daily energy expenditure and cost of activity in a free-living mammal. *Oecologia* **51**, 253–259.

Karasov, W.H. (1983). Wintertime energy conservation by huddling in Antelope ground squirrels (*Ammospermophilus leucurus*). *J. Mammal.* **64**, 341–345.

Karasov, W.H. (1992). Daily energy expenditure and the cost of activity in mammals. *Am. Zool.* **32**, 238–248.

Karasov, W.H. and Diamond, J. (1985). Digestive adaptations for fueling the cost of endothermy. *Science* **228**, 202–204.

Kasting, N.W., Adderley, S.A.L., Safford, T. and Hewlett, K.G. (1989). Thermoregulation in beluga (*Delphinapterus leucas*) and killer (*Orcinus orca*) whales. *Physiol. Zool.* **62**, 687–701.

Kenagy, G.J. (1987). Energy allocation for reproduction in the golden-mantled ground squirrel. *Symp. Zool. Soc. Lond.* **57**, 259–273.

Kenagy, G.J., Sharbaugh, S.M. and Nagy, K.A. (1989). Annual cycle of energy and time expenditure in a golden-mantled ground squirrel population. *Oecologia* **78**, 269–282.

Kenagy, J.G., Masman, D., Sharbaugh, S.M. and Nagy, K.A. (1990). Energy expenditure during lactation in relation to litter size in free-living golden-mantled ground squirrels. *J. Anim. Ecol.* **59**, 73–88.

Kennedy, G.C., Pearce, W.M. and Parrott, D.M.V. (1958). Liver growth in the lactating rat. *J. Endocrinol.* **17**, 158–160.

King, J.R. (1974). Seasonal allocation of time and energy resources in birds. In: *Avian Energetics* (Ed. by R.A. Paynter), pp. 4–70. Nuttal Ornithological Club 15, Cambridge, Massachusetts.

Kinnear, A. and Sheild, J.W. (1975). Metabolism and temperature regulation in marsupials. *Comp. Biochem. Physiol.* **52A**, 235–245.

Kirkwood, J.K. (1983). A limit to metabolisable energy intake in mammals and birds. *Comp. Biochem. Physiol.* **75A**, 1–3.

Klaus, S., Heldmaier, G. and Riquier, D. (1988). Seasonal acclimation of bank voles and wood mice: non-shivering thermogenesis and thermogenic properties of brown adipose tissue mitochondria. *J. Comp. Physiol.* **185**, 157–164.

Kleiber, M. (1932). Body size and metabolism. *Hilgardia* **6**, 315–353.

Kleiber, M. (1975). *The Fire of Life: An Introduction to Animal Energetics*. Wiley, New York.

Koopman, K. (1994). *Handbook of Zoology. Vol. VIII: Mammalia. Part 60: Chiroptera systematics*. De Gruyer, Berlin.

Koskela, E., Jonsson, P., Hartikainen, T. and Mappes, T. (1998). Limitation of reproductive success by food availability and litter size in the bank vole *Clethrionomys glareolus*. *Proc. R. Soc. [B]* **265**, 1129–1134.

Koteja, P. (1987). On the relation between basal and maximum metabolic rate in mammals. *Comp. Biochem. Physiol.* **87A**, 205–208.

Koteja, P. (1991). On the relation between basal and field metabolic rates in birds and mammals. *Funct. Ecol.* **5**, 56–64.

Koteja, P. (1996a). Limits to the energy budget in a rodent, *Peromyscus maniculatus*: the central limitation hypothesis. *Physiol. Zool.* **69**, 981–993.

Koteja, P. (1996b). Limits to the energy budget in a rodent, *Peromyscus maniculatus*: does gut capacity set the limit? *Physiol. Zool.* **69**, 994–1020.

Koteja, P. (1996c). Measuring energy metabolism with open-flow respirometric systems? Which design to choose? *Funct. Ecol.* **10**, 675–677.

Koteja, P. and Weiner, J. (1993). Mice, voles and hamsters: metabolic rates and adaptive strategies in muroid rodents. *Oikos* **66**, 505–514.

Koteja, P., Krol, E. and Stalinski, J. (1994). Maximum cold- and lactation-induced rate of energy assimilation in *Acomys cahirinus*. *Polish Ecological Studies* **20**, 369–374.

Konarzewski, M. and Diamond, J. (1994). Peak sustained metabolic rate and its individual variation in cold stressed mice. *Physiol. Zool.* **67**, 1186–1212.

Konarzewski, M. and Diamond, J. (1995). Evolution of basal metabolic rate and organ masses in laboratory mice. *Evolution* **49**, 1239–1248.

Krzanowski, A. (1961). Weight dyamics of bats wintering in a cave of Pulawy (Poland). *Acta theriologica* **4**, 242–264.

Krebs, H.A. (1950). Body size and tissue respiration. *Biochim. Biophys. Acta* **4**, 249–269.

Krebs, J.R. (1978). Optimal foraging: decision rules for predators. In: *Behavioural Ecology: An Evolutionary Approach* (Ed. by J.R. Krebs and N.B. Davies), pp. 23–64. Blackwells, Oxford.

Krog, H., Monson, M. and Irving, L. (1954). Influence of cold upon metabolism and body temperature of wild rats, albino rats and albino rats conditioned to the cold. *J. Appl. Physiol.* **7**, 349–354.

Kunz, T.H., Robson, S.K. and Nagy, K.A. (1998a). Economy of harem maintenance in the greater spear-nosed bat, *Phyllostomus hastatus*. *J. Mammal.* **79**, 631–642.

Kunz, T.H., Wrazen, J.A. and Burnett, C.D. (1998b). Changes in body mass and fat reserves in pre-hibernating little brown bats (*Myotis lucifugus*). *Ecoscience* **5**, 8–17.

Kurta, A. (1991). Torpor patterns in food deprived *Myotis lucifugus* (Chiroptera: vespertilionidae) under simulated roost conditions. *Can. J. Zool.* **69**, 255–257.

Kurta, A., Bell, G.P., Nagy, K.A. and Kunz, T.H. (1989). Energetics of pregnancy and lactation in free-ranging little brown bats, *Myotis lucifugus*. *Physiol. Zool.* **62**, 804–818.

Kurta, A., Kunz, T.H., and Nagy, K.A. (1990). Energetics and water flux of free-ranging big brown bats (*Eptesicus fuscus*) during pregnancy and lactation. *J. Mamm.* **71**, 59–65.

LaBarbera, M.L. (1989). Analysing body size as a factor in ecology and evolution. *Ann. Rev. Ecol. Syst.* **20**, 97–117.
Lack, D. (1956). *Swifts in a Tower.* Metheun London.
Larochelle, J. (1998). Comments on a negative appraisal of taxidermic mounts as tools for studies of ecological energetics. *Physiol. Zool.* **71**, 596–598.
Lasiewski, R.C. (1963). Oxygen consumption of torpid, resting, active and flying hummingbirds. *Physiol. Zool.* **36**, 122–141.
Lifson, N. and McClintock R. (1966). Theory of use of the turnover rates of body water for measuring energy and material balance. *J. Theor. Biol.* **12**, 46–74.
Lifson, N., Gordon, G.B. and McClintock, R. (1955). Measurement of total carbon dioxide production by means of $D_2^{18}O$. *J. Appl. Physiol.* **7**, 704–710.
Lovegrove, B.G. (1996). The low basal metabolic rates of marsupials: the influence of torpor and zoogeography. In: *Adapations to the Cold: Tenth International Hibernation Symposium.* (Ed. by F. Geiser, A.J. Hulbert and S.C. Nicol), pp. 141–151. University of New Engalnd Press, Armidale, Australia.
Lucas, M.C. and Armstrong, J.D. (1991). Estimation of meal energy intake from heart rate records of pike, *Esox lucius* L. *J. Fish. Biol.* **38**, 317–319.
Lucas, M.C. and Priede, I.G. (1992). Utilisation of metabolic scope in relation to feeding and activity by individual and group housed zebrafish *Brachydanio rerio* (Hamilton Buchanan) *J. Fish. Biol.* **41**, 175–190.
Luckett, W.P. and Hartenberger, J.L. (1985). *Evolutionary Relations Among Rodents: A Multidimensional Analysis.* Plenum, New York.
Lyman, C.P. (1954). Activity, food consumption and hoarding in hibernators. *J. Mammal.* **35**, 545–552.
Marsh, R.L. (1983). Adaptations of the gray catbird *Dumetella carolinensis* to long-distance migration: energy stores and substrate concentrations in plasma. *Auk* **100**, 170–179.
Margaria, R., Cerretelli, P., Aghemo, P. and Sassi, G. (1963). Energy cost of running. *J. Appl. Physiol.* **18**, 367–370.
Martin, A.W. and Fuhrman, F.A. (1955). The relationship between summated tissue respiration and metabolic rate in the mouse and dog. *Physiol. Zool.* **28**, 18–34.
Martins, E.P. and Garland, T. (1991). Phylogenetic analysis of the correlated evolution of continuous characters—a simulation study. *Evolution* **45**, 534–557.
Masman, D. and Klaassen, M. (1987). Energy expenditure during free-flight in trained and free living Eurasian kestrels (*Falco tinnunculus*). *Auk* **104**, 603–616.
Masman, D., Dijkstra, C., Daan, S. and Bult, A. (1989). Energetic limitation of avian parental effort: field experiments in the kestrel (*Falco tinnunculus*). *J. Evol. Biol.* **2**, 435–455.
Meyer, P.J., Lange, C.S., Bradley, M.U. and Nichols, W.W. (1991). Gender differences in age-related decline in DNA double strand break damage and repair in lympocytes. *Ann. Hum. Biol.* **18**, 405–415.
McArthur, R.H. and Pianka, E.L. (1966). On the optimal use of a patchy environment. *Am. Nat.* **100**, 603–609.
McDevitt, R.M. and Andrews, J.F. (1997). Seasonal variation in brown adipose tissue mass and lipid droplet size of *Sorex minutus*, the pygmy shrew: the relation between morphology and metabolic rate. *J. Therm. Biol.* **22**, 127–135.
McDevitt, R.M. and Speakman, J.R. (1994a). Central limits to sustained metabolic rate have no role in cold acclimation of the short-tailed field vole (*Microtus agrestis*). *Physiol. Zool.* **67**, 1117–1139.
McDevitt, R.M. and Speakman, J.R. (1994b). Limits to sustainable metabolic rate during transient exposure to low temperatures in short-tailed field voles (*Microtus agrestis*). *Physiol. Zool.* **67**, 1103–1116.

McGillvery, R.W. (1971). *Biochemistry: A Functional Approach*. Saunders, Philadelphia.
McLean, J.A. and Speakman, J.R. (1995). Elimination rate of Zn-65 as a measure of food intake—a validation study in the mouse (*Mus* sp.). *J. Appl. Physiol.* **79**, 1361-1369.
McMillan, R.E. and Hinds, D.S. (1992). Standard, cold-induced, metabolism of rodents and exercise induced. In: *Mammalian Energetics: Interdisciplinary Views of Metabolism and Reproduction* (Ed. by T.E. Tomasi and T.H. Horton), pp. 16–33. Comstock, Ithaca.
McNab, B.K. (1963). A model of the energy budget of a wild mouse. *Ecology* **44**, 521-532.
McNab, B.K. (1978). The comparative energetics of neotropical marsupials. *J. Comp. Physiol. [B.]* **125**, 115-128.
McNab, B.K. (1980). Food habits, energetics and the population biology of mammals. *Am. Nat.* **116**, 106-124.
McNab, B.K. (1983). Energetics, body size and the limits to endothermy. *J. Zool.* **199**, 1-29.
McNab, B.K. (1986a). Food habits, energetics and the reproduction of marsupials. *J. Zool.* **208**, 595-614.
McNab, B.K. (1986b). The influence of food habits on the energetics of eutherian mammals. *Ecol. Monogr.* **56**, 1-19.
McNab, B.K. (1987a). The reproduction of marsupial and eutherian mammals in relation to energy expenditure. *Symp. Zool. Soc. Lond.* **57**, 29–41.
McNab, B.K. (1987b). Basal rate and phylogeny. *Funct. Ecol.* **1**, 159-167.
McNab, B.K. (1988). Food habits and the basal rate of metabolism in birds. *Oecologia* **77**, 343-349.
McNab, B.K. (1997). On the utility of uniformity in the definitions of basal rate of metabolism. *Physiol. Zool.* **70**, 718-720.
McNab, B.K. and Morrison, P. (1963). Body temperature and metabolism in subspecies of *Peromyscus* from arid and mesic environments. *Ecol. Monogr.* **33**, 63-82.
McNeil Alexander, R.M. (1998). When is migration worthwhile for animals that walk, swim or fly? *J. Avian Biol.* **29**, 387-394.
McNeil Alexander, R.M. (1999). *Energy for Animal Life*. Cambridge University Press, Cambridge.
Meerlo, P., Bolle, L., Visser, G.H., Masman, D. and Daan, S. (1997). Basal metabolic rate in relation to body composition and daily energy expenditure in the field vole, *Microtus agrestis*. *Physiol. Zool.* **70**, 362-370.
Merritt, J.F. (1986). Winter survival adaptations of the short-tailed shrew (*Blarina brevicauda*) in an Appalachian montane forest. *J. Mammal.* **67**, 450–464.
Migula, P. (1969). Bioenergetics of pregnancy and lactation in European common vole. *Acta Theriol.* **14**, 167-179.
Monadjem, A. and Perrin, M.R. (1998). The effect of supplemental food on the home range of the multi-mammate mouse, *Mastomys natalensis*. *South Afr. J. Wildl. Sci.* **28**, 1-3.
Morhadt, J.E. and Morhadt, S.S. (1971). Correlation between heart rate and oxygen consumption in rodents. *Am. J. Physiol.* **221**, 1580-1586.
Mullen, R.K. (1970). Respiratory metabolism and body water turnover rates of *Perognathus formosus* in its natural environment. *Comp. Biochem. Physiol.* **32**, 259-265.
Mullen, R.K. (1971a). Note on the energy metabolism of *Peromyscus crinitus* in its natural environment. *J. Mammal.* **52**, 633-635.

Mullen, R.K. (1971b). Energy metabolism and body water turnover rates of two species of free-living kangaroo rats, *Dipodomys merriami* and *Dipodomys microps*. *Comp. Biochem. Physiol.* **39A**, 379–380.

Mullen, R.K. and Chew, R.M. (1973). Estimating the energy metabolism of free-living *Perognathus formosus*: a comparison of direct and indirect methods. *Ecology* **54**.

Munger, J.C. and Karasov, W.H. (1989). Sublethal parasites and host energy budgets: tapeworm infection in white footed mice. *Ecology* **70**, 904–921.

Munks, S.A. and Green, B. (1995). Energy allocation for reproduction in a marsupial arboreal folivore, the common ringtail possum (*Pseudocheirus peregrinus*). *Oecologia* **101**, 94–104.

Mutze, G.J., Green, B. and Newgrain, K. (1991). Water flux and energy use in wild house mice (*Mus domesticus*) and impact of seasonal aridity on breeding and population levels. *Oecologia* **4**, 529–538.

Millar, J.S. (1979). Energetics of lactation in *Peromyscus maniculatus*. *Can. J. Zool.* **57**, 1015–1019.

Myrcha, A. (1964). Variations in the length and weight of the alimentary tract of *Clethrionomys glareolus* (Schreber 1780). *Acta Theriol.* **9**, 139–148.

Nagy, K.A. (1980). CO_2 production in animals: analysis of potential errors in the doubly labeled water method. *Am. J. Physiol.* **238**, R466-R473.

Nagy, K.A. (1987). Field metabolic rate and food requirement scaling in mammals and birds. *Ecol. Monogr.* **57**, 111–128.

Nagy, K.A. (1994). Field bioenergetics of mammals: what determines field metabolic rates? *Aust. J. Zool.* **42**, 43–53.

Nagy, K.A. and Gruchacz, M.J. (1994). Seasonal water and energy metabolism of the desert-dwelling kangaroo rat (*Dipodomys merriami*). *Physiol. Zool.* **67**, 1461–1478.

Nagy, K.A. and Montgomery, G.G. (1980). Field metabolic rate, water flux, and food consumption in three-toed sloths (*Bradypus varieatus*). *J. Mammal.* **61**, 465–472.

Nagy, K.A. and Suckling, G.C. (1985). Field energetics and water balance of sugar gliders, *Petaurus breviceps* (Marsupialia: Petauridae). *Aust. J. Zool.* **33**, 683–691.

Nagy, K.A., Seymour, R.S., Lee, A.K. and Braithwaite, R. (1978). Energy and water budgets of free-living *Antechinus stuartii* (Marsupilia: Dasyuridae). *J. Mammal.* **59**, 60–68.

Nagy, K.A., Lee, A.K., Martin, R.W. and Fleming, M.R. (1988). Field metabolic-rate and food requirement of a small dasyurid marsupial, *Sminthopsis crassicaudata*. *Aust. J. Zool.* **36**, 293–299.

Nagy, K.A., Bradley, A.J. and Morris, K.D. (1990). Field metabolic rates, water fluxes, and feeding rates of Quokkas, *Settonix brachyurus* and Tammars, *Macropus eugenii*, in western Australia. *Aust. J. Zool.* **37**, 553–560.

Nagy, K.A., Bradshaw, S.D. and Clay, B.T. (1991). Field metabolic rate, water flux and food requirements of short-nosed bandicoots, *Isoodon obesulus* (Marsupialia: Peramelidae). *Aust. J. Zool.* **39**, 299–305.

Nagy, K.A., Meienberger, C., Bradshaw, S.D. and Wooler, R.D. (1995). Field metabolic rate of a small marsupial mammal, the honey possum (*Tarsipes rostratus*). *J. Mammal.* **76**, 862–866.

Nagy, K.A., Girard, I.A. and Brown, T.K. (1999). Energetics of free-ranging mammals, reptiles and birds. *Ann. Rev. Nutr.* **19**, 247–277.

Nagy, T.R. and Negus, N.C. (1993). Energy acquisition and allocation in male collared lemmings (*Dicrostonyx groenlandicus*): effects of photoperiod, temperature, and diet quality. *Physiol. Zool.* **66**, 537–560.

Nicoll, M.E. and Thompson, S.D. (1987). The energetics of reproduction in therian mammals: didelphids and tenrecs. *Symp. Zool. Soc. Lond.* **51**, 1–27.

Nolet, B.A., Butler, P.J., Masman, D. and Woakes, A.J. (1992). Estimation of daily energy expenditure from heart rate and doubly labeled water in exercising geese. *Physiol. Zool.* **65**, 1188–1216.

Norberg, U.M. (1986). Evolutionary convergence in foraging niche and flight morphology in insectivorous aerial-hawking birds and bats. *Ornis Scand.* **17**, 253–260.

Norberg, U.M. (1990). *Vertebrate Flight.* Springer, Berlin.

Novacek, M.J. (1992). Mammalian phylogeny—shaking the tree. *Nature* **356**, 121–125.

Novacek, M.J. (1994). Morphological and molecular inroads to phylogeny. In: *Interpreting the Hierarchy of Nature: From Systematic Patterns to Evolutionary Process Theories* (Ed. by L. Grande and L. Rieppel), pp. 85–131, Academic Press, New York.

Oliver, J.E. and Fairchild, R.W. (1984). *The Encyclopedia of Climatology.* Van Nostrand Reinhold, New York.

Padley, D. (1985). Do life history parameters of passerines scale to metabolic rate independently of body mass? *Oikos* **45**, 285–287.

Pagel, M.D. and Harvey, P.H. (1988). Recent developments in the analysis of comparative data. *Q. Rev. Biol.* **63**, 413–440.

Paladino, F.V. and King, J.R. (1984). Thermoregulation and oxygen consumption duirng terrestrial locomotion by white crowned sparrows *Zonotrichia leucophrys gambelli*. *Physiol. Zool.* **57**, 226–236.

Pasanen, S. (1971). Seasonal variation in interscapular brown fat in three species of small mammals wintering in an active state. *Aquilo Ser. Zool.* **11**, 1–31.

Pasquis, P., Lacaisse, A. and Dejours, P. (1970). Maximal oxygen uptake in four species of small mammals. *Resp. Physiol.* **9**, 298–309.

Pearson, O.P. (1947). The rate of metabolism of some small mammals. *Ecology* **28**, 127–145.

Pennycuick, C.J. (1969). The mechanics of bird migration. *Ibis* **111**, 525–556.

Pennicuick, C.J. (1989). *Bird Flight Performance.* Oxford University Press, Oxford.

Perrigo, G. (1987). Breeding and feeding strategies in deer mice and house mice when females are challenged to work for their food. *Anim. Behav.* **35**, 1298–1316.

Peters, E.L., Shawki, A.I., Tracy, C.R., Whicker, F.W. and Nagy, K.A. (1995). Estimation of the metabolic rate of the desert iguana (*Dipsosaurus dorsalis*) by a radio-nuclide technique. *Physiol. Zool.* **68**, 16–342.

Peters, R.H. (1983). *The Ecological Implications of Body Size.* Cambridge University Press, Cambridge.

Peterson, C.C., Nagy, K.A. and Diamond J. (1990). Sustained metabolic scope. *Proc. Natl. Acad. Sci. USA* **87**, 2324–2328.

Peterson, R.M. and Batzli, G.O. (1975). Activity patterns in natural populations of the brown lemming (*Lemmus trimucronatus*). *J. Mammal.* **56**, 718–720.

Peterson, R.M. Jr., Batzli, G.O. and Banks, E.M. (1986). Activity and energetics of the brown lemming in its natural habitat. *Arctic Alpine Res.* **8**, 131–138.

Pettigrew, J.D., Jamieson, B.G.M., Robson, S.K., Hall, L.S., McAnally, K.I. and Cooper, H.M. (1989). Phylogenetic relations between microbats, megabats and primates (Mammalia: Chiroptera and primates). *Phil. Trans. R. Soc. [B]* **325**, 489–559.

Pettigrew, J.D. (1995). Flying primates: crashed or crashed through? In Racey, P.A. and Swift, S.M. Ecology, evolution and behaviour of bats. *Symp. Zool. Soc. Lond* **67**, 3–26.

Pianka, E.L. and Parker, W.S. (1975). Age specific reproductive tactics. *Am. Nat.* **109**, 453–464.

Poehlman, E.T., Toth, M.J. and Webb, G.D. (1993). Sodium–potassium pump activity contributes to the age related decline in resting metabolic rate. *J. Clin. Endocrinol. Metab.* **76**, 1054–1057.

Poppitt, S.D., Speakman, J.R. and Racey, P.A. (1993). The energetics of reproduction in the common shrew, *Sorex araneus*. *Physiol. Zool.* **66**, 964–982.

Poppitt, S.D., Speakman, J.R. and Racey, P.A. (1994). The energetics of reproduction in the hedgehog tenrec (*Echinops telefari*). *Physiol. Zool.* **67**, 976–994.

Priede, I.G. (1977). Natural selection for energetic efficiency and the relationship between activity level and mortality. *Nature* **267**, 610–611.

Priede, I.G. (1983). Heart rate telemetry from fish in the natural environment. *Comp. Biochem. Physiol.* **76A**, 515–524.

Priede, I.G. and Tytler, P. (1977). Heart rate as a measure of metabolic rate in teleost fishes: *Salmo gairdneri*, *Salmo trutta* and *Gadus morhua*. *J. Fish Biol.* **10**, 231–242.

Pucek, Z. (1970). Seasonal and age changes in shrews as an adaptive process. *Symp. Zool. Soc. Lond.* **26**, 189–207.

Racey, P.A. and Speakman, J.R. (1987). The energetics of pregnancy and lactation in heterothermic bats. *Symp. Zool. Soc. Lond.* **54**, 101–129.

Racey, P.A. and Stephenson, P.J. (1996). Reproductive and energetic differentiation of the Tenrecidae of Madagascar. *Biogeographie de Madagascar* **1996**, 307–319.

Randolph, P.A. (1980). Daily energy metabolism of two rodents (*Peromyscus leucopus* and *Tamias striatus*) in their natural environment. *Physiol. Zool.* **53**, 70–81.

Rayner, J.M.V. (1979). A new approach to animal flight mechanics. *J. Exp. Biol.* **80**, 17–54.

Rayner, J.M.V. (1985). Linear relations in biomechanics—the statistics of scaling functions. *J. Zool. Lond.* **206**, 415–439.

Reynolds, P.S. and Lee, R.M. (1997). Phylogenetic analysis of avian energetics: passerines and non-passerines do not differ. *Am. Nat.* **147**, 735–759.

Ricklefs, R.E., Konarzewski, M. and Daan, S. (1996). The relation between basal metabolic rate and daily energy expenditure in birds and mammals. *Am. Nat* **147**, 1047–1071.

Riska, B. (1991). Regression models in evolutionary allometry. *Am. Nat.* **138**, 283–299.

Rogowitz, G.L. (1996). Trade-offs in energy allocation during lactation. *Am. Zool.* **36**, 197–204.

Rogowitz, G.L. (1998). Limits to milk flow and energy allocation during lactation of the hispid cotton rat (*Sigmodon hispidus*). *Physiol. Zool.* **71**, 312–320.

Rogowitz, G.L. and McClure, P.A. (1995). Energy export and offspring growth during lactation in cotton rats (*Sigmodon hispidus*). *Funct. Ecol.* **9**, 143–150.

Root, T. (1988). Energy constraints on avian distribution and abundances. *Ecology* **69**, 330–339.

Rosenman, M.P., Morrison, P. and Feist, D. (1975). Seasonal changes in the metabolic capacity of red-backed voles. *Physiol. Zool.* **48**, 303–310.

Rothe, H.J., Biesel, W. and Nachtigall, W. (1987). Pigeon flight in a wind tunne: II. Gas exchange and power requirements. *J. Comp. Physiol.* **157B**, 99–109.

Rowe-Rowe, D.T., Green, B. and Crafford, J.E. (1989). Estimated impact of feral house mice on sub-antarctic invertebrates at Marion Island. *Polar Biol.* **9**, 457–460.

Ruben, J.A. (1995). The evolution of endothermy in mammals and birds: from physiology to fossils. *Ann. Rev. physiol.* **57**, 69–76.

Salsbury, C.M. and Armitage, K.B. (1994). Resting and field metabolic rates of adult male yellow-bellied marmots, *Marmota flaviventris*. *Comp. Biochem. Physiol.* **108A**, 579–588.

Salsbury, C.M. and Armitage, K.B. (1995). Reproductive energetics of adult male yellow-bellied marmots, *Marmota flaviventris*. *Can. J. Zool.* **73**, 1791–1797.

Sarich, V.M. (1985). Rodent macromolecular systematics. In: *Evolutionary Relationships Among Rodents*. (Ed. by W.P. Luckett and J.L. Hartenberger), pp. 423–452. Plenum, New York.

Scantlebury, D.M., Butterwick, R.F. and Speakman, J.R. (in press) Lactation in domestic dog (*Canis familiaris*) breeds of two species. Comparative Biochemistry and Physiology.

Schartz, R.L. and Zimmerman, J.L. (1971). The time and energy budget of the male dickcissel (*Spiza americana*). *Condor* **73**, 65–76.

Schmidt-Hempel, P. and Wolf, T. (1988). Foraging effort and lifespan of workers in a social insect. *J. Anim. Ecol.* **57**, 509–522.

Schmidt Nielsen, K. (1975). *Animal Physiology: Adaptation and Environment*. Cambridge University Press, Cambridge.

Schoener, T.W. (1971). Theory of feeding strategies. *Ann. Rev. Ecol. Syst.* **2**, 369–404.

Scott, I. and Evans, P.R. (1992). The metabolic output of avian (*Sturnus vulgaris, Calidris alpina*) adipose tissue, liver and skeletal muscle: implications for BMR/body mass relations. *Comp. Biochem. Physiol.* **103A**, 329–332.

Seymour, R.S., Withers, P.C. and Weathers, W.W. (1998). Energetics of burrowing, running and free-living in the namib desert golden mole (*Eremitalpa namibensis*) *J. Zool. Lond.* **244**, 107–117.

Shoemaker, V.H., Nagy, K.A., and Costa, W.R. (1976). Energy utilisation and temperature regulation by jackrabbits (*Lepus californicus*) in the Mojave desert. *Physiol. Zool.* **49**, 364–375.

Simmons, N.B. (1998). A reappraisal of interfamilial relationships of bats. In: *Bats: Phylogeny, Morphology, Echolocation and Conservation Biology* (Ed. by T.H. Kunz and P.A. Racey), pp. 3–27. Smithsonian Institute Press, Washington, DC.

Simmons, N.B. and Geisler, J.H. (1998). Phylogenetic relationships of *Icaronycteris, Archaeonycteris, Hassianycteris*, and *Paleochiropteryx* to extant bat lineages, with comments on the evolution of echolocation and foraging strategies in microchiroptera. *Bull. Am. Mus. Nat. Hist.* **235**, 1–182.

Skutch, A.F. (1949). Do tropical birds rear as many young as they can nourish? *Ibis* **91**, 430–455.

Smith, A.P., Nagy, K.A., Fleming, M.R. and Green, B. (1982). Energy requirements and water turnover in free-living Leadbeater's possums, *Gymnobelideus leadbeateri* (Marsupialia: Petauridae). *Aust. J. Zool.* **30**, 717–749.

Speakman, J.R. (1987). Calculation of CO_2 production in doubly-labelled water studies. *J. Theor. Biol.* **126**, 101–104.

Speakman, J.R. (1995). Energetics and the evolution of body size in small terrestrial mammals. *Symp. Zool. Soc. Lond.* **69**, 63–81.

Speakman, J.R. (1997a). *Doubly-labelled Water: Theory and Practice*. Chapman and Hall, London.

Speakman, J.R. (1997b). Factors influencing daily energy expenditure of small mammals. *Proc. Nutr. Soc.* **56**, 1119–1136.

Speakman, J.R. (1998). Doubly-labeled water: history and theory of the technique. *Am. J. Clin. Nutr.* **68**, S932–938.

Speakman, J.R. and McQueenie, J. (1996). Limits to sustained metabolic rate: the link between food intake, BMR, and morphology in reproducing mice (*Mus musculus*). *Physiol. Zool.* **69**, 746–769.

Speakman, J.R. and Racey, P.A. (1988). The doubly-labelled water technique for measurement of energy expenditure in free-living animals. *Sci. Prog. (Oxford)* **72**, 227–237.

Speakman, J.R. and Racey, P.A. (1987). The energetics of pregnancy and lactation in the brown long-eared bat, *Plecotus auritus*. In: *Recent Advances in the Study of*

Bats (Ed. by M.B. Fenton, P.A. Racey and J.M.V. Rayner), pp. 368–393. Cambridge University Press, Cambridge.
Speakman, J.R. and Racey P.A., (1989). Hibernal ecology of the pipistrelle bat: energy expenditure, water requirements and mass loss, implications for survival and the function of winter emergence flights. *J. Anim. Ecol.* **58**, 797–814.
Speakman, J.R. and Racey, P.A. (1991). No cost of echolocation for bats in flight. *Nature* **350**, 421–423.
Speakman, J.R. and Rowland, A. (1999). Preparing for inactivity: how insectivorous bats deposit a fat store for hibernation. *Proc. Nutr. Soc.* **58**, 123–131.
Speakman, J.R., Anderson, M.E. and Racey, P.A. (1989). The energy cost of echolocation in pipistrelle bats (*Pipistrellus pipistrellus*). *J. Comp. Physiol. [A]* **165**, 679–685.
Speakman, J.R., McDevitt, R.M. and Cole, K. (1993). Measurement of basal metabolic rates: don't lose sight of reality in the quest for comparability. *Physiol. Zool.* **66**, 1045–1049.
Stearns, S.C. (1976). Life history tactics: a review of the ideas. *Q. Rev. Biol.* **51**, 3–47.
Stearns, S.C. (1983). The impact of size and phylogeny on patterns of co-variation in the life history traits of mammals. *Oikos* **41**, 173–187.
Stearns, S.C. (1993). *The Evolution of Life Histories.* Oxford University Press, Oxford.
Stephenson, P.J. and Racey, P.A. (1993a). Reproductive energetics of Terecidae (Mammalia: Insectivora). I. Large eared Tenrec *Geogale aurita*. *Physiol. Zool.* **66**, 643–663.
Stephenson, P.J. and Racey, P.A. (1993b). Reproductive energetics of Terecidae (Mammalia: Insectivora). II. The shrew tenrecs (*Microgale* spp.). *Physiol. Zool.* **66**, 664–685.
Stephenson, P.J. and Racey, P.A. (1994). Seasonal variation in resting metabolic rate and body temperature of streaked tenrecs, *Hemicentetes nigriceps* and *H. spinosus* (Insectivora: Tenrecidae). *J. Zool. (Lond).* **232**, 285–294.
Stephenson, P.J. and Racey, P.A. (1995). Resting metabolic rate and reproduction in the Insectivora. *Comp. Biochem. Physiol.* **112A**, 215–223.
Stephenson, P.J., Speakman, J.R. and Racey, P.A. (1994). Field metabolic rate in two species of shrew-tenrec *Microgale dobsoni* and *M. talazaci*. *Comp. Biochem. Physiol.* **107A**, 283–287.
Stephenson, R., Butler, P.J. and Woakes, A.J. (1986). Diving behaviour and heart rate in tufted ducks (*Aythya fuligula*). *J. Exp. Biol.* **126**, 341–359.
Strelkov, P.P. (1969). Migratory and stationary bats (Chiroptera) of the European part of the Soviet Union. *Acta Zool. Crac.* **16**, 393–440.
Sutherland, W.G. (1982). Do oystercatchers select the most profitable cockles? *Anim. Behav.* **30**, 857–861.
Swaminathan, R., Burrows, G. and McMurray, J. (1982). Energy cost of sodium pump activity in man: an *in vivo* study of metabolic rate in human subjects given digoxin. *IRCS Med Sci* **10**, 949.
Swaminathan, R., Chan, L.P.C., Sin, X.L., King, N.S., Fun, N.S. and Chan, A.Y.S. (1989). The effect of ouabain on metabolic rate in guinea pigs: estimation of the energy cost of sodium pump activity. *B. J. Nutr.* **61**, 467–473.
Szarski, H. (1983). Cell size and the concept of frugal and wasteful evolutionary strategies. *J. Theor. Biol.* **105**, 201–209.
Taigen, T.L. (1983). Activity metabolism of anuran amphibians: implications for the evolution of endothermy. *Am. Nat.* **121**, 94–109.
Tarkkonen, T.H. (1971). Effect of repeated short term cold exposure on BAT of mice. *Ann. Zool. Fenn.* **8**, 434.
Tatner, P. and Bryant, D.M. (1989). The double labelled water technique for measuring energy expenditure. In: *Techniques in Comparative Respiratory*

Physiology (Ed. by C.R. Bridges and P.J. Butler), pp. 77–112. Cambridge University Press, Cambridge.

Taylor, C.R. (1974). Exercise and thermoregulation. In: *MTP International Review of Science. Environmental Physiology,* Series 1, Vol. 7 (Ed. by D. Robert-Shaw), pp. 163–184. Butterworths, London.

Taylor, C.R. (1981). Scaling limits of metabolism to body size: implications for animal design. In: *A Companion to Animal Physiology* (Ed. by C.R. Taylor, K. Johansen and L. Bolis), pp. 161–170. Cambridge University Press, Cambridge.

Tedman, R. and Green, B. (1987). Water and sodium fluxes and lactational energetics in suckling pups of Weddell seals (*Leptonychotes weddellii*). *J. Zool. (Lond.)* **212**, 29–42.

Thomas, D.W. (1985). Data attributed to Thomas, presented in Fleming, T.H. (1988). *The Short-tailed Fruit Bat. A Study in Plant–Animal Interactions.* University of Chicago Press, Chicago.

Thomas, S.P. (1975). Metabolism during flight of two species of bats, *Phyllostomus hastatus* and *Pteropus gouldii. J. Exp. Biol.* **63**, 273–292.

Thompson, S.D. (1992). Gestation and lactation in small mammals: basal metabolic rate and the limits of energy use. In: *Mammalian Energetics. Interdisciplinary Views of Metabolism and Reproduction* (Ed. by T.E. Tomasi and T.H. Horton), pp. 213–259. Comstock, Ithaca.

Thompson, S.D. and Nicoll, M.E. (1986). Basal metabolic rate and energetics of reproduction in therian mammals. *Nature* **321**, 690–693.

Townsend, C.R. and Calow, P. (1981). *Physiological Ecology: An Evolutionary Approach to Resource Use.* Blackwells, Oxford.

Travis, J. (1982). A method for statistical analysis of time-energy budgets. *Ecology* **63**, 19–25.

Tucker, V.A. (1966). Oxygen consumption of a flying bird. *Science* **154**, 150–151.

Utter, J.M. and LeFebvre, E.A. (1973). Daily energy expenditure of purple martins (*Progne subis*) during the breeding season: estimates using $D_2^{18}O$ and time budget methods. *Ecology* **54**, 597–603.

Vogt, F.D. and Lynch, G.R. (1982). Influence of ambient temperature, nest availability, huddling, and daily torpor on energy expenditure in the white footed mouse *Peromyscus leucopus. Physiol. Zool.* **55**, 56–63.

von Helversen, O. and Reyer, H.U. (1984). Nectar intake and energy expenditure in a flower visiting bat. *Oecologia* **63**, 178–184.

Wallis, I.R. and Green, B. (1992). Seasonal field energetics of the rufous rat-kangaroo (*Aepyprymnus rufescens*). *Aust. J. Zool.* **40**, 279–290.

Wallis, I.R., Green, B. and Newgrain, K. (1997). Seasonal field energetics and water fluxes of the long-nosed potoroo (*Potorus tridactylus*) in Southern Victoria. *Aust. J. Zool.* **45**, 1–11.

Walsberg, G.E. and Wolf, B.O. (1996). An appraisal of operative temperature mounts as tools for study of ecological energetics. *Physiol. Zool.* **69**, 658–681.

Wang, L.H.C. (1978). Factors limiting maximum cold induced heat production. *Life Sci.* **23**, 2089–2098.

Weatherly, A.H., Rogers, S.C., Pincock, D.G. and Patch, J.R. (1982). Oxygen consumption of active trout, *Salmo gairdneri* Richardson, derived from electromyograms obtained by radio-telemetery. *J. Fish Biol.* **20**, 479–489.

Weathers, W.W. and Nagy, K.A. (1980). Simultaneous doubly labeled water ($^3HH^{18}O$) and time–budget estimates of daily energy-expenditure in *Phainopepla nitens. Auk* **97**, 861–867.

Weathers, W.W., Buttemer, W.A., Hayworth, A.M. and Nagy, K.A. (1984). An evaluation of time-budget estimates of daily energy expenditure in birds. *Auk* **101**, 459–472.

Weber, J.M. (1992). Pathways for oxidative fuel provision to working muscles: ecological consequences of maximal supply limitations. *Experientia* **48**, 557–564.

Webster, M.D. and Weathers, W.W. (1990). Heat produced as a by-product of foraging activity contributes to thermoregulation in verdins, *Auriparus flaviceps*. *Physiol. Zool.* **63**, 777–794.

Weihn, J. and Korpimaki, E. (1997). Food limitation and brood size: experimental evidence in the Eurasian kestrel. *Ecology* **78**, 2047–2050.

Weindruch, R. and Walford, R.L. (1988). *The Retardation of Aging and Disease by Dietary Restriction*. C.C. Thomas, Springfield, Illinois.

Weiner, J. (1987). Maximum energy assimilation rates in the djungarian hamster (*Phodopus sungorus*). *Oecologia* **72**, 297–302.

Weiner, J. (1989). Metabolic constraints to mammalian energy budgets. *Acta Theriol.* **34**, 3–35.

Weiner, J. (1992). Physiological limits to sustainable energy budgets in birds and mammals: ecological implications. *TREE* **7**, 384–388.

Westerterp, K.R. and Bryant, D.M. (1984). Energetics of free-existence in swallows and martins (Hirundinidae) during breeding: a comparative study using doubly-labelled water. *Oecologia* **62**, 376–381.

Winter, Y. (1998). Energetic cost of hovering flight in a nectar feeding bat measured with fast response respirometry. *J. Comp. Physiol.* **168**, 434–444.

Winter, Y. and von Helversen, O. (1998). The energy cost of flight: do small bats fly more cheaply than birds? *J. Comp. Physiol.* **168**, 105–111.

Winter, Y., Voight, C. and von Helversen, O. (1998). Gas exchange during hovering flight in a nectar-feeding bat *Glossophaga soricina*. *J. Exp. Biol.* **201**, 237–244.

Withers, P.C. (1977). Measurements of Vo_2, Vco_2 and evaporative water loss with a flow through mask. *J. Appl. Physiol.* **42**, 120–123.

Wolf, T. and Schmidt-Hempel, P. (1989). Extra loads and foraging life span in honeybee workers. *J. Anim. Ecol.* **58**, 943–954.

Wunder, B.A., Dobkins, D.S. and Gettinger, R.D. (1977). Shifts of thermogenesis in the prarie vole (*Microtus ochrogaster*): strategies for survival in a seasonal environment. *Oecologia* **29**, 11–26.

Yu, B.P. (1994). *Modulation of Aging Processes by Dietary Restriction*. CRC Press, Boca Raton, Florida.

Zebra, E. and Walsberg, G.E. (1992). Exercise-generated heat contributes to thermoregulation by Gambel's quail in the cold. *J. Exp. Biol.* **171**, 409–422.

Zubaid, A. and Gorman, M.L. (1991). The diet of wood mice *Apodemus sylvaticus* living in a sand dune habitat in north-east Scotland. *J. Zool. (Lond.)* **225**, 227–232.

APPENDIX A

Measurement of Oxygen Consumption by Indirect Calorimetry

To measure oxygen consumption animals are usually confined in chambers through which a flow of dry air is passed. The animals consume some of the incoming oxygen and replace it with carbon dioxide. Thus the O_2 content of gas in the outflow differs from that in the inflow. Generally the O_2 consumption of the animal (Vo_2) is estimated by:

$$Vo_2 = V_{in} \cdot pO_{2in} - V_{out} \cdot pO_{2out}$$

where V_{in} and V_{out} are the volumes of air flowing into and out of the chamber respectively and pO_{2in} and pO_{2out} are the proportion of the gas flowing in and out that comprises oxygen. Because the amount of CO_2 added to the air by the animal is generally not exactly equal to the O_2 it removes, there is a discrepancy in the flow rate entering and leaving the chamber ($V_{in} = V_{out}$). The effect on the flow rate is quite small (usually less than 0.5%) but this has a knock-on effect on the estimated proportional O_2 content of the outflow which is more significant (up to 6%). This difference may be further exacerbated if the CO_2 is removed from the outflow before measuring the O_2 content of the excurrent gas. In respirometry it is unusual for both upstream and downstream flow rates to be determined, and often the inflowing O_2 content is also not quantified but assumed to be equal to the atmospheric average (0.2095). The exact calculation of O_2 consumption depends on exactly where the flow is measured (i.e. upstream or downstream of the chamber), and the treatment of the gases before measurement of flow and oxygen content (with or without CO_2 removed). Several authors (Depocas and Hart, 1957; Hill, 1972; Withers, 1977; Gessaman, 1987) have reviewed the alternative equations for determination of O_2 consumption in these different situations.

The system that minimizes the error in the oxygen consumption estimate, if RQ is unknown, is to absorb the CO_2 in the excurrent stream and to measure the flow downstream of the CO_2 absorber (condition 3; Hill, 1972). Because the CO_2 is completely absorbed in this configuration, the measurement is completely independent of the assumption of an RQ. The worst system for estimating O_2 consumption, if the RQ is unknown, is to measure the flow upstream of the chamber and combine this with a measurement of the downstream O_2 content without absorbing the CO_2. This system effectively assumes an RQ of 1.0 and errors occur if the actual RQ deviates from 1.0. The maximum error occurs if the animal is metabolizing fat (RQ = 0.7) and in this circumstance the error is approximately 6%. However, Koteja (1996c) has shown that, while this latter system may result in erroneous estimates of O_2 consumption, when the RQ is lower than 1.0, the error in the estimated O_2 content cancels exactly with the error in the conversion of O_2 consumption to energy expenditure for the same RQ. Consequently the resultant estimate of energy expenditure with this configuration is independent of the assumed RQ. In contrast the estimated energy expenditure when converting estimates of O_2 consumption derived from condition 3 are not independent of RQ. Therefore, the best system configuration for determination of O_2 consumption (when RQ is unknown) is *not* the best system for accurate determination of energy expenditure.

This may be confusing because it has generally been assumed that minimizing error in measurement of O_2 consumption will automatically minimize error in estimated energy expenditure. Many workers are familiar with the advice from Hill (1972) and Withers (1977) that measurements of Vo_2 are best made by absorbing CO_2 and measuring flow downstream of the absorber. In my experience (with referees for papers) these researchers tend to respond negatively when presented with studies which use the worst possible protocol for measurement of V_{O_2}, on the basis that this minimizes error in the estimated energy expenditure. To clarify why minimizing error in the measurement of O_2 consumption does not also minimise the error in the estimated energy expenditure, consider the following example.

In this example we have an animal in a respirometry chamber consuming O_2 at a rate of 10 ml min^{-1}. The animal is metabolizing carbohydrate and therefore has an RQ of 1.0. The flow rate into the chamber is 1000 ml min^{-1}. Because the RQ is 1.0 the flow rate out of the chamber is also 1000 ml min^{-1}. The inflowing gas consists of 20.95% O_2 and the outflowing gas consists of 19.95% O_2. The missing 1% O_2 has been replaced by 1% CO_2 in the outflow. The energy expenditure of the animal depends on the oxycalorific equivalent for carbohydrate utilization, which is 20.92 J ml^{-1}. The energy expenditure is therefore 3.4867 W (= (20.92 × 10)/60). Imagine that half-way through a measurement session the animal switches its metabolic substrate to fat. The RQ changes to 0.7. We will assume that, when the substrate utilization changes, the O_2 consumption remains stable at 10 ml min^{-1} and that the energy expenditure of the animal changes. The oxycalorific equivalent for fat and an RQ of 0.7 is 19.66 J ml^{-1}. Consequently the energy expenditure falls to (19.66 × 10)/60 = 3.2766 W.

Given this basic system we will evaluate what happens when we make measurements of O_2 consumption and energy expenditure using three different standard configurations. First, consider what happens when we have a flowmeter upstream of the chamber metering the dry gas entering the chamber and we measure the O_2 content of the dry gas leaving the chamber *without* absorbing the CO_2 before analysis. In this situation, at the start of the experiment, we measure 1000 ml min^{-1} entering the chamber. The gas leaving the chamber each minute has a volume of 1000 ml and an O_2 content of 199.5 ml, thus the measured percentage O_2 content of the outflow is 19.95%. Assuming the inflow gas content is standard dry atmospheric with an O_2 content of 20.95%, the O_2 consumption estimated at the start of the experiment is 1000 × (0.2095 − 0.1995) = 10 ml min^{-1}. If we assume an RQ of 1.0, the estimated energy expenditure is 3.486 W. Both these estimates (O_2 consumption and energy expenditure) are exactly correct. However, consider what happens after the substrate change. Now there is a slight discrepancy between the inflow and outflow. Because the RQ is now 0.7, rather than 1000 ml min^{-1} leaving the chamber there is only 997 ml min^{-1} (i.e. 10 ml O_2 is consumed per min but only 7 ml CO_2 is put back each minute). The error in the estimated outflow is only 0.03%. The O_2 consumption remains constant, thus 3.2775 ml oxygen is in the outflow. However, the analyser measures this as a percentage of the entire flow. Thus the actual measured O_2 content of the outflow is 199.5/997 = 20.01%. The estimated O_2 consumption in this situation thus becomes $1000 \times (0.2095 - 0.2001) = 9.4$ ml min^{-1}. Despite the error in the outflow of the chamber being only 0.03%, the resultant error in the estimated O_2 consumption becomes 6%. Consider, however, the estimated energy expenditure during this second part of the experiment. Using the same assumed RQ of 1.0, the energy expenditure estimate is $(9.4 \times 20.92)/60 = 3.2674$ W. This is only 0.03% different from the actual energy expenditure.

In this situation, measuring flow upstream of the chamber and the O_2 content downstream without absorbing the CO_2 and assuming an RQ of 1.0, the estimated O_2 consumption may be in error by up to 6%. The estimated energy expenditure, however, using the same RQ assumption of 1.0 is effectively correct, independent of the true RQ. If the assumption that RQ = 1.0 is replaced with another assumed RQ, the estimated energy expenditure estimate is also compromised. For an assumed RQ = 0.85 the actual expenditure is underestimated by 3% in all conditions, independent of the true RQ, and for an assumed RQ of 0.7 the underestimate in all conditions, irrespective of the real RQ, increases to 6%.

Consider now what happens in this system if we do not measure flow going into the chamber but rather measure the dried outflow. As with the above scenario the O_2 content of the outflow is measured without absorbing the CO_2. At the start of the experiment the outflow meter records 1000 ml min^{-1} and the O_2 content of the outflowing gas is 19.95%. The estimated O_2 consumption is $1000 \times 0.2095 - (0.1995) = 10$ ml min^{-1}. In the second part of the experiment, however, the outflow metered is lower at 997 ml min^{-1} and the recorded O_2 content in the outflow is 20.01% (see above for explanation of the changes between the first and second parts of the experiment). Thus the estimated O_2 consumption in the second part of the experiment becomes $997 \times (0.2095 - 0.2001) = 9.37$ ml min^{-1}. This is 6.3% lower than the actual O_2 consumption. Using an RQ of 1.0 and an oxycalorific coefficient of 20.92 J ml^{-1} gives estimated energy expenditures of 3.4866 and 3.2674 W, which have errors of 0% and 0.3% respectively, compared with the true values. Changing the RQ assumption to 0.85 gives a systematic error of 3% in the estimated energy expenditure, and changing it to 0.7 gives a systematic error of 6%.

When measuring O_2 content of the excurrent gas without absorbing CO_2 first, it appears to be marginally better to meter the flow upstream of the chamber rather than downstream. This gives a marginally lower error when estimating O_2 consumption and a very slightly more accurate estimate of the resultant energy expenditure. In practice, however, measurement errors in both flow and O_2 content of the gases makes the difference between the methods negligible. There is, however, a slight difference in the positioning of the flowmeter for the consequences of leaks in the chamber. (Because the chamber must accommodate an entry point for the animal, the chamber seal is the most likely source of leaks in the entire system.) This also depends on the positioning of the air pump that is driving or sucking air through the system. When the pump is upstream, the chamber is under positive pressure. Slight leaks in the seals will lead to air being pushed out of the chamber. When the pump is downstream, there is a slight negative pressure in the system and small leaks in the chamber seals lead to air being sucked into the chamber. Consider what happens with a pump sucking air through the chamber. If the flow into the chamber is measured, small leaks in the chamber seals will lead to errors in the estimated O_2 consumption. This is because the true flow rate coming into the chamber will be greater than that being measured (as a small amount of unmetered air will be sucked in through the seals). The error is basically in direct proportion to the contribution to the total flow made by the leak. If the leak provides 10% of the inflow the estimated O_2 content will be 10% too low. In contrast, if one meters downstream of the chamber (with a sucking pump), a slight leak in the chamber seal will make no difference because the outflow is being monitored and the leak contributes only to the inflow. On the other hand, if the pump is blowing air through the system, air will be lost outwards through leaking seals. In this situation the error is on the outflow, so metering the air into the chamber will be unaffected but metering the outflow will involve an error because some of the outflow has been lost. In general, therefore, meter position (upstream or downstream of the chamber) makes a negligible impact on the accuracy of the measurement (when CO_2 is not absorbed). However, one should always position the meter upstream if an upstream blow through pump is being employed and downstream if a downstream suck through pump is being used since this minimizes any errors that might occur due to a leaky chamber. (In both cases, however, a better solution is to eliminate the leaks, because the above interpretations of the impact of leaks make the assumption that any leaking air is fully mixed with the chamber contents, either before it leaks out, or after it leaks in. In practice this will depend on the exact location of the leak relative to the main inflow and outflow pipes in the system.) In my laboratory we typically use blow-through systems that are metered upstream of the chamber and we do not absorb CO_2 before measurement of O_2 content.

Finally, let us consider what happens in this system if we use the configuration recommended for the most accurate determination of O_2 consumption. In this scenario the dried outflow has its CO_2 removed before the flow being metered and the O_2 content measured. Because 10 ml O_2 in the flow is being consumed and all CO_2 is being absorbed before measurement of the gas flow and O_2 content, the measured flow rate at the start of the experiment is 990 ml min^{-1}. The O_2 content measured is $199.5/990 = 20.15\%$. Using the equation provided by Withers (1977) for this condition, the measured O_2 consumption is:

$$V_{O_2} = (990 \times (0.2095 - 0.2015))/(1 - 0.2015)$$
$$= 9.92 \text{ ml min}^{-1}$$

The error in this estimate is 0.8% compared with the true O_2 consumption of 10 ml min^{-1}. When the RQ changes to 0.7, the CO_2 in the outflow from the chamber changes and the flow rate at the chamber outflow also changes. However, the different amount of CO_2 in the outflow has no impact on this system since all outflowing CO_2 is absorbed. Thus, after the change, the measured flow remains the same and the measured O_2 consumption remains 9.91 ml min^{-1} compared with the true value of 10 ml min^{-1}. This system for measuring O_2 consumption is far more robust to changes in RQ compared with the previous two systems. However, consider now what happens when we convert to energy expenditure. If we assume an RQ of 1.0, the initial energy expenditure estimate is 3.4788 W which has an error of 0.8%. However, after the change in the real RQ to 0.7, the estimated expenditure becomes 3.1988 W, which is an error of 6%. In contrast, assuming an RQ of 0.7 throughout reverses the errors. In this situation the initial error is 6% and the final error is only 0.8%. Making an assumption of RQ which is midway between the extremes leads to an error of about 3% in the energy expenditure estimate throughout the experiment (initially 3% lower, then 3% higher).

These examples should clarify that minimizing error in O_2 consumption estimates does not lead to the most accurate estimate of energy expenditure. When aiming to measure accurate O_2 consumption of animals, one should absorb CO_2 downstream of the chamber, and meter the flow of the resultant gas and its O_2 content. Oxygen consumption is estimated using equation (4) from Withers (1977). The resultant estimated O_2 consumption is independent of the actual RQ and accurate to better than 1%. If an estimate of energy expenditure is subsequently required from these estimates, the error in this derived estimate is minimized by assuming an RQ of around 0.85. Maximum errors in the derived energy expenditure will normally be less than 3%. If, from the outset, the intention is to measure accurate estimates of energy expenditure, one should meter the dry inflow to the chamber (for blow-through systems) or the dry outflow (for suck-through systems) and measure the O_2 content of the outflow without absorbing the CO_2. Oxygen consumption is estimated as the flow multiplied by the fractional O_2 content change between inflow and outflowing gases. This O_2 content estimate may be in error by up to 6%. This should be converted to energy expenditure using an oxycalorific equivalent to an RQ of 1.0. Because of the cancelling errors, the resultant estimate of energy expenditure is independent of the actual RQ and accurate to better than 0.5%.

It is perhaps important to note that, while the errors in the estimates of energy expenditure cancel when using O_2 consumption, this is not the case for estimates based on CO_2 production. Using the above scenario again as an example, if air flow is measured upstream of the chamber, and CO_2 content is measured downstream, at the start of the experiment, when the RQ is 1.0, the CO_2 content of the exhaust stream will be 1%. The estimated CO_2 production will be 10 ml min⁻¹. (This assumes that the inflowing stream has atmospheric CO_2 scrubbed from it or all the chamber readings are referred to a pre-measurement baseline.) In the second part of the experiment the CO_2 production falls to 7 ml min⁻¹ and the total flow falls to 997 ml min⁻¹. The measured CO_2 content in the outflow is 0.702% and the estimated CO_2 production is 7 ml min⁻¹. Assuming an RQ of 1.0 gives an estimated energy expenditure of 3.483 W, in the first part of the experiment (correct), but an estimate of 2.87 W in the second part, which is 34% too low. Errors are minimized using an intermediate RQ of 0.85, but the estimated energy expenditure may still be in error by up to 16% (in either direction), compared with an error of less than 0.5% for measurements based on O_2 consumption using the same configuration.

These examples clarify that in practice, in most scenarios, whichever method is employed the error in the estimated O_2 consumption or energy expenditure is at most 6%, and more normally less than 3%.

APPENDIX B
Raw Data

The data are listed chronologically by date of publication. The Latin name of the species and common name, author and date of the publications are followed by data for distinct groups of the given species, for example different seasons, sites or sexes. For each group the body mass (g), field metabolic rate (kJ per day), approximate latitude of the study site and ambient temperature T(°C) are given.

Species	Common name	Body mass (g)	Metabolic rate (kJ day⁻¹)	Latitude (°)	T(°C)	Reference
Perognathus formosus	Pocket mouse	18.60	24.30	35.00	23.7	Mullen (1970)
		18.60	29.20	35.00	21.9	
		18.60	46.60	35.00	13.3	
		18.60	39.30	35.00	9.7	
Dipodomys merriami	Kangaroo rat	35.90	61.80	35.00	–	Mullen (1971b)
Dipodomys microps		58.60	117.40	35.00	–	Mullen (1971b)
Peromyscus crinitus		13.40	40.10	35.00	–	Mullen (1971a)
Perognathus formosus	Pocket mouse	16.20	23.10	35.00	22.8	Mullen and Chew (1973)
		16.40	28.30	35.00	19.7	
		16.40	44.70	35.00	10.6	
		17.30	37.80	35.00	7.1	
		16.70	47.90	35.00	0.6	
		19.20	53.20	35.00	5.5	
		19.20	55.40	35.00	12.8	
		21.50	62.50	35.00	–	
		16.40	21.50	35.00	29.1	
Lemmus trimucronatus	Lemming	62.00	191.40	70.00	0.0	Peterson et al. (1976)
Lepus californicus	Jack rabbit	1800.00	1416.00	35.00	30.0	Shoemaker et al. (1976)
		1800.00	1175.00	35.00	30.0	

continued

Species	Common name	Body mass (g)	Metabolic rate (kJ day^{-1})	Latitude (°)	T(°C)	Reference
Antechinus stuartii	Antechinus	25.70	72.00	36.00	8.0	Nagy et al. (1978)
Bradypus variegatus	Three-toed sloth	3830.00	589.82	8.00	30.0	Nagy and
		4220.00	489.52	8.00	30.0	Montgomery
		4450.00	738.00	8.00	30.0	(1980)
Tamias striatus	Chipmunk	93.10	99.33	38.00	19.0	Randolph (1980)
Peromyscus leucopus	White-footed mouse	19.40	36.54	38.00	19.0	Randolph (1980)
Ammospermophilus leucurus	Antelope ground squirrel	97.50	130.60	35.00	30.0	Karasov (1981)
		85.50	104.30	35.00	30.0	
		79.90	79.20	35.00	20.0	
		82.10	79.60	35.00	25.0	
Clethrionomys rutilus	Red-backed vole	13.60	64.90	65.00	23.0	Holleman et al. (1982)
		18.30	58.00	65.00	−15.0	
		15.00	66.10	65.00	8.0	
		16.10	64.30	65.00	−14.0	
Gymnobelideus leadbeateri	Leadbeater's possum	129.00	226.00	7.00	7.5	Smith et al. (1982)
Ammospermophilus leucurus	Antelope ground squirrel	96.10	82.64	5.00	0.0	Karasov (1983)
Dasyurus viverrinus	Eastern quoll	1029.00	793.00	2.00	12.0	Green and Eberhard (1983)
		1102.00	169.00	42.00	5.0	
Thomomys bottae	Pocket gopher	99.40	126.60	33.00	20.0	Gettinger (1984)
		108.20	127.70	33.00	5.0	
		103.90	135.90	33.00	15.0	
Arvicola terrestris	Vole	85.80	118.90	47.00	–	Grenot et al. (1984)
Anoura caudifer	Glossopgagine bat	11.50	51.90	0.00	30.0	von Helversen and Reyer (1984)
Petaurus breviceps	Sugar glider	135.00	192.00	37.00	1.5	Nagy and Suckling (1985)
		112.00	153.00	37.00	11.5	
Carollia perscipillata	Short-tailed fruit bat	19.50	79.30	10.00	–	Thomas (1985)
Macrotus californicus	Californian big-eared bat	12.90	22.80	35.00	25.0	Bell et al. (1986)
Sekeetamys calurus		41.20	44.00	29.00	–	Degen et al. (1986)
Acomys cahirinus	Spiny mouse	38.30	51.80	29.00	–	Degen et al. (1986)
Acomys russatus	Spiny mouse	45.00	47.60	29.00	–	Degen et al. (1986)
Mus musculus	House mouse	13.00	9.80	–	–	Nagy (1987)
Pseudocheirus peregrinus	Ring-tailed possum	717.00	556.00	–	–	Nagy (1987)
Plecotus auritus	Brown long-eared bat	8.50	27.00	57.00	10.0	Speakman and Racey (1987)
Pipistrellus pipistrellus	Pipistrelle bat	7.60	29.30	57.00	12.0	Racey and Speakman (1987)
Pseudomys albocinereus		32.60	62.20	–	–	Nagy (1987)
Praeomys natalensis	Multi-mammate mouse	57.30	86.60	26.00	27.0	Green and Rowe-Rowe (1987)
Sminthopsis crassicaudata	Fat-tailed dunnat	16.60	68.70	36.00	20.0	Nagy et al. (1988)
Pseudocheirus herbertensis		1103.00	446.00	8.00	20.0	Goudberg (1988)
Hemibeldideus lemuroides		1026.00	675.00	18.00	20.0	Goudberg (1988)
Peromyscus maniculatus	Deer mouse	18.40	68.40	35.00	7.6	Hayes (1989a)
		17.60	46.30	35.00	21.0	
Mus musculus	House mouse	19.30	65.10	46.00	7.0	Rowe-Rowe et al. (1989)
Myotis lucifugus	Little brown bat	8.45	27.60	43.00	0.0	Kurta et al. (1989)
Peromyscus leucopus	White-footed mouse	18.20	53.10	45.00	12.0	Munger and Karasov (1989)
		19.60	58.30	45.00	12.0	
		20.20	62.20	45.00	12.0	

Species	Common name	Body mass (g)	Metabolic rate (kJ day^{-1})	Latitude (°)	T(°C)	Reference
Peromyscus maniculatus	Deer mouse	20.60	56.10	35.00	–	Hayes (1989a)
Bassariscus astutus	Ring-tailed cat	752.00	472.00	–	–	Chevalier (1989)
Phascogale calura	Wambenger	33.50	61.90	37.00	–	Green et al. (1989)
Bettongia penicillata	Short-nosed rat kangaroo	1100.00	593.00	35.00	–	Green (1989)
Spermophilus saturatus	Golden-mantled ground squirrel	224.00	232.00	47.00	5.0	Kenagy et al. (1989)
		231.00	263.00	47.00	15.0	
		256.00	248.00	47.00	22.0	
		174.00	159.00	47.00	5.0	
		215.00	289.00	47.00	15.0	
		241.00	205.00	47.00	22.0	
Antechinus swainsonii	Dusky antechinus	43.00	74.00	36.00	15.0	Green and Crowley (1989)
		43.00	162.00	36.00	5.0	
		40.00	130.00	36.00	–2.0	
		57.00	222.00	36.00	10.0	
		60.00	175.00	36.00	10.0	
Antechinus stuartii	Antechinus	25.80	65.18	36.00	5.0	Green and Crowley (1989)
		22.80	96.17	36.00	–2.0	
		30.00	160.20	36.00	10.0	
Phascogale calura	Wambenger	36.00	43.00	32.00	8.0	Green et al. (1989)
		34.00	110.00	32.00	11.0	
		35.00	70.00	32.00	22.0	
		30.00	54.00	32.00	22.0	
Petauroides volans	Greater glider	1018.00	336.00	32.00	25.0	Foley et al. (1990)
Spermophilus saturatus	Golden-mantled ground squirrel	232.00	336.00	47.00	2.0	Kenagy et al. (1990)
Setonix brachyurus	Quokka	1900.00	548.00	32.00	27.0	Nagy et al. (1990)
Eptesicus fuscus	Big brown bat	20.80	47.60	43.00	20.0	Kurta et al. (1990)
		17.40	75.30	43.00	20.0	
Microtus arvalis	Field vole	20.00	90.00	52.00	15.0	Daan et al. (1990b)
Clethrionomys glareolus	Bank vole	23.40	88.00	52.00	15.0	Daan et al. (1990)
Microtus agrestis	Short-tailed field vole	27.20	83.00	52.00	15.0	Daan et al. (1990)
Spermophilus parryi	Arctic ground squirrel	630.00	817.00	65.00	6.0	Daan et al. (1990)
Mus domesticus	House mouse	14.00	45.10	36.00	–	Mutze et al. (1991)
Psammomys obesus	Fat sand rat	175.70	184.50	30.00	9.7	Degen et al. (1991)
		165.60	146.30	30.00	25.3	
Isoodon obesulus	Short-nosed bandicoot	1230.00	644.00	38.00	17.0	Nagy et al. (1991)
Antechinus stuartii	Antechinus	33.00	86.40	37.00	5.0	Green et al. (1991)
Antechinus stuartii	Antechinus	16.20	45.40	35.00	25.0	Green et al. (1991)
		22.80	54.05	5.00	20.0	
		55.60	133.94	35.00	15.0	
		25.10	89.75	35.00	15.0	
		52.30	104.12	35.00	15.0	
		24.80	74.27	35.00	15.0	
		28.80	94.17	35.00	20.0	
		26.70	78.81	35.00	20.0	
Vulpes cana	Blanford's fox	874.00	568.10	31.00	2.3	Geffen et al. (1992)
		1016.00	640.20	31.00	17.5	
Gerbillus allenbyi	Allenby's gerbil	22.80	35.60	29.00	30.0	Degen et al. (1992)
Gerbillus pyramidium	Greater Egyptian gerbil	31.80	45.20	29.00	30.0	Degen et al. (1992)
Aepyprymnus rufescens	Rat kangaroo	2850.00	1372.00	28.00	25.0	Wallis and Green (1992)
		2870.00	1353.00	28.00	25.0	
		2800.00	1551.00	28.00	10.0	
		2980.00	1439.00	28.00	10.0	

continued

Species	Common name	Body mass (g)	Metabolic rate (kJ day^{-1})	Latitude (°)	T(°C)	Reference
Sorex areneus	Common shrew	9.00	104.80	57.00	12.0	Poppitt et al. (1993)
		6.00	90.30	57.00	2.0	
Microgale dobsoni	Shrew tenrec	42.60	77.10	18.10	30.0	Stephenson et al. (1994)
Microgale talazaci	Shrew tenrec	42.80	66.50	18.30	30.0	Stephenson et al. (1994)
Marmota flaviventris	Yellow-bellied marmot	3240.00	2434.40	38.50	0.0	Salsbury and Armitage (1994)
Dipodomys merriami	Kangaroo rat	28.70	34.29	35.00	24.0	Nagy and Gruchacz (1994)
		32.90	40.78	35.00	35.0	
		32.70	40.76	35.00	29.0	
		32.50	36.00	35.00	24.0	
		32.30	50.89	35.00	25.0	
		32.60	58.70	35.00	11.0	
		33.60	58.27	35.00	6.0	
		36.20	56.24	35.00	12.0	
		36.90	57.58	35.00	12.0	
		37.90	44.44	35.00	14.0	
		33.10	35.79	35.00	22.0	
		34.70	32.24	35.00	31.0	
		35.10	36.40	35.00	32.0	
		34.60	34.56	35.00	33.0	
		35.00	42.16	35.00	15.0	
Isoodon auratus	Bandicoot	307.00	76.00	21.00	30.0	Bradshaw et al. (1994)
		333.00	257.00	21.00	30.0	
Pseudomys nanus	Barrow island mouse	40.30	18.10	21.00	30.0	Bradshaw et al. (1994)
		41.50	36.40	21.00	30.0	
		51.50	79.30	21.00	30.0	
Zyzomys (argurus?)	Rock rat	46.80	15.60	21.00	30.0	Bradshaw et al. (1994)
Tarsipes rostratus	Honey possum	9.90	14.40	30.00	8.0	Nagy et al. (1995)
Marmota flaviventris	Yellow-bellied marmot	3210.00	2307.00	38.58	0.0	Salsbury and Armitage (1995)
Pseudochirus peregrinus	Ring-tailed possum	968.00	561.30	40.00	17.0	Munks and Green (1995)
		968.00	550.60	40.00	13.1	
		1059.00	698.10	40.00	17.0	
		993.00	759.20	40.00	22.0	
		994.00	643.30	40.00	17.0	
Microtus pennsylvanicus	Meadow vole	32.30	77.50	45.00	17.3	Berteaux et al. (1996a)
Microtus pennsylvanicus	Meadow vole	33.20	91.30	45.00	17.6	Berteaux et al. (1996b)
		32.30	92.05	45.00	17.6	
Vulpes velox	Swift fox					Covell et al. (1996)
		1990.00	1488.00	40.00	−2.6	
		2220.00	2079.00	40.00	14.2	
Microtus pennsylvanicus	Meadow vole	35.60	100.40	45.00	19.6	Berteaux et al. (1997)
Potorous tridactylus	Long-nosed potoroo	868.00	629.00	37.00	12.0	Wallis et al. (1997)
		757.00	512.00	37.00	12.0	
		824.00	453.00	37.00	16.0	
		852.00	473.00	37.00	16.0	
Marmosa robinsoni	Mouse opposum	28.00	53.00	7.00	32.0	Green (1997)
Parantechinus apicalis		51.00	64.00	–	–	Green (1997)
		47.00	92.00	–	–	
		59.00	127.00	–	–	
		49.00	94.00	–	–	
Bettongia penicillata	Bettong	1100.00	524.00	–	–	Green (1997)
		1100.00	570.00	–	–	
		1100.00	695.00	–	–	
Bettongia gaimardi	Tasmanian bettong	1700.00	892.00	44.00	12.0	Johnson et al., in Green (1997)
		1700.00	856.00	44.00	12.0	

Species	Common name	Body mass (g)	Metabolic rate (kJ day^{-1})	Latitude (°)	T(°C)	Reference
Lagochestes hirsuitus	Rufous hare wallaby	1453.00 1351.00	661.00 532.00	23.00 23.00	20.0 35.0	lundie-Hjenkins, in Green (1997)
Isoodon obesulus	Bandicoot	486.00 517.00	366.00 459.00	– –	– –	Green (1997)
Macrotis lagotis	Great bilby	1208.00 1132.00	626.00 455.00	25.00 25.00	20.0 35.0	Southgate, in Green (1997)
Microtus agrestis	Hort-tailed field vole	26.50	72.70	53.00	15.0	Meerlo *et al.* (1997)
Phyllostomus hastatus	Spear-nosed bat	87.10 74.40	59.70 132.20	10.00 10.00	30.0 30.0	Kunz *et al.* (1998a)
Eremitalpa namibensis	Desert golden mole	20.70	12.46	22.00	32.1	Seymour *et al.* (1998)
Apodemus sylvaticus	Woodmouse	22.00 18.90	58.10 54.70	57.00 57.00	7.0 5.0	Corp *et al.* (1999)
Syconycteris australis	Common blossom bat	17.40	76.90	29.00	16.0	Coburn and Geiser (1999)
Talpa europaea	Mole	87.70	173.00	54.00	11.1	Frears *et al.* (unpublished results)

APPENDIX C

Phylogeny used in the Construction of the Phylogenetically Independent Contrasts

Part 1: Marsupials

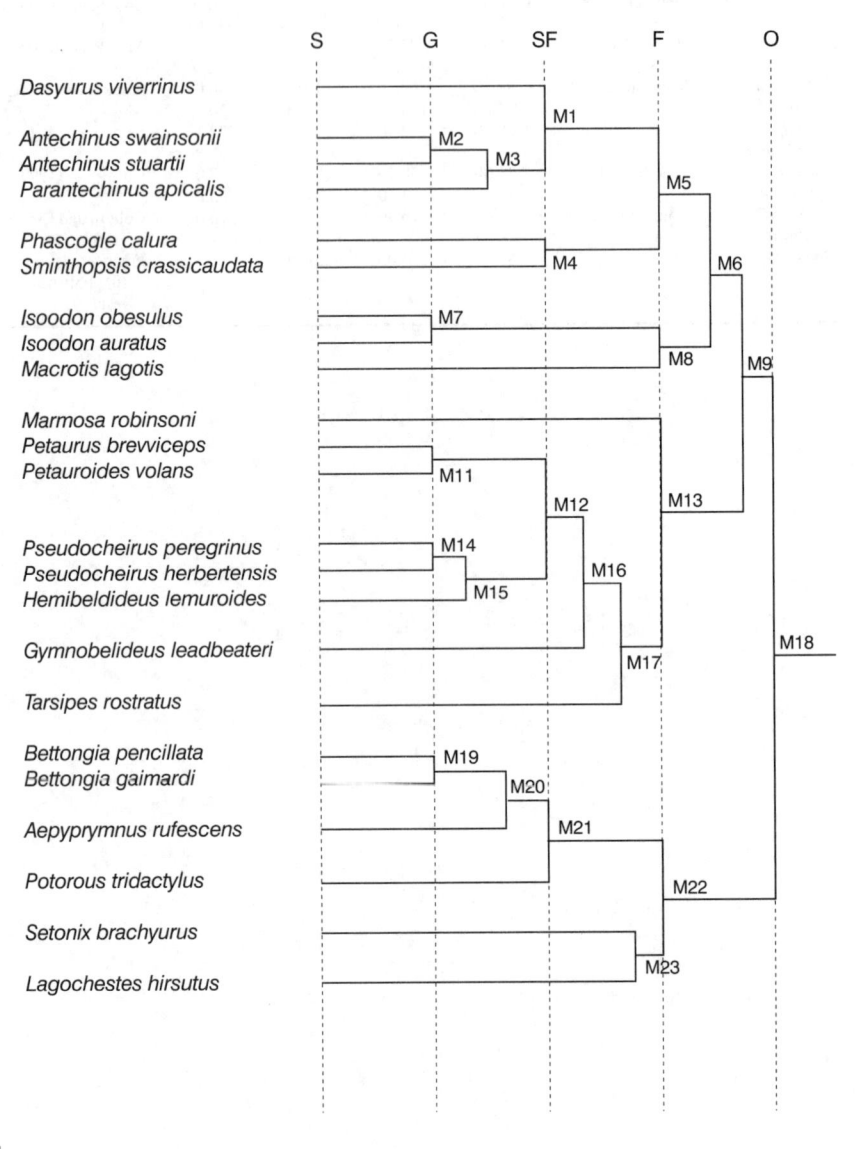

Dasyurus viverrinus
Antechinus swainsonii
Antechinus stuartii
Parantechinus apicalis
Phascogle calura
Sminthopsis crassicaudata
Isoodon obesulus
Isoodon auratus
Macrotis lagotis
Marmosa robinsoni
Petaurus breviceps
Petauroides volans
Pseudocheirus peregrinus
Pseudocheirus herbertensis
Hemibeldideus lemuroides
Gymnobelideus leadbeateri
Tarsipes rostratus
Bettongia pencillata
Bettongia gaimardi
Aepyprymnus rufescens
Potorous tridactylus
Setonix brachyurus
Lagochestes hirsutus

key
S = Species   G = Genus   SF = Sub-family
F = Family   O = Order   C = Class

Part 2: Bats

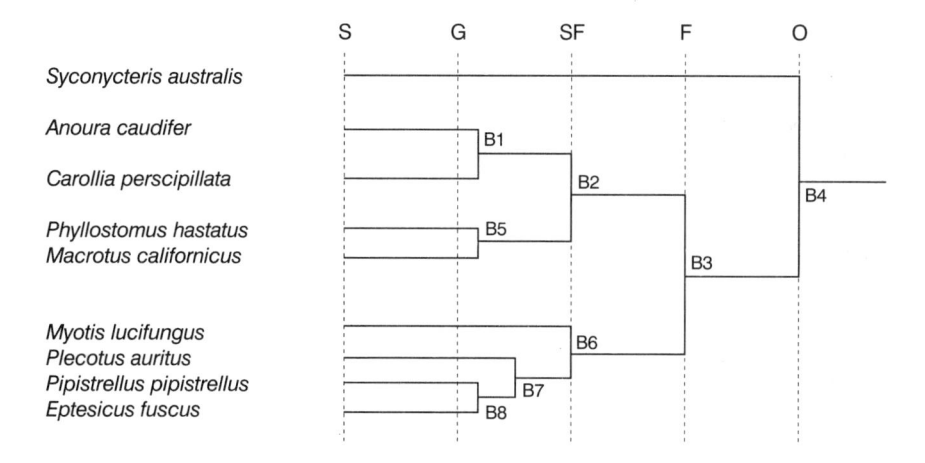

	S	G	SF	F	O
Syconycteris australis					
Anoura caudifer		B1			
Carollia perscipillata			B2		B4
Phyllostomus hastatus		B5			
Macrotus californicus				B3	
Myotis lucifungus			B6		
Plecotus auritus					
Pipistrellus pipistrellus		B7			
Eptesicus fuscus		B8			

Part 3: Insectivora

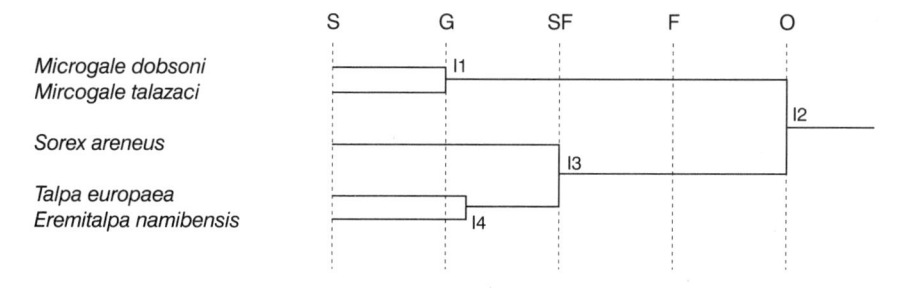

	S	G	SF	F	O
Microgale dobsoni		I1			
Mircogale talazaci					I2
Sorex areneus			I3		
Talpa europaea					
Eremitalpa namibensis		I4			

key
S = Species   G = Genus   SF = Sub-family
F = Family   O = Order   C = Class

Part 4: Rodents and Lagomorphs (Glires)

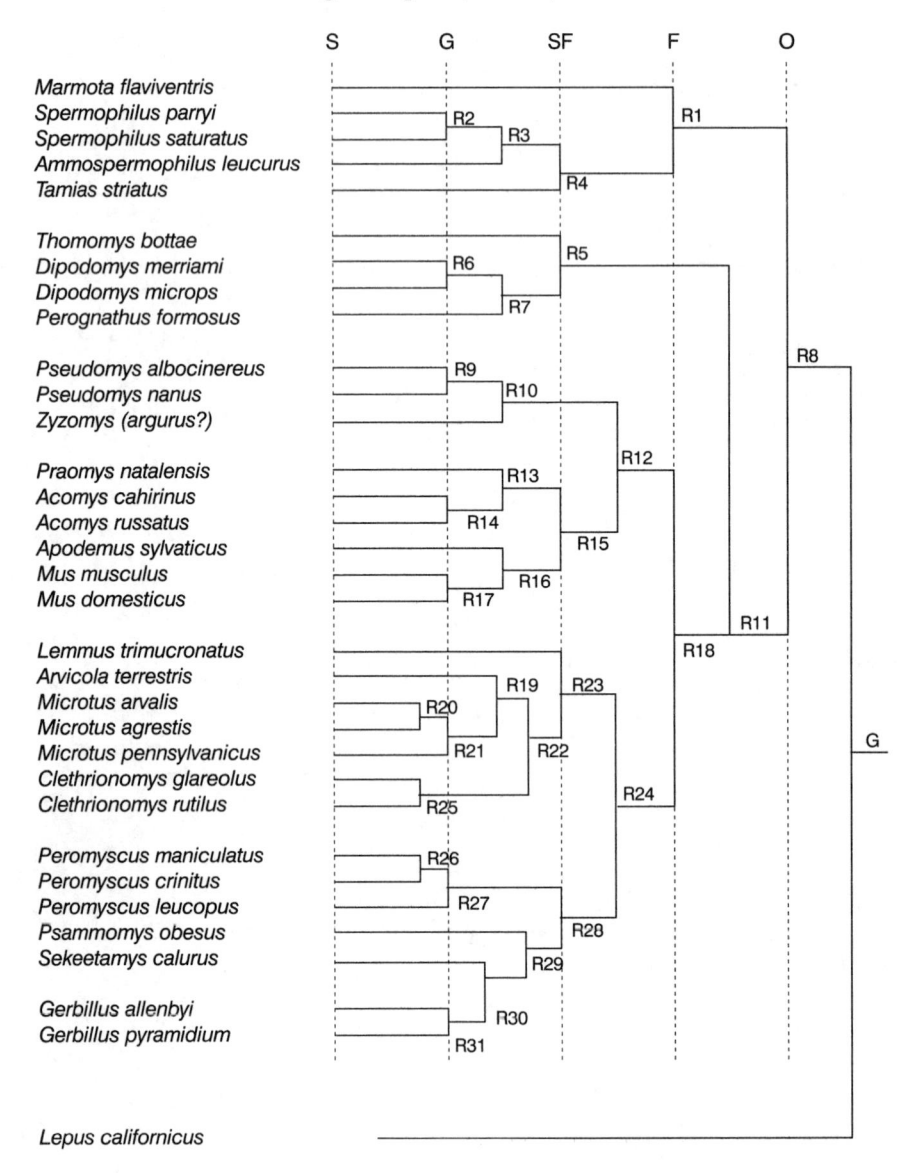

	S	G	SF	F	O

- Marmota flaviventris
- Spermophilus parryi
- Spermophilus saturatus
- Ammospermophilus leucurus
- Tamias striatus

- Thomomys bottae
- Dipodomys merriami
- Dipodomys microps
- Perognathus formosus

- Pseudomys albocinereus
- Pseudomys nanus
- Zyzomys (argurus?)

- Praomys natalensis
- Acomys cahirinus
- Acomys russatus
- Apodemus sylvaticus
- Mus musculus
- Mus domesticus

- Lemmus trimucronatus
- Arvicola terrestris
- Microtus arvalis
- Microtus agrestis
- Microtus pennsylvanicus
- Clethrionomys glareolus
- Clethrionomys rutilus

- Peromyscus maniculatus
- Peromyscus crinitus
- Peromyscus leucopus
- Psammomys obesus
- Sekeetamys calurus

- Gerbillus allenbyi
- Gerbillus pyramidium

- Lepus californicus

key
S = Species   G = Genus   SF = Sub-family
F = Family   O = Order   C = Class

Part 5: *Carnivores, Edentates and Higher Nodes*

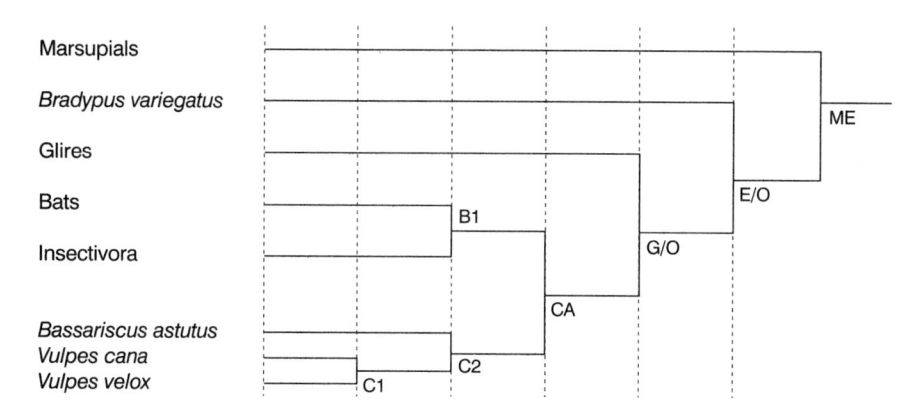

Marsupials

Bradypus variegatus

Glires

Bats

Insectivora

Bassariscus astutus
Vulpes cana
Vulpes velox

ME, E/O, G/O, CA, B1, C2, C1

key
S = Species   G = Genus   SF = Sub-family
F = Family   O = Order   C = Class

Manipulative Field Experiments in Animal Ecology: Do They Promise More Than They Can Deliver?

D. RAFFAELLI AND H. MOLLER

I. Summary 299
II. Introduction 300
III. Designs of Published Field Experiments 302
    A. Plot Sizes, Replication and Duration 302
    B. Why Are Some Experimental Designs So Weak? 312
    C. For How Long Should Experiments Run? 320
    D. Issues of Scale 324
    E. Meta-analysis: A Way Forward? 327
IV. Conclusions 328
Acknowledgements 330
References 330

I. SUMMARY

Manipulative field experiments have the potential to be the most rigorous and persuasive tests of hypotheses in ecology. Community press experiments have been especially successful in this respect and there have been increasing calls for application of the experimental approach to large-scale management issues. However, the designs of large-scale experiments are often constrained by limited resources (person-power, funding, area available), and are thus limited in their statistical power and ability to deliver convincing outcomes. These issues are explored through a comparison of community press experiments in aquatic and terrestrial systems with reference to plot size, replication, duration and scale. Insights into the underlying reasons for the choice of particular experimental designs are gained through interviews with researchers. These highlight the greater use of pragmatic and logistic criteria in the design of terrestrial as opposed to aquatic experiments. Several areas are identified where more research effort is required, including the development of 'stopping rules' for field experiments and issues of appropriate spatial and temporal scales. Finally, we discuss the potential of meta-analysis for evaluating the generality of outcomes of independent field experiments.

II. INTRODUCTION

In order to convince scientific peers, environmental managers, politicians and the public of the plausibility of processes underlying a pattern, evidence is required. The kind of evidence marshalled to support arguments may include (in order of strength): the frequent repetition of anecdotes, logical argument, mathematical modelling, observation and experimental tests of hypotheses (Hairston, 1989; McArdle, 1996; Underwood, 1997). Perhaps not surprisingly, the past 20 years have seen a significant increase in the use of and reliance on experimental tests, in part as a reaction against the rather uncritical use of other less strong forms of evidence to support and develop ecological arguments (Lawton, 1996; Underwood, 1997). In addition, funding agencies and journals have appeared less likely to support or publish research that does not incorporate an experimental hypothesis-testing approach (Lawton, 1996).

However, arguments based on experimental tests of hypotheses are only as robust as the design of the experiment. Institutions, especially applied agricultural and rangeland research organizations, with in-house statistical expertise and a culture of involvement with biometricians at the design and planning stages, have always delivered rigorous experimental science. In contrast, less applied and more academic science has taken a surprisingly long time to incorporate experimental approaches into field studies (Paine, 1994), sadly often without the benefit of competent statistical advice at both the planning and analysis stages. Recipes for the proper design, analysis and interpretation of experiments have lined the mathematics and statistics sections of library shelves for decades, but these have been largely inaccessible to all but the most ardent experimental ecologist. Only relatively recently have sympathetically written texts become available, specifically targeted at ecologists taking a less applied, fundamental approach (e.g. Hairston, 1989; Underwood, 1997). As a consequence, many of the early field experiments stimulated by the dramatic and compelling outcomes of marine intertidal studies (e.g. Connell, 1961; Paine, 1966) did not pay sufficient attention to design and analytical considerations (Underwood, 1981; Hurlbert, 1984).

The majority of field experiments in the literature have been carried out on relatively small spatial and short temporal scales (Kareiva and Anderson, 1989) but some of the most interesting and challenging ecological processes operate at much larger scales (May, 1994; Carpenter *et al.*, 1995; Lawton, 1996). These large-scale processes are the very ones in which society has a stake and for which managers require guidance, if not answers (Moller and Raffaelli, 1998). Clearly there is a need to work at much larger scales than is presently the case and several research teams have recently made forceful and compelling arguments for the funding of large ecosystem-scale experiments (Carpenter *et al.*, 1995). However, large-scale experiments may not always deliver sensible and informative answers: the experimental designs are necessarily constrained and compromised by the trade-off between the demands of carrying out experiments at those large scales and the resources that can be made available (see below). If the design is sufficiently compromised, then the strength of evidence acquired may be no better than that which derives from anecdote or logical argument.

In the present review we consider the potential of manipulative field experiments for addressing ecological questions at larger scales, particularly with respect to scale-dependent constraints imposed on the experimental design. We illustrate this by reviewing the designs of published experiments extracted from the mainstream ecological literature over the past few years. All the experiments involve either the removal of animals, often predators, from plots, or addition of animals to those plots. They are, therefore, classical press experiments, where the perturbation is maintained at the same level for the duration of the experiment (Bender *et al.*, 1984). There are two distinct and separate data sets or lists of studies. The second data set (see Table 2) includes 16 freshwater (mainly lentic) studies as well as 22 marine and 42 terrestrial studies. These experiments are an 'unbiased' selection, in that they represent all of such studies retrievable on the BIDS (Bath Information and Data Services) database. Although the selection process was unbiased, this approach inevitably leads to over-and under-representation of particular habitats or taxa within freshwater, marine and terrestrial systems. For instance, there are many more studies involving small mammals than birds in terrestrial systems, and more intertidal than sublittoral studies in marine systems. Given the small sample sizes of habitat types or taxa, it has not been possible to carry out finer-grained comparisons within each of the three major habitat types.

The first set comprises 56 experiments, (two freshwater, 14 marine and 40 terrestrial studies) previously compiled by Moller and Raffaelli (1998). These studies were originally selected by the authors to illustrate particular issues but were deemed representative of the majority of marine and terrestrial experiments in the literature.

We have not pooled these two data sets as they represent two separate time periods and because it was of interest to establish whether they revealed similar or different patterns. Combining them would have obscured any differences.

Whilst there have already been a number of reviews of field experiments in the literature (e.g. Underwood, 1981; Connell, 1983; Schoener, 1983; Hurlbert, 1984; Sih *et al.*, 1985; Kareiva and Anderson, 1989; Tilman, 1989), those reviews focused exclusively on the prevalence of particular kinds of interactions, such as competition and predation, and different focal taxa, such as plants, on the strength of field experiments in relation to alternative approaches, or the pitfalls of design and statistical analysis. None of the above reviews specifically considered the broader issues discussed here: why ecologists persist in carrying out experiments they know to be poorly designed, whether there may be cultural differences in experimental approach between ecologists working in different habitat-types (terrestrial versus marine versus freshwater), and whether it is worth attempting experiments at all for some kinds of question. We have touched on some of these issues elsewhere in relation to the feasibility of evaluating the potential community-wide impact of biological control of European rabbits by a pathogen, calicivirus (Moller and Raffaelli, 1998). The aim of the present review is to make these issues more accessible to a wider ecological audience and to extend and amplify our analyses.

III. DESIGNS OF PUBLISHED FIELD EXPERIMENTS
A. Plot Sizes, Replication and Duration

Whilst there is a great diversity of designs and approaches in the 136 experiments analysed (Tables 1 and 2), several features should be basic to all designs and therefore common to all studies. It is these design features that we wish to compare for experiments done in terrestrial, freshwater and marine systems.

Table 1
Experiments reviewed in Moller and Raffaelli (1998)

Aaser, et al. (1995)	Moss et al. (1996)
Akbar and Gorman (1993)	Norrdahl and Korpimaki (1995)
Bell and Coull (1978)	Oksanen and Moen (1994)
Brown (1994)	Paine (1974)
Brown (In press)	Patterson and Fuchs (1996)
Caley (1995)	Patterson and Laing (1995)
Clifton-Hadley et al. (1995)	Power (1990)
Crowder et al. (1994)	Proulx et al. (1996)
Dilks and Wilson (1979)	Quammen (1984)
Erlinge (1987)	Raffaelli et al. (1989)
Eves (1993)	Redfield et al. (1978)
Floyd (1996)	Sanders (1996)
Hartley (In press)	Schimmel and Granstrom (1996)
Jacquet and Raffaelli (1989)	Taitt and Krebbs (1983)
Krebs et al. (1995)	Thrush et al. (1995)
Lodge (1948)	Thrush et al. (1996)
McCarthy (1993)	Watson et al. (1984)
Moller et al. (1991)	Wilesmith et al. (1982)

Table 2
New experiments reviewed in the present paper

Abramsky et al. (1996)	Kuhlmann (1994)
Abramsky et al. (1990)	Lagos et al. (1995)
Alterio and Moller (1997)	Lagrange et al. (1995)
Baines et al. (1994)	Leibold (1989)
Batzer (1998)	Lewis (1986)
Bazly and Jeffries (1986)	Malhotra and Thorpe (1993)
Belovsky and Slade (1993)	Marsh (1986)
Benedettie-Cecchi and Cinelli (1997)	Marshall and Keogh (1994)
Berlow and Navarrete (1997)	Martin et al. (1994)
Berteaux et al. (1996)	Mattila and Bonsdorff (1989)
Beukers and Jones (1998)	Moen and Oksanen (1998)
Biondini and Manske (1996)	Montgomery et al. (1997)
Bowers (1993)	Moran and Hurd (1994)
Branch et al. (1996)	Ouellet et al. (1994)
Bukaveckas and Shaw (1998)	Parmenter and Macmahon (1988)
Chase (1996)	Pastor et al. (1993)
Churchfield et al. (1991)	Pennings (1990)
Dale and Zbigniewicz (1997)	Petersen et al. (1994)
Edgar and Robertson (1992)	Posey and Hines (1991)
Farrell (1991)	Pringle and Hamazaki (1998)
Flecker (1992)	Sarnelle (1993)
Gascon (1992)	Scheibling and Hamm (1991)
Hargeby (1990)	Schmitz (1998)
Hermony et al. (1992)	Schulte-Hostedde and Brooks (1997)
Hershey (1985)	Spiller and Schoener (1994)
Heske et al. (1994)	Steele (1998)
Huang and Sih (1991)	Todd and Keogh (1994)
Hubbs and Boonstrana (1997)	Turchin and Ostfeld (1997)
Hulme (1996)	Tyler (1995)
Hurd and Eisenberg (1990)	Valentine and Heck (1991)
James and Underwood (1994)	van Buskirk and Smith (1991)
Joern (1992)	Virtanen (1998)
Joern (1986)	Virtanen et al. (1997)
Juliano (1998)	Walde and Davies (1984)
Kennedy (1993)	Walde (1994)
Kerans et al. (1995)	Wilson (1989)
Kielland et al. (1997)	Witman (1985).
Klemola et al. (1997)	

1. Size of Plots Manipulated

We define a plot as an area of habitat perturbed which the authors (or in some cases ourselves) consider constitutes a true replicate treatment or control based on accepted criteria (Hurlbert, 1984). For some lentic freshwater studies, the plot size was reported as a volume, probably because the response organisms move in a three-dimensional environment. If linear dimensions of the volume were provided by the authors, we have taken the plan area as the plot size. In reality this is no different from the plot areas specified by sediment, soil and forest ecologists, all of whom are essentially dealing with a volume of habitat that has been conveniently delimited by a plan area. For aquatic studies where conversion to plan area was not possible, for instance when the plot size was given in litres, this particular datum was excluded from the analysis. The shape of plots, as defined by length–width ratios, was not considered an issue, as almost all studies used square or circular plots.

2. Number of Plots (Replicates)

The number of plots was usually clearly defined by the authors. Where pseudo-replication clearly occurred, or where repeated sampling over time of individual plots took place, we have attempted to clarify what was the actual number of replicates present in the experimental design. In many cases, pseudo-replication had not been recognized by the author, as has previously been reported by Hurlbert (1984), Underwood (1981) and McArdle (1996), and the inferences the authors drew from the statistical analysis of their experiment are probably in many cases unjustified. Pseudo-replication is usually due to constraints placed on the experimental design by external non-scientific factors, and it is those constraints we wish to highlight, rather than any niceties of statistical analysis. Where authors have reported pseudo-replicates as replicates, we have deduced the correct degree of replication ourselves and used this in our analyses.

Whilst the present review will not deal explicitly with the consequences of pseudo-replication for statistical analyses, it is worth reiterating a point made by McArdle (1996) concerning the confusion over pseudo-replication and statistical independence in field experiments, because it is probably the most frequent basis for rejecting experimental papers and thus assigning them to the great mass of unpublished material on field experiments (see below). Hurlbert's (1984) paper on pseudo-replication and the design of field experiments had been cited by an astonishing 1543 journal papers at the time of writing (BIDS database). The paper has performed a valuable service for ecology by inserting rigour into the design of field experiments and by alerting reviewers and editors to the difficulties of drawing inference from certain kinds of experimental design. However, as McArdle (1996) has pointed out, Hurlbert's original definition of pseudo-replication '… the use of inferential statistics to test for treatment effects with data from experiments where either treatments are not replicated or replicates are not statistically independent …', conflates two quite separate issues: the purpose of replication and the independence of the sampling units. Treatment replication is necessary because it allows us confidently to separate treatment effects which are inherently variable. Experimenters are guilty of pseudo-replication only if they make claims that their analysis demonstrates a much more general effect when, in fact, the error terms used in the statistical analysis reflect the variation between samples within a treatment replicate. Thus, authors cannot be accused of the sin of pseudo-replication if they are content to make inferences only about the variation within a single plot and they refrain from generalizing beyond that area. Whilst such limited inferences are rarely deemed by journals to be sufficiently interesting to attract a large audience, some of the most convincing field experiments have entered the mainstream literature despite this handicap (see below). Also, at the largest scales of all, such as an entire estuary, lake or forest, replication may be impossible or even inappropriate if it is the particular area that is of intense interest.

On the issue of independence of sampling units, Hurlbert's (1984) statement perpetuates a fallacy that is clearly of irritation to biometricians (McArdle, 1996). Non-independent replicates can be analysed as long as non-independence is explicitly incorporated into the analysis used, and McArdle (1996) and Underwood (1997) provide appropriate guidance and references on how to go about this. In fact, it is the *errors* after fitting the usual statistical models (*t* tests, analysis of variance) routinely applied to the analysis of field experiments that must be independent.

3. The Duration of the Experiment

We only consider experiments where the treatment was effectively applied for the entire duration of the experiment. In the terminology of Bender *et al.* (1984) these are 'press' experiments, as opposed to 'pulse' experiments where the treatment is applied momentarily and the processes involved in the return to equilibrium are the main focus of the study. For most of the studies we review, the duration was clearly stated, although it is not always clear whether the duration was fixed *a priori* or whether the period analysed was selected *a posteriori*, perhaps because the integrity of the experimental set-up was becoming questionable. In experiments involving repeated measures within the same plots at several time intervals, and where treatment effects were apparent only in particular seasons, it was not always obvious what the duration of the experiment should be. In such cases we have excluded that experiment from this aspect of the analysis.

4. Trends in Experimental Designs

Our earlier analysis (Moller and Raffaelli, 1998) showed that experiments with the highest degree of replication have a relatively small plot size (< 10 m^2) (Figure 1), for both treatment and control plots and for both aquatic (almost all marine) and terrestrial systems. Several studies had no treatment replication (only one treatment plot), and this was more frequent in experiments employing large plot sizes (Figure 1). A similar pattern is seen for control replicates (Figure 2),

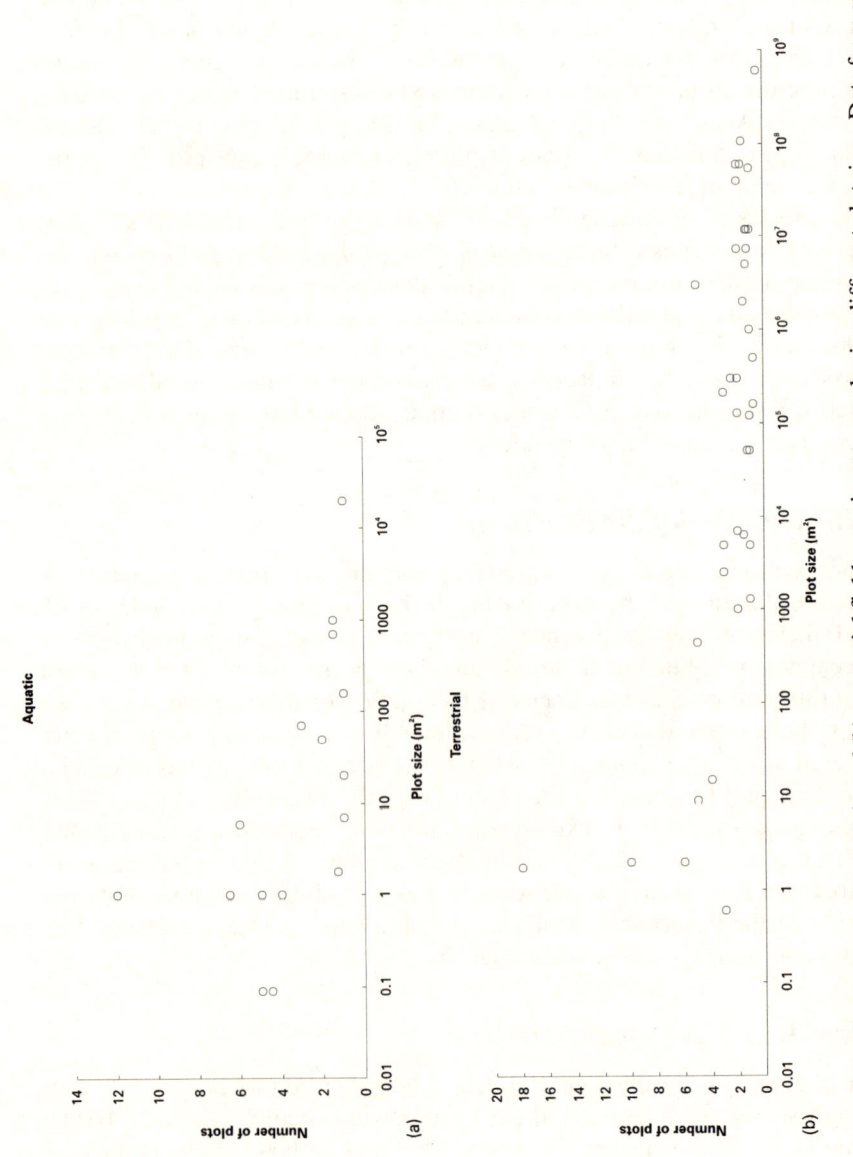

Fig. 1. Number of treatment plots in (a) aquatic and (b) terrestrial field experiments employing different plot sizes. Data from experiments reviewed in Moller and Raffaelli (1998), Table 1.

with five aquatic (marine) and three terrestrial experiments having no control plots. Many of the experimental designs were unbalanced, for both habitat types, in that there were different numbers of control and treatment plots (31% of aquatic and 35% of terrestrial experiments), but there was no obvious consistent trend towards more plots of one type (Figure 3). Experiments employing a relatively small plot size were mostly aquatic (marine), whereas those using a larger plot size were nearly all terrestrial (Figures 1 and 2).

The choice of experiments in our earlier analysis (Moller and Raffaelli, 1998) was idiosyncratic, based in part on our familiarity with research on shallow-water marine systems (D.R.) and in wildlife management (H.M.). Our analysis of 80 further experiments (Table 2), selected more objectively (see above), confirm that the majority of experiments carried out in marine systems, involve plot sizes of < 10 m^2 whilst plot sizes of terrestrial experiments extend to much larger sizes, the majority being > 10 m^2 in size (Figures 4 and 5). Freshwater experiments (which could not be treated as a distinct group in Moller and Raffaelli (1998) because of their small sample size) included a greater proportion with plot sizes < 1 m^2, compared with marine, but the range of plot sizes is similar to those used in marine systems and quite different from that in terrestrial systems (Figures 4 and 5). We recognize that some kinds of habitats permit only relatively small plot sizes (e.g. tide pools, narrow streams and ponds), and this may account for the tendency towards small plots in many of the freshwater studies. However, this factor cannot entirely account for the between-system differences shown in Figures 4 and 5, for larger areas of freshwater benthic habitat and for the extensive areas available for experimentation in intertidal flats, seagrass beds and benthic sublittoral.

All three systems had a similar range (1–6) of treatment and control replication (Figure 4), with a high degree of treatment replication in one marine ($n = 15$) and one terrestrial ($n = 11$) system. Balanced experimental designs (equal numbers of control and treatment replicates) were found for only a proportion of studies in any of the three systems: 18% of freshwater, 24% of marine and 33% of terrestrial experiments (Figure 6). Several experiments lacked controls altogether, but this was more common for terrestrial than aquatic systems (Figures 4 and 6).

In the studies reported in our original analysis (Moller and Raffaelli, 1998), small-plot experiments were run for a much greater range of duration than large-plot experiments (Figure 7), although most small-plot experiments lasted for weeks or months rather than months and years. This overall pattern is also apparent in the most recent analysis (Figures 8 and 9), although the relationships are not statistically significant (Spearman's rank) when each of the systems is analysed separately.

In summary, analysis of two independent data sets indicates that large-plot (> 10m^2) experiments tend to be less well replicated than those employing smaller plots, many larger plot experiments having no replication of treatment

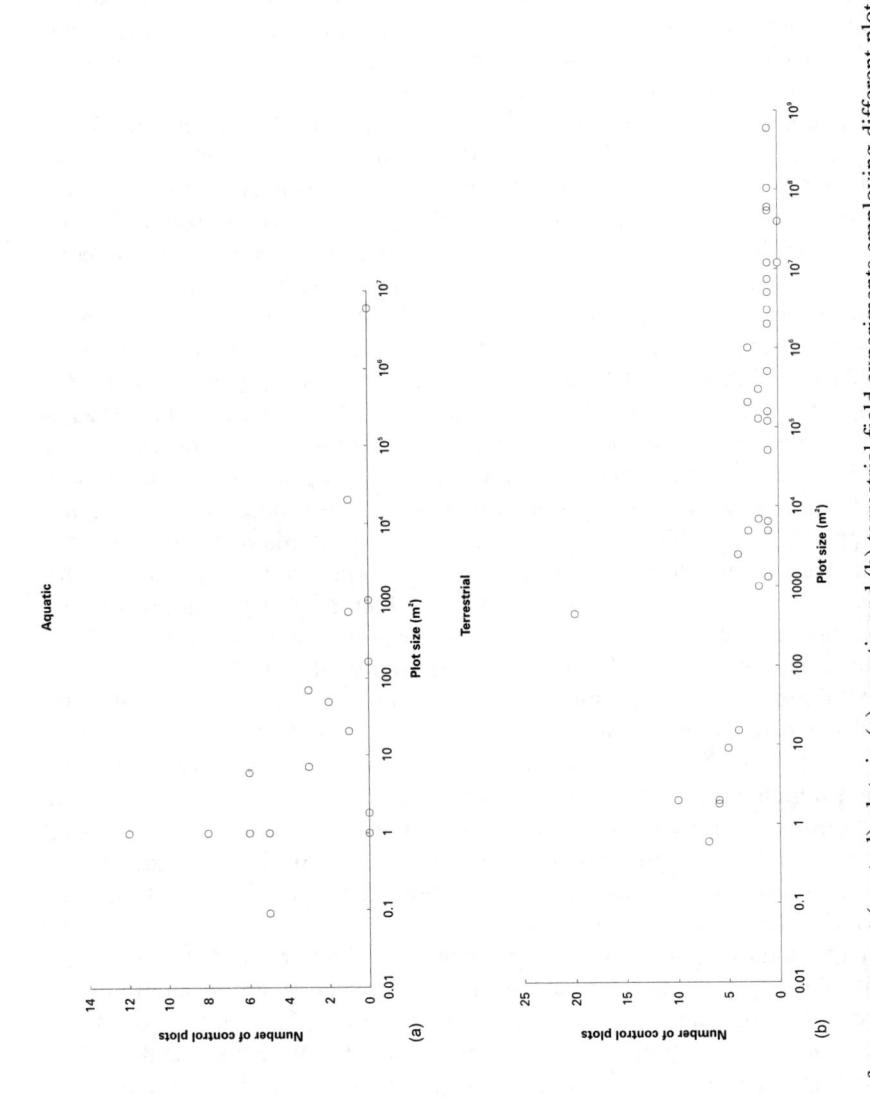

Fig. 2. Number of non-treatment (control) plots in (a) aquatic and (b) terrestrial field experiments employing different plot sizes. Data from experiments reviewed in Moller and Raffaelli (1998), Table 1.

Aquatic

(a)

Terrestrial

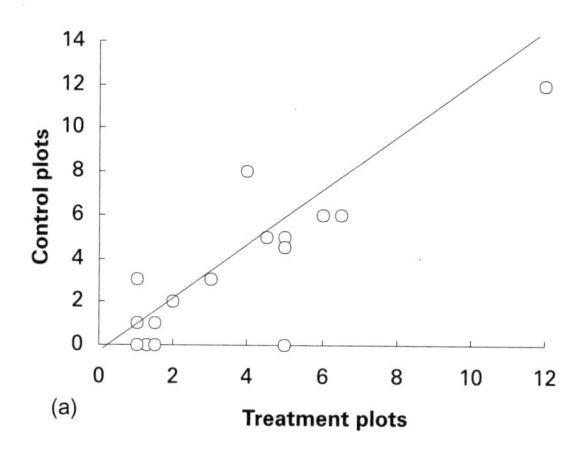

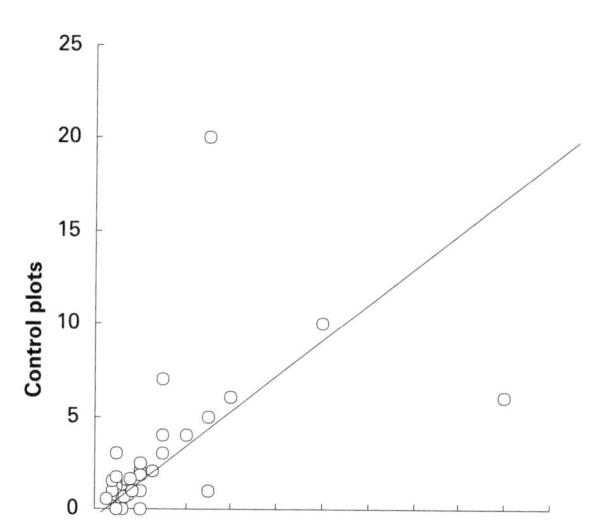

(b)

Fig. 3. Number of treatment and control plots employed in (a) aquatic and (b) terrestrial field experiments. For clarity, data points for several of the smaller plot-size experiments have been displaced on the graphs. Data from experiments reviewed in Moller and Raffaelli (1998), Table 1.

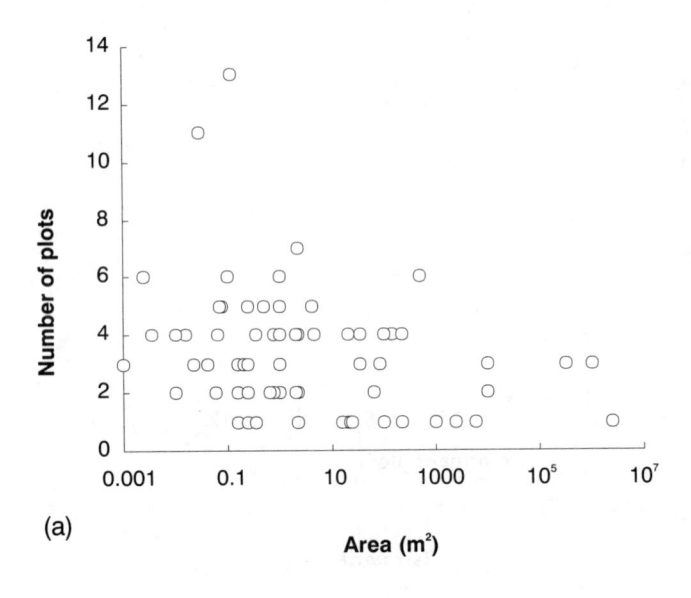

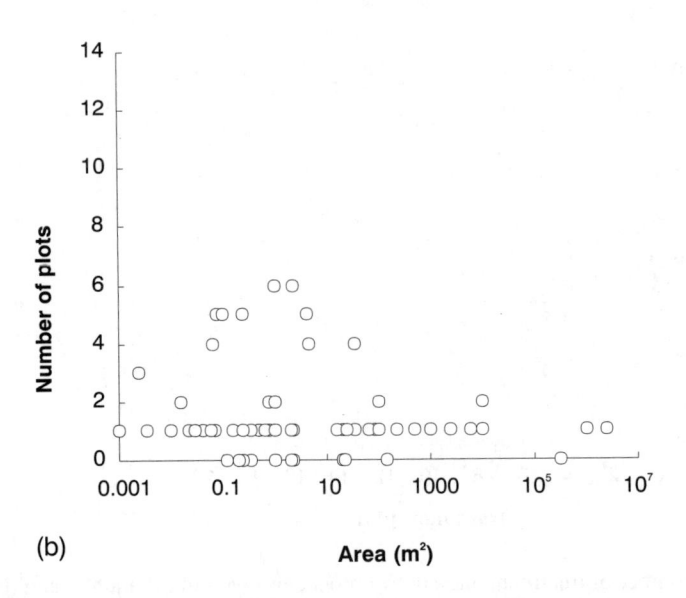

Fig. 4. Number of (a) treatment and (b) control plots in field experiments employing different plot sizes. Data from published experiments listed in Table 2.

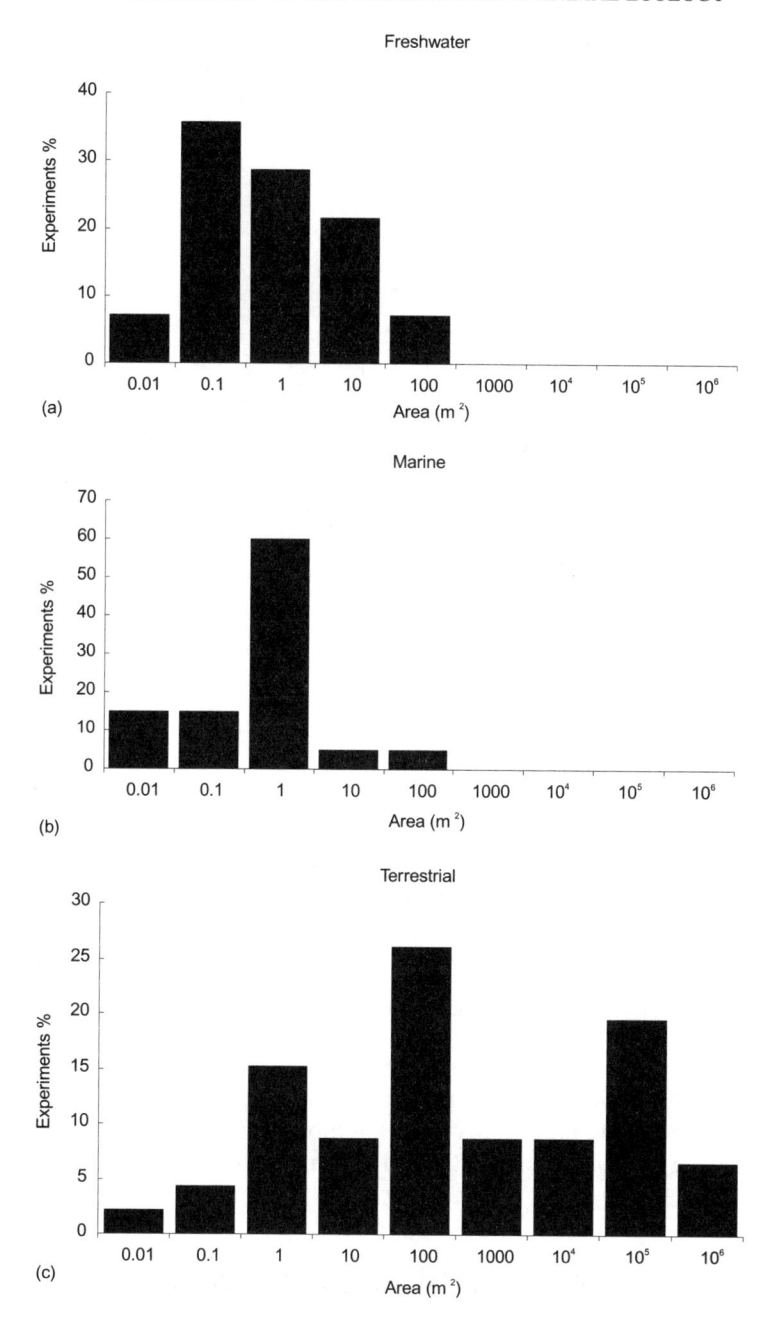

Fig. 5. Distribution of plot sizes employed in freshwater, marine and terrestrial field experiments listed in Table 2.

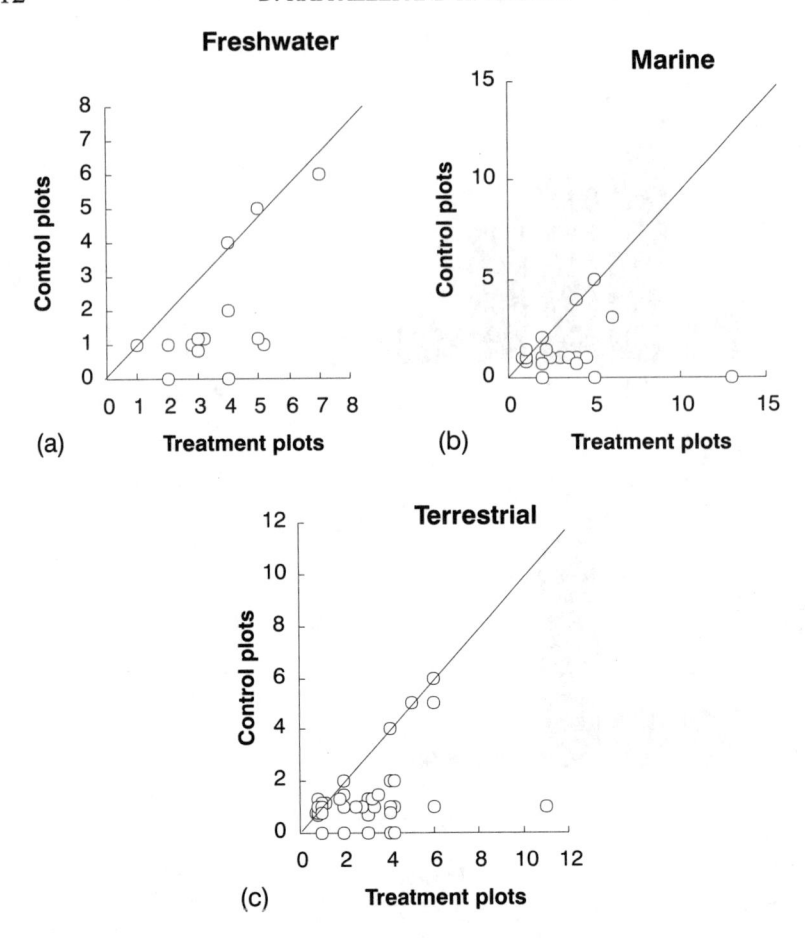

Fig. 6. Number of treatment and control plots employed in (a) freshwater, (b) marine and (c) terrestrial field experiments listed in Table 2. For clarity, data points for several of the smaller plot-size experiments have been displaced on the graphs.

plots and, sometimes, no control or reference plots at all. On average, experiments in aquatic systems employ smaller plots than those in terrestrial systems, and large-plot experiments tend to run for longer than small-plot experiments.

B. Why Are Some Experimental Designs So Weak?

Whilst these analyses provide a quantitative statement of the differences between approaches to experimental design in terrestrial (mostly large-plot) and aquatic (mostly small-plot) systems, they will probably come as no

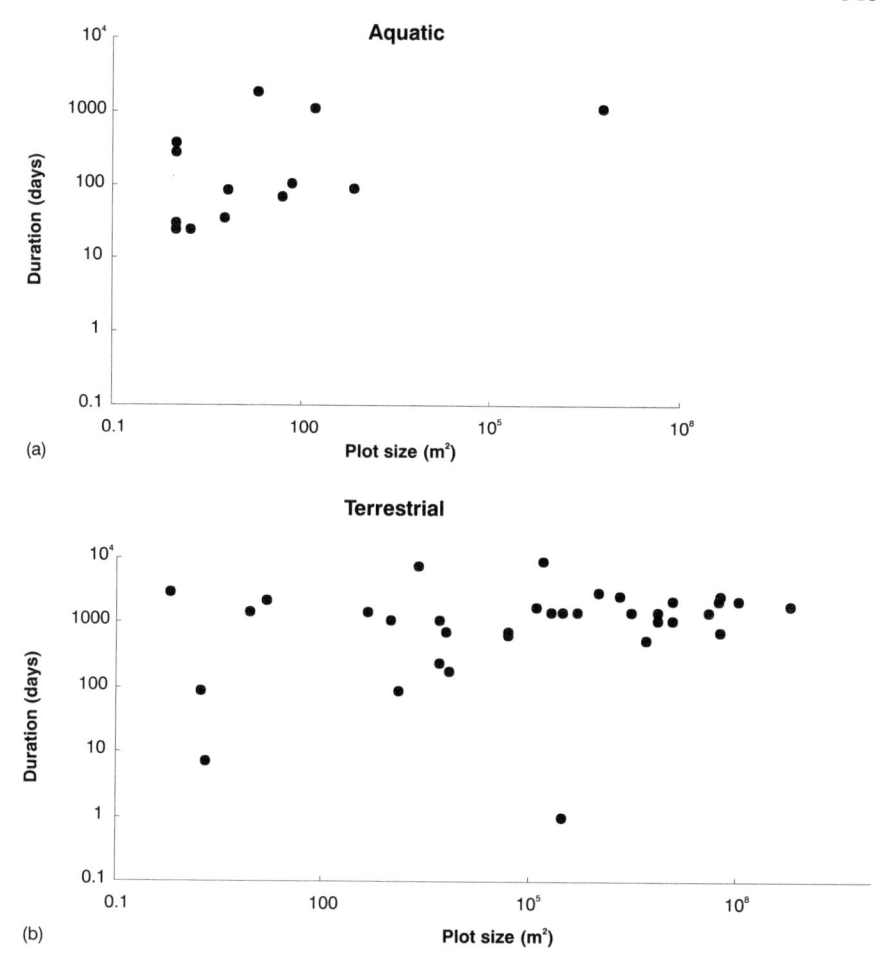

Fig. 7. Relationship between plot size and duration of field experiments in (a) aquatic and (b) terrestrial systems. Data from experiments reviewed in Moller and Raffaelli (1998), Table 1.

surprise to most ecologists. Many questions in terrestrial animal ecology will clearly demand working with larger areas simply because of the larger size, lower density and larger home-range sizes of the focal species, whether this be the organism manipulated or the response species. Thus, if a raptor forages over several square kilometres on small mammal prey which have a home-range of several hundred square metres, an experimental design employing exclosures of only a few square metres is unlikely to be taken seriously by reviewers. In our earlier analysis (Moller and Raffaelli, 1998) there was a good correlation between plot size selected and the density and home range of the

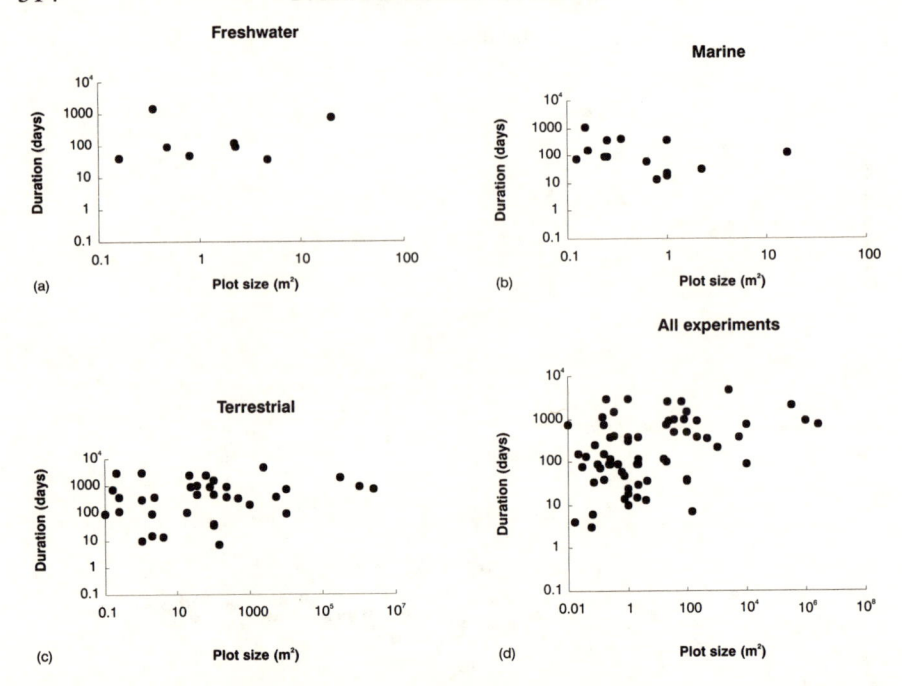

Fig. 8. Relationship between plot size and duration of field experiments in (a) freshwater, (b) marine, (c) terrestrial and (d) all systems combined. Data from experiments listed Table 2.

response species, indicating some attempt at scaling the experiment appropriately (but see below). Nevertheless, a large proportion of large-plot terrestrial experiments are unreplicated or have only one replicate with respect to treatment plots or lack control plots altogether, and the experiments have very limited power. Several large-plot aquatic experiments also suffer from the same weaknesses, and these are discussed below. In this section, we focus on experiments in terrestrial systems for which we have additional information on why particular designs were chosen.

1. Interviews with Authors of Field Experiment Papers

The reasons for designing experiments with weak power are complex, reflecting differences in culture, training, resource availability (especially funding and the extent of habitat available) and whether the question addressed was of a pure or applied nature. In discussing the relative contribution of these factors and the interactions between them, we are aware that we bring prejudices from our own marine (D.R.) and terrestrial (H.M.)

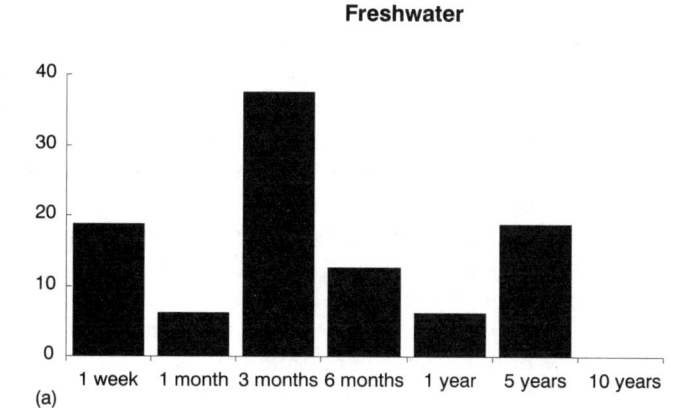

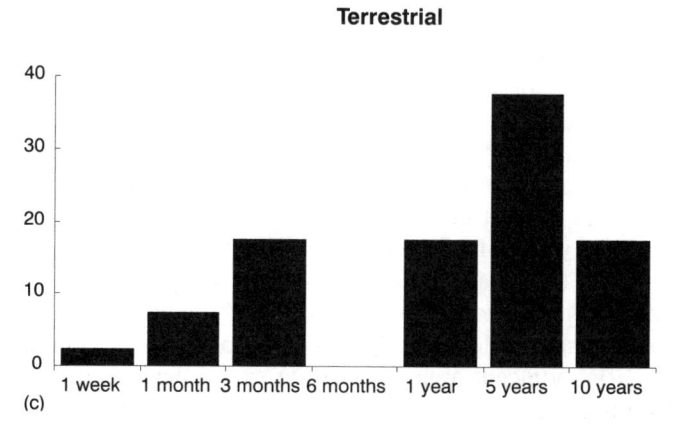

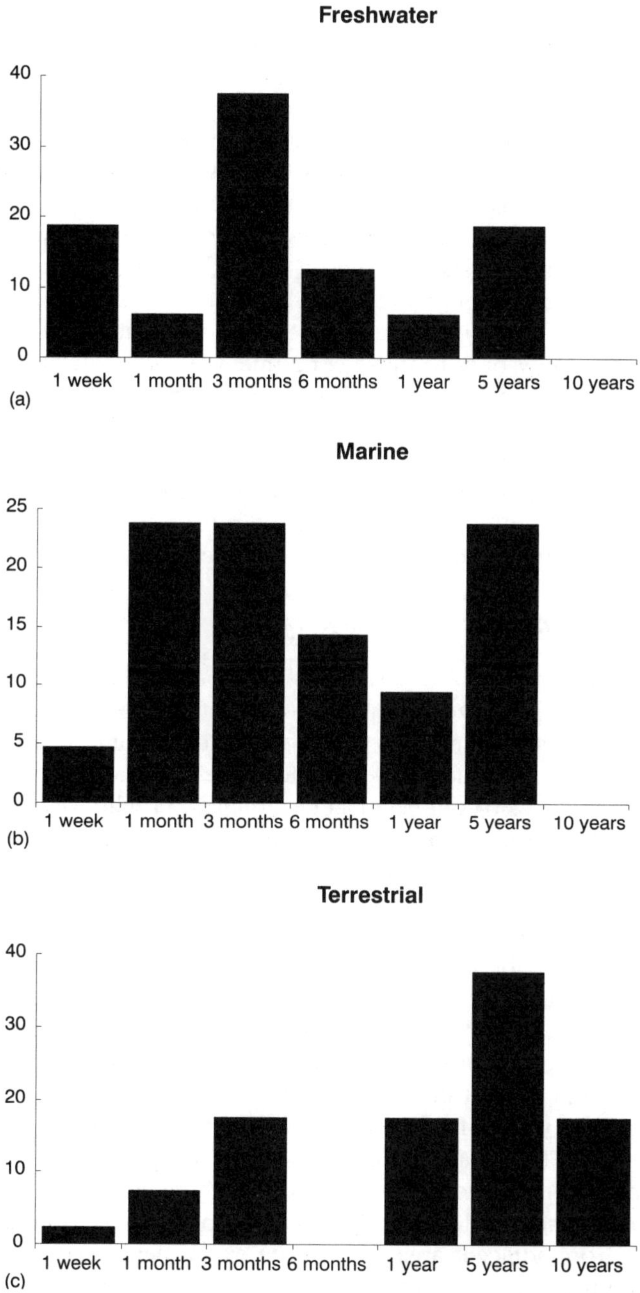

Fig. 9. Frequency distribution of durations of field experiments in (a) freshwater, (b) marine and (c) terrestrial systems. Data from experiments listed in Table 2.

backgrounds. We therefore chose to gain insights into the underlying reasons for the choice of a particular design by directly interviewing the authors of the large-plot terrestrial experiments listed in our original analyses (Table 1) or asking them to complete a questionnaire. We confined the interview to authors of large-plot (> 10 m^2) experiments because it is these experiments that seem to have the most weaknesses. Given the relatively small number of such experiments in the literature, and the fact that several authors declined to be interviewed or to complete an anonymous questionnaire, our analysis is based on a small number of responses ($n = 26$). This represents a significant proportion (78%) of the large-plot terrestrial experiments listed in Table 1. The interview–questionnaire analysis results strictly apply only to those experiments in Table 1, but we believe they are not unrepresentative of large-plot terrestrial experiments in general. It was a condition of the questionnaire and interviews that the responses remain anonymous and this is respected in the present review, unless the authors have agreed to be identified for clarification of a particular point. A full analysis of these interviews can be found in Moller and Raffaelli (1998) and only a summary is given here.

2. Reasons for Choosing a Plot Size

Few researchers explicitly described in their published account why they chose a particular plot size, irrespective of whether the plots were small or large, or whether the experiments were done in aquatic or terrestrial systems. (Incidently, this is also true for the second, more recent, data set in Table 2.) Nearly all of those who completed a questionnaire or who were interviewed (i.e. authors of large-plot terrestrial experiments) stated that plot size was the first and most important feature of their design, with the number of replicates being a secondary, consequential, consideration. However, no single factor emerged as the basis for choice of plot size. Most respondents ($n = 21$) had decided on a plot size using pragmatic criteria, including convenient landscape boundaries (islands, peninsulas, habitat boundaries), all the habitat that was available, mimicking similar work done elsewhere, or because this was all that resources (funds, helpers or equipment) would run to. These criteria are not obviously related to the dynamics of the species manipulated. Only five of the 26 interviewees chose a plot size based on the characteristics of the focal (perturbed or response) species or the magnitude of the response anticipated. For instance, one team calculated that plots of 35 ha would be needed to exclude wasps (by poisoning) in order to avoid re-invasion from adjacent areas during the experiment. Another team estimated the minimum area that would hold enough individuals of the response variable (Kiwi chicks) following exclusion of introduced mammalian predators (Moller and Raffaelli, 1998).

3. Reasons for Choosing the Number of Replicates

Having fixed on a plot size, most interviewees then estimated the number of replicates of these plots that they could manage given the funding, logistic and spatial resources available. Again, these are pragmatic, not ecological, considerations. Only one interviewee had carried out a formal power analysis before the experiment to estimate the number of replicate plots required. However, the results of that analysis was then ignored because the number of plots suggested would have been beyond the resources available. Encouragingly, all interviewees stated they would have gone for greater replication if resources had permitted.

4. Repetition of Experiments

When replication is limited, perhaps because of the short-term cycle of resource funding or lack of suitable habitat, a convincing case for an effect can sometimes be made if the experiment is repeated on several occasions or at several locations contemporaneously and a consistent outcome obtained. Whilst it is not straightforward formally to bring together data from repeated experiments as replicates in a ANOVA-type model (McKone and Lively, 1993; Lively and McKone, 1994), statistical approaches exist that can be used to establish whether the same kind of experiment repeated several times generates a similar outcome (Hedges and Olkin, 1985; Hunter and Schmidt, 1990). This meta-analysis approach is discussed further below.

It should be noted that the power of the individual experiments is estimated in meta-analyses, so that these procedures are not a panacea for poorly replicated experiments. Nevertheless, experiments need to be repeated several times if their outcomes are to be judged convincing (Walters and Holling, 1990). Repeated experiments can be executed concurrently or with different starting times and conditions, but good examples employing either approach are rare. Underwood and Chapman (1992) and Brown (1994) provide two fine exceptions. One of the most convincing demonstrations of the need to repeat experiments comes from a series of rocky intertidal experiments in New South Wales carried out by Fairweather and Underwood (1991). They repeated the same removal experiment contemporaneously at different locations to evaluate the generality of the effect of removing a predatory whelk on the dynamics of intertidal assemblages. The magnitude of the effects differed markedly between sites, and a significant predation effect was only one of a series of possible outcomes. If the experiment had been performed only once at a single site, Fairweather and Underwood (1991) might have demonstrated a non-effect, in itself an important result when attempting to understand the factors organizing this intertidal assemblage, or a large effect, an outcome that would probably have had a much higher chance of being accepted by a journal.

The difference in acceptability of zero-effect and strong-effect outcomes by the authors of field experiments (who are less likely to attempt to publish non-effect papers), and later by editors and reviewers, potentially limits formal meta-analysis: the generality of an effect can be properly evaluated only if experiments are both repeated and reported (but see also below).

None of the large-plot experiments in the survey carried out by Moller and Raffaelli (1998) had been repeated. Furthermore, none of our interviewees had any future plans to repeat their experiments. In 40% of cases this was because they considered the outcomes 'proven' and further experiments were therefore unnecessary. This viewpoint is by no means confined to terrestrial ecologists, but the need to establish the generality of an outcome may be more urgent and critical for large-plot experiments directed at providing advice for expensive conservation and management programmes. The remaining respondents (60%) had decided not to repeat their experiment for pragmatic and logistical reasons, such as cessation of funding or because the work was too difficult or too boring (Moller and Raffaelli, 1998). One reasonable constraint on the repetition of large-plot experiments is that they might have to be repeated at the same location. Whilst this is undesirable (the experiments may not be considered independent), it was not stated as a reason for not repeating experiments.

This lack of repetition is disappointing, but not unexpected given the constraints on funding, the expectations of funding agencies and peer reviewers for researchers to address increasingly novel issues, and perhaps the failure of scientists themselves to appreciate the need for repeat experiments.

5. When Poorly Designed Experiments are Accepted

We do not believe that the diversity of reasons revealed by our interviews and questionnaire for choosing the design of field experiments is limited to terrestrial ecologists. Similar reasons underlie many of the experimental designs found in the aquatic literature. However, in many kinds of aquatic systems there is the opportunity to use small plots which can be replicated cheaply in order to manipulate animals with a small body size, short generation time and small home range, so that aquatic ecologists can at least aspire to a rigorous experimental design within a given resource base. For the same reasons, it is relatively easy for aquatic ecologists to carry out an *a priori* power analysis when designing experiments and to refine the experiment accordingly (Raffaelli and Hawkins, 1996; Underwood, 1997), since there are sufficient time and resources available to carry out pilot experiments within the usual 3–5-year funding cycle.

Despite the obvious advantages offered by aquatic systems, there are numerous examples in the marine and freshwater literature of experimental designs that are far from satisfactory, especially in many of the earlier studies.

In some cases, experiments with no replication of treatments or control plots are above criticism, because the authors have refrained from using inferential statistics, or accepted that their analyses cannot be generalized (cf. McArdle, 1996). More interestingly, some of the most convincing and exciting field experiments carried out to date had no treatment replication and/or no equivalent control areas. These include the now classic large-plot removal experiments carried out to assess the effects of starfish (Paine, 1974) and limpets (Lodge, 1948) on the organization of marine intertidal assemblages.

Bob Paine's extensive work on predator removals in shallow-water marine systems has made a major contribution to ecological science over the past 30 years (reviewed in Paine, 1980, 1994; Raffaelli and Hawkins, 1996), and has given birth to lasting concepts, such as keystone species. Although in danger of becoming heresies in the wrong hands (Mills *et al.,* 1993; Raffaelli and Hawkins, 1996; Hurlbert, 1997; Simberloff, 1998), these concepts have attained paradigm status in mainstream ecology. Yet close reading of Paine's earlier papers which formed the basic of the keystone paradigm shows that the initial experiment involving removal of the starfish *Pisaster ochraceous* at Mukkaw Bay, Washington State, USA, had a single experimental plot of 20.3 m^2 and a single control plot of 13.8 m^2 (Paine, 1974); that is, neither control nor experimental areas were replicated and they were of very different size. In a second experiment at Tatoosh Island, there was a single experimental (starfish removal) area of 160 m^2, but no obvious control. Similarly, Lodge's (1948) much earlier limpet (*Patella vulgata*) removal experiment on the Isle of Man, UK, involving clearing a single 10-m wide strip of shore from the high to the low tide level, provided many insights into the dynamics of zonation, succession and the organization of intertidal assemblages (Raffaelli and Hawkins, 1996). Here, too, there was a single experimental plot and no obvious pre-defined control area.

Despite scant attention to the niceties of experimental design and problems of pseudo-replication, both studies have maintained their standing in the ecological literature and few researchers would be prepared to argue over the results obtained. The monopolization of primary space by the mussel *Mytilus californianus* and a reduction in species richness of other primary space occupiers in Washington State, and a dramatic 'greening' of the shore by opportunistic benthic algae on the Isle of Man, are now well accepted outcomes of removing starfish and limpets, respectively, on these shores. Yet it is difficult to imagine that these experiments would be accepted for publication today because of their seemingly poor experimental design. The power of these experiments to convince us lies not in their experimental designs, but in their dramatic outcomes, which are so completely at odds with the 'normal' pattern known to occur over large spatiotemporal scales. In other words, the normal zonation patterns and community composition that had been painstakingly and repeatedly described by earlier workers on rocky shores served as the appropriate reference or control plots for these two studies. The combination of a large effect size and the departure of the manipulated system from an accepted reference point makes the inferences of these two experiments as compelling as the results of any maticulously designed experiment.

However, the experiments of Paine and Lodge are unusual: often the 'normal' ecological pattern is not so well documented and experimental effects not so dramatic. If inferences from poorly designed large-plot terrestrial experiments are to be convincing, the effect size must be large and the effect clearly at odds with a well documented reference situation. Sadly, neither of these conditions is usually met; effect sizes are often much smaller and probably within what most terrestrial ecologists would expect as normal background variation. Sometimes, however, large-plot experiments in terrestrial systems are as convincing as those of Paine or Lodge. For example, Moss et al. (1996) removed territorial male grouse (*Lagopus scotticus*) from a single (i.e. unreplicated) area of moorland in the Scottish Highlands in order to examine the hypothesis that the status of male birds was associated with grouse population cycles. The effect size in this experiment was large (the decline phase of the cycle was clearly disrupted) and the outcome was quite different from the normal (and well accepted) ecological pattern (a well defined temporal cycle). The reader is therefore totally convinced of Moss et al.'s (1996) results, despite the limited experimental design.

In discussing these three experiments, we are not making a case for ignoring the requirements of a sound experimental design, but are merely pointing out that occasionally experiments are convincing *despite* their design. One should always attempt to design an experiment so that it will provide a valid test of a hypothesis, so that any inferences regarding the outcome will convince the wider community. One wonders how sympathetic the ecological research community would have been if the effect sizes recorded by Paine (1974), Lodge (1948) and Moss et al. (1996) had been small, and the authors had inferred there had been no detectable effects of the removals. Arguing the case for a zero effect will be a particular problem for large-plot terrestrial experiments, where the design may well be compromised. Experiments done to explore contentious management issues, such as whether non-target species will be affected by the introduction of a pathogen to control a pest (Moller and Raffaelli, 1998) or the removal of a wildlife host species to prevent disease transmission to farm-stock (Krebs, 1998), will be difficult to interpret and defend to the wider public if a zero-effect outcome is claimed. At the very least, an *a posteriori* power analysis should be carried out (e.g. Hall et al., 1990; Raffaelli and Hall, 1992) in order to establish whether that design could have detected an effect if one had occurred.

C. For How Long Should Experiments Run?

Power analysis can be performed on the results of a pilot experiment to estimate the likelihood of detecting a stated effect size for a given number of replicates. A full-scale experiment can then be designed appropriately or the idea abandoned altogether if the conclusions of the power analyses are too daunting. Such analyses indicate whether the experimental plots are sufficiently well replicated. A much less tractable issue in experimental design is for how long to run the experiment. In our interviews with those engaged in large-plot terrestrial field experiments (Moller and Raffaelli, 1998), more than half of the interviewees did not state why they had chosen a particular duration. (This is also true for many of the studies in Table 2.) Most of the remainder gave pragmatic reasons (e.g. length of funding cycle or a management obligation to continue) but, encouragingly, six researchers based the duration of their experiment on some aspect of the population dynamics of the focal species, (i.e. an ecological criterion). We suspect that a similar diversity of reasons would be given by aquatic ecologists, although in aquatic systems the length of the funding cycle (3–5 years) may well be considered appropriate for the dynamics of the focal experimental species chosen. Our own extensive experience with the marine shallow-water research community suggests that the duration of an experiment may be decided on the basis of previous work, the designs used by other workers, the length of time for which a manipulated species can be expected to survive within an enclosure, and the likelihood of the experimental structure, such as a cage, surviving storms or suffering artefacts, such as clogging of the cage mesh, which would confound interpretation of the data. Occasionally, the duration is decided by nature itself, as when cages are damaged or lost altogether. We also suspect that aquatic ecologists are not alone in sometimes adding artificially high densities of animals to enclosures in an attempt to obtain a larger effect over a shorter time.

1. Do Short- and Long-term Effects Differ?

Whilst few researchers state why a particular duration was chosen, the length of an experiment is likely to be critical for its interpretation since the direction and magnitude of a response to perturbation may be quite different in the short and long term (Yodzis, 1988). Operational definitions of 'short' and 'long' in this sense must always remain subjective, although Yodzis (1988) has produced a 'rule-of-thumb' for predicting the time to final equilibrium following a press perturbation. This stopping rule is essentially twice the summation of the generation times of all the species in the longest trophic path which links the species being perturbed (e.g. a predator) and the species of interest (e.g. a prey). If this rule is followed, the time to equilibrium for aquatic systems will, on average, be relatively fast, since experimenters have tended to choose relatively short-lived species and the interaction of interest is usually direct, involving one or two trophic links at the most. Marine shallow-water experiments tend to have a rather limited interaction web, probably due to a combination of the small areas (plots) manipulated and the inherently simple structure of these assemblages. In contrast, animals of interest in terrestrial experiments tend to be longer lived and embedded within a more complex interaction web, so that the number of links between the perturbed species and the response species of interest may be greater. We acknowledge that these situations represent extremes of a continuous range of longevities and web complexities, but the point is well illustrated by examples from our own research.

Moller and Raffaelli (1998) considered the design of a pilot experiment aimed at assessing the impact of removing the European rabbit from a New Zealand food web (Figure 10). The response organism of interest is the giant skink, an indigenous species. Summation of the generation times of all the species in the longest trophic pathway between the perturbed and response species (i.e. rabbits, tussock grass, insects and skinks) gave an estimate of 25 years. Applying Yodzis' rule-of-thumb, the experiment would need to run for 50 years to ensure reaching post-perturbation equilibrium. Needless to say, we did not attempt the experiment. In another experiment, Raffaelli *et al.* (1989) evaluated the effects of predation by the shore-crab *Carcinus maenas* on its invertebrate prey (e.g. *Macoma balthica*, *Hydrobia ulvae* and *Corophium volutator*) living in mudflats of the Ythan estuary, Aberdeenshire. The plots were small ($\simeq 0.1 \text{m}^2$), but of the appropriate scale for this interaction web (Figure 11). The trophic pathway in all these interactions is direct (one link) and the summation of the generation times is about 3 years. Applying Yodzis' rule-of-thumb, the experiment would have to run for only 5–6 years, an order of magnitude less than in our New Zealand terrestrial example. Evaluating the effects of shorebirds

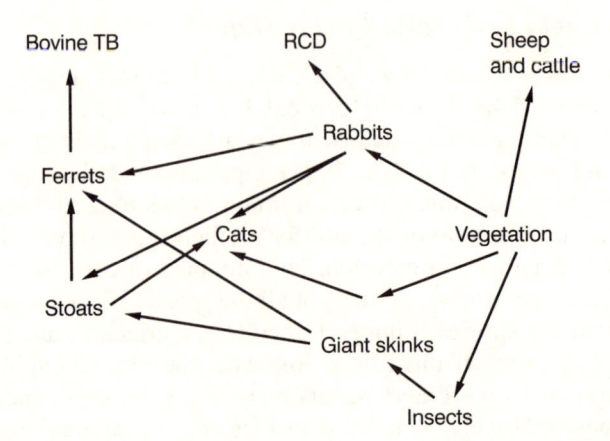

Fig. 10. Interaction web of a putative field experiment designed to detect the effects of removing rabbits on the abundance of giant skinks in a New Zealand pastoral system. From Moller and Raffaelli (1998). (RCD, Rabbit Calicivirus Disease; TB, Tuberculosis).

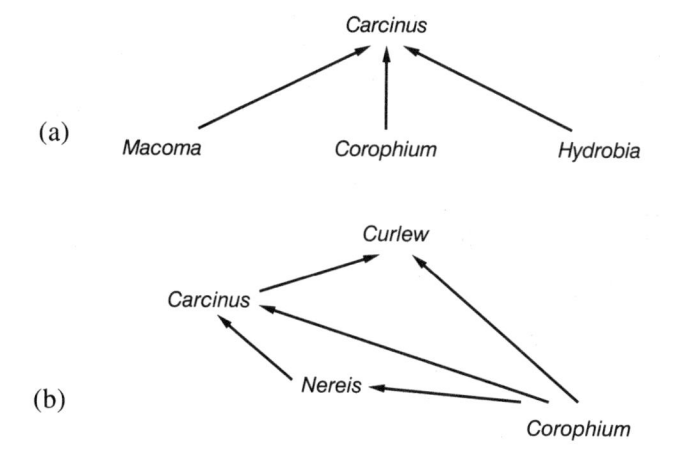

Fig. 11. Interaction webs from two field experiments on the Ythan estuary, Scotland, investigating (a) the effects of the shore-crab (*Carcinus*) on mudflat invertebrates (Raffaelli *et al.*, 1989) and (b) the effects of a shorebird, the curlew (*Numenius arquata*), on invertebrates (Raffaelli and Milne, 1987).

on prey abundance in the Ythan food web (Raffaelli and Milne, 1987) inevitably demands larger plots, and the interaction web will be more complex (Figure 11). Here, the longest trophic pathway is curlew (*Numenius arquata*), crab (*C. maenas*), *Nereis*, and *Corophium*, and the summation of the generation times for these species is about 15 years, so that an experiment on curlew exclusion should run for about 30 years.

Applying Yodzis' rule respectively to our Ythan experiments has been a depressing experience. In all our predator manipulations in this system the experiments were carried out for at least an order of magnitude *less* than the time demanded by the rule-of-thumb (Raffaelli and Hall, 1992, 1995). According to Yodzis' rule, none of our experiments could be expected to have reached post-perturbation equilibrium. The durations of our experiments are fairly typical of those in the literature, but this is cold comfort. At the most, we can only make inferences concerning short-term effects, a point also made by Tilman (1989). He noted that of the 180 field experiments he reviewed, only 7% had a duration of more than 5 years, whilst 40% ran for less than 1 year. It is sometimes argued that Yodzis' rule-of-thumb may be unrealistic, but in the absence of evidence to the contrary it is the only guide we have.

Of course, if the short-term and longer-term effects of an experimental manipulation are similar, then we need not wring our hands. The outcome of short-term experiments will be at least qualitatively similar to those from long-term experiments. However, there are examples where this is clearly not the case. Paine's (1974) starfish removal experiments revealed that, if the perturbation is maintained for sufficiently long, the response species (mussel) is allowed to grow into a prey size-refuge, at which point the starfish is unable to alter the assemblage. Before the point in time when mussels enter this size-refuge, the assemblage can be viewed as globally stable: relaxation of the perturbation would probably result in the assemblage reverting to its initial (reference) structure and composition. Once mussels are too large for starfish to predate, the system has moved to an alternative stable state and recovery to the original (reference) assemblage is possible only if large mussels are removed by another perturbation, such as a storm. Thus, the short- and long-term effects of starfish removal can be quite different. Similar ecological changes seem to have occurred through a natural perturbation in South African shallow-water assemblages (Barkai and McQuaid, 1988). Here, removal of rock lobsters (*Jasus*) by an unknown natural factor from one island allowed their usual prey, carnivorous whelks, to grow into a size-refuge. Attempts to reintroduce rock lobsters to this island consistently failed, not because the whelks were not too large to eat, but because they were now sufficiently large and numerous to kill the rock lobsters (Barkai and McQuaid, 1988). Few field experiments run for sufficiently long for the kinds of effects observed by Paine (1974) and Barkai and McQuaid (1988) to become apparent, although there is compelling evidence that species removals in other systems could flip that system into an alternative stable state (e.g. Dublin *et al.*, 1990). Clearly, it may not be wise to assume that short-term experiments will have similar outcomes to those with a much longer duration.

D. Issues of Scale

Only relatively recently have serious attempts been made formally to assess scaling issues with respect to the design and interpretation of ecological experiments (Englund and Olsson, 1996; Legendre, 1993; Thrush *et al.*, 1995, 1996, 1997a,b, in preparation; Schneider *et al.*, 1997), although the general problems that might arise have been acknowledged for some time (Legendre and Fortin, 1989; Wiens, 1989; Levin, 1992; Dayton, 1994; Giller *et al.*, 1994; Hall *et al.*, 1994; May, 1994; Schneider, 1994). Whilst ecological scale is explicitly referred to or implicitly indicated in many of the experiments we reviewed, it is clear that different authors use the term 'scale' in different ways. First, most discussions relate to spatial scale, whilst issues of temporal scale are ignored. We have argued above that issues of temporal scale may be very significant.

Second, spatial scale has at least three components (grain, lag and extent; Schneider, 1994), which may be relevant to the design and sampling of field experiments. Grain is the size of the sample unit used to record, for example, the abundance of a response species within a predator exclosure. Note that this is rarely equivalent to the size of the exclosure itself, especially in large-plot experiments. Lag is the distance between the sample units. Extent is the area over which the sampling or experiment occurs. Analogous definitions could be made for the lag, extent and grain of sampling or monitoring over time. It is rarely stated which of these three elements are implied in discussions of ecological scale, although most authors appear to be referring to spatial extent.

Third, and more worrying, is a tendency to confuse scale with plot size. Larger plots do not necessarily imply larger scale. Scale must be referenced to the size of organism and its behavioural attributes, such as movement rates, home-range size or foraging area (Raffaelli et al.,1994). Thus, a 10 m^2 plot can be considered a large scale for a tiny invertebrate, whereas a 1000 m^2 plot would be a small scale for a large rare vertebrate. In this respect, many of the terrestrial experiments reported in the literature are not necessarily done on larger scales than those in aquatic systems; they merely tend to employ larger plot sizes, and the ecological scales may in fact be very similar. An analogous argument could be made for short-and long-term temporal scales. In the experiments analysed by Moller and Raffaelli (1998), there were good correlations between the size of the experimental plots and the size, density, generation time and home range of the response species (Figure 12; Table 3), suggesting that, on average, large-plot (mostly terrestrial) experiments at least acknowledge the larger sizes, generation times and home ranges of the response animals. (However, if separate analyses are made for the aquatic and terrestrial experiments in Moller and Raffaelli (1998), the correlations are weaker for many of the relationships (Table 3).

Whether the trends shown in Figure 12 reflect the choice of the *appropriate* scale is a different question. Choosing the appropriate scale of an experiment is not straightforward because simple rules, recipes and guidelines do not yet exist (Schneider et al., 1997). The grain, lag and extent of an experimental design (including the monitoring of the response species) may all affect the variance in the data, and hence the power of the experiment. For instance, in marine systems, between-sample variation (a factor in power estimation) in the density of contagiously distributed benthic animals will generally be greater for small as opposed to large sample units. The extent and lag chosen should also be appropriate for the spatial heterogeneity of the response species that is monitored. Most species have heterogeneous distributions so that closely-sited samples are likely to be spatially autocorrelated (Thrush, 1991; Legendre, 1994). It is important that this heterogeneity is identified and quantified before setting out the experimental blocks in order to ensure proper independence of subsequent samples or plots (Legendre, 1994). Decisions on plot size should at least take into account the potentially confounding effects of movement of the response species out of the treatment area (Englund and Olsson, 1994; Englund, 1997).

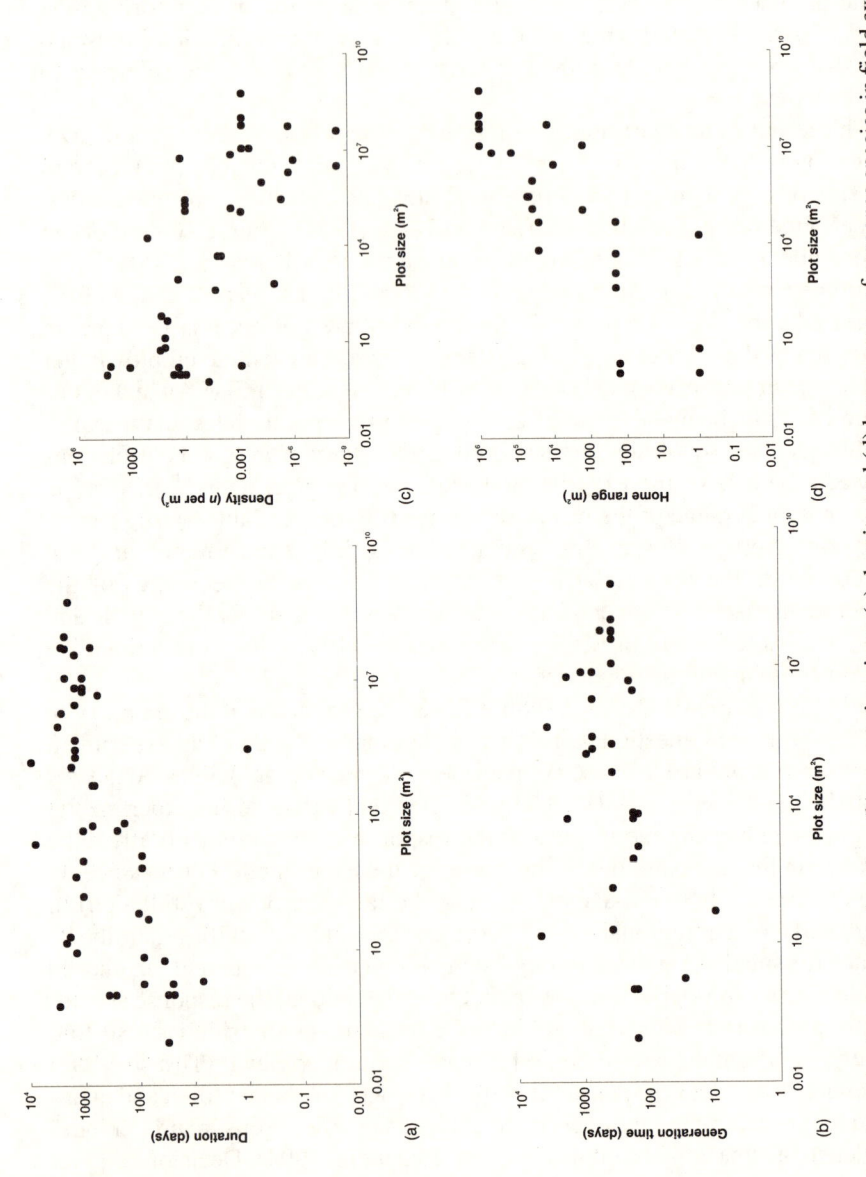

Fig. 12. Relationships between (a) duration, (b) generation time, (c) density and (d) home range of response species in field experiments reviewed by Moller and Raffaelli (1998) and listed in Table 1.

Table 3.
Spearman's rank correlation between plot size and experimental variables in experiments reviewed by Moller and Raffaelli (1998)

Plot size x	All experiments	Aquatic	Terrestrial
Duration of experiment	***	***	not significant
Generations of response species	***	not significant	not significant
Size of response species	***	–	***
Density of response species	***	*	*
Home range of response species	***	**	***

$*P = 0.05$, $**P = 0.01$, $***P = 0.001$; n.s., not significant.

E. Meta-analysis: A Way Forward?

Given the difficulty of carrying out large-plot (and large-scale) experiments with sufficient replication, techniques that permit integration of the results of repeated or similar experiments have considerable potential for gaining insights into ecological processes. ANOVA approaches have been suggested (McKone and Lively, 1993; Lively and McKone, 1994) but these have recently been questioned by Underwood (1997). A more elegant approach to meta-analysis has been described by Hunter and Schmidt (1990), Hedges and Olkin (1985), Rosenthal (1991) and Mullen (1989), whilst Gurevitch *et al.* (1992), Arnqvist and Wooster (1995) and Osenberg and St Mary (1998) have discussed its application to the analysis of field experiments. The technique allows three basic questions to be addressed to a selected set of studies (Arnqvist and Wooster, 1995). (1) What is the combined magnitude of the effect under study? (2) Is this overall effect significantly different from zero? (3) Do any characteristics of the studies influence the magnitude of the observed effect (Arnqvist and Wooster, 1995). The procedures are elegantly illustrated by Gurevitch *et al.* (1992) and by Arnqvist and Wooster, (1995); the statistics are relatively simple to compute; and a ready-made package, *MetaWin*, is available (Rosenberg *et al* 1996) to provide ecologists with a gateway for carrying out such analyses.

Whilst the potential of the approach cannot be doubted, it has been criticized at a number of levels, although most of these criticisms apply equally well to written reviews. First is the problem of potential bias in the selection of experiments to be included in a set, as these may not be representative of all experiments conducted. There is a tendency for journals to publish experiments that show significant effects, so there is likely to be an unknown number of unpublished experiments showing a zero-effect. Arnqvist and Wooster (1995) show how it is possible to estimate the likely magnitude of this unpublished zero-effect mass and to set this against the meta-analysis tolerance for null results. In other words, it permits an assessment of the 'threat' posed by unpublished studies to the analysis outcome and hence the likely robustness of the analysis. There is no equivalent process for written reviews.

The second criticism levelled at meta-analysis is that it may mix 'apples and oranges' if there is sufficient variation between the selected studies in, for instance, the habitats in which the experiments were carried out. Any generalizations that emerge will therefore be at the level of 'fruit' rather than apples or oranges (Rosenthal, 1991). Again, meta-analysis is not alone in this respect. In the present review we have been constrained by the small sample sizes of habitat divisions finer than marine, freshwater and terrestrial. This need not be a disincentive to applying meta-analysis (or making synthetic written reviews) as long as one is aware of the limitations and these are clearly stated. However, as more experiments are added to the analysis, it is likely that any inferences will be more generic and less specific, although it is still possible to evaluate differences between subgroups of studies (Gurevitch *et al.*, 1992; VanderWerf, 1992). The quite different issue of mixing 'good' and 'bad' experiments in the same analysis is to an extent coped with by estimating the power of each study and by accounting for differences in reliability across studies (Arnqvist and Wooster, 1995).

A third criticism is the perceived danger of making available such a powerful tool to researchers who may not appreciate the above and other limitations (Osenberg and St Mary, 1998). In particular, the choice of metric for estimating effect size can have a dramatic influence on the outcome of the analysis (Figure 13). Osenberg and St Mary (1998) are concerned that the choice of metric for estimating effect size in commercial packages like MetaWin is too limited and they fear that many biologists will select metrics because of their accessibility rather than whether they best quantify an effect in relation to the hypothesis being tested. They argue that the most frequently used metrics are statistical in nature and that more biologically relevant metrics should be devised (Figure 13). Finney (1995) has similar misgivings concerning the limited metrics available in computer packages. Clearly, the concerns and fears of these authors relate specifically to the potential misuse of a powerful technique, rather than a critique of the meta-analysis approach *per se*. We concur with Osenberg and St Mary (1998) that meta-analysis will have a large impact in biology, especially in evaluating the outcomes of community press experiments of the kind reviewed here, provided that sensible and appropriate metrics are used.

IV. CONCLUSIONS

Our review of a range of manipulative field experiments in animal ecology has highlighted a number of issues. It is clear that large-plot experiments tend to be more design-constrained than small-plot experiments and, in the absence of repeated experiments, it may be better not to proceed with an experimental approach for some large-plot or large-scale questions. The inferences from

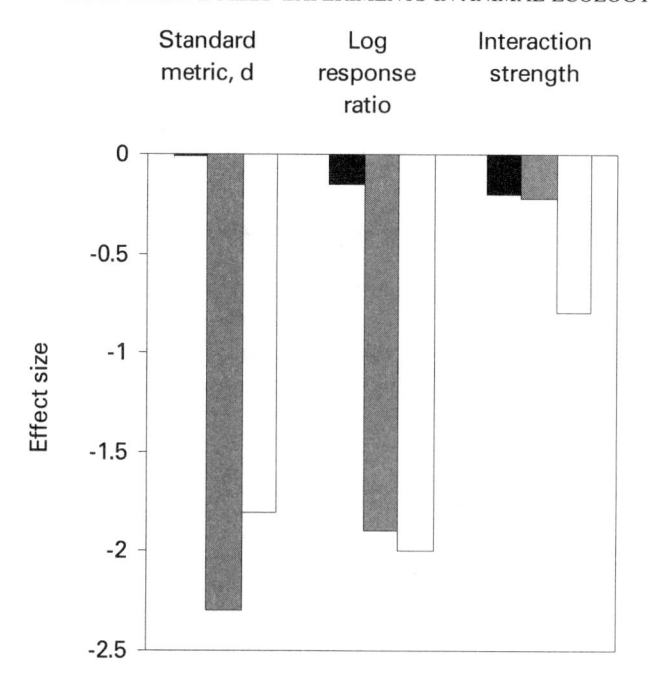

Fig. 13. Consequences of metric choice on the interpretation of meta-analysis of experiments on the effects of fish predation on snail abundance. Note that the three metrics (the standard metric, d, log response ratio and interaction strength) used to define effect size produce divergent results, in particular for mixtures (▨) of generalist (■) and specialist (☐) species. From Osenberg and St Mary (1998).

such experiments are likely to be weak and unlikely to convince either scientific peers or, in the case of experiments that address management issues, politicians and the public. In the absence of rigorous experimental examination of propositions, the researcher must fall back on weaker lines of evidence and argument, but, if these are consistent in direction, it may still be possible to present a compelling case. A second issue identified in this review is the need to design experiments at the correct spatial and temporal scales. It is not yet possible to assess the degree to which the inferences of the experiments described here can be translated across even limited spatiotemporal scales or even whether they were carried out at the correct scale for the hypothesis under investigation. In particular, the duration (extent of temporal scale) of many experiments is probably far too short (at least an order of magnitude) for the documentation of longer-term effects, which may be quite different from those recorded in the short term. Meta-analysis appears to have considerable potential for assessing the generality of effects in field experiments and may provide a valuable tool for those attempting to make inferences from experiments of variable quality. However, we are mindful that a poorly designed experiment may not be worth doing at all, and that reliance on an experimental approach alone for examining ecological processes is unwise; field experiments should be seen only as one of a larger suite of approaches to be brought to bear on ecological issues.

ACKNOWLEDGEMENTS

The authors thank Lindsey Hewitson for efforts in retrieving many of the data on the field experiments listed in Table 2 and all those authors who kindly agreed to be interviewed or who completed the questionnaire on experimental design. The manuscript benefited greatly from the constructive criticism of an anonymous reviewer.

REFERENCES

Aaser, H.F., Jeppesen, E. and Sondergaard, M. (1995). Seasonal dynamics of the mysid *Neomysis integer* and its predation on the copepod *Eurytemora affinis* in a shallow hypertrophic brackish lake. *Mar. Ecol. Prof. Ser.* **127**, 47–56.

Abramsky, Z., Rozenweig, M.L., Pinshowe, B., Brown, J.S., Kotler, B. and Mitchell, W.A. (1990). Habitat selection, an experimental field test with 2 gerbil species. *Ecology* **71**, 2358–2369.

Abramsky, Z., Strauss, E., Subachn, A., Kotler, B. and Reichman, A. (1996). The effect of barn owls on the activity and microhabitat selection of *Gerbillus allenbyi* and *G. pyramidum*. *Oecologia* **105**, 313–319.

Akbar, Z. and Gorman, M.L. (1993). The effect of supplementary feeding upon the demography of a population of woodmice *Apodemus sylvaticus*, living in a system of maritime sand-dunes. *J. Zool. (Lond.)* **230**, 609–617.

Alterio, N. and Moller, H. (1997). Daily activity of stoats (*Mustela erminea*), feral ferrets (*Mustela furo*) and feral house cats (*Felis catus*) in coastal grassland, Otago Peninsula, New Zealand. *N. Z. J. Ecol.* **21**, 89–95.

Arnqvist, G. and Wooster, D. (1995). Meta-analysis — synthesising research findings in ecology and evolution. *Tr. Ecol. Evol.* **10**, 236–240.

Baines, D., Sage, R.B. and Baines, M.M. (1994). The implications of red deer grazing to ground vegetation and invertebrate communities of Scottish native pinewods. *J. Appl. Ecol.* **31**, 776–783.

Barkai, A. and McQuaid, C. (1988). Predator–prey reversal in a marine benthic ecosystem. *Science* **242**, 62–64.

Batzer, D.P. (1998). Trophic interactions among detritus, benthic midges and predatory fish in a freshwater marsh. *Ecology* **79**, 1688–1698.

Bazly, D.R. and Jeffries, R.L. (1986). Changes in the composition and standing crop of salt marsh communities in response to the removal of a grazer. *J. Ecol.* **74**, 693–706.

Bell, S.S. and Coull, B.C. (1978). Field evidence that shrimp predation regulates meiofauna. *Oecologia* **35**, 141–148.

Belovsky, G.E. and Slade, J.B. (1993). The role of vertebrate and invertebrate predators in a grasshopper community. *Oikos* **96**, 193–201.

Bender, E.A., Case, T.J. and Gilpin, M.E. (1984). Perturbation experiments in community ecology: theory and practice. *Ecology* **65**, 1–13.

Benedetti-Cecchi, L. and Cinelli, F. (1997). Confounding in field experiments: direct and indirect effects of artifacts due to the manipulation of limpets and algae. *J. Exp. Mar. Biol. Ecol.* **209**, 171–184.

Berlow, E.L. and Navarrete, S.A. (1997). Spatial and temporal variation in rocky intertidal community organization: lessons from repeating field experiments. *J. Exp. Mar. Biol. Ecol* **214**, 195–229.

Berteaux, D., Bergeron, J.M., Thomas, D.W. and Lapierre, H. (1996). Solitude versus gregariousness: do physical benefits drive the choice in overwintering meadow voles? *Oikos* **76**, 330–336.

Beukers, J.S. and Jones, G.P. (1998). Habitat complexity modifies the impact of piscivores on a coral reef fish population. *Oecologia* **114**, 50–59.

Biondini, M.E. and Manske, L. (1996). Grazing frequency and ecosystem processes in a northern mixed prairie, USA. *Ecol. Appl.* **6**, 239–256.

Bowers, M.A. (1993). Influence of herbivorous mammals on an old field plant community, years 1–4 after disturbance. *Oikos* **67**, 129–141.

Branch, L.C., Villarreal, D., Hierro, J.L. and Portier, K.M. (1996). Effects of local extinction of the plains vizcacha on vegetation patterns in semi-arid scrub. *Oecologia* **106**, 389–399.

Brown, J.H. (1999). The desert granivory experiments at Portal. In: *Issues and Perspectives in Experimental Ecology* (Ed. by J. Resetarits and W. Bernrado. Oxford University Press, Oxford.

Brown, V.K. (1994). Herbivory: a structuring force in plant communities. In: *Individuals, Populations and Patterns in Ecology* (Ed. by S.R. Leather, A.D. Watt, N.J. Mills and K.F.A. Walters), Intercept, Andover, UK.

Bukaveckas, P. and Shaw, W. (1998). Phytoplankton responses to nutrient and grazer manipulations among northeast lakes of varying pH. *Can. J. Fish. Aq. Sci.* **55**, 958–966.

Caley, P. (1995). Effects of ferret control on cattle reactor incidence and on the rate of increase of rabbit populations. *Year One Progress Report*. Landcare Research Contract Report LC9596/56, Wellington.

Carpenter, S.R., Chisholm, S.W., Krebs, C.J., Schindler, D.W. and Wright, R.F. (1995). Ecosystem experiments. *Science* **269**, 324–327.

Chase, J.M. (1996). Abiotic controls of trophic cascades in a simple grassland food chain. *Oikos* **77**, 495–506.

Churchfield, S., Holier, J. and Brown, V.K. (1991). The effects of small mammal predators on grassland invertebrates investigated by a field exclosure experiment. *Oikos* **60**, 283–290.

Clifton-Hadley, R.S., Wilesmith, J.W., Richards, M.S., Upton, P. and Johnson, S. (1995). The occurrence of *Mycobacterium bovis* infection in cattle in and around an area subjected to extensive badger (*Meles meles*) control. *Epidemiol. Infect.* **114**, 179–193.

Connell, J.H. (1961). The influence of interspecific competition and other factors on the distribution of the barnacle *Chthamalus stellatus*. *Ecology* **42**, 710–773.

Connell, J.H. (1983). On the prevalence and relative importance of interspecific competition: evidence from field experiments. *Am. Nat.* **122**, 661–696.

Crowder, L.B., Wright, R.A., Rose, K.A., Martin, T.H. and Rice, J.A. (1994). Direct and indirect effects of flounder predation on a spot population: experimental and model analyses. In: *Theory and Application in Fish Feeding Ecology* (Ed. by D. J. Stouder, K.L. Fresh and R.J. Fellers), University of South Carolina Press, Columbia.

Dale, M.R.T. and Zbigniewicz, M.W. (1997). Spatial pattern in boreal shrub communities: effects of a peak in herbivore density. *Can. J. Bot.* **75**, 1342–1348.

Dayton, P.K. (1994). Community landscape: scale and stability in hard-bottom communities. In: *Aquatic Ecology: Scale, Pattern and Process* (Ed. by P. Giller, A. Hildrew and D. Raffaelli), pp. 289–332. Blackwell Science, Oxford.

Dilks, P.J. and Wilson, P.R. (1979). Feral sheep and cattle and royal albatrosses on Campbell Island; population trends and habitat changes. *N. Z. J. Zool.* **6**, 127–139.

Dublin, H.T., Sinclair, A.R.E. and McGlade, J. (1990). Elephants and fire as causes of multiple stable states in the Serengeti-Mara woodlands. *J. Am. Ecol.* **59**, 1147–1164.

Edgar, G.J. and Robertson, A.L. (1992). The influence of seagrass structure on the disribution and abundance of mobile epifauna pattern and process in a western Australian amphibolis bed. *J. Exp. Mar. Biol. Ecol.* **160**, 13–31.

Englund, G. (1997). Importance of spatial scale and prey movements in predator caging experiments. *Ecology* **78**, 2316–2325.

Englund, G. and Olsson, T. (1996). Treatment effects in a stream fish exclosure experiment: influence of predation rate and prey movements. *Oikos* **77**, 519–528.

Erlinge, S. (1987). Predation and noncyclicity in a microtine population in southern Sweden. *Oikos* **50**, 347–352.

Eves, J. (1993). The East Offaly Badger Research Project: an interim report. In: *The Badger.* (Ed. by T.J. Hayden), Royal Irish Academy, Dublin.

Fairweather, P.G. and Underwood, A.J. (1991). Experimental removals of a rocky intertidal predator: variations within two habitats in the effects on prey. *J. Exp. Mar. Biol. Ecol.* **154**, 29–75.

Farrell, T.M. (1991). Models and mechanisms of succession, an example from a rocky intertidal community. *Ecol. Monogr.* **61**, 95–113.

Finney, D.J. (1995). A statistician looks at meta-analysis. *J. Clin. Epidemiol.* **48**, 87–103.

Flecker, A.S. (1992). Fish trophic guilds and the structure of a tropical stream, weak direct versus strong indirect effects. *Ecology* **73**, 927–940.

Floyd, T. (1996). Top-down impacts on creosote bush herbivores in a spatially and temporally complex environment. *Ecology* **77**, 1544–1555.

Gascon, C. (1992). Aquatic predators and tadpole prey in central amazonian field data and experimental manipulations. *Ecology* **73**, 971–980.

Giller, P., Hildrew, A. and Raffaelli, D. (1994). *Aquatic Ecology: Scale, Pattern and Process.* Blackwell Scientific, Oxford.

Gurevitch, J., Morrow, L.L., Wallace, A. and Walsh, J.S. (1992). Meta-analysis of competition in field experiments. *Am. Nat.* **140**, 539–572.

Hairston, N.G.Sr. (1989). *Ecological Experiments: Purpose, Design, and Execution.* Cambridge University Press, Cambridge.

Hall, S.J., Raffaelli, D. and Turrell, W. (1990). Predator caging experiments. A re-examination of their value. *Am. Nat.* **136**, 657–672.

Hall, S.J., Raffaelli, D. and Thrush, S. (1994). Patchiness and disturbance in shallow water benthic assemblages. In: *Aquatic Ecology: Scale, Pattern and Process* (Ed. by P. Giller, A. Hildrew and D. Raffaelli), pp. 333–376. Blackwell Science, Oxford.

Hargeby, A. (1990). Macrophyte associated invertebrates and the effect of habitat permanence. *Oikos* **57**, 338–346.

Hartley, S.E. (1999). The effects of grazing and nutrient inputs on grass–heather competition. *B. A. J. Scot.* **47**.

Hedges, L.V. and Olkin, I. (1985). *Statistical Methods for Meta-analysis.* Academic Press, San Diego.

Hermony, I. Shachak, M. and Abramsky, Z. (1992). Habitat distribution in the desert snail. *Oikos* **64**, 516–522.

Hershey, A.E. (1985). Effects of predatory sculpin on the chironmid communities in an arctic lake. *Ecology* **66**, 1131–1138.

Heske, E.J., Brown, J.H. and Minstry, S. (1994). Long term experimental study of a chihuahuan desert rodent communitiy: 13 years of competition. *Ecology* **75**, 438–445.

Huang, C. and Sih, A. (1991). An experimental study on the effects of salamander larvae on isopods in stream pools. *Freshw Biol.* **25**, 451–460.

Hubbs, A. and Boonstrana, R. (1997). Population limitation in Arctic ground squirrels: effects of food and predation. *J. Anim. Ecol.* **I66**, 527–541.

Hulme, P.E. (1996). Herbivores and the performance of grassland plants: a comparison of arthropod mollusc and rodent herbivory. *J. Ecol.* **84**, 43–51.

Hunter, J.E. and Schmidt, F.L. (1990). *Methods of Meta-analysis: Correcting Error and Bias and Research Findings*. Sage, California.

Hurd, L.E. and Eisenberg, R.M. (1990). Arthropod community responses to manipulation of a bitrophic predator guild. *Ecology* **71**, 2107–2114.

Hurlbert, S.H. (1984). Pseudoreplication and the design of ecological field experiments. *Ecol. Monogr.* **54**, 187–211.

Hurlbert, S.H. (1997). Functional importance vs. keystoness: reformatting some questions in theoretical biocenology. *Aust. J. Ecol.* **22**, 369–382.

James, R.D. and Underwood, A.J. (1994). Influence of colour of substratum on recruitment of spirobid tubeworms to different types of intertidal boulders. *J. Exp. Mar. Biol. Ecol.* **181**, 105–115.

Jaquet, N. and Raffaelli, D. (1989). The ecological importance of the sand goby *Pomatoschistus minutus (Pallas). J. Exp. Mar. Biol. Ecol.* **128**, 147–156.

Joern, A. (1986). Experimental study of avian predation on co-existing grasshopper populations *Orthoptera acrididae* in a sandhills grassland. *Oikos* **46**, 243–249.

Joern, A. (1992). Variable impact of avian predation on grasshopper assemblages in sandhills grassland. *Oikos* **64**, 458–463.

Juliano, S.A. (1998). Species introduction and replacement among mosquitos: interspecific resource competition or apparent competition. *Ecology* **79**, 255–268.

Kareiva, P. and Anderson, M. (1989). Spatial aspects of species interactions: the wedding of models and experiments. In: *Community Ecology* (Ed. by A. Hastings). Springer, New York.

Kennedy, A.D. (1993). Minimal predation upon meiofauna by endobenthic macrofauna in the Exe Estuary, south west England. *Mar. Biol.* **117**, 311–319.

Kerans, B.L., Peckarsky, B.L. and Anderson, C. (1995). Estimates of mayfly mortality. Is stonefly predation a significant source? *Oikos* **7**, 315–323.

Kielland, K., Bryant, J. and Ruess, R.W. (1997). Moose herbivory and caron turnover of early successional stands in interior Alaska. *Oikos* **80**, 25–30.

Klemola, T., Koivula, M., Korpimaki, E. and Norrdahl, K. (1997). Small mustelid predation slows population growth of *Microtus* voles: a predator reduction experiment. *J. Anim. Ecol.* **66**, 607–614.

Krebs, C.J., Boutin, S., Boonstra, R., Sinclair, A.R.E., Smith, J.N.M., Dale, M.R.T., *et al.* (1995). Impact of food and predation on the snowshoe hare cycle. *Science* **269**, 1112–1115.

Krebs, J.R. (1998). *Bovine Tuberculosis in Cattle and Badgers*. MAFF Publications, London.

Kuhlmann, M.L. (1994). Indirect effects of a predatory gastropod in a seagrass community. *J. Exp. Mar. Biol. Ecol.* **183**, 163–178.

Lagos, V.O., Contreras, L.C., Meserve, P.L. and Jaksic, F.M. (1995). Effects of predation risk on space use by small mammals: a field experiment with a neotropical rodent. *Oikos* **74**, 259–264.

Lagrange, T.G., Hansen, J.L., Andrews, R.D, Hancock, A.W. and Kienzler, J.M. (1995). Electric fence predator exclosure to enhance duck nesting: a long term case study in Iowa. *Wildl. Soc. Bull.* **23**, 261–266.

Lawton, J. (1996). Patterns in ecology. *Oikos* **75**, 145–147.

Legendre, P. (1993). Spatial autocorrelation: trouble or new paradigm? *Ecology* **74**, 1659–1673.

Legendre, P. and Fortin, M.J. (1989). Spatial pattern and ecological analysis *Vegetatio* **80**, 107–138.

Leibold, M.A. (1989). Resource edibility and the effects of predators and productivity on the outcome of trophic interactions. *Am. Nat.* **134**, 922–949.

Levin, S.A. (1992). The problem of scale in ecology. *Ecology* **73**, 1943–1967.

Lewis, S.M. (1986). The role of herbivorous fishes in the organization of a caribbean reef community. *Ecol. Monogr.* **56**, 183–200.

Lively, C.M. and McKone, M.J. (1994). Choosing an appropriate ANOVA for experiments conducted at a few sites. *Oikos* **69**, 335.

Lodge, S.M. (1948). Algal growth in the absence of *Patella* on an experimental strip of foreshore, Port St Mary, Isle of Man. *Proc. Trans. Live Biol. Soc.* **56**, 78–83.

Malhotra, A., Thorpe, R.S. (1993). An experimental field study of a eurytopic anole *Anole oculatus*. *J. Zool.* **229**, 163–170.

Marsh, C.P. (1986). Rocky intertidal community organization the impact of avian predators on mussel recruitment. *Ecology* **67**, 771–786.

Marshall, P.A. and Keogh, M.J. (1994). Asymmetry in intraspecific competition by the limpet *Cellana tramoserica*. *J. Exp. Mar. Biol. Ecol.* **177**, 121–138.

Martin, A.J., Seaby, R.M. and Young, J.O. (1994). Food limitation in lake dwelling leeches: field experiments. *J. Anim. Ecol.* **63**, 93–100.

Mattila, J. and Bonsdorff, E. (1989). The impact of fish predation on shallow soft bottoms in brackish waters South West Finland: an experimental study. *Neth. J. Sea Res.* **23**, 69–81.

May, R.M. (1994). The effects of spatial scale on ecological questions and answers. In: *Large-scale Ecology and Conservation Biology* (Ed. by P.J. Edwards, R.M. May, and N.R. Webb). pp. 1–17. Blackwell Science, Oxford.

McArdle, B.H. (1996). Levels of evidence in studies of competition, predation and disease. *N. Z. J. Ecol.* **20**, 7–15.

McCarthy, J. (1993). The badger vaccination trial in west Cork: progress report. In: *The Badger* (Ed. by T.J. Hayden), Royal Irish Academy, Dublin.

McKone, M.J. and Lively, C.M. (1993). Statistical analysis of experiments conducted at multiple sites. *Oikos* **67**, 184–186.

Mills, L.S., Soule, M.E. and Doak, D.F. (1993). The keystone concept in ecology and conservation. *BioScience* **43**, 219–224.

Moen, J. and Oksanen, L. (1998). Long term exclusion of folivorous mammals in 2 arctic-alpine plant communities: a test of the hypothesis of the exploitation hypothesis ecosystems. *Oikos* **82**, 333–346.

Moller, H. and Raffaelli, D. (1998). Predicting risks from new organisms: the potential of community press experiments. In: *Statistics in Ecology and Environmental Monitoring: Risk Assessment and Decision Making in Biology* (Ed. by D.J. Fletcher, L. Kavalieris and B.J.F. Manly), pp. 131–156. Otago University Press, Dunedin.

Moller, H., Tilley, J.A.V., Thomas, B.W. and Gaze, P.D. (1991). Effects of introduced social wasps on the standing crop of honeydew in New Zealand beech forests. *N. Z. J. Zool.* **18**, 171–179.

Montgomery, W.I., Wilson, W.L. and Elwood, R.W. (1997). Spatial regulation and population growth in the woodmouse *Apodemus sylvaticus*: experimental manipulations of males and females in natural populations. *J. Anim. Ecol.* **66**, 755–768.

Moran, M.D. and Hurd, L.E. (1994). Short term responses to elevated predator densities: non-competitive intraguild interactions and behaviour. *Oecologia* **98**, 269–273.
Moss, R., Watson, A. and Parr, R. (1996). Experimental prevention of a population cycle in red grouse. *Ecology* **77**, 1512–1530.
Mullen, B. (1989). *Advanced BASIC Meta-analysis.* Lawrence Erlbaum.
Norrdahl, K. and Korpimaki, E. (1995). Effects of predator removal on vertebrate prey populations: birds of prey and small mammals. *Oecologia* **103**, 241–248.
Oksanen, L. and Moen, J. (1994). Species-specific plant responses to exclusion of grazers in three Fennoscandian tundra habitats. *Ecoscience* **1**, 31–39.
Osenberg, C.W. and St Mary, C.M. (1998). Meta-analysis: synthesis or statistical sugjugation? *Integr. Biol.* **1**, 37–41.
Ouellet, J., Boustinstan, P. and Heard, D.C. (1994). Responce to simulated grazing and browsing of vegetation available to caribou in the Arctic. *Can. J. Zool.* **72**, 1426–1435.
Paine, R.T. (1966). Food web complexity and species diversity. *Am. Nat.* **100**, 65–75.
Paine, R.T. (1974). Intertidal community structure. Experimental sudies on the relationship between a dominant predator and its principal predator. *Oecologia* **15**, 93–120.
Paine, R.T. (1980). Food webs: linkage, interaction strength and community infrastructure. *J. Anim. Ecol.* **49**, 667–685.
Paine, R.T. (1994). *Marine Rocky Shores. An Experimentalist's Perspective.* Nordbruite, Ecology Institute. Oldendorf, Germany.
Parmenter, R.R. and Macmahon, J.A. (1988). Factors influencing species composition and population sizes in a ground beetle community. Carabidae predation by rodents. *Oikos* **52**, 350–356.
Pastor, J., Dewey, B., Naiman, R.J., Mcinnes, P.F. and Cohen, Y. (1993). Moose browsing and soil fertility in the boreal forests of Isle Royale National Park. *Ecology* **74**, 467–480.
Patterson, I.J. and Fuchs, R.M.E. (1996). Management of grassland for grazing waterfowl at the Loch of Strathbeg, Grampian: 1995/96. Unpublished report to Royal Society Protection of Birds, Edinburgh.
Patterson, I.J. and Laing, R.M. (1995). Nesting distribution and nesting success of eiders on Forvie National Nature Reserve, 1995. Unpublished report to Scottish Natural Heritage.
Pennings, S.C. (1990). Size related shifts in herbivory specialization in the sea hare *Aplysia californica. J. Exp. Mar. Biol. Ecol.* **142**, 43–62.
Petersen, C.G., Weibel, A.C., Grimm, N.B. and Fisher, S.G. (1994). Mechanisms of benthic algal recovery following spates: comparison of stimulated and natural events. *Oecologia* **98**, 280–290.
Posey, M.H. and Hines, A.H. (1991). Complex predator prey interactions within an estuarine benthic community. *Ecology* **72**, 2155–2169.
Power, M.E. (1990). Effects of fish in river food webs. *Science* **250**, 811–814.
Pringle, C.M. and Hamazaki, T. (1998). The role of omnivory in a neotropical stream: Separating diurnal and nocturnal effects. *Ecology* **79**, 269–280.
Proulx, M., Pick, F.R., Mazumber, A., Hamilton, P.B. and Lean, D.R.S. (1996). Effects of nutrients and planktivorous fish on the phytoplankton of shallow and deep aquatic systems. *Ecology* **77**, 1556–1572.
Quammen, M.L. (1984). Predation by shorebirds, fish and crabs on invertebrates in inter-tidal mudflats: an experimental test. *Ecology* **65**, 529–537.
Raffaelli, D. and Hall, S.J. (1992). Compartments and predation in an estuarine food web. *J. Anim. Ecol.* **61**, 551–560.

Raffaelli, D. and Hall, S.J. (1995). Assessing the relative importance of trophic links in food webs. In: *Food Webs: Patterns and Processes* (Ed. by G. Polis and K. Winemuller). pp. 185–191. Chapman and Hall, New York.

Raffaelli, D. and Hawkins, S. (1996). *Intertidal Ecology*. Chapman and Hall, London.

Raffaelli, D.G. and Milne, H. (1987). An experimental investigation of the effects of shorebirds and flatfish on estuarine invertebrates. *Est. Coast. Shelf Sci.* **24**, 1–13.

Raffaelli, D.G., Conacher, A., McLachlan, H. and Emes, C. (1989). The role of epibenthic crustacean predators in an estuarine food web. *Est. Coast. Shelf Sci.* **28**, 149–160.

Raffaelli, D., Hildrew, A. and Giller, P. (1994). Scale, pattern and process in aquatic systems: concluding remarks. In: *Aquatic Ecology: Scale, Pattern and Process* (Ed. by P. Giller, A. Hildrew and D. Raffaelli). pp. 601–605. Blackwell Science, Oxford.

Redfield, J.A., Taitt, M.J. and Krebs, C.J. (1978). Experimental alteration of sex ratios in populations of *Microtus townsendii*, a field vole. *Can. J. Zool.* **56**, 17–27.

Rosenberg, M.S., Adams, D.C. and Gurevitch, J. (1996). *MetaWin: Statistical Software for Meta-analysis with Resampling Tests*. Version 1.0. Sinaur Associates, Sunderland, Massachusetts.

Rosenthal, R. (1991). *Meta-analytic Procedures for Social Research Synthesis*. Bussel Sage Foundation.

Sanders, M.D. (1996). Effects of fluctuating lake levels and habitat enhancement on black stilts (*Himantopus novaezelandiae* Gould, 1841). *PhD thesis*, University of Canterbury.

Sarnelle, O.J. (1993). Herbivore effects on phytoplankton succession in a eutrophic lake. *Ecol. Monogr.* **63**, 129–149.

Scheibling, R.E. and Hamm, J. (1991). Interactions between sea urchins and their predators in field and laboratory experiments. *Mar. Biol.* **110**, 105–116.

Schimmel, J. and Granstrom, A. (1996). Fire severity and vegetation response in a boreal Swedish forest. *Ecology* **77**, 1436–1450.

Schmitz, O.J. (1998). Direct and indirect effects of predation and predation risk in old field interaction webs. *Am. Nat.* **151**, 327–342.

Schneider, D.C. (1994). Scale-dependent patterns and species interactions in marine nekton. In: *Aquatic Ecology: Scale, Pattern and Process*. (Ed. by P. Giller, A. Hildrew and D. Raffaelli), pp. 441–468. Blackwell Science, Oxford.

Schneider, D.C., Walters, R., Thrush, S.F., Dayton, P.K. (1997). Scale-up and ecological experiments: density variation in the mobile bivalve *Macomona liliana J. Exp. Mar. Biol. Ecol.* **216**, 219–152.

Schoener, T. (1983). Field experiments on interspecific competition. *Am Nat.* **122**, 240–285.

Schulte-Hostedde, A.I. and Brooks, R.J. (1997). An experimental test of habitat selection by rodents of Algoquin Park. *Can. J. Zool* **75**, 1989–1993.

Sih, A., Crowley, P., McPeek, M., Petranka, J. and Strohmeier, K. (1985). Predation, competition, and prey communities: a review of field experiments. *Ann. Rev. Ecol. Syst.* **16**, 269–311.

Simberloff, D. (1998). Flagships, umbrellas and keystones: is single-species management passe in the landscape era? *Biol. Cons.* **183**, 247–257.

Spiller, D.A. and Schoener, T.W. (1994). Effects of top and intermediate predators in a terrestrial food web. *Ecology* **75**, 182–196.

Steele, M.A. (1998). The relative importance of predation and competition in 2 reef fishes. *Oecologia* **115**, 222–232.

Taitt, M.J. and Krebs, C.J. (1983). Predation, cover, and food manipulations during a spring decline of *Microtus townsendii*. *J. Anim. Ecol.* **52**, 837–848.

Thrush, S.F. (1991). Spatial patterns in soft bottom communities. *Tr. Ecol. Evol.* **6**, 75–78.
Thrush, S.F., Hewitt, J.E., Cummings, V.J. and Dayton, P.K. (1995). The impact of habitat disturbance by scallop dredging on marine benthic communities: what can be predicted from the results of experiments? *Mar. Ecol. Prog. Ser.* **129**, 141–150.
Thrush, S.F., Whitlach, R.B., Pridmore, R.D., Hewitt, J.E., Cummings, V.J. and Wilkinson, M.R. (1996). Scale-dependent recolonization: the role of sediment stability in a dynamic sandflat habitat. *Ecology* **77**, 2472–2487.
Thrush, S.F., Pridmore, R.D., Bell, R.G., Cummings, V.J., Dayton, P.K., Ford, R. *et al.* (1997a). The sandflat habitat: scaling from experiments to conclusions. *J. Exp. Mar. Biol. Eco.* **216**, 1–9.
Thrush, S.F., Schneider, D.C., Legendre, P., Whitlach, R.B., Dayton, P.K., Hewitt, J.E. *et al.* (1997b). Scaling–up from experiments to complex ecological systems: where to next? *J. Exp. Mar. Biol. Eco.* **216**, 243–254.
Thrush, S.F., Hewitt, J.E., Cummings, V.J., Green, M.O., Funnell, G.A. and Wilkinson, M.R. (1998). Improving the generality of field experiments: the interaction of processes operating over different spatial scales on intertidal sandflats. (In Prep).
Tilman, D. (1989). Ecological experimentation: strengths and conceptual problems. In: *Long-term Studies in Ecology. Approaches and Alternatives.* (Ed. by G.E. Likens). Springer, New York.
Todd, C.D. and Keogh, M.J. (1994). Larval settlement in hard substratum epifaunal assemblages: a manipulative field study of the effects of substratum filming and the presence incumbents. *J. Exp. Mar. Biol. Ecol.* **181**, 159–187.
Turchin, P. and Ostfeld, R.S. (1997). Effects of density and season on the population rate of change in the meadow vole. *Oikos* **78**, 355–361.
Tyler, C.M. (1995). Factors contributing to post fire seedling establishment in chapparal: direct and indirect effects of fire. *J. Ecol.* **83**, 1009–1020.
Underwood, A.J. (1981). Techniques in analysis of variance in experimental marine biology and ecology. *Ann. Rev. Oceanogr. Mar. Biol.* **19**, 513–605.
Underwood, A.J. (1997). *Experiments in Ecology: Their Logical Design and Interpretation Using Analysis of Variance.* Cambridge University Press, Cambridge.
Underwood, A.J. and Chapman, M.G. (1992). Experiments on topographic influences on density and dispersion of *Littorina unifasciata* in New South Wales. In: *Proceedings of the Third International Symposium on Littorinid Biology* (Ed. by J. Grahame, P.J. Mill and D.G. Reid). The Malacological Society of London, London.
Valentine, J.F. and Heck, K.L. Jr. (1991). The role of sea urchin grazing in regulating subtropical seagrass meadows, evidence from field manipulations in the northern gulf of Mexico. *J. Exp. Mar. Biol. Ecol.* **154**, 215–230.
van Buskirk, J. and Smith, D.C. (1991). Density dependent population regulation in a salamander. *Ecology* **72**, 1747–1756.
VanderWerf, E. (1992). Lack's clutch size hypothesis — an examination of the evidence using meta-analysis. *Ecology* **73**, 1699–1705.
Virtanen, R. (1998). Impact of grazing and neighbour removal on a heath plant community transplanted onto a snowbed, NW Finnish Lapland. *Oikos* **81**, 359–367.
Virtanen, R., Risto, R., Henttonen, H. and Laine, K. (1997). Lemming grazing and structure of a snowbed plant community, a long term experiment at Kipisjarrvi, Finnish Lapland. *Oikos* **79**, 155–166.
Walde, S.J. (1994) Immigration and the dynamics of a predator–prey interaction in biological control. *J. Anim. Ecol.* **63**, 337–346.
Walde, S.J. and Davies, R.W. (1984). Invertebrate predation and lotic prey communities evaluation of *in-situ* enclosure–exclosure experiments. *Ecology* **65**, 1206–1213.

Walters, C.J. and Holling, C.S. (1990). Large-scale management experiments and learning by doing. *Ecology* **71**, 2060–2068.

Watson, A., Moss, R. and Parr, R. (1984). Effects of food enrichment on numbers and spacing behaviour of red grouse. *J. Anim. Ecol.* **53**, 663–678.

Wiens, J.A. (1989). Spatial scaling in ecology. *Funct. Ecol.* **3**, 385–397.

Wilesmith, J.W., Little, T.W.A., Thompson, H.V. and Swan, C. (1982). Bovine tuberculosis in domestic and wild mammals in an area of Dorset. I. Tuberculosis in cattle. *J. Hyg. (Camb.)* **89**, 195–210.

Wilson, W.H. Jr. (1989). Predation and the mediation of intraspecific competition in an infaunal community in the Bay of Fundy, New Brunswick, Nova Scotia, Canada. *J. Exp. Mar. Biol. Ecol.* **132**, 221–245.

Witman, J.D. (1985). Refuges, biological disturbance and rocky subtidal community structure in New England, USA. *Ecol. Monogr.* **55**, 421–446.

Yodzis, P. (1988). The indeterminacy of ecological interactions as perceived through perturbation experiments. *Ecology* **69**, 508–515.

Fractal Properties of Habitat and Patch Structure in Benthic Ecosystems

P.E. SCHMID

I.	Summary	339
II.	Introduction	341
III.	Fractal Dimension and its Measurement	342
	A. Fractal Dimension of Boundary and Profile Lines	342
	B. Fractal Dimension of Deposition Processes	348
	C. Fractal Dimension of Surfaces	349
IV.	Fractal Properties in Riverine Ecosystems	358
	A. Large Spatial Scales	358
	B. Intermediate to Small Spatial Scales	363
	C. Fractal Dimension Across Several Spatial Scales	373
V.	Fractal Random Walks	380
VI.	Fractal Properties of Biotic Structures	387
	A. Coral Reefs	387
	B. Mussel Beds	388
	C. Colonization of Artificial Pond Weeds	388
VII.	Fractal Coexistence and Species Abundance Distribution	390
VIII.	Concluding Remarks	395
	Acknowledgements	396
	References	397
	Appendix A	401
	Appendix B	401

I. SUMMARY

Fractal geometry provides a quantitative approach that deals explicitly with spatial irregularities and scale dependency in ecology. Fractals occupy a borderline between linear geometry and complete randomness. A fractal set is composed of geometric objects whose dimension is not an integer, therefore exceeding the topological dimension. Mathematical and natural fractals are shapes whose roughness and fragmentation remains essentially unchanged at any scale. Subsets of mathematical fractal objects are infinitesimally subdivisional, each subset, however small, containing no less detail than the complete set. Unlike mathematical fractals, natural fractals are not infinitely self-similar but statistically self-similar, because natural structures are truncated at certain scales.

How the physical structure of habitats influences the organisms residing in them is a question of abiding interest to ecologists. Habitat structure has effects on aspects of community structure and population dynamics such as species diversity, population abundances, dispersal and body sizes of residing organisms. It is the complex product of factor interactions such as habitat type, surface relief, hydrology and biological activity that act over different spatial scales causing variations at the level of individuals and populations.

Habitats such as gravel sediments can be viewed as a random two-phase system consisting of a pore and a grain space. Gravel streams, for instance, are composed of sediment particles with various diameters, more numerous in proportion as the diameter decreases. Furthermore, cross-sectional profiles of substrate surfaces display relief features of different sizes and roughness due to processes such as erosion and epilithic growth. The relief features and pore/grain spaces in benthic habitats can range over several orders of magnitude, thus displaying their fractal character.

This review considers evidence for habitat-related fractal properties in both freshwater and marine benthic environments. Different approaches in applying fractal techniques will be discussed, such as methods for the measurement of the dimension of habitat surfaces. The latter is analysed either by dealing with the habitat surface in its entirety or by examining in a lower dimension the intersection of the surface with horizontal and/or vertical planes.

An important ecological question is how the composition and surface heterogeneity of habitats may affect the movement, patch dynamics and size distribution of organisms. Data are presented that elucidate whether the distribution of abundances of organisms of different size relate to the fractal dimension of gravel sediments. Species' perception of habitat structure may depend on the physical patchiness as well as the different spatiotemporal scales at which species interact, because of differences in vagility and in types of resources used. Distances moved by motile organisms may depend on both animal size and the fractal dimension of habitat surfaces. Results obtained from laboratory experiments suggest that mobility pathways of stream invertebrates are fractal and that the fractal dimension of these trajectories may relate to habitat structure and body size distribution.

Patches formed by organisms, which are defined by boundary conditions, are themselves heterogeneous and hierarchical, for they can be divided into aggregations of greater or lesser density. Patch structure may display fractal pattern, allowing the determination of 'random boundaries' between species in dynamic environments. This review examines the proposition that patch structure and body size distribution in stream sediments are fractal.

Fractal scaling appears as a ubiquitous property of nature. It may both describe habitat and patch structure and provide a way to quantify similarities or differences in how organisms of different body size, life history or vagility respond to environmental heterogeneity.

II. INTRODUCTION

Fractal geometry is an extension of conventional euclidean geometry. A unique difference between fractal objects and euclidean objects is that the length, when measured, depends on the scale, which denotes the resolution within the range of the measured quantity. Euclidean objects are defined by their constant length, regardless of the length of the measuring device, whereas fractal contours increase with increasing length of the measuring device because of the addition of more detail seen at larger magnification scales. This dependence of the measured length on the measuring scale is expressed as a dimension measure and reflects the scale invariance of the object.

Before introducing the fractal dimension, the concept of dimension should be defined. It is known empirically that the dimension of a point is 0, of a line and a plane are respectively 1 and 2, and that organisms move in a three-dimensional space. Conventionally, we can consider integer dimensions which are exponents on length, such as surface = length^2 or volume = length^3. These integer dimensions are inadequate in describing the highly complex natural forms and dynamical physiological processes that do not have specific scales of length and time. To define the topological characteristics of dynamical structures and processes adequately, fractional power dimensions are necessary. Fractal geometry allows the measurement of objects in a non-integer or fractional way when the unit of measurement changes, hence the term fractal (Mandelbrot, 1983). Scale-dependent processes, such as dispersal and scale-dependent patterns in the shape and form of organisms, and the physical structure of the habitats can be visualized in fractal dimensions rather than euclidean dimensions (Frontier, 1987; Sugihara and May, 1990). Fractal geometry provides a tool for analysing dynamic forms and functions of tissue cells, organs and their physiological states (West et al., 1997; Maina, 1998). Fractals have been used to analyse non-linear spatial and temporal phenomena, and can also be extended to abstract objects developing in phase-space, such as strange attractors of dynamic complex systems (Burrough, 1981; Frontier, 1987). An attractor that has a positive maximal Lyapunov exponent and, consequently, a high sensitivity to initial conditions, often has a very complex and chaotic structure. The trajectories of such an attractor may have fractal structure with the properties of a Cantor set (Holden and Muhamad, 1987).

Mathematical fractals are shapes whose roughness and fragmentation remains essentially unchanged at smaller scales. Subsets of mathematical fractal objects are infinitesimally subdivisional, each subset, however small, containing no less detail than the complete set (self-similar). Simple examples, such as the recursively constructed Cantor dust, Koch curve or Apollonian gasket, iterate themselves under discrete scale changes, having the property of self-similarity. Unlike mathematical fractals, natural or random fractals are only self-similar (scale invariant) in a statistical sense, because natural structures are truncated at certain scales. Thus, by relating the size of the random variations to the scale, enlargements of small parts of an object have the same statistical distribution as the whole set. Often, random sets, such as the outline of a detritus particle, are statistically self-similar not only for a given value of the scaling ratio (λ), but for all scaling ratios above some lower breakpoint (the microscale) and some upper breakpoint (the macroscale). Thus, natural fractals look qualitatively the same over several scales, but the dimension of these fractal objects is restricted to ranges of scales.

The scale transformation for self-similar fractals is isotropic, which means that dilation increases the size of the object uniformly in every spatial direction. Many ecological studies deal with sets that are not self-similar. For instance, when we trace the movement of an organism, the positions of the individual in time are different physical quantities and, thus, scale with different ratios. These objects are self-affine rather than self-similar and can be re-scaled using an anisotropic transformation (Barabási and Stanley, 1995). An affine transformation is a combination of a translation, rotation and dilation of an object. Unlike similarities, affine transformations contract with different ratios in different directions. Consequently, the measurement technique that can be applied to self-affine objects differs from techniques that can be applied to statistically self-similar objects (Falconer, 1990). More details on self-affine and self-similar sets are found in Mandelbrot (1983), Feder (1988) and Falconer (1992).

In aquatic ecosystems water flow, generally governed by dynamic non-linearity, is one of the major structuring forces of benthic habitats, affecting the spatial and temporal distribution of communities. Fractal patterns are typically observed in systems that develop far from equilibrium and fractals may be looked upon as a model for spatial and/or temporal stochasticity (Mandelbrot, 1983; Falconer, 1985). This review will address random patterns observed for movement paths, spatiotemporal distribution and patch structure of benthic assemblages that are characterized by fractal properties.

III. FRACTAL DIMENSION AND ITS MEASUREMENT

A. Fractal Dimension of Boundary and Profile Lines

Leaves and their particulate fractions form a major structuring element as well as food source in the streambed interstices. Thus, a leaf fragment, as found in the streambed sediments weeks after leaf fall (Figure 1), is used as an example to demonstrate the scale-dependent extent of micro-habitat boundary outlines using the following three procedures.

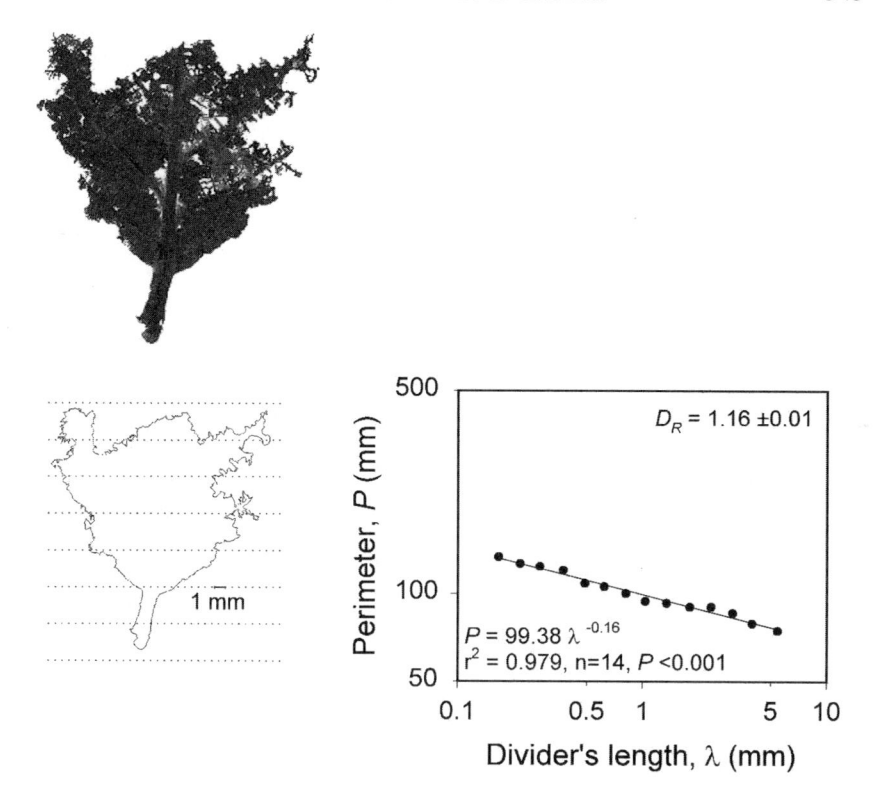

Fig. 1. Leaf fragment of an ash tree (*Alnus glutinosa* L.) as found in the streambed sediments of the Afon Mynach (North Wales) in November 1996, weeks after leaf-fall. To characterize the fractal dimension of a rugged boundary, the leaf outline is superimposed by parallel dotted lines, which are used to link up intersection points on the profile boundary to form a polygon. The Richardson dimension (D_R) was calculated over 14 divider's lengths. See text for further details.

1. Richardson Procedure

The principle of the Richardson procedure is given by the outline of the object, which is converted to polygonal approximations using different side lengths (Richardson, 1961). The total number of sides $N(\lambda)$ multiplied by the length gives an estimate of the perimeter of the object (P), which increases as the step size (λ) becomes smaller. The number of steps and the length of the final side length will vary for a given particle boundary line depending on the starting point. Thus, several different starting points are used and the results averaged to estimate the dimension. It is also necessary to decide how the step is to be made when the same side length can intercept the boundary line of the object in more than one location. The choices are to take the nearest neighbouring point or the furthest point. An unbiased polygon construction algorithm is to lay a set of parallel lines across the boundary and use the intercepts of the parallel lines with the boundary line as a set of points to be linked by the polygon (Kaye, 1989). If the spacing of the lines is λ (Figure 1), the length of the polygon side spanning two intersection points can be calculated using the Pythagoras theorem as:

$$d^2_n = (x_{n+1} - x_n)^2 + (y_{n+1} - y_n)^2$$

where x, y are the cartesian co-ordinates of the boundary line. The importance of this method using pythagorean polygons lies in its general application and programmability using image analysis techniques (Schwarz and Exner, 1980; see section III). Plotting the measured perimeter (P) against the value of the step length λ using a logarithmic axis gives a negative linear regression which may cover several orders of magnitude (Figure 1). The increase in perimeter length as the measuring scale is reduced is due to the boundaries of leaf particles being composed of irregularities at all scales. As the magnification is increased, more of the boundary roughness is revealed, evidencing self-similarity. The slope (b) of the line lies between 0 and 1. The fractal dimension of the boundary line is the sum of the magnitude of the slope and the topological dimension of a line (Figure 1).

2. Kolmogorov or Box-counting Dimension

The profile or boundary of the object to be evaluated is overlaid by a square grid. This process is illustrated in Figure 2, which shows both the boundary of a leaf fragment and its surface fragmentation. By placing a cartesian lattice over the profile, the image of the object is transformed into a mosaic of squares. The number of squares through which any part of the line or curve passes is counted. This process is repeated with random placement of grids of progressively smaller square size on the profile, and the number of squares overlaying the profile is plotted against the length of the square side on a log–log scale (Figure 2). For each box size, the grid should be overlaid in such a way that the minimum number of boxes is occupied, which can be accomplished by consecutively rotating the grid over the boundary. The Kolmogorov or capacity dimension is defined as the exponent D_B in the relationship:

$$N(\lambda) \propto 1/\lambda^{D_B}$$

where $N(\lambda)$ is the number of squares of side length λ necessary to cover a set of points distributed in a two-dimensional plane. The basis of this method is that, for objects that are euclidean, $N(\lambda)$ defines their dimension. A number of boxes proportional to $1/\lambda$ is needed to cover a set of points lying on a smooth

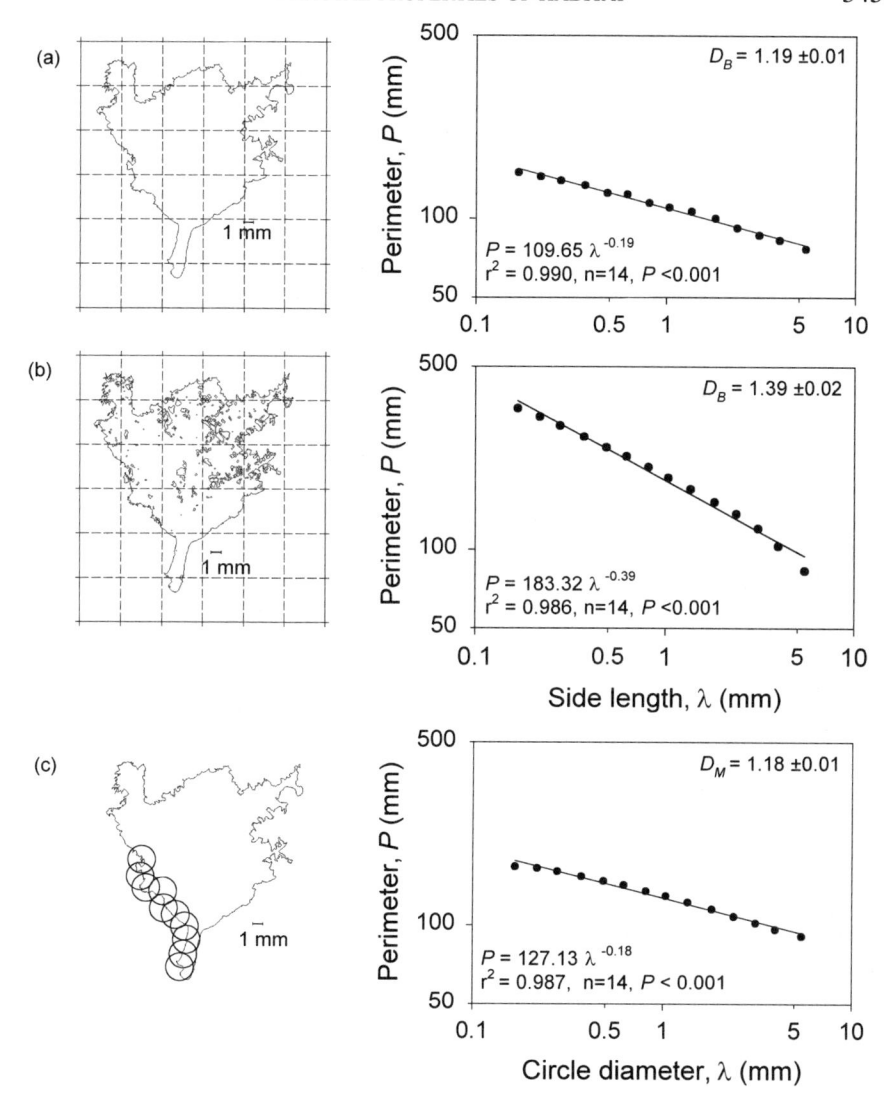

Fig. 2. Kolmogorov method and Minkowski logic used to characterize the rugged outline of a leaf fragment. In the Kolmogorov method a square lattice is placed over the boundary line and the number of squares through which any part of the line passes is counted. To obtain unbiased estimates of the profile boundary, the lattice was rotated each time by 15°. In the Minkowski method a circle is continuously swept along the line and the area covered is determined. (a) The Kolmogorov dimension (D_B) is based on the log–log relation between side length of the squares and the perimeter of the leaf outline. (b) Kolmogorov dimension (D_B) estimated for the fragmented inner and outer boundary lines of the leaf, demonstrating an increase in fractal dimension with boundary fragmentation. (c) The Minkowski dimension (D_M) for the outline of the leaf fragment based on the log–log relation between circle diameter λ and estimated leaf perimeter (P).

line, and proportional to $1/\lambda^2$ to cover a set of points evenly distributed on a plane. To estimate D_B, the number of squares of linear size λ necessary to cover the set for a range of values of λ is counted and the logarithm of $N(\lambda)$ versus the logarithm of λ is plotted. If the set is fractal, this plot will follow a straight line with a negative slope that equals $-D_B$. To obtain evenly spaced points in log–log space, it is best to choose side lengths (λ) that follow a geometric progression. However, values of λ chosen should consider the lower and upper limit of resolution, such as above and below 0.25 the object's length.

3. Minkowski Logic

Many fractal boundaries are fuzzy, which means that the exact location of each point of the boundary line cannot be determined. The technique for evaluating the fractal structure of a fuzzy boundary and self-affine profiles is based on the mathematical logic to estimate the length of continuous curves with undefined lengths and derivatives, originally proposed by Minkowski (1901). A circle is drawn around each point on the fuzzy boundary and these circles merge to form a ribbon overlaying the boundary (Figure 2c). The area of this merged set of circles is measured and the length of the ribbon is divided by the diameter of the underlying circles to obtain an estimate of the boundary or profile length. The log of each resulting area divided by the circle diameter is plotted against the log of the circle diameter. The Minkowski dimension of a fractal object (D_M) is related to the slope (b) of this log–log relation as:

$$D_M = 1 - b.$$

The circle diameter is the resolution parameter at which to estimate the length of the indeterminate boundary. Thus, in applying the Minkowski logic, the boundary covered by the circle can be sharp or fuzzy. The major difference between the Minkowski logic and the Richardson procedure is that the circle is moved so that its centre lies in every point of the boundary or profile line.

4. Dilation Methods

Digitized images of objects can be displayed as mosaics in the form of picture elements (pixel). By using a high-resolution digitizing tablet, the object boundary is traced with a stylus, where the co-ordinates are quantified in the x and y dimensions. Moreover, raster scan imaging devices, such as flat-bed image scanners for objects and their photographs, produce images that initially have several hundreds to thousands of pixels across each horizontal and vertical line with a similar number of lines. I have used images with each pixel having grey values, which range from white to black over a range of 2^8 to 2^{12} steps. Based on brightness or local texture classification, the square pixels are separated into foreground and background picture elements. To define the boundary or profile, a threshold setting was established which lies in the range of brightness values between the peaks of the brightness phases, such as white and black. Objects were placed on a non-reflecting black background and scanned *in situ* or photographed with a digital camera using light sources normal to the object surface. However, the use of pixel images imposes some limitations on the measurement of the fractal dimension of the boundary. Image resolution and the size of the pixel, for instance, sets the lower limit of dimension for any kind of plot.

Dilation is a modification of the Minkowski method, using images of the object (Flook, 1978; Smith *et al.*, 1989). The process of dilation is based on the addition of pixels, around the outline of a boundary. This can be accomplished by a convolution operation with binary disks of different pixel diameter. The rate at which the total surface area of the boundary line grows as a function of the diameter of the convolution kernel depends on the fractal dimension D. The log of each resulting area divided by the disk diameter is plotted against the log of the disk diameter. The dilation dimension of a fractal object (D_{MD}) is related to the slope (b) of this log–log relation as:

$$D_{MD} = 1 - b.$$

Mosaic amalgamation logic is a technique similar to the Kolmogorov procedure for the evaluation of fractal structures (Kaye, 1978). The image of the object to be evaluated is transformed into a mosaic at the smallest pixel size available. The original image gives the finest scale approximation to the boundary length, and progressive coarsening of the image representation is used to construct log–log plots of total profile length versus measurement scale.

As with the Kolmogorov method, the profile or boundary to be evaluated is overlaid by a rectangular grid of a particular pixel size. To estimate the length of the boundary, the number of squares that include any fraction of the boundary is counted. The area of the band covering the actual profile is calculated as $A = n\lambda^2$, where λ is the length of the grid spacing and n the number of squares. An unbiased estimate of the area is the number of squares falling within the profile plus 50% of the squares on the boundary. The pixels from the original image are averaged together to form a new image in which pixels are larger in area. The larger pixels smooth the details of the profile line and the measured perimeter will decrease. To evaluate the fractal dimension of the boundary, the profile is converted into a series of mosaics of increasing square size and the perimeter is plotted against the width of the respective pixel size on a log–log scale.

Euclidean distance mapping (EDM) is a method to estimate the fractal dimension of a black and white image object which produces a grey-scale result in which each pixel has a brightness value equal to its distance to the nearest background or boundary feature point (Danielsson, 1980). Thresholding the EDM for either the boundary features or the background produces uniform erosion (pixel removed) and dilation (pixel addition) to any distance from the original boundary. After generating the EDM, the image can be thresholded at any brightness level to select pixels with various distances from the original boundary, which produces a Minkowski plot. The width of circles is uniform regardless of whether the boundary is isotropic or anisotropic. Increasing grey-scale levels and the resulting thresholded area are plotted on a log–log scale and the fractal dimension (D_{EDM}) is calculated from the slope of this relation as:

$$D_{EDM} = 2 - b$$

B. Fractal Dimension of Deposition Processes

1. Flocculated Dispersion

When attraction forces become larger than repulsion and brownian motion, particles both in the water column and in sediments can aggregate on contact (Macosko, 1994). The resulting loosely structured aggregates, or flocs, reveal a complex structure that has no homogeneous internal structures. The growth kinetics of flocs have been studied in depth and simulated by Goodarz-Nia (1975). Figure 3 shows an image of flocculated particles as occasionally found in the streambed sediments of the mountain stream Afon Mynach, North Wales. Different degrees of flocculated aggregation are evident, ranging from single detritus particles to complex aggregates. The right panel of Figure 3 shows three different types of flocs found at benthic substrate surfaces and in the bed sediments. The simplest description of such a physical deposition process is by diffusion-limited aggregation (DLA) in which particles perform a random walk until they encounter the surface of another particle. The DLA model represents the succession of fine particulate matter released sequentially from a distance, forming a growing floc aggregate. Wherever a particle touches, it has a certain probability of sticking to the surface. Aggregates, such as flocs of particulate matter, can be characterized by the mass fractal dimension. The mass within a radius λ of the floc centre increases as:

$$M \propto \lambda^{D_{MASS}}$$

where D_{MASS} gives the mass fractal dimension. As the detritus aggregate becomes more compact and more particles cluster together, the mass fractal dimension increases (Figure 4). The presence of clustered flocs of particulate matter in benthic systems may increase the complexity of substrate surfaces and available surface area, particularly for microbenthic and meiobenthic

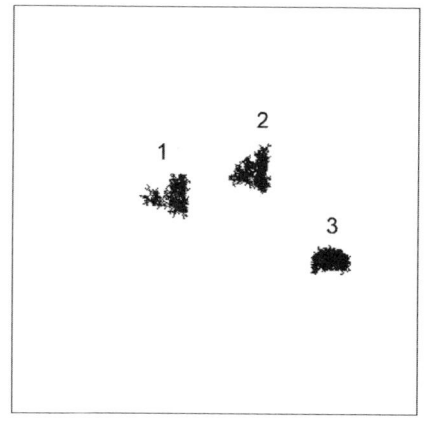

Fig. 3. Flocculated aggregates of particulate matter as found at the streambed surface of the gravel stream Afon Mynach (North Wales; 52°57′N, 3°38′W; National Grid Reference SH909407) in July 1996. The right panel shows three floc types generated by diffusion-limited aggregation with 100% (type 1), 50% (type 2) and 1% (type 3) sticking probability.

organisms. Aggregates formed by flow are expected to be dense and complex (type 3; Figure 3), because flow may cause breakdown and detachment of the outer floc regions and the formation of dense particle aggregates (Macosko, 1994). Looser and more open aggregates of particulate matter (type 1, 2; Figure 3) may be found in the sheltered inner zones of moss patches and in 'death zones' of exposed boulders in stream systems. Therefore, a greater number of smaller-sized organisms are expected to occur in denser fractal aggregates than in open aggregates of particulate matter.

C. Fractal Dimension of Surfaces

Surfaces of benthic features, such as biotic and abiotic substrates, can be evaluated in terms of roughness by direct analogy with the evaluation of boundary or profile lines. The most common approach to measuring surface fractals is to reduce the dimensionality and measure the fractal dimension of the boundary line produced by intersecting the surface with a sample plane. This may be produced by cross-sectioning or embossing the surface to produce islands. In either case, the rougher the surface, the more irregular the outline of islands. Profiles across the surface can be used to estimate surface roughness and fractal dimension, whether they lie along the x or y axes, or in some other direction. Passing a plane through the surface in a vertical direction produces

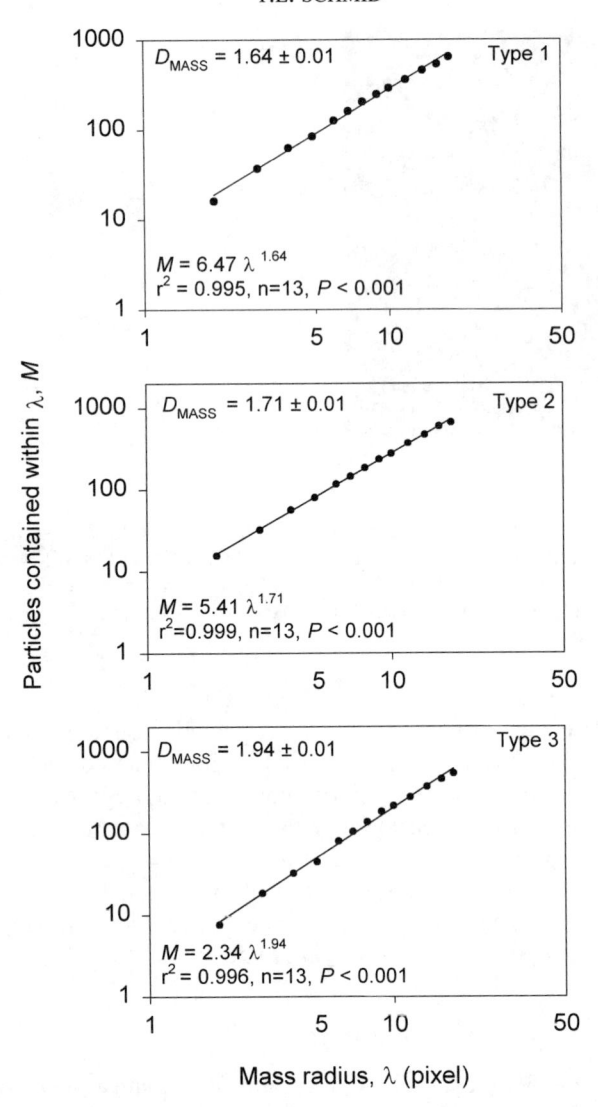

Fig. 4. Measurement of the mass fractal dimension (D_{MASS}) of each of three floc types (see Figure 3). The regressions between different mass radii λ in pixels and the number of particles contained within each radius, M, are given.

an elevation profile, which is self-affine rather than self-similar. Passing a plane through the surface in a horizontal direction produces an intersection profile where all points on the profile have the same values of z, which is the direction in which the scaling factor is different (Russ, 1994).

1. Direct Measurement of Benthic Surface Fractals

Surface absorption isotherms have long been used to measure the surface area of catalysts and similar surfaces. The Brunauer–Emmett–Teller (BET) method (Brunauer *et al.*, 1938) is used to coat surfaces with monolayers of gas molecules of the same or different size. The principle of the measurement is based on the application of two absorption vessels with the same volume; one vessel is filled with the object to be measured and the other remains empty. Both vessels are filled under the same atmospheric pressure with the gas, for instance liquid nitrogen at room temperature, and cooled to –196°C. The absorbed film on surfaces causes a pressure difference between the filled and empty vessel, which is registered by a differential manometer. The surface area determined by absorption isotherms both of single or different-sized molecules allows an estimate of the surface fractal dimension.

Three methods are used to estimate the fractal dimension:

(1) A single absorbate molecule is applied to objects (particles) of different size. The surface area per unit mass S is related to the surface fractal dimension D and particle radius λ_R by $A_M \propto \lambda_R^{D-3}$.
(2) The surface sample is covered by a series of monolayers of different gases whose cross-sectional dimension d varies. The number of molecules varies as $N \propto d^{(-D/2)}$.
(3) The distribution of pore space from gas absorption is used to determine the surface fractal dimension D from the change of cumulative void space P with respect to pore radius λ_R, where $-dP/d\lambda_R \propto \lambda_R^{2-D}$.

The advantage of these three techniques for estimating the surface fractal dimension of objects is that they can assess all the fine details of the surface, pores, etc., which may be undetected by other direct methods that record single elevation values at x, y co-ordinates (see below).

Profilers, physical contact measurement devices, are widely used methods for measuring surface roughness, particularly of sediment surfaces. Profilometers are used to scan the elevation profile either along a linear transect (De Jong and Ergenzinger, 1995) or in a specific sampling area (Gore, 1978; Schmid-Araya, 1993). However, finer surface features which may be ecologically important remain undetected with this method. The fractal dimension of elevation profiles, such as those measured by a profiler, can be estimated using, for instance, the root-mean-square (RMS) deviation or roughness method. For a self-affine profile, the standard deviation (SD) or RMS deviation, measured at windows of size λ, is related to the Hurst exponent as $SD(\lambda) = \lambda^H$. For a given window width λ, the profile is subdivided into several intervals of length λ, and the RMS is calculated in each window after detrending, by subtracting a best-fit straight line. The estimate is taken to be $SD(\lambda)$, the average of the RMS measured in each window. By repeating this procedure for different λ, the values of the standard deviation are plotted against the window width. The slope of the log–log plot equals the Hurst exponent H. The fractal dimension of the profile can be calculated from the Hurst exponent H as:

$$D_{RMS} = 2 - H$$

where D_{RMS} is the fractal dimension of the RMS method.

The RMS method can also be applied to profiles and traces that do not have constant sampling intervals, as the estimates of RMS should not depend on the number of samples in the window, provided that each window contains a minimum of five samples.

2. Indirect Measurements of Benthic Surface Fractals

Interstitial sediments and resulting void space are dominant habitat features in benthic environments. At a smaller scale (e.g. < 100 μm) the surface structure of biofilms may play an important role in structuring available microspace. Two examples from streambed sediments illustrate the estimation of surface fractal dimensions using images of habitat features. The first example (Figure 5) is a subsample of streambed sediments collected from depths between 10 and 20 cm using the freeze-core method in the Afon Mynach (North Wales) in March 1994. The second example (Figure 6) shows two randomly selected sections of a biofilm surface covering sediment particles collected in the gravel stream Oberer Seebach (Lower Austria). To estimate the fractal dimension of these abiotic and biotic surface structures, these images were subjected to texture pattern analysis.

Brightness pattern analysis, representing the brightness of light scattered from the surface or the production of secondary electrons in a scanning electron micrograph (SEM), is performed on images of surface structures. Imaging systems, which rely on the re-emission of light from surfaces, generate signals that are related to the surface slope of the object. Scattered light from a fractal surface produces a brightness pattern, which is mathematically fractal. In these images, the roughness or texture of the surface is revealed. The texture values, estimated as the slope of the graph of mean pixel brightness difference versus pixel separation distance, do not agree numerically with the physical dimension generated from measured elevation profiles (Russ, 1994). However, measuring surfaces of similar material using the same illumination allows a good qualitative comparison of the brightness (texture) fractal dimension of light scattered from the surface. A rather constant illumination is obtained by scanning the surface structure, where the light source is placed normal to the nominal surface orientation.

1000 μm

Fig. 5. Streambed substrate collected from sediment depths between 10 and 20 cm using the freeze-core method in the Afon Mynach (North Wales) in March 1994. The sample consists of grain sizes < 2.5 mm and covers a planar surface area of 1 cm².

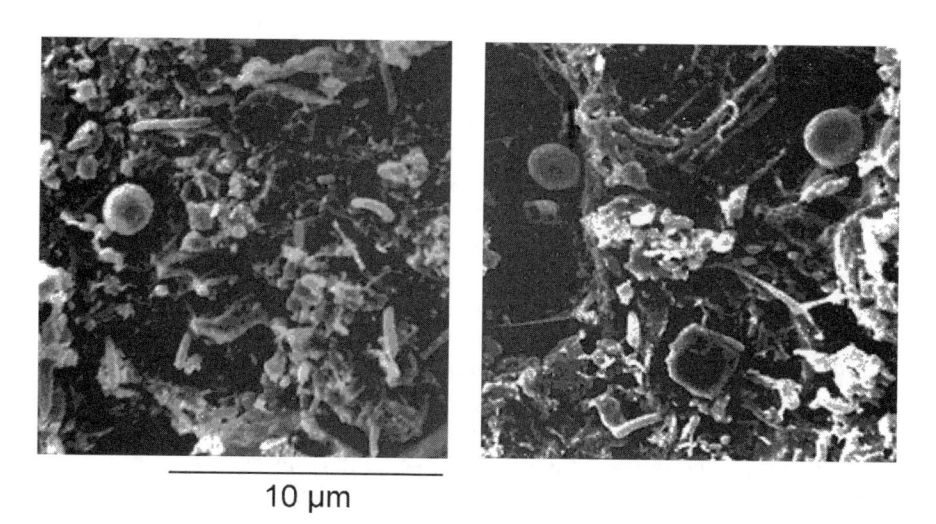

10 μm

Fig. 6. Electron micrograph of two 230 μm² sections of biofilm covering sediment particles in the Oberer Seebach (modified after Leichtfried, 1991).

The methods used to estimate the fractal texture dimension are applied to range or elevation images, which are self-affine. An estimate of fractal texture dimension, which can be interpreted in terms of the physical surface dimension, is based on the Fourier analysis in two dimensions. This method is insensitive to the presence of noise in images due to the operation of the measuring device (Russ, 1992). The Fourier transform consists of magnitude and phase data covering a two-dimensional array, where each point represents a frequency and an orientation in the original image of the surface structure. The Fourier transform is defined as the spatial domain function $f(x)$, where x is a real variable representing distance in one dimension across the image, and the transform F as the frequency space function:

$$F(n) = \int_{-\infty}^{+\infty} f(x) \exp(-2\pi i n x) \, dx$$

where the exponential notation relies on the identity

$$\exp(-2\pi i n x) = \cos(2\pi n x) - i \sin(2\pi n x)$$

where i is the imaginary number and n represents the continuous frequency component added together to construct the spatial domain function. Thus, this function $F(n)$ describes the amount of each frequency term that must be added together to produce sine and cosine terms of increasing frequency, and can be rewritten as:

$$f(x) = \int_{-\infty}^{+\infty} F(n) \exp(2\pi n x) \, dn$$

where $f(x)$ is a real function such as spatially varying image brightness. The complex transform function $F(n)$ is the sum of a real part R and an imaginary part i giving $F(n) = R(n) + i(n)$. This can be rewritten in polar rather than cartesian form as $F(n) = |F(n)| \exp[i\phi(n)]$, where $|F|$ is the amplitude and ϕ is the phase of the spectrum. The square of the amplitude is referred to as the power spectrum or spectral density of $f(x)$. In image analysis the integrals from minus to plus infinity are reduced to a summation of terms of increasing frequency (discrete Fourier transform), limited by the finite spacing of the sampled points in the image. This summation is performed over n terms up to $n/2$ the dimension of the image in number of pixels. The direct extension from the one-dimensional function to two- or three-dimensional ones can be done by performing the above transformation separately in each horizontal and vertical direction of the surface image. Reviews and comprehensive discussions on Fourier transform can be found elsewhere (Feder, 1988; Hastings and Sugihara, 1993). To estimate the fractal dimension of an image using the two-dimensional discrete Fourier transform (2DFT), the log-power spectrum (amplitude^2) is plotted against log-frequency. For a surface, the dimension D is related to the slope b of the plot as $D = (6 + b)/2$.

Figure 7 shows the surface fractal dimension of the Fourier analysis for a mixture of sediment particles and voids scanned in a horizontal direction. The two-dimensional estimate of surface fractal dimension was obtained by calculating the slope of the amplitude^2/frequency relation for all major directions over the image. The resulting slopes were plotted as a rose plot (Figure 7a) and the average slope of the image was calculated. The estimated fractal surface dimension, based on the mixture of grains and interstitial voids, is 2.57 (Figure 7a). This estimate falls into the range of values for fractal surface dimensions obtained for sediments at smaller scales (< 350 μm) based on SEM images (e.g. Krohn, 1988). To compare the fractal value obtained by two-dimensional Fourier analysis with analysis in one dimension, the relation between the power spectrum and frequency was calculated both for horizontal and vertical profiles across the surface (Figure 7b). The results in one dimension confirm the value obtained for the surface fractal dimension of these sediments.

Moreover, Figure 8 shows the results obtained for the fractal dimension on one-dimensional profiles across horizontal and vertical directions of the SEM biofilm images (see Figure 6). Both examples indicate a similar surface structure with a fractal dimension in the range 1.51–1.54 (Figure 8). It should be pointed out that a fractal dimension of 1.5 for a profile is equivalent to an ideal random walk. This phenomenon, called $1/f$ 'pink' noise, is common to many physical processes, and suggests random processes with no temporal correlation as a surface structuring force.

3. Differences Between Fractal Measures

In addition to the methods described above, there are other methods, such as the perimeter-area and Korcak dimension, which can be applied to boundary lines. Details on these and other methods can be found in the sections below, and in Mandelbrot (1983) and Hastings and Sugihara (1993). The various methods discussed here for estimating the fractal dimension are not mathematically identical. Formally, only the Hausdorff dimension (Mandelbrot, 1983), which results from the Richardson procedure, or the mass fractal is technically a dimension. The Minkowski logic and Korcak method result in fractal values that are equal to or greater than the Hausdorff dimension, while the Kolmogorov dimension estimates the upper limit of the Minkowski dimension (Falconer, 1990). The procedure to be used may depend on the form in which the data are acquired and the relative efficiency of performing the measurement. For instance, the box-counting method should be applied only to objects, which are statistically self-similar, whereas the Minkowski logic can also be used for self-affine structures.

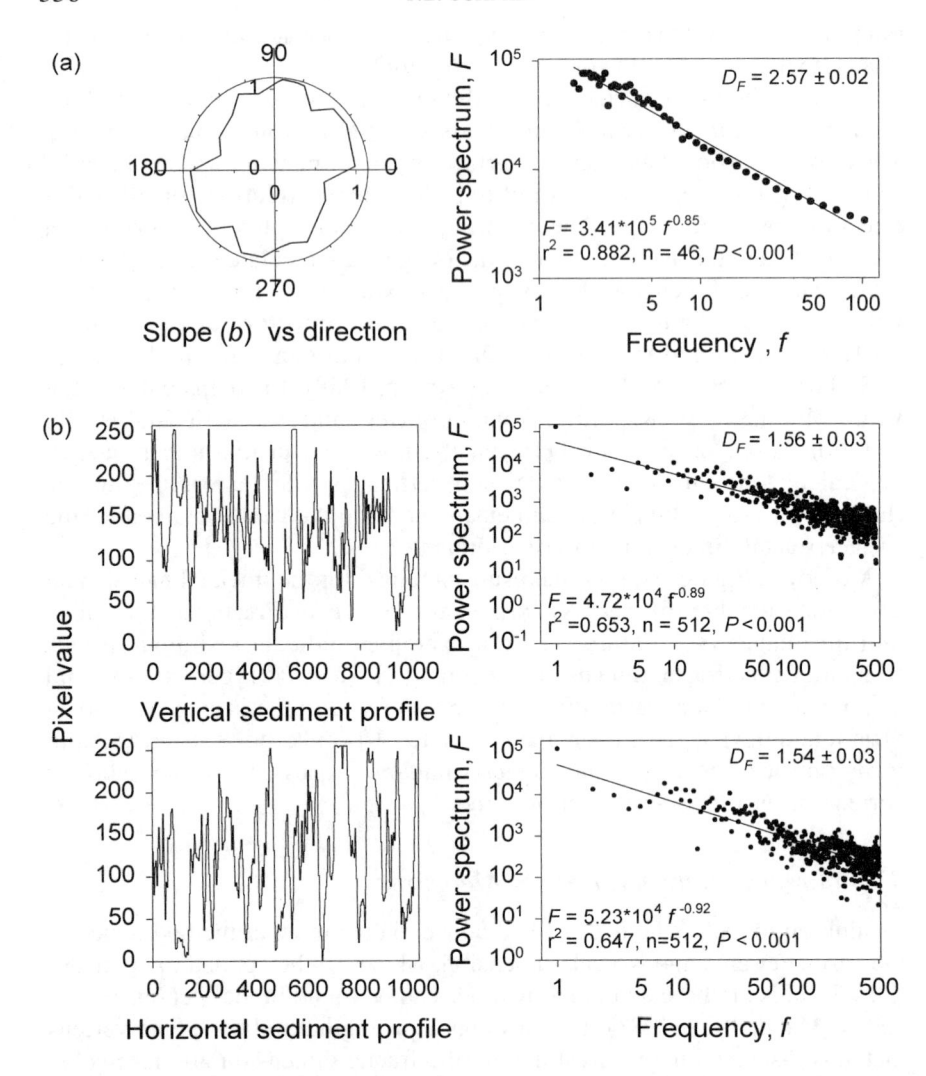

Fig. 7. Fourier analysis of the particle–void surface from interstitial sediments collected in the Afon Mynach (North Wales). (a) Rose plot of the slope (b) of the log power spectrum–log frequency relationship in all directions across the particle–void texture surface and log–log relationship between values of the power spectrum averaged over all directions and frequency. (b) Brightness values of vertical and horizontal profiles across sediment particles/voids and the log–log relation between power spectrum and frequency for vertical and horizontal profiles, resulting in a similar fractal dimension (D_F).

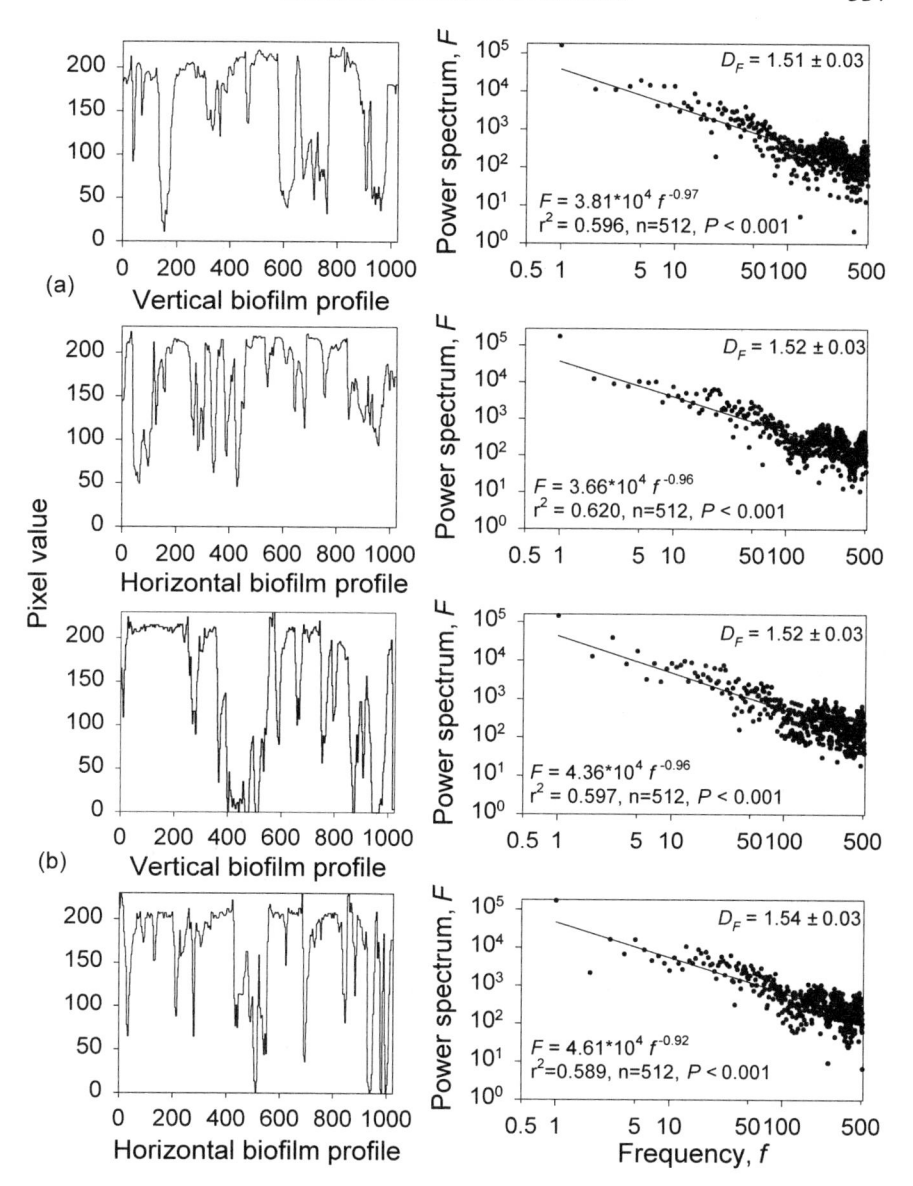

Fig. 8. Fourier analysis of two biofilm surface sections from interstitial sediments collected in the Oberer Seebach (Lower Austria). (a) Brightness pattern of vertical and horizontal profiles across the biofilm image of section A and (b) of section B.

4. Multifractals

Following Feder's (1988) definition of multifractals, one may distinguish a measure (of probability, or some physical quantity) from its geometric support, which may or may not have fractal geometry. Then, if the measure has different fractal dimension on different parts of the support, the measure is a multifractal. In contrast, Hastings and Sugihara (1993) distinguish multifractals from multi-scaling fractals, which have different fractal dimensions at different scales (e.g. show a break in slope in a divider's plot, or some other power law).

Mathematically, multifractal measures are related to the study of a distribution of physical or biotic quantities on a support such as a surface or a volume which could itself be fractal (Rammal, 1984). An example of a measure γ in sediment interstices V may be the quantity of leaf patches among sediment voids. For each subset V of the interstices, the measure is a function of $\gamma(V)$, which is the amount of leaf patches in V. If we divide the void space into equally sized volumes V_1 and V_2, it is evident that their respective leaf patch quantities $\gamma(V_1)$ and $\gamma(V_2)$ will differ. This subdivision could be carried through to the smallest voids, where some voids are filled with leaf fragments and others are empty. Other quantities, such as the densities of benthic species, may exhibit the same behaviour; that is, the quantity γ, the amount of an object within V, is an example of a measure that is irregular at all scales. Thus, the individuals within benthic populations may display a distribution pattern of fractal clustering. Another example of multifractal behaviour is found in the context of DLA of particle flocs (see section III.B). A particle cluster has a mass fractal dimension, and the growth of each location in the particle floc is different for each point in the cluster. Thus, each of these locations scales with a power law, having a different dimension (Meakin, 1987). The dimension of each subset is different, hence it behaves like a multifractal, and the basic multifractal measures are obtained as limits of products of random factors (Mandelbrot, 1984).

IV. FRACTAL PROPERTIES IN RIVERINE ECOSYSTEMS

There are several phenomena in riverine ecosystems that are examples of fractals, ranging from river networks, which possess features of self-similarity over a considerable range of scales (Tarboton *et al.*, 1988; Tarboton, 1996), to discharge (Mandelbrot, 1983), to the flow and deposition processes in porous media (Adler, 1996).

A. Large Spatial Scales

1. Catchment Area

The shape of a whole river system can also be described in fractal terms. The catchment area (A_B) of a river increases with increasing measurement scale.

River systems with a complex network of tributaries, such as the Amazon, exhibit a Kolmogorov dimension of 1.85, whereas the catchment area of the River Nile has a fractal dimension of 1.40 (Takayasu, 1990). The fractal complexity of river networks, therefore, appears to increase with increasing precipitation (Takayasu, 1990).

In order to analyse the relation between length scale and catchment area of river systems several orders of magnitude smaller than the above examples, I chose the River Ybbs, a tributary of the River Danube in Lower Austria (Figure 9), and the stream Oberer Seebach, a tributary of the Ybbs. I used the Kolmogorov method to estimate changes in catchment area (A_B) with increasing resolution. The relation between length scale and drainage basin of

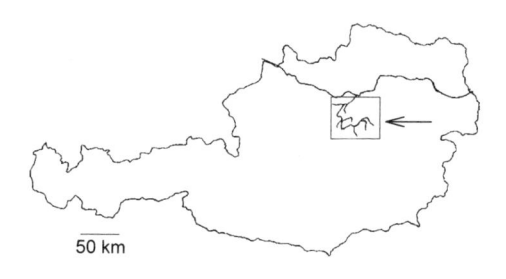

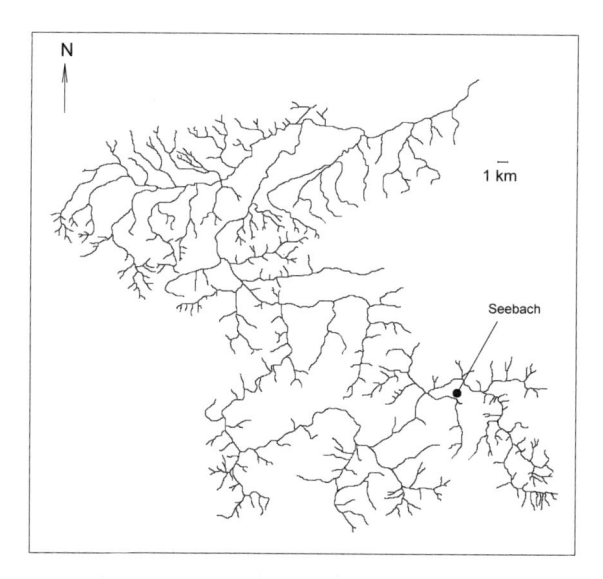

Fig. 9. Catchment area of the River Ybbs, a sixth-order tributary of the River Danube, in Lower Austria (48°10′N, 15°06′E). The symbol (●) indicates the location of the gravel stream Oberer Seebach (47°51′N, 5°04′E).

the River Ybbs resulted in two distinct regimes with a transition at a scale of 1 km (Figure 10). For scales less than 1 km, the relationship between length scale and estimated A_B remained rather constant, as a result of the resolution limit of the network map (Figure 9). I estimated the fractal properties of the drainage basin of the River Ybbs for scales greater than 1 km, which resulted

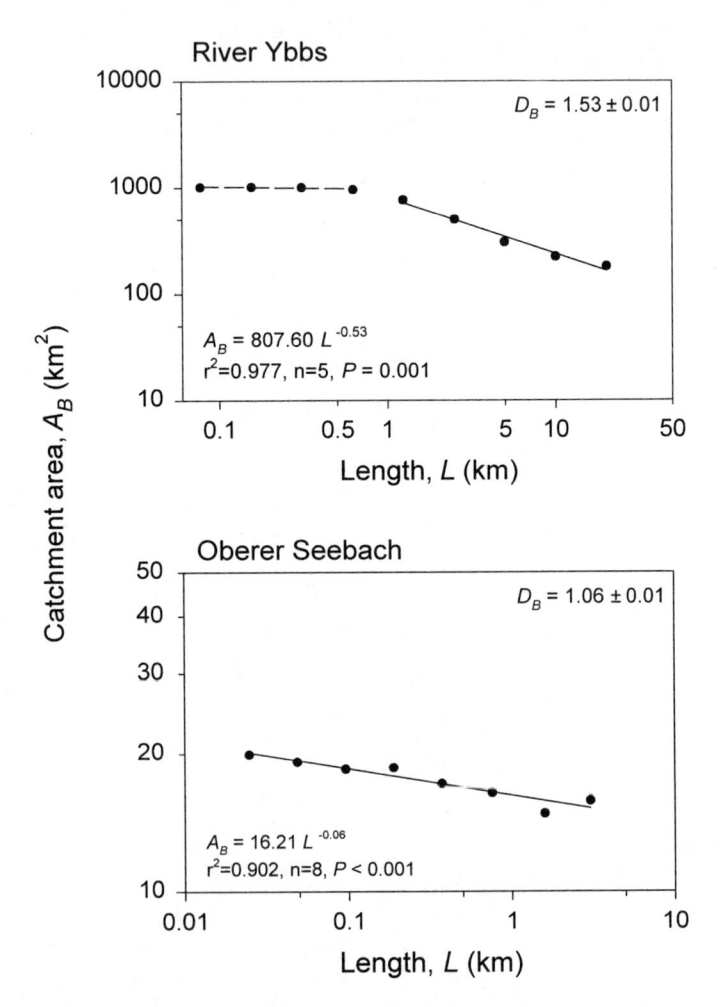

Fig. 10. Log–log relation between different side-lengths (L) of square lattices covering the river network and the resulting estimates of catchment area (A_B) of the River Ybbs and the Oberer Seebach. The River Ybbs is characterized by two regimes with a transition at a scale > 1 km. Scales < 1 km indicate the resolution limit. The Kolmogorov dimension (D_B) of the drainage basin of the River Ybbs was calculated over spatial scales > 1 km and for the Oberer Seebach over scales > 0.02 km.

in a Kolmogorov dimension of 1.53 and an estimated catchment area of 807.6 km² (Figure 10). In contrast, the relation between scale and catchment area of the Oberer Seebach resulted in a Kolmogorov dimension of 1.06, and an estimated area of 19.9 km² for the smallest resolution (Figure 10). The results of the above contrasting cases suggest that differences in the fractal dimension of catchment areas depend not only on the frequency of rainfall but also on other factors such as the geomorphology of the drainage basin. Moreover, it can be argued that river networks with a high fractal dimension may be characterized by a variety of habitat types and, consequently, a higher species diversity, compared with networks with lower complexity.

2. Mainstream Length and Drainage Area Relationship

Owing to the sinuosity of streams, the empirical power law relationship between mainstream length (L_M) and the area of the drainage basin, A_B (Hack, 1957; Leopold et al., 1964)

$$L_M \propto A_B^{0.6}$$

could be interpreted as the mainstream length having a fractal dimension D, (Mandelbrot, 1983; Tarboton, 1996). Combining the assumptions that the standard relation $L \propto A^{1/2}$, and that rivers as well as their drainage basins are mutually similar, thus, $L^{1/D} \propto A_B^{1/2}$, the above relation can be rewritten as $A_B^{1/2} \propto L_M^{(1/1.2)}$. Hence, the fractal dimension D of the mainstream–basin relationship approximates 1.2. As a general rule for all rivers, values of D of the relation mainstream length versus drainage area fall in the range 1.1–1.4, centred around 1.15 (Mandelbrot, 1983; Tarboton, 1996).

3. Channel Length

Stream length influences sediment transport and consequently the amount of potentially available habitat area in a drainage basin. Formally, channel length is recognized as a fractal measure (Tarboton, 1996). Using maps at a single scale, the mainstream is stepped off repeatedly with dividers set at different lengths. In this manner, a stream length (L) is obtained for each divider's length (λ). The apparent channel length measured at scale λ is calculated as:

$$L(\lambda) = c\lambda^{1-D}$$

where c is the fractal stream length, which is a measure independent of scale and finiteness of measurement. The length of the stream diverges as $\lambda \to 0$. Figure 11 shows the relationship between stream length and divider's length for the gravel stream Oberer Seebach (47°51′N, 5°04′E; Lower Austria),

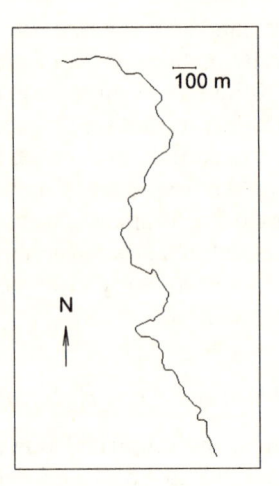

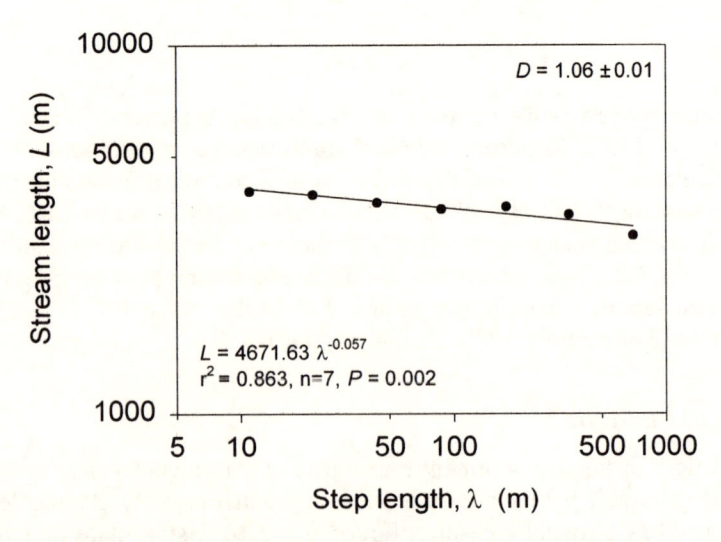

Fig. 11. Stream course of the Oberer Seebach, flowing in roughly a south–north direction, between the outflow of the lake Lunzer Mittersee and its mouth forming lake Untersee (47°50′N, 15°04′E). Richardson plot of the log–log relation of measured step length (λ) and estimated total stream length (L), resulting in an estimated fractal stream length of 4.7 km.

which gives a fractal dimension of 1.06, similar to the dimension obtained for its catchment area (see section IV.A). Estimates of D of various streams, calculated in this way, are in the range 1–1.2 (Robert and Roy, 1990; Nikora, 1991).

B. Intermediate to Small Spatial Scales

Empirical and theoretical research on porous media, such as sediments, revealed how their geometric structure in sediments displays fractal properties (Turcotte, 1986; Hildgen *et al.*, 1997). Krohn (1988) showed that at a microscale, the pore–rock interface for sandstone and carbonates is a fractal, with surface fractal dimensions ranging from 2.27 to 2.89.

Below, I deal with methods for estimating fractal dimension within streambed sediments, followed by aspects of body-size relations of a larval chironomid community and their possible dependence on habitat structure. I examine the hypothesis that the fractal dimension of the pore–particle boundary and the surface fractal of particles in a gravel stream relates to the body-size distribution, population densities and species richness found in streambed sediments and, therefore, influences community structure.

1. Fractal Dimension of the Pore–Particle Boundary

Bretschko and Leichtfried (1988) investigated *in situ* the interstitial spaces in the gravel stream Oberer Seebach. They carefully injected a mixture of coloured cement and an epoxy resin (Araldite) into the streambed sediments. After the cement had set, blocks of sediments were extracted from different depths (down to 60 cm) and cut into horizontal and vertical cross-sections (Figure 12). I estimated the fractal dimension of the boundary line between sediment particles and voids for both horizontal and vertical cross-sections of the streambed sediments using the Kolmogorov and Minkowski dilation method. The outline of sediment particles ($\varnothing \geq 1$ mm; $\varnothing$ is the diameter of sediment particles), which were clearly distinguishable from cement-filled interstices, was traced with a stylus. The images of digitized outlines for 20 sediment sections (10 × 10 cm) were superimposed by a square lattice, with side lengths ranging from 2 to 56 pixels at increments of 2 (Figure 13a). The lattice was repositioned, by rotating the grid, and the slope of the log–log relation between the size of the square and the number of squares covering the outline of the particle–void interface was calculated to estimate the fractal dimension D_B (Figure 13a). Similarly, the Minkowski dilation was applied on the same images using 28 different pixel sizes at increments of 2 (Figure 13b).

For a transect across a surface (a line), the fractal dimension must by definition lie in the range $1 \leq D \leq 2$ (Mandelbrot, 1983). The mean value of the fractal dimension based on horizontal and vertical sections was 1.68 (95%CL:1.57–1.79) for the Kolmogorov dimension and 1.65 (95%CL:1.56–1.78) for the Minkowski dimension. Assuming a mean dimension (D_{BM}) of 1.67 for the sediment boundaries of the total sediment column, heuristic lower and upper limits on D_{BM} for the particle surfaces are 2.7 and 3.3, respectively. The tentative lower and upper bound of the expected

Fig. 12. Photographic image of a vertical section (10 × 15 cm) through the streambed sediments of the Oberer Seebach. Dark background shows the cement-filled interstitial spaces between sediment particles.

increase in potentially perceived fractal surface area, when scale decreases by an order-of-magnitude, is obtained by adding 1 and by squaring the linear increase in distance, respectively (Morse *et al.*, 1985; Lawton, 1986; Sugihara and May, 1990).

2. *Fractal Dimension of Particle Surfaces*

The surface area of particles in the streambed sediments of the Oberer Seebach was measured using the BET method based on the surface absorption of N_2 molecules (Brunauer *et al.*, 1938; Leichtfried, 1985). The surface area per unit mass S is related to the surface fractal dimension D_S and particle radius λ_R as $S \propto \lambda_R^{(D_S-3)}$. If $D_S = 2$, the surface is smooth in a three-dimensional embedding space, and the surface area of a particle of size λ scales as λ^2. For fractally rough particle surfaces, the surface area can scale with an exponent lying in the interval $2 \leq D_S \leq 3$. The theoretical minimum surface area per unit mass was obtained by assuming particles of a specific density to be spherical in

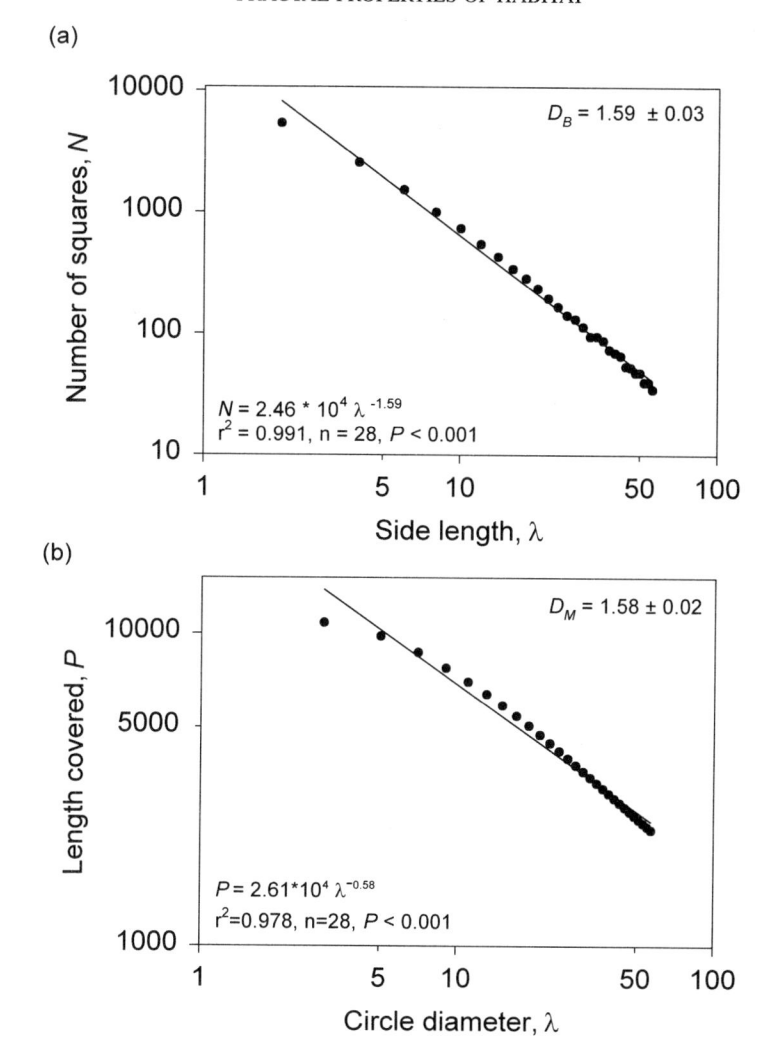

Fig. 13. Fractal dimension of the vertical section through the streambed sediments of the Oberer Seebach. (a) Kolmogorov dimension (D_B) based on the log–log relation between 28 different square side-lengths (λ) and number of squares in a lattice (N) covering the outline of the sediment–void interface. (b) Minkowski dimension (D_M) based on the log–log relation between 28 different circle diameters (λ) and the length of the sediment–void interface covered (P) by each circle diameter, using the dilation logic.

shape. Figure 14 shows the log–log relation between particle size and the measured particle surface area as well as the theoretical minimum surface area in gravel sediments. The result is a fractal dimension (D_S) of 2.7 for data based on BET isotherms.

Particle-size composition and porosity of each of seven depth layers (0–70 cm) was assessed from single freeze-cores taken on each of six sampling occasions between 1984 and 1985 (P.E. Schmid, unpublished data). However, only 32 of the 42 sampling units could be analysed, because the amount of sediment in the remaining samples was insufficient. To estimate the percentage of water-filled interstitial spaces among sediment particles, the sediment porosity of each frozen sample unit (depth layer) was measured using the displacement method (Gordon et al., 1992). Sample units were oven-dried, weighed and dry-sieved using 12 screen sizes (from > 0.01 mm to >100 mm) following the method of Leichtfried (1985).

The mean surface fractal dimension of streambed particles ranged between 2.53 and 2.76 during six sampling occasions in 1984–1985 (Figure 15). A pairwise comparison of all fractal exponents revealed significant differences between the highest fractal dimension and the two lowest values in October 1984 and April 1985 (Tukey–Kramer test: 0.234 and 0.165

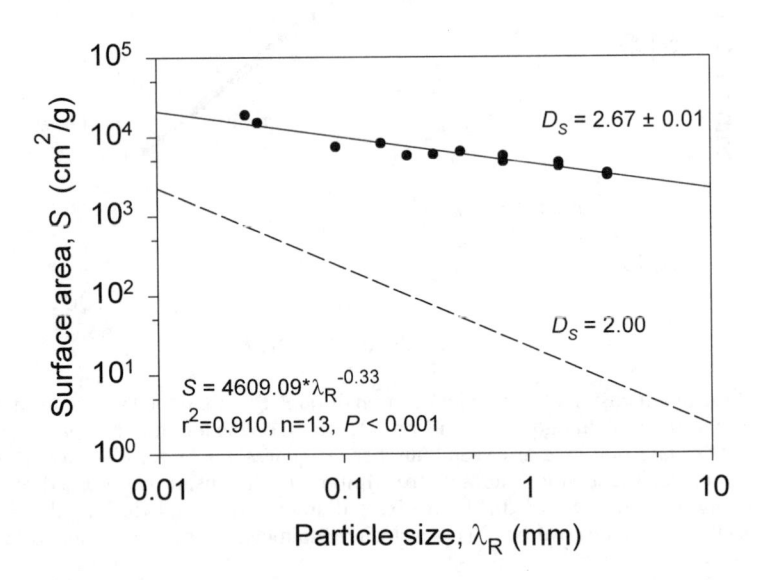

Fig. 14. Relation between log particle size (λ_R) and log particle surface area (S) cm² g⁻¹ dry weight (DW). The full line shows the regression fitted to data obtained by BET isotherms, resulting in an expected surface fractal dimension (D_S) of 2.7. The broken line shows the regression of the theoretical relation between spherical particles of a specific density and their expected surface area. Further details are given in section III.C.

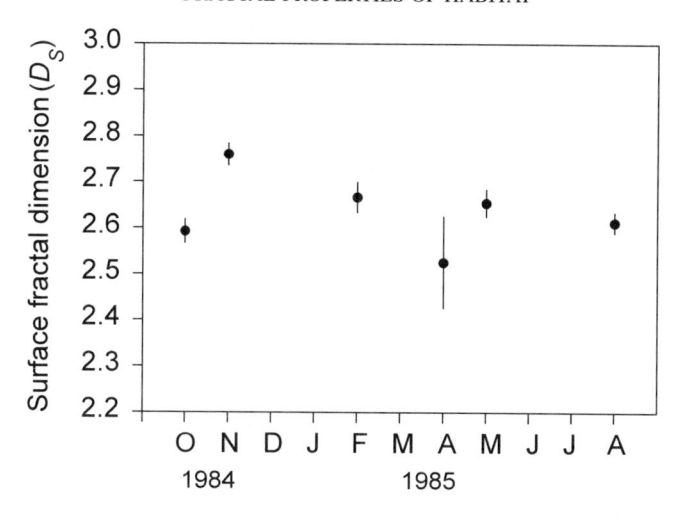

Fig. 15. Bias adjusted mean (± 1 SE) surface fractal dimension (D_S) of streambed sediments collected with freeze-cores (0–70 cm) in the Oberer Seebach on six sampling occasions during 1984 and 1985. Bias-adjusted mean and SE are based on 5 × 10⁴ bootstrap replications.

respectively; $P < 0.05$). The highest value of surface dimension coincided with periods of leaf-fall (November 1985). Thus, differences in fractal dimension may be due to seasonal changes in habitat complexity. Moreover, the heuristic estimates for the surface dimension based on particle–pore boundaries coincided with values based on the BET method (Figures 13 and 14; section IV.B).

3. Relation Between Body-size Distribution and Fractal Dimension

Using data from Schmid (1992, 1993), I tested the hypothesis that the number of larval individuals and species of chironomids found in the streambed sediments might be limited by available particle surface area. I assumed that there should be more small-sized individuals than larger ones, because each small-sized species uses less space. If animal body length is used as divider's length, and if body length is related to the resource use by animals (May, 1978), it is possible to combine body length with number of individuals (Morse et al., 1985; Gunnarsson, 1992). Assuming that the mean metabolic rate of insects scales as the 0.75 power of body mass (Peters, 1983; West et al., 1997) and that population densities may be proportional to the reciprocal of metabolic rate, Morse et al. (1985) suggested that population density (N) scales approximately with body length (L) as $N \propto (L^3)^{-0.75}$. Based on the empirically established linear relationship between the exponent b of the length–weight relationship and the maximum body length (Mackey, 1977; Nolte, 1991) of 10 chironomid species (Appendix A), the exponent b was interpolated for all other chironomid species within the length range of the species weighed (Appendix B).

To test both the 'energetic equivalence' and the inefficiency in energy transfer hypotheses, the relationship between log body size and log population abundance for all chironomid species was analysed using four different regression approaches (Blackburn et al., 1993). The relationship between log body mass and log population abundance for 64 species is shown in Figure 16. Although body size is a poor predictor of abundance (only 10% explained), it is evident that population abundance declines in a linear band with increasing body mass. Table 1 summarizes the different slope values calculated using different regression methods. Both calculated slopes, that based on ordinary least squares (OLS) and that based on the upper bound slope method (UBS_{OLS}), did not depart significantly from either -0.75 or -1.0 ($P > 0.05$; Tukey–Kramer test). In contrast, significant departures from the energy equivalence and inefficiency hypotheses were found for slope values based on reduced major axis (RMA) regressions, while the UBS_{RMA} slope did not depart significantly from -1.0 but departed from -0.8 (Table 1; $P > 0.05$; Clarke's T test). These results demonstrate that the assemblage slopes of the log mass–log abundance relation depend on the regression model used (Blackburn et al., 1993). The OLS slope

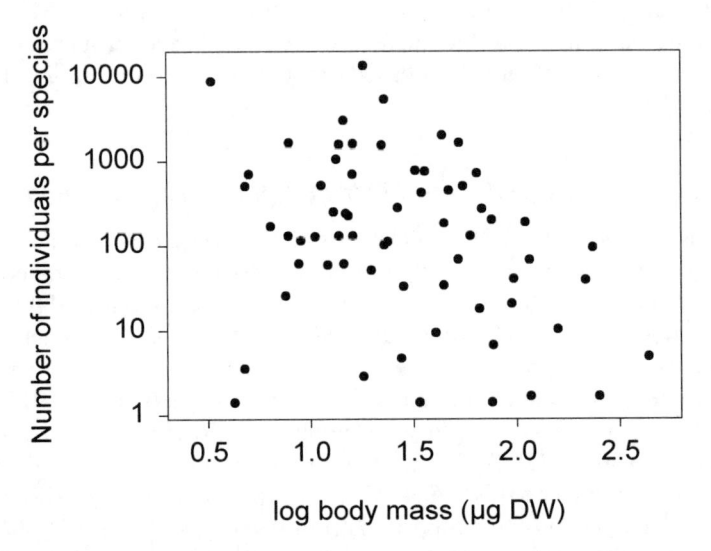

Fig. 16. Plot of log body size against log total number of individuals for each chironomid species found in 560 freeze-core samples of the streambed sediments in the Oberer Seebach during 1984 and 1985.

Table 1
Regression slope for the relationship between body size and abundance per species in a larval chironomid assemblage in the Oberer Seebach

n	r^2	b_{OLS}	b_{RMA}	$b_{UBS-OLS}$	$b_{UBS-RMA}$
64	0.104	−0.663 (−1.224 to −0.101)	−2.058*† (−2.611 to −1.612)	−1.367 (−2.281 to −0.452)	−1.561* (−2.487 to −0.980)

OLS, ordinary least squares; RMA, reduced major axis; UBS, upper bound slope. Body size is given as dry weight (μg). r^2 is the variance explained by the correlation of body size with all species in the assemblage. b is the regression slope through all species in the assemblage, calculated using either OLS, RMA or UBS regression. Ninety-five per cent confidence limits are given for OLS and RMA (McArdle, 1988). *OLS regression coefficients significantly different from −0.75, or RMA slopes significantly different from 0.80 (Clarke, 1980; $P \leq 0.05$); †OLS, RMA regression coefficients significantly different from −1 ($P \leq 0.05$, T test).

is nearer to − 0.75 than to −1.0 as predicted by the 'energy equivalence' hypothesis, whereas the value of the RMA slope indicates that small species control a greater proportion of available resources. The upper boundary of the relationship between mass and abundance is where resource use is most likely to be limiting (Lawton, 1989). However, the data on the chironomid assemblage yield weak evidence for resource constraints on abundances, particularly of small species.

The relation between sediment fractals and scaling of body lengths with number of individuals is obtained by combining the increase in surface area due to D_{BM} with the expected increase in the number of individuals with decreasing body length. It follows from $L(\lambda) \propto \lambda^{1-D}$ that, for a decrease in body length from 10 to 1 mm, the potentially available surface area for benthic organisms increases between 4.68 and 21.88 times. If body length of larval chironomids scales with population density as $(L^3)^{-0.75}$, a decrease in body length by a factor of 10 may result in a density increase by a factor of 178. Thus, the predicted slopes in the streambed sediment range between −2.92 and −3.59. This prediction was tested using the frequency distribution of logarithmic length classes of chironomids in the total sediment column (0–70 cm; Figure 17). The slope of the log body length–log density regression was estimated for all size classes to the right of the highest frequency class. The slope was −2.95 (± SE 0.25) and fell within the range of predicted slopes (Figure 17) based on the fractal dimension of the particle–pore boundary (section IV.B).

The hypothesis that variation in larval densities and species richness may be related to changes in habitat complexity was examined. The total number of individuals found at each of six sampling occasions was significant and curvilinear related to the estimated surface fractal dimension D_S (Figure 18). Moreover, a significant positive curvilinear relation was found between the number of species and values of fractal dimension (Figure 18). Further evidence that habitat complexity and not habitat space may be the main

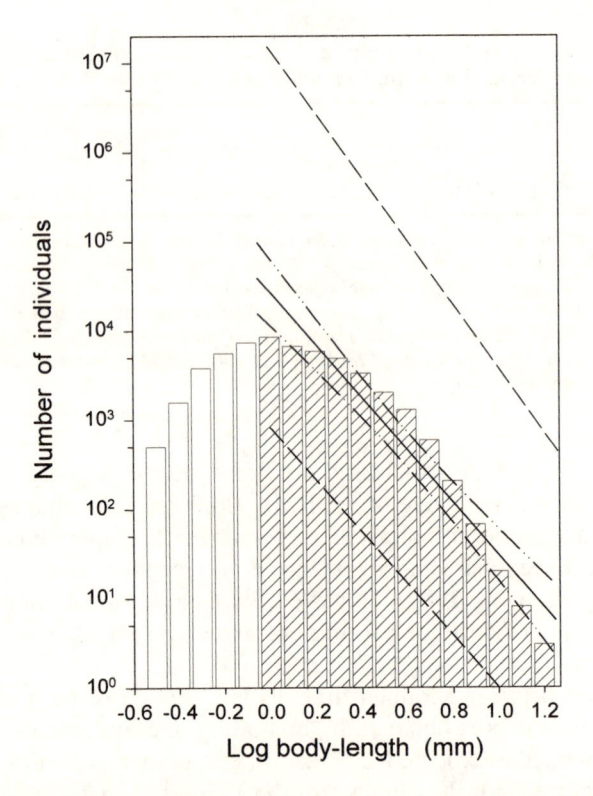

Fig. 17. Body length distribution, in millimetres on a log scale, of all chironomids in the total sediment column (0–70 cm) of the stream Oberer Seebach. Regression (± 95% confidence intervals) fitted through the histograms, forming the upper tail of the distribution (diagonal hatched), approximates the decline of individuals with body length and gives a slope of –2.95. The broken lines give the upper (–3.59) and lower (–2.92) bound prediction of slopes in the bed sediments.

structuring force is given by a weak, but significant and negative, correlation between the percentage of interstitial space and D_S ($r_s = -0.524$, $t_{30} = -3.37$, $P = 0.002$). As the interstitial space decreases, smaller particles fill the space between the larger grains, thus increasing the surface fractal dimension and habitat diversity.

In the Oberer Seebach, particles in the size range > 0.01 to > 100 mm displayed a similar surface fractal dimension (D_S) to those for limestone sediments reported by Turcotte (1986), Wong et al. (1986) and Krohn (1988). The fractal dimension of particle surface areas (D_S) was highest in the period of leaf-fall (see Figure 15), coinciding with higher larval densities and species richness compared with other seasons. There is further evidence that higher

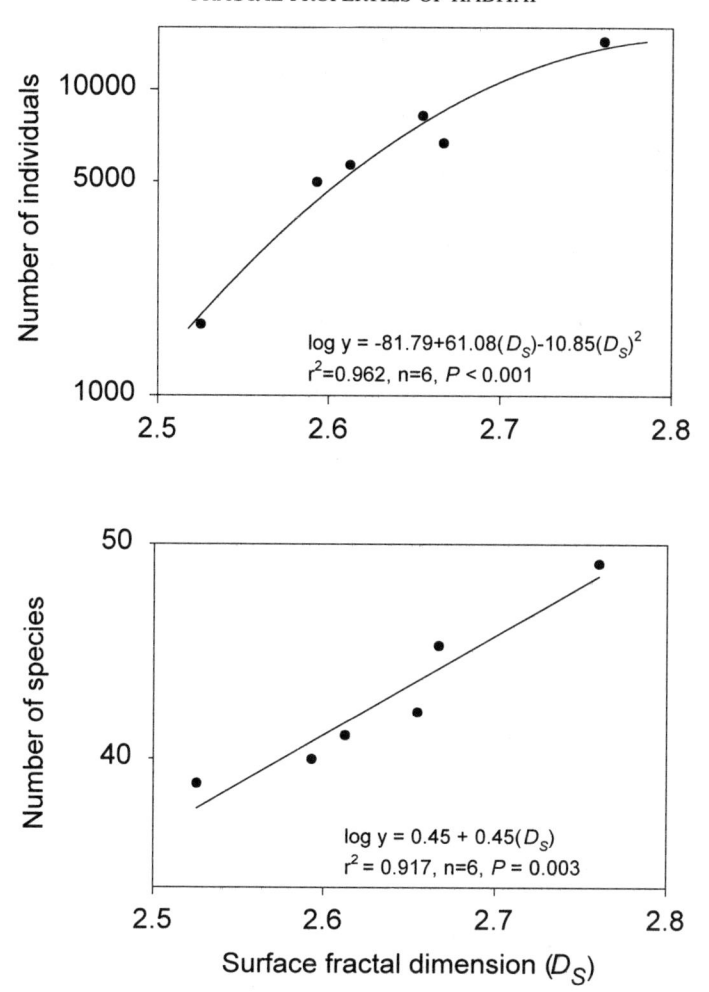

Fig. 18. Total number of larval individuals and chironomid species found in the streambed sediments on each of six sampling occasions during 1985 and 1986, plotted against the bias-adjusted mean surface fractal dimension (D_S).

fractal dimension of sediment surfaces is related to an increased habitat diversity from a study conducted in an Australian stream (B. Robson, personal communication). Robson used fractal geometry to describe the physical structure of both bedrock and cobble-bed riffles over several horizontal scales, ranging from 10 cm to 7 m. As expected, cobble-bed riffles were significantly more complex than bedrock riffles at most of these scales. Examination of the body-size distribution of benthic invertebrates in these habitats showed a broader range of maximum body sizes in cobble-bed riffles, along with a larger number of taxa than in bedrock riffles (B. Robson, personal communication).

The values of fractal dimension empirically estimated for the outline of particle–void interfaces in vertical and horizontal sections of the streambed sediments of the Oberer Seebach showed a similar range to those of plants (Morse et al., 1985; Shorrocks et al., 1991; Gunnarsson, 1992), but with a higher mean value. Assuming that a mean D_{BM} value of 1.67 is representative for cross-sections of the bed sediments, the qualitative prediction about scaling of body size with number of individuals was supported for length classes of larval chironomids larger than 1.0 mm. The observed slope of the log abundance–log length relation of all larval chironomids (Figure 17), found at depths down to 70 cm, was similar to the expected range for both D_S and D_{BM}. In contrast, if the habitat does not have a fractal character, then $D_{BM} = 1$ and $D_S = 2$, and the expected slope of the chironomid body-size frequency distribution would be 2.3.

The negative linear relationship between density and body size may break down if species use different volumes of space. Species that use the environment in three dimensions show distinct variation around the regression coefficients of abundance–body weight relations. The scattered, weakly correlated, pattern found for the abundance–body mass relationship of larval chironomids implies that, for a given body size, species that inhabit the streambed sediments in three dimensions exhibit a wide range of densities. Thus, there is no evidence that chironomid abundance is strongly allometrically related to body length. Moreover, Tokeshi (1990) showed that the population density of chironomids on the macrophyte *Myriophyllum spicatum* was independent of body size. The dichotomy in the log abundance–log body mass relation implies both energy equivalence and inequality within the chironomid assemblage. Guilds of larger-sized species might show a stronger tendency for equivalence than small-sized species. Previous results on an assemblage of small-sized chironomid species showed that stochastic factors (i.e. interstitial water through-flow variations) may influence the species composition (Schmid, 1993, 1997) and, therefore, the length–frequency distribution of larvae.

A weak point in the hypothesis proposed by Morse et al. (1985) lies in the energetic argument about how population densities scale with body size (Gunnarsson, 1992). Most chironomid populations in streams rich in fine particulate organic matter may not be energy limited, and competition for food and space may play a minor role in structuring those assemblages (Schmid and Schmid-Araya, 1997). Moreover, population densities appear to be independent of body size, suggesting that energy use by most species in the chironomid assemblage is rather inequitable, a view expressed for communities in general (Griffiths, 1998).

C. Fractal Dimension Across Several Spatial Scales

Many factors, such as hydrophysical (e.g. hydraulic shear stress), chemical (e.g. biofilm availability) and biotic (e.g. competition, predation), influence patch structure and, therefore, the spatiotemporal distribution of benthic organisms. Benthic environments are composed of a mosaic of patches, but these are not self-evident and must be defined in relation to the phenomenon under consideration. Patches can be represented as discrete areas (i.e. spatial domain) or periods (i.e. temporal domain) of relatively homogeneous environmental conditions where the patch boundaries are distinguished by discontinuities in the variables. Patches may be appropriately defined by non-random distribution of resource use among environmental units, as recognized in the concept of 'grain response' (Wiens, 1976). Patches are dynamic and occur on a variety of spatial and temporal scales, and might therefore demonstrate fractal properties. A patch at a given scale has an internal structure that is a reflection of patchiness at finer scales, and the mosaic containing that patch has a structure that is determined by patchiness at broader scales (Kotliar and Wiens, 1990). Thus, regardless of the basis for defining patches, benthic environments comprise a hierarchy of patch mosaics across a range of scales.

To study the patch structure and scales of spatial heterogeneity of benthic species, three randomly chosen cross-sectional transects were sampled in the gravel stream Oberer Seebach in July 1992 (J.M. Schmid-Araya, S. Wiedenbrüg and P.E. Schmid, unpublished data). The first cross-sectional transect (T1) passed across a slow-flow riffle zone (mean ± 1 SE: 24.6 ± 10.3 cm s^{-1}), the second transect (T2), 40 m downstream from the first, passed across a slow-flow area (29.7 ± 8.7 cm s^{-1}). The third transect (T3) was located 30 m downstream of T2 and passed across a fast-flow riffle zone (83.8 ± 34.8 cm s^{-1}). Quantitative samples were taken from the sediment surface of the streambed using a modified Hess sampler (0.024 m^2; mesh net 80 μm). A total of 70 samples were taken across the flooded area of the streambed from bank to bank by carefully positioning the samplers simultaneously one beside each other. Substrate heterogeneity was measured using a modified profiler, which consists of a plate with steel needles vertically positioned at 2-cm intervals across the sediment surface. The profiler is set parallel at a constant distance to the substrate surface and the steel needles are pushed downwards until they contact the sediment surface. The data obtained from the profiler were used to calculate the relative roughness or heterogeneity of the substrate surface using a three-point moving average similar to the approach of De Jong and Ergenzinger (1995). In addition, the direction of water flow at each sampling position was measured with a compass meter.

In the laboratory, larval chironomids and both immature stages (larvae and nymphs) and imagines of Hydrachnellidae, which numerically dominated at the sediment surface, were sorted and identified to species level. The body-length and width of all specimens were measured to the nearest 0.01 mm using an eyepiece micrometer.

To summarize the variance on sediment roughness and the distribution of organisms as a function of scale, the semivariance or structure function of each variable was calculated as:

$$\gamma(h) = n(h)\, 0.5 \sum_{i=1}^{n(h)} (z(i) - z(i+h))^2$$

where h is the distance (cm), n is the number of pairs of points separated by distance λ, $z(i + h)$ is the density of an organism or the value of a hydrophysical variable separated from point i by distance h. Lags up to 50% of transect length where included in the analysis, because the number of point pairs decrease at large separation distances. The behaviour of the variogram $2\gamma(h)$ and covariance near the origin provides information on the continuity properties of the spatial random process $z(i)$. The possible discontinuity of γ at the origin is the nugget effect (c_0), which is caused by microscale variations. Both the variance of the white-noise process, the inherent site variability, and the measurement-error variance represent the nugget effect (Cressie, 1991). The sum of the nugget effect and structural variance (c) defines the sill, beyond which data are spatially independent of one another. To analyse the transition between stationary and non-stationary patterns of $z(i)$, isotropic semivariogram models were fitted to the biotic and abiotic data (Table 2). These models differ by the importance of stochastic, as opposed to period, components (Cressie, 1991).

Substrate roughness displayed a pattern of small-scale linearity and spatial independence at larger scales in each of the three transects (Figure 19). Figure 19 shows that, at larger scales, sediment roughness appeared to be rather homogeneous in transect 1. Transects 2 and 3 were characterized by a more heterogeneous surface sediment and stronger spatial dependence on a scale less than 28 cm (Table 2). Surface roughness fitted significantly to the exponential semivariogram model at distance intervals of both 2 and 17.5 cm in each of three cross-sections (Table 3; $P < 0.001$).

Figure 20 summarizes the semivariograms for the spatial distribution pattern of the two most abundant species, the detritivorous *Parorthocladius nudipennis* K. and *Corynoneura lobata* E., at the sediment surface. The semivariograms of the three transects displayed constant slopes across the scales considered (Figure 20). Moreover, both abundant species exhibited distribution patterns of statistical self-similarity, which imply scale independence across streambed habitats ranging between from centimetres to several metres. The semivariance across transects for *P. nudipennis* and *C. lobata* followed the

Table 2

Fitted semivariogram models and fractal dimension for sediment surface roughness, water current direction and the distribution pattern of four abundant benthic invertebrate species (72.8%) in the Oberer Seebach in July 1992

	Model	RSS	Sill	D	r^2	P_L	D_P	r^2_S
Roughness								
T1–2 cm	$\gamma(h) = 0.023 + 0.034[1 - e^{(-h/120.16)}]^a$	4.67×10^{-4}‡	0.057	1.54 ± 0.09	0.916‡	62	1.75 ± 0.23	0.677‡
T2–2 cm	$\gamma(h) = 0.007 + 0.041[1 - e^{(-h/26.70)}]^a$	2.55×10^{-3}‡	0.049	1.46 ± 0.15	0.931‡	12	1.30 ± 0.09	0.985‡
T3–2 cm	$\gamma(h) = 0.001 + 0.056[1 - e^{(-h/27.90)}]^a$	1.61×10^{-2}‡	0.057	1.47 ± 0.09	0.954‡	18	1.35 ± 0.07	0.983‡
T1	$\gamma(h) = 0.021 + 0.034[1 - e^{(-h/123.62)}]^a$	3.43×10^{-4}‡	0.056	1.88 ± 0.18	0.864‡	88	1.85 ± 0.41	0.867*
T2	$\gamma(h) = 0.010 + 0.031[1 - e^{(-h/36.40)}]^a$	9.67×10^{-5}‡	0.041	1.89 ± 0.28	0.939†	53	1.89 ± 0.36	0.992*
T3	$\gamma(h) = 0.003 + 0.052[1 - e^{(-h/31.80)}]^a$	1.61×10^{-3}‡	0.056	1.72 ± 0.34	0.948*	35	n.a.	n.a.
Direction								
T1	$\gamma(h) = 0.108 + 0.027 h^{0.60}]^b$	5.89×10^{-2}‡	—	1.74 ± 0.13	0.921‡	53	1.47 ± 0.01	0.999‡
T2	$\gamma(h) = 0.001 + 0.003[1 - e^{(-h^2/1.7 \times 10^4)}]^c$	2.75×10^{-7}‡	0.003	1.59 ± 0.12	0.950‡	105	1.68 ± 0.34	0.893†
T3	$\gamma(h) = 0.267 + 2.175[1 - e^{(-h^2/5.3 \times 10^3)}]^c$	6.54×10^{-1}‡	2.442	1.52 ± 0.09	0.979‡	70	1.42 ± 0.03	0.999‡
Parorthocladius nudipennis								
T1	$\gamma(h) = 0.064 + 0.304[1 - e^{(-h^2/1.7 \times 10^5)}]^c$	2.34×10^{-3}‡	0.368	1.88 ± 0.54	0.687‡	123	1.86 ± 0.61	0.619*
T2	$\gamma(h) = 0.023 + 0.083[1 - e^{(-h^2/3.7 \times 10^5)}]^c$	1.69×10^{-4}‡	0.106	1.69 ± 0.15	0.938‡	105	1.76 ± 0.41	0.689*
T3	$\gamma(h) = 0.052 + 0.038[1 - e^{(-h^2/2.9 \times 10^5)}]^c$	2.98×10^{-3}†	0.090	1.87 ± 0.31	0.860‡	70	1.85 ± 0.67	0.857*
Corynoneura lobata								
T1	$\gamma(h) = 0.023 + 0.082[1 - e^{(-h/90.30)}]^a$	3.30×10^{-3}‡	0.105	1.86 ± 0.58	0.767*	53	n.s.	n.s.
T2	$\gamma(h) = 0.066 + 0.251[1 - e^{(-h^2/4.8 \times 10^4)}]^c$	3.99×10^{-3}‡	0.317	1.76 ± 0.43	0.662†	105	n.s.	n.s.
T3	$\gamma(h) = 0.212 h^{0.279} d$	2.63×10^{-3}‡	—	1.87 ± 0.54	0.666*	35	n.a.	n.a.
Nilotanypus dubius								
T1	$\gamma(h) = 0.083 + 0.080[1 - e^{(-h^2/2.9 \times 10^5)}]^c$	6.43×10^{-3}‡	0.163	1.88 ± 0.40	0.564‡	140	n.s.	n.s.
T2	$\gamma(h) = 0.107 + 0.180[1 - e^{(-h^2/5.5 \times 10^5)}]^c$	3.54×10^{-3}‡	0.287	1.87 ± 0.62	0.505*	70	1.84 ± 0.18	0.998*
T3	$\gamma(h) = 0.110 + 0.554[1 - e^{(-h^2/6.8 \times 10^5)}]^c$	9.21×10^{-4}†	0.664	1.85 ± 0.48	0.786*	35	n.a.	n.a.
Torrenticola elliptica								
T1	$\gamma(h) = 0.081 + 0.209[1 - e^{(-h^2/4.5 \times 10^5)}]^c$	8.97×10^{-3}‡	0.290	1.75 ± 0.18	0.844‡	123	1.83 ± 0.70	0.728*
T2	$\gamma(h) = 0.032[1 - \sin(0.135h/(0.135h)]^e$	7.09×10^{-4}*	0.032	n.s.	n.s.	35	n.a.	n.a.
T3	$\gamma(h) = 0.001 + 0.236[1 - e^{(-h/51.26)}]^a$	3.69×10^{-3}‡	0.236	1.72 ± 0.25	0.889‡	70	1.60 ± 0.19	0.973*

Models were fitted using the Marquardt method with a maximum number of 10^3 iterations. Models presented are the most significant fit of the data based on RSS (reduced or error sums of squares). The Sill represents the spatially independent variance; P_L is the patch length (cm); D is the fractal dimension (± 1 SE) of the stationary part of transect length and D_P is the fractal dimension (± 1 SE) of patch length. [a]Exponential; [b]power; [c]gaussian; [d]fractional brownian; [e]hole (Cressie, 1991). T1 to T3 are the three transects sampled at 17.5-cm intervals; T1–2 cm to T3–2 cm represent the surface roughness measured at 2-cm intervals in each of three transects. *$P < 0.05$, †$P < 0.01$, ‡$P < 0.001$; n.s. not significant; n.a. value cannot be calculated owing to small sample size (< 4).

Surface roughness

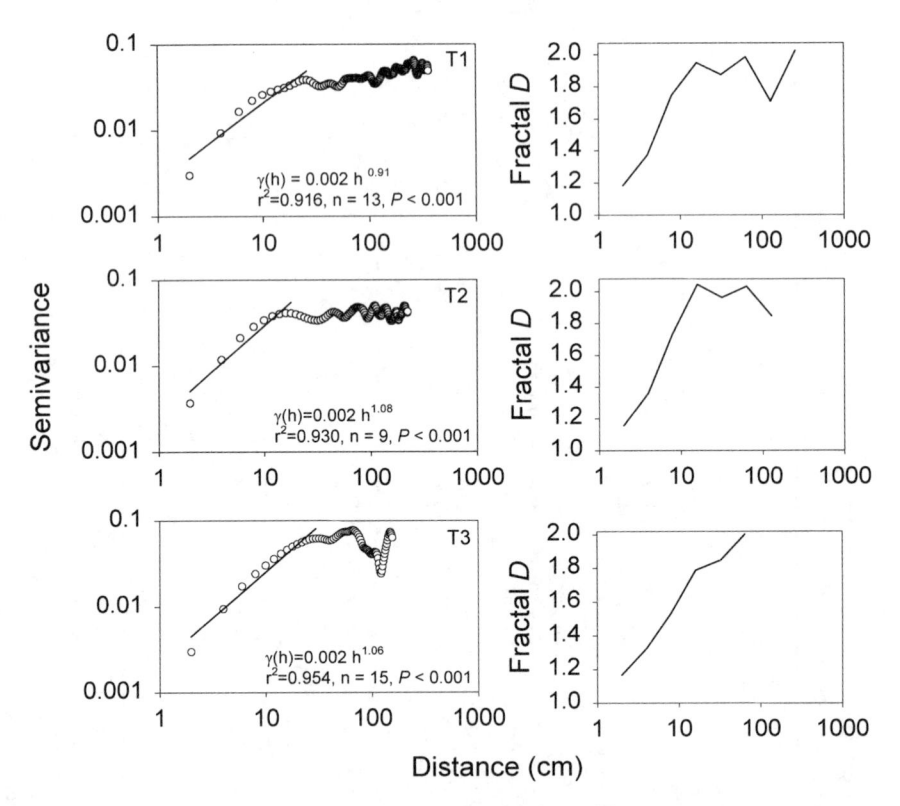

Fig. 19. Semivariograms (left panel) and fractograms (right panel) on variations of sediment surface roughness observed across three transects (T1–T3) in the Oberer Seebach in July 1992. Fractograms are shown for spatial scales ranging over two to three orders-of-magnitude depending on transect length. Best-fit regressions are given for the stationary part of transects on log distance versus log semivariance. Fractal dimension (D) of the sediment structure is given in Table 2.

gaussian model, which is similar to the exponential model but assumes a gradual rise in reaching the sill. The exponential and gaussian models are characterized by the importance of the stochastic, compared with the periodic, component (Robert and Richards, 1988).

The spatial pattern of current direction and other abundant species across transects exhibited a variable site-specific pattern, which could not be predicted by a single model (Table 3). A pattern of small-sized ($\leq$ 35 cm), roughly equidistant patches was found for the hydrachnellid species *Torrenticola elliptica* M. in the cross-section characterized by slow-flow

Table 3
Results of Kruskal–Wallis tests for differences of density patch length (P_L), fractal D and D_P values between benthic taxa, species and transects in the Oberer Seebach in July 1992

	d.f.	P_L T	P_L P	D_P T	D_P P	D T	D P
Transects	2	12.14	0.001	0.78	0.682	3.03	0.235
Species	13	16.67	0.193	9.98	0.583	14.39	0.358
Taxa	1	1.72	0.188	1.15	0.291	0.16	0.702

Probability estimates are based on 10^4 Monte Carlo simulations.

conditions. This pattern resulted in a significant fit to the Hole or Wave model ($P < 0.05$), which is dominated by a periodic component (Table 2).

Following Burrough (1983), the generalized fractal dimension D can be estimated from the slope b of the log–log relation between $z(i)$ and h by the function $D = (4 - b)/2$. In non-linear models, the values below the sill were used to fit least-square regressions. Semivariograms generally tend to deteriorate with increasing spatial lag, so that a criterion is needed for linear semivariogram models when deciding upon an appropriate range of lags to include in the regression. The values of lags that maximized the coefficient of determination of least-square regressions were used for linear models. For these one-dimensional transects, D ranges between 1 and 2; larger D values reflect the dominance of small-scale effects and persistently larger-scale variations result in smaller D values. The fractal dimension (D_P) was also calculated over the estimated patch length (see Sinsabaugh et al., 1991). Patch length (P_L) was defined as the lag at which the spatial autocorrelation coefficient is close to zero. Changes in fractal dimension when shifting between scales may delineate transition zones separating different scale ranges, where changes in environmental properties or constraints acting upon organisms or habitat structure are apparent (Frontier, 1987). Multifractal patterns may be apparent where surface roughness and/or distribution patterns of organisms vary with dimension. To analyse possible scale-dependent changes in dimension, the fractal D was plotted as the change in variance of surface roughness and densities with increasing inter-point distances (see Palmer, 1988).

The fractal dimension of surface roughness ($\geqslant 2$ cm intervals) increased from 1.2 at the smallest spatial scale to 2.0 at a scale larger than the estimated substrate patch length, which ranged from 12 to 26 cm (Figure 19; Table 3). At scales larger than 12 cm the fractal D values were a constant function of scale and, therefore, spatial structure of surface roughness appeared to be statistically self-similar. The fractograms on the spatial structure of the two abundant chironomid species are given in Figure 20. The fractal dimension varied between and within cross-sections, and only the D values of the spatial

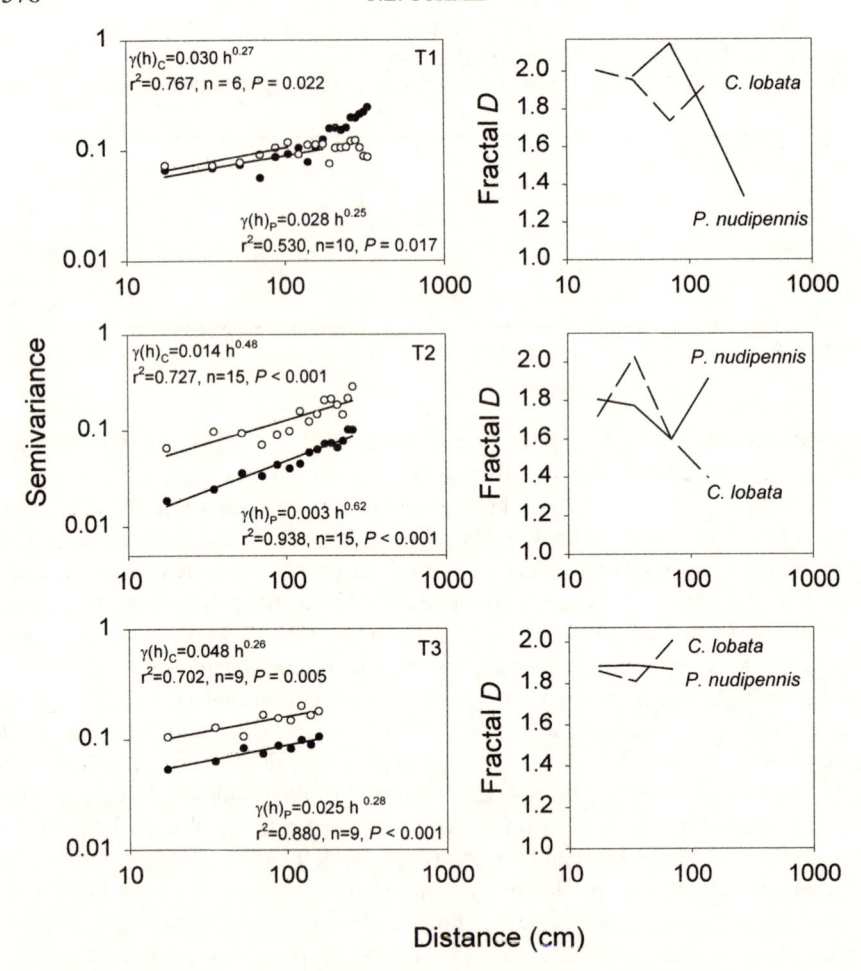

Fig. 20. Semivariograms (left panel) and fractograms (right panel) of the spatial density structure of the two most abundant invertebrate species *Parorthocladius nudipennis* (●) and *Corynoneura lobata* (○) observed along three transects (T1–T3) in the Oberer Seebach in July 1992. Fractograms are shown for spatial scales ranging over one to two orders of magnitude, depending on transect length. Best-fit regressions on log distance versus log semivariance, $\gamma(h)_C$ and $\gamma(h)_P$ are given for *C. lobata* and *P. nudipennis* respectively. Fractal dimension (D) of the spatial density structure is given in Table 2.

structure of *P. nudipennis* in the riffle zone were a constant function of scale (T3; Figure 20). In general, the D values of the spatial structure at scales < 1 m were closer to 2 than to 1 for both species, implying weak spatial dependence. At increasing scales (> 1 m) the spatial structure of *P. nudipennis* appeared to follow a distinct gradient of increasing surface roughness in T1 (Figure 20).

At the spatial scale considered, patch lengths (P_L) of the 14 common species varied from 35 to 123 cm, and were always larger than the single sampling area (> 17.5 cm). No apparent differences in P_L were found between larval chironomids and mites (Table 3; $P > 0.1$). However, a comparison of P_L values between transects showed that patches extended over a significantly wider range in T1 compared with the other cross-sections (Table 3; $P < 0.010$). The transect passing across the riffle zone (T3) exhibited the shortest patch lengths both for substrate and densities (Table 3), which could be attributed to higher variations in water flow (greater turbulences). Fractal D values ranged between 1.68 and 1.98 (mean ± 1 SE: 1.86 ± 0.01) and were similar between taxa, species and transects (Table 3; $P > 0.1$). Similarly, D_p values of patch structure ranged between 1.55 and 1.98 (mean ± 1 SE: 1.80 ± 0.03), without apparent differences between taxa, species and transects (Table 3; $P > 0.1$). The D values were on average 0.099 higher than D_p, which suggests that complexity increased due to spatial random variations between patches (Schmid, 1993).

The spatial density structure of three hydrachnellid and 11 chironomid species was as complex as the substrate structure across transects at scales ≥ 17.5 cm (permutation test: $\Delta = -1.449$, 39 d.f., $P_{\text{2-sided MC}} = 0.149$). The apparent similarity in fractal D values suggests that the spatial distribution pattern at the sediment surface may be at least partly mediated by habitat complexity. In contrast, density patch structure (D_p) of the 14 species was less complex than substrate patch structure (permutation test: $\Delta = -2.471$, 21 d.f., $P_{\text{2-sided MC}} = 0.011$).

Density variations among species of different body size may result in variations of spatial density structure. Mean body length was significantly and negatively correlated with the D values of spatial density structure across the transects ($r_s = -0.767$; $t_{12} = -4.14$, $P = 0.001$). Thus, smaller-sized individuals appear to form a more complex spatial pattern than larger-sized species at the sediment surface of the Oberer Seebach. A similar complexity of surface roughness and density structure at scales ≥ 17.5 cm implies that heterogeneous habitats are more likely to be inhabited by more smaller-sized than larger-sized species.

At a smaller spatial scale, epilithic microbial communities have a central role in the transfer of matter and energy. These communities might serve to elucidate the fine-scale organization by empirically analysing the patch structure over a range of scales. Sinsabaugh et al. (1991) studied the epilithon patch structure in a fourth-order stream in northern New York State to identify the scale at which the major processes responsible for epilithon patch structure operate. Benthic and suspended sampling units, which consisted of 120 cm long strands of 5 mm glass beads oriented parallel to the current, were placed in a riffle and a slow-flowing channel 2 months before sampling. Some 30 benthic and 12 suspended sampling units were collected during a spring, summer and winter period. The contour of the sediment surface beneath the strands of glass beads was mapped, and the patch development was examined at a 1 cm resolution assaying bead pairs for either chlorophyll or alkaline phosphatase activity. Patch structure for each sampling unit was examined as patch length, patch amplitude and fractal dimension.

Patch length, estimated by Moran's I, was 10 ± 8 cm, roughly half the length of the cobble and boulder substratum (22 ± 9 cm), with no apparent differences between sites and seasons. Patch amplitude, based on chlorophyll concentrations and alkaline phosphatase activity, did show spatial and temporal patterns that were related to water-flow and season, respectively. The average fractal dimension of chlorophyll concentration and alkaline phosphatase activity was greater ($D = 1.81$) than that of the substratum ($D = 1.25$), suggesting that epilithon patch structure emerged primarily from small-scale processes such as turbulence and macroinvertebrate activity (Sinsabaugh *et al*., 1991).

The data of Sinsabaugh *et al.* (1991) and of J.M. Schmid-Araya *et al.* (unpublished results) suggest that, across transects of riverine substrates, fractal dimension provides information on the spatial configuration of process–pattern interactions over several spatial scales. Changes in fractal dimension when shifting between scales, as observed across transects in the Oberer Seebach, may demarcate transition zones separating scale ranges dominated by different physical processes, where, for instance, hydrophysical constraints acting upon organisms change.

V. FRACTAL RANDOM WALKS

Considering patterns of patch dynamics observed in benthic environments, the aggregated patterns observed for motile organisms may be a consequence of their movement towards higher food concentrations. Motile meiobenthic and macrobenthic species display movements including straight trajectory paths which alternate with patterns similar to those of brownian motion (Figure 21). Food and microhabitat patches may occur randomly (Schmid and Schmid-Araya, 1997), thus the search path of a motile benthic organism may be described as random walk through interstitial voids. As soon as food patches are located, the larger-sized exploration path may be replaced by a convoluted movement pattern within a smaller spatial range (Figure 21).

In the laboratory, single meiobenthic (< 1 mm) and macrobenthic (1.0–4.5 mm) specimens were placed in 1 ml (10 cm^2) and 5 ml (40 cm^2) graticules, respectively. Fine particulate organic matter (FPOM; size range 0.05–1 mm) was added to the experimental units. The 'interstitial space' available was between 38% and 51%, compared with the total surface area. The movement of five meiobenthic and macrobenthic species was traced continuously under a microscope using video recording (SVHS 3-CCD JVC camera) for 17 min.

Conchapelopia pallidula

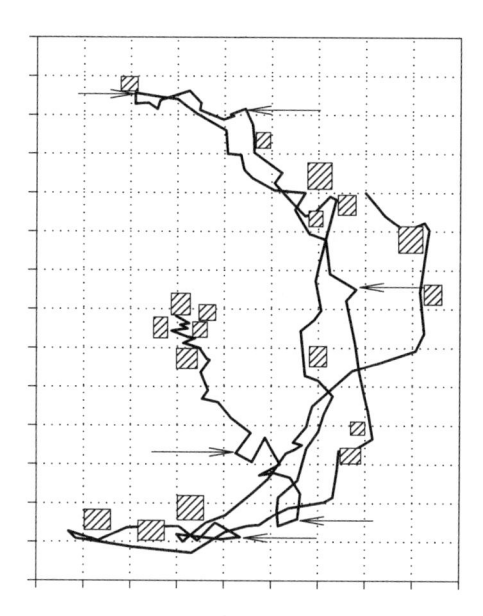

Fig. 21. Trajectory path of a larval individual of the predatory chironomid species *Conchapelopia pallidula* recorded over an observation period of 17 min. Arrows indicate locations of prey encounter. ▨, Particle patches near the trajectory.

Figure 22 shows the trajectory paths for both the filter-feeder bdelloid *Rotaria tardigrada* Ehrb. and one of its major predatory chironomid species *Conchapelopia pallidula* M. (Schmid and Schmid-Araya, 1997). The position of the individual is shown for every 8 s. Both species exhibit straight interpatch trajectories and short-step within-patch trails. In contrast to the tanypod, the trace of the bdelloid species shows strongly convoluted intersections, particularly in food-patch areas (FPOM aggregations). Shorter within-patch trails and sharp turns were observed for the predator at periods of prey encounter. In addition, the small-sized detritivorous orthoclad *C. lobata* E. showed a similar feeding path, with sharp turns in food-patch zones and straight interpatch trajectories (Figure 23). The monogonont *Cephalodella gibba* Ehrb., another common rotifer in the interstices of streambed sediments, showed a strongly convoluted trajectory, associated with particle patches (Figure 23).

To establish whether these trajectories represent random walks, the position of an individual within a lattice was recorded for each second. The displacement length was measured for each individual from the starting point (co-ordinate (0/0); Figures 22 and 23) of the trace in time. This process was

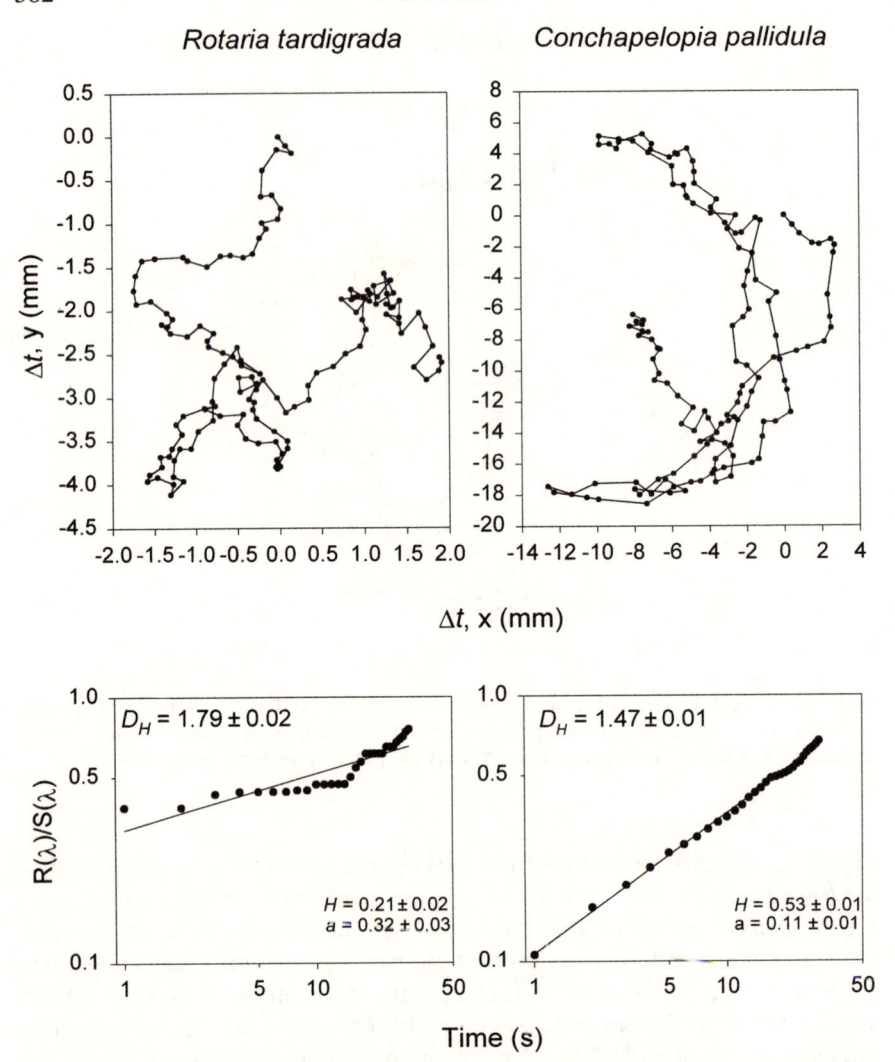

Fig. 22. Trajectory paths of a bdelloid rotifer (*Rotaria tardigrada*) and one of its predator species the tanypod *Conchapelopia pallidula* (here shown as instar II). Tracing started at the lattice point (0/0) and was followed continuously for 17 min. For clarity, dots represent the position of the individual after every 8 s. The regressions are the fitted curve of the rescaled range analysis: $R/S = (a\lambda)^H$, where λ is the lag in seconds and a is the intercept, and H is the Hurst coefficient. The trajectory path of the bdelloid and the tanypod individual fitted significantly to the model of fractal brownian motion ($r^2 = 0.787$ and 0.997 respectively, $P < 0.01$).

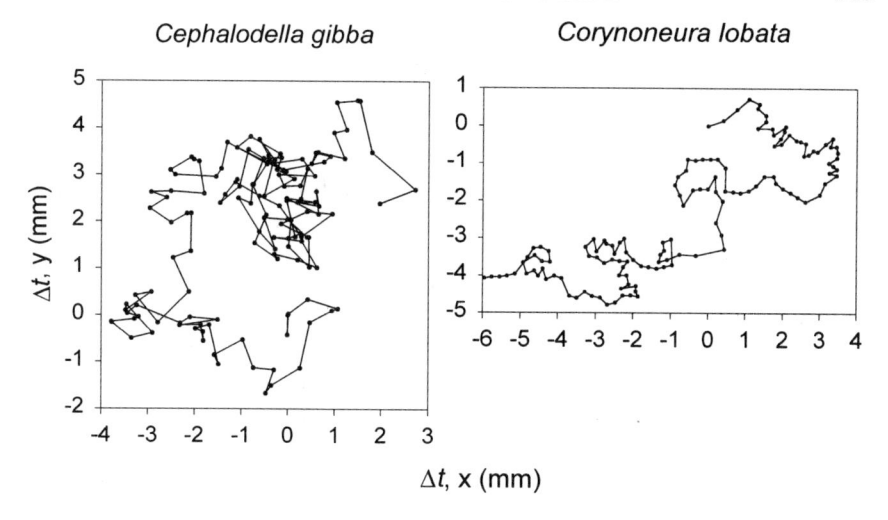

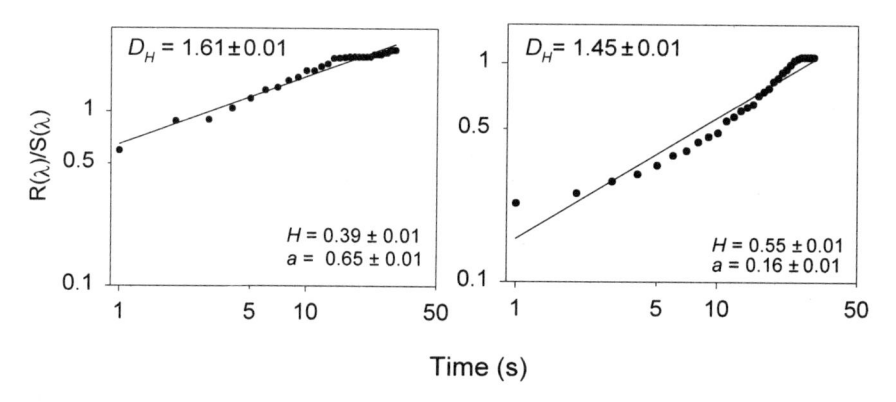

Fig. 23. Trajectory paths of one of the most common monogononth benthic rotifer species *Cephalodella gibba* and the most common detritivorous chironomid species *Corynoneura lobata* (here shown as instar II). The trajectory path of the rotifer and the chironomid individual fitted significantly to the model of fractal brownian motion ($r^2 = 0.969$ and 0.944 respectively, $P < 0.001$). Details are given in the legend for Figure 22.

repeated 1025 times (17 min) and the length of the displacement vector (B) plotted for each second.

The trajectory path of organisms is essentially a self-affine profile, because time and position are scaled with different ratios. The method of rescaled range (R/S) or Hurst analysis provides a simple tool for measuring temporal patterns. Considering a temporal interval of lag λ in a trace, we can define within that lag the range taken by the y values (B) in that interval as $R(\lambda)$. This range is a random function with scaling property:

$$R(\lambda) \approx \lambda^H$$

where H is the Hurst exponent. Because of self-affinity, the range taken by the values of B in an interval λ should be proportional to the lag to a power equal to the Hurst exponent H. This range is defined as the minimum and maximum neighbour value along the time axis and is similar to the dilation used in estimating the Minkowski dimension for a self-affine profile (section III.A). However, R/S analysis differs from the Minkowski method, because it is the greatest difference between the maximum and minimum values among all data, not the total area swept out, that is plotted. To normalize the vertical scale and make it independent of the size of the data set, the difference is rescaled by dividing it by the standard deviation $S(\lambda)$ of the data. It follows that the rescaled range R/S is given by:

$$R(\lambda)/S \approx \lambda^H$$

where the fractal dimension of the trace D_H can be derived from H as $2 - H$. For comparative purposes, in addition to the R/S method, the RMS logic was applied to 61 trajectory paths of the five representative benthic species. Displacement vectors of similar length sequence in time would result in a fractal dimension close to 1, whereas higher values of fractal dimension reflect the temporal variability in spatial displacement.

Comparison of the four examples given in Figures 22 and 23, and the results of the trajectories of a predatory microturbellar *Stenostomum* sp. (Table 4), revealed that the movement path of the rotifer species had a higher fractal dimension than the two chironomid and microturbellar species. Thus, for benthic rotifers, the range of displacement lengths varies more in time and the search path is closer to a random walk ($D_H \approx 1.5$) compared with the other species. Trajectory paths show a high variability within species (Table 4) and the fractal dimension of the trajectories differed significantly between species and method estimations (Kruskal–Wallis test with Monte Carlo estimate: $T_4 = 18.31$; $P_{MC} < 0.001$ and $T_1 = 34.08$, $P_{MC} < 0.001$, respectively). Although these two methods are mathematically similar, each method measures slightly different aspects of the trajectory. The R/S method measures the vertical range of the profile, whereas the RMS method (section III.C) measures the standard deviation between two points of the trajectory. The Hurst R/S method produces D values that tend to be higher than those of the RMS method, and appear to be generally too high (Russ, 1994). However, the fractal dimension of the trajectory paths calculated for the RMS and R/S method was significantly

Table 4
Mean fractal dimension (95% CL) for the displacement vector based on trajectory paths obtained in laboratory experiments for five meio- and macrobenthic species. Movement paths were monitored for each individual per second over an observation period of 17-min. The fractal exponent was calculated using the root-mean-square deviation (RMS) and Hurst (R/S) method. n gives the number of individuals traced. L is the mean body length (95% CL) in mm. Bias adjusted mean values and 95% bootstrap CL are given based on 10^4 bootstrap replications.

Species	n	L	D_{RMS}	$D_{R/S}$
Cephalodella gibba Ehrb.	9	0.17 (0.09–0.34)	1.58 (1.49–1.67)	1.55 (1.48–1.61)
Rotaria tardigrada Ehrb.	12	0.39 (0.36–0.43)	1.53 (1.44–1.61)	1.58 (1.42–1.75)
Stenostomum sp.	6	0.56 (0.48–0.62)	1.48 (1.38–1.58)	1.48 (1.42–1.52)
Corynoneura lobata E.	12	1.02 (0.91–1.14)	1.34 (1.29–1.38)	1.46 (1.32–1.61)
Conchapelopia pallidula M.	22	2.20 (1.82–2.61)	1.21 (1.07–1.34)	1.35 (1.30–1.41)

and negatively correlated with individual body lengths ($r_s = -0.568$, $t_{59} = -5.29$, $P < 0.001$ and $r_s = -0.274$, $t_{59} = -2.19$, $P = 0.032$, respectively). Consequently, the mobility path of smaller species is characterized by more strongly convoluted trajectories, whereas larger-sized species are differentiated by a higher probability for linear spatial displacements.

The step length (λ) varied between 0.045 and 0.500 mm for *R. tardigrada* and between 0.250 and 2.625 mm for the tanypod *C. pallidula*. I used an approach similar to that of Weiss and Murphy (1988) to model the trajectory path over a surface between two reference points of maximum projected length (L_M), 1 m apart. I applied the structured walk model of Richardson as:

$$L_A(\hat{\lambda}) = \hat{\lambda}^{1-D}$$

where $\hat{\lambda}$ is the mean step length of a species and $L_A(\hat{\lambda})$ is the mean distance moved between two reference points. The fractal dimension of the path is given as:

$$D = 1 + |b| \qquad 1 < D < 2$$

where b is the slope of the power relation. Under laboratory conditions, the actual distance moved by individuals of *R. tardigrada* and *C. pallidula* relates to a linear displacement of given length by a factor of 3–5. Figure 24 shows the results based on the laboratory experiment and model extrapolations to natural habitats of different fractal dimension D. The fractal dimension was 1.12 and 1.16 for the rotifer and tanypod individual, respectively. Assuming fractal dimensions for sediment structures between 1.6 and 1.9, the actual distance moved linearly between two points 1m apart increases by a factor of 100–1000 for the bdelloid and by a factor of 30–200 for the chironomid species. Thus, distances moved by mobile organisms in the streambed sediments depend upon both the step length of the organism and habitat structure.

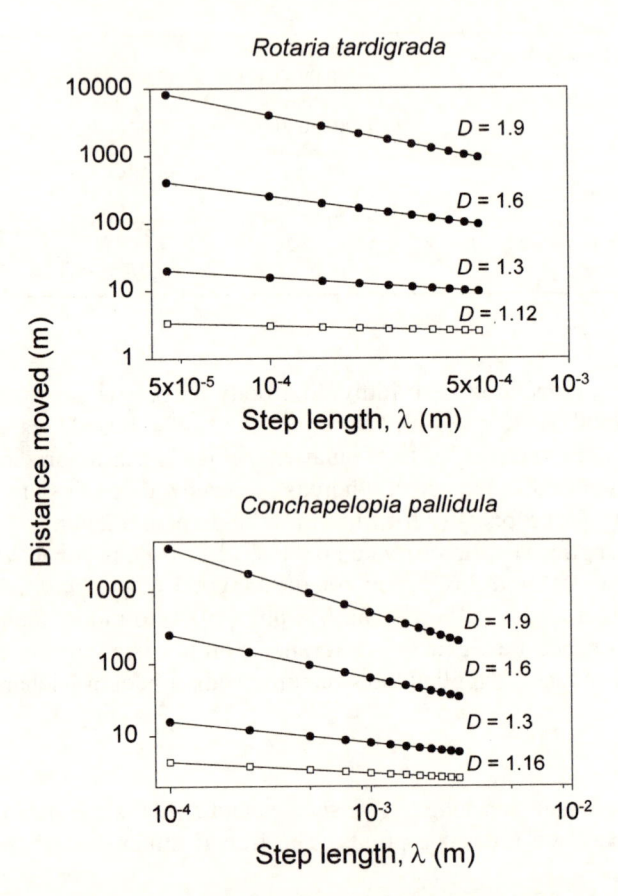

Fig. 24. Estimated distance moved (m) by the bdelloid rotifer *Rotaria tardigrada* and the tanypod *Conchapelopia pallidula* over a linear surface distance of 1 m with different habitat complexity. Regressions on step length (λ) *versus* distance moved (δ) represent the modelled trajectory following $\delta = \lambda^{1-D}$. Relationship between step length and distance moved (□), based on laboratory observations, resulted in fractal estimates (D) of habitat complexity, of 1.12 for *R. tardigrada* and 1.16 for *C. pallidula*. Modelled relation between step size and distances moved between two points linearly 1 m apart (●) is given for different fractal dimensions of habitat.

VI. FRACTAL PROPERTIES OF BIOTIC STRUCTURES

Besides physical features of the habitat, biotic structures may exhibit complex growth-related fractal properties. In aquatic environments, many sessile, often colony-forming, organisms form habitats for other organisms, and environmental heterogeneity may often be closely associated with growth and activity processes of those biotic features. However, few studies have been conducted in aquatic environments to determine the structure and scale-dependent properties of sessile organisms. The examples given below, therefore, cover a wide range of scale-related aspects.

A. Coral Reefs

Bradbury *et al.* (1984) investigated the fractal dimension D of an Australian coral reef crest. Using data on transects across the reef surface, they applied the divider's method to study the within- and between-scale variation of D. In order to examine the local variability of D at each of seven length scales, two interlaced subsequences were formed by selecting alternate observations from the original sequence of data. The interlaced results are shown in Figure 25 and imply that within-scale variation in D is small in the Myrmidon reef crest. Thus, scale-invariant dynamics within each size class generated the boundary features. They found that D declines from a value of about 1.1 at the finest scale (10 cm) to approximately 1.05 for intermediate scales (from 20 to 200 cm), and

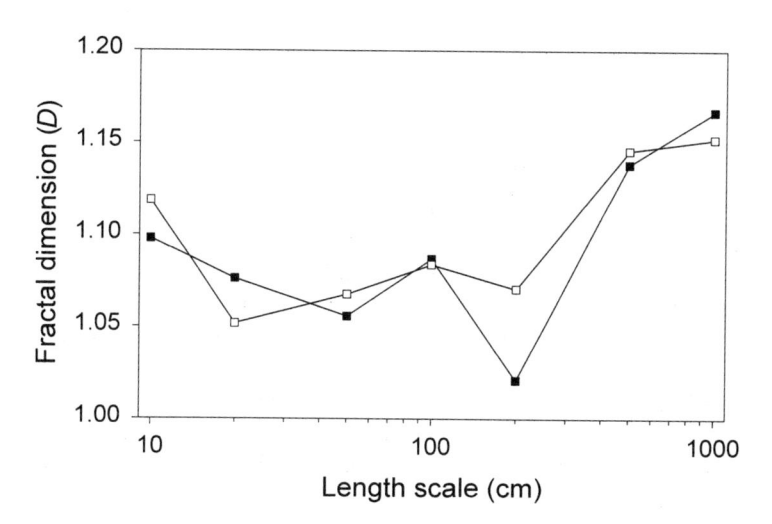

Fig. 25. Fractal dimension for two interlaced subsequences (within-scale variation) at each of seven length scales (between-scale variation) on Myrmidon coral reef (modified from Bradbury *et al.*, 1984).

rises markedly to approximately 1.15 at the largest scales (from 5 to 10 m). The finest scale of observation corresponds to the size of the branches and convolutions within individual coral colonies, whereas the boundary features of living colonies corresponds to the intermediate scale. The largest scales correspond to the major morphological structures such as groves and buttresses. The shifts in fractal dimension at different scales appear to show transition zones in the hierarchical organization of the coral reef.

B. Mussel Beds

The blue mussel *Mytilus edulis* L. forms dense beds that can cover extensive benthic marine areas, and these beds and the spaces between mussel individuals serve as habitat for other organisms. To study the spatial complexity and distribution of the blue mussel, Snover and Commito (1998) took areal-view photographs 1 m above the mussel beds in a population at Bob's Cove, an intertidal flat in Jonesboro, Maine. A quadrat (0.5 × 0.5 m) was placed in the field of each photograph for scale. Using the boundary-grid method, they estimated the outline of the mussels within each of 25 randomly chosen quadrats. The values of the box-counting dimension ranged from 1.36 to 1.86 and were significantly and curvilinearly related to mussel percentage cover and densities within the quadrats (Figure 26; $r^2 = 0.92$, $P < 0.01$; Snover and Commito, 1998). The highest fractal dimension was found at intermediate levels of both percentage cover and densities (Figure 26). The fractal dimension of the mussel outline declined significantly and linearly with increasing randomness value (Morisita's index). These results indicate that fragmented patch structure, which relates to a higher fractal dimension of the mussel bed, may be caused by physical disturbances such as storms and ice scouring.

C. Colonization of Artificial Pond Weeds

Aquatic macrophytes are an important and often abundant structural feature in benthic environments. They form habitat and are a food source for a variety of organisms. The structural features of macrophytes may alter food supply, and colonization patterns of mobile organisms may vary as invertebrates recognize differences between plant structures (Rooke, 1986). Jeffries (1993) tested the hypothesis that the abundance of invertebrate taxa and individuals will increase with increasing fractal complexity, by conducting invertebrate colonization experiments using artificial pond weed, which mimicked the shape of real plants. Artificial pond weed designs with the same surface area but different fractal dimensions were used in two field experiments. The first colonization trial was conducted using four types of artificial pond weed placed randomly among macrophytes in an old aquaculture pond. After roughly 7 months of exposure the artifical clumps were retrieved. The total number of

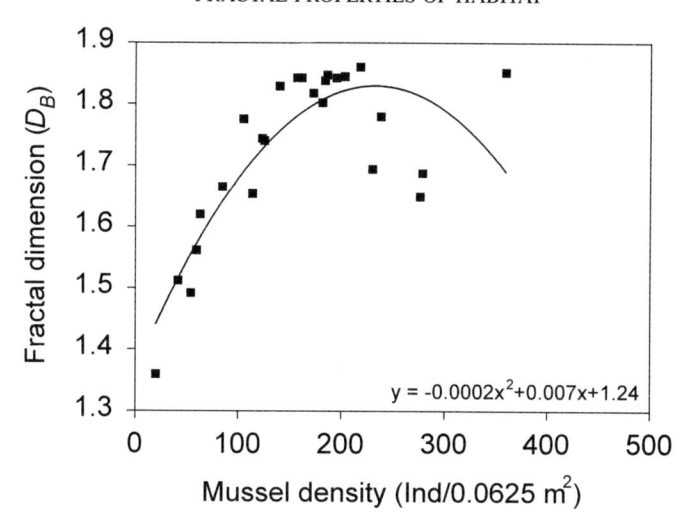

Fig. 26. Curvilinear relation between the fractal dimension (*D*) of the mussel outline and the density of *Mytilus edulis* L. in an intertidal flat of sand mud in Jonesboro (modified from Snover and Commito, 1998).

individuals, taxa and number of several species increased markedly with increasing fractal complexity. Figure 27 shows the significant polynomial relationship between total number of individuals and fractal dimension of the four weed designs ($r^2 = 0.339$, $P < 0.001$; Jeffries, 1993).

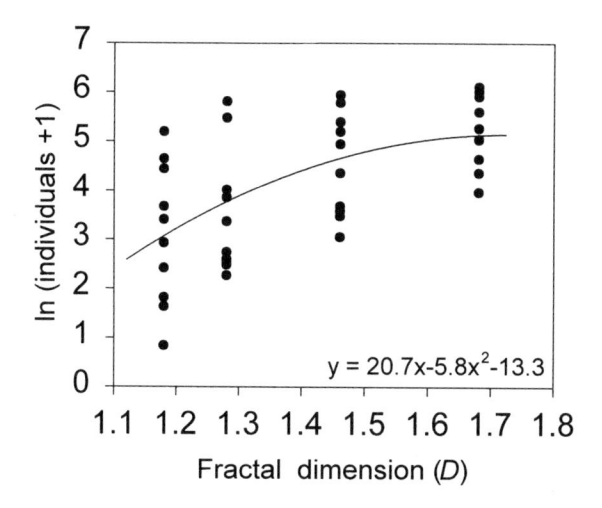

Fig. 27. Curvilinear relation between total number of individuals found on artificial weed stands and fractal dimension (modified from Jeffries, 1993).

In a second experiment, the colonization pattern of the beetle *Cyphon* sp. was monitored in 30 pools in freshwater marshes (East Lothian, Scotland). This trial included artificial pond weed of two different designs and no pond weed. A significant higher number of beetle larvae occurred on the more complex weed design. These results confirm that habitat complexity is an important factor determining invertebrate numbers on biotic habitat features such as freshwater macrophytes, the number of individuals increasing with increasing fractal dimension of the habitat surface.

VII. FRACTAL COEXISTENCE AND SPECIES ABUNDANCE DISTRIBUTION

The fractal nature of resource use by organisms may mediate a body size-related coexistence of species. Fractal resource use would imply that a resource space is more fragmented and patchy in occurrence for smaller species than larger ones. Tokeshi (1999) envisaged this idea in an assemblage of sedentary species, each holding a specific habitat area for its exclusive use. The close packing of such a two-dimensional habitat by one species leaves unused areas that can be exploited by other species (Figure 28a). Assuming that the body size or activity range is roughly a circular area, the body length or activity radius of the smaller species finding space among the larger would be a constant proportion of the size of the predecessor.

I used the mathematical model of apollonian packing (Mandelbrot, 1983), applying Soddy's equation (Coxeter, 1961) to calculate the potential activity radius of smaller species among larger ones. Considering, for instance, a feeding range of 40 mm around the tube of a sedentary soft-bottom species such as *Chironomus* sp., it follows that further sedentary species could theoretically coexist without resource partitioning if the activity ranges of, for instance, five further species decrease in size from 6.2, 2.5, 1.4, 0.9 to 0.6 mm (Figure 28a).

Extending the model of species packing to that of an apollonian gasket with subsets of two-dimensional habitat (curvilinear triangles) would imply that similar-sized individuals could coexist without resource overlap. (Figure 28b). The fractal dimension D for the apollonian gasket, a fractal curve, is a measure of fragmentation (Mandelbrot, 1983). Thus, the species assemblage model yields a fractal dimension of roughly 1.3 (Boyd, 1973). This apollonian model can also be interpreted as a model of species coexistence in habitats such as benthic interstices (here shown as circular triangles) where species occupy a fractal habitat corresponding to their body mass.

Body size differences among closely related species can be considered as one of the mechanisms that facilitate coexistence, resulting in a differential use of resources among species. Models of species abundance patterns attempt to reflect different ecological processes of community formation and the

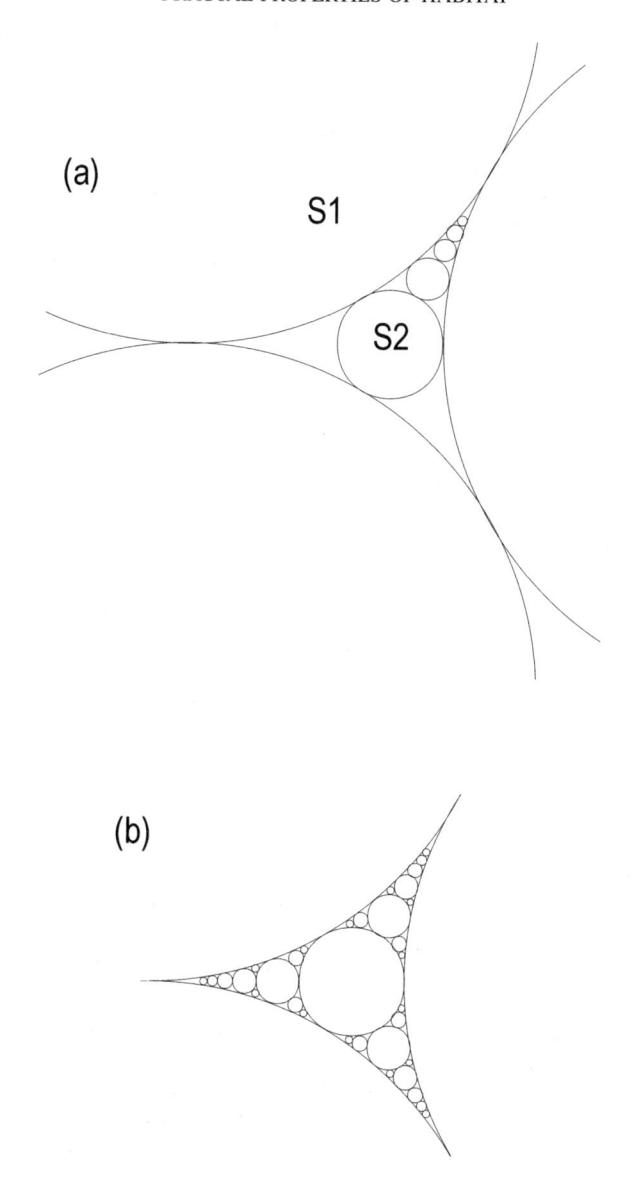

Fig. 28. Models of species packing and coexistence. (a) Pattern of size-related fractal coexistence among sessile species (*see* Tokeshi, 1999). (b) Apollonian model of fractal coexistence of motile species in interstitial habitats.

subsequent patterns of relative abundance (Tokeshi, 1993). These models aim to describe patterns as well as clarify processes of assemblage structure. To analyse aspects of size-related coexistence and resource use, I used chironomid biomass data based on 560 samples taken from the streambed sediments of the Oberer Seebach (Schmid, 1993). Building on previous results for a five-species assemblage (Schmid, 1997), I compared an assemblage of 16 chironomid 'species' (99% of the total chironomid biomass) with different resource-oriented models of rank abundance. To test the conformity to any of these models, I used a process-oriented approach, where resources of different size can be occupied by any species. Thus, species were ranked according to biomass independent of species identity.

Random assemblage models such as Random Assortment (RA; Tokeshi, 1993) assume that species use resources independently of one another on a quasi-random basis. The total resource space available is rarely completely filled by species, because species continuously redistribute.

Random resource apportionment models refer to models that assume a direct coupling between resource use and species abundance. A sequential breakage model stipulating that the total resource space is divided at random among species is the MacArthur Fraction (MF) model (Tokeshi, 1993). This model assumes that a new species colonizes the resource space of a more abundant species with a higher probability than a less abundant one.

The Random Fraction (RF) model (Tokeshi, 1993) presumes that species colonize and select resource space randomly, obtaining a random fraction of the resource space occupied by one of the existing species.

To consider both spatial and temporal aspects of possible resource allocation (resource filling) among species, I conceptualized further derivations of resource apportionment models (Figure 29). The Random Resource Allocation (RM) model envisages a situation in which common species (i.e. small-sized highly mobile generalists) colonize any available resource space (Figure 29). Rare species occupy a smaller fraction of resource space and the resources allocated to them may depend on the fraction occupied by common species. Rarer, often larger-sized, species, which occupy smaller resource space, may display patterns of speciation such as particular feeding habits. The RM model stresses that all except the rare species have the same probability of selecting a random fraction of the resource space available. The major distinction between the RF and RM models is that resource allocation of rare species is assumed to be dependent on the resources left by common species.

For each of the random models, 10^4 simulations were performed to create 10^4 species assemblages. To judge whether the observed relative abundance of the chironomids is in agreement with one of the random models, 95% confidence limits were calculated for the mean abundances of each species from the simulated assemblages (Tokeshi, 1993; Schmid, 1997). If, for all

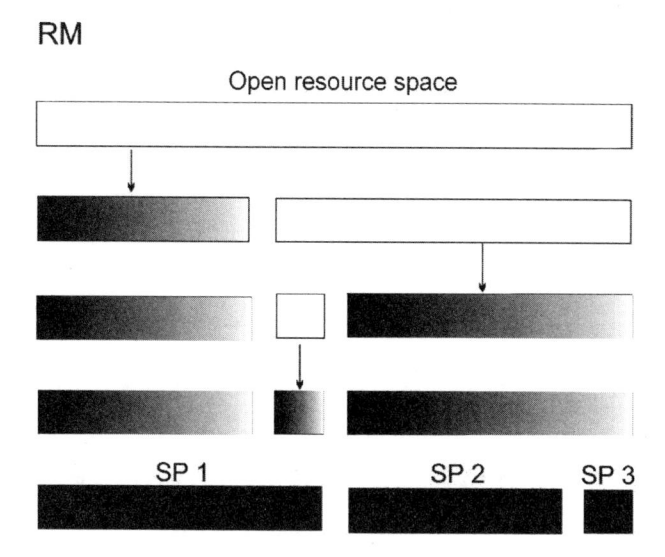

Fig. 29. Schematic representation of the Random Resource Allocation (RM) model. Details are given in the text.

species, the observed species abundance values fell within the corresponding 95% confidence limits from one of the models, the observed pattern did not significantly depart from the model's prediction ($P < 0.05$).

The observed species abundance pattern on biomass departed significantly, except for the first four species ranks, from the RA model (Figure 30). Among models, which cover a more equitable distribution of biomass among available resource space, only the RM model fitted significantly to the observed data (Figure 30). Both the MF and RF models departed from the observed pattern of relative abundance, particularly for rarer species (Figure 30). Therefore, abundant, mostly small-sized, species (i.e. genus *Corynoneura*, *Thienemanniella*) invade any random fraction of open resource space in the sediments of the Oberer Seebach. Larger-sized species appear to select the remaining unoccupied resource space. Large-sized larval individuals of the xenophage genus *Brillia*, for instance, occur in much lower numbers, inhabiting zones of woody debris located at the uppermost sediment surface layer.

In contrast to the above results, using a species-oriented approach, numerical data on the species abundance distribution of the five most abundant chironomid species fitted significantly to both RA and RF models (Schmid, 1997). This suggests that similar to mass-related abundance patterns, single species may disperse independently from one another selecting a random fraction of the available resource space.

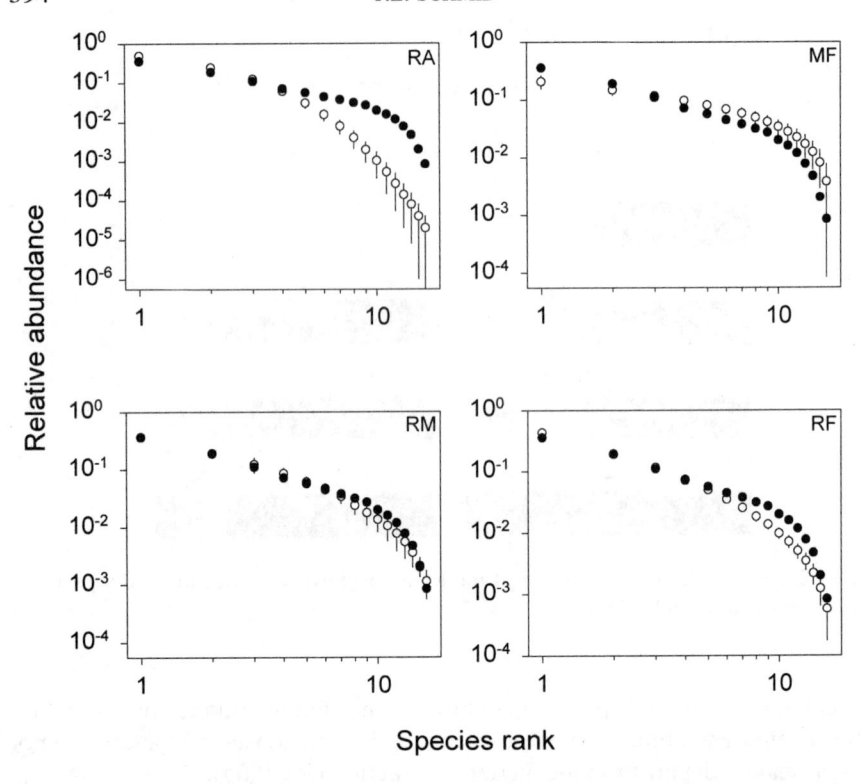

Fig. 30. Pattern of observed relative abundance in terms of biomass (●) in a chironomid assemblage (99% of total chironomid biomass) and expected values from four random resource models (○). Vertical lines are the expected 95% confidence intervals for 10^4 simulated 16-species assemblages. Models are represented by Random Assortment (RA) and three random resource apportionment models given in order of declining evenness: MacArthur Fraction (MF), Random Resource Allocation (RM) and Random Fraction (RF). RA, MF and RF models are after Tokeshi (1993) and details on the RM model are outlined in the text.

The spatial distribution of motile species, particularly in ecosystems mediated by physical disturbances, may be governed by available resource space, its temporal persistence, the structural complexity of the environment and, to a lesser extent, biotic interactions. Therefore, most species may respond to random divisions of resource space independently of one another. Resource apportionment can be envisaged as resource fragmentation processes with different degrees of convergence towards even distribution of constituent species. Assemblages with a complex patterns of random resource fragmentation fit more closely to the MF model and, with increasing uneven resource use, species abundance may fit to the RM and RF models.

Considering that the evenness of species assemblages has the form of the fractal dimension of a Cantor set (Frontier, 1985), the fractal structure of the species distribution declines from the MF towards RF model predictions. Species assemblages that are loosely structured may be characterized by a lower fractal dimension compared with assemblages with a complex pattern of resource fragmentation.

VIII. CONCLUDING REMARKS

Fractal geometry is foremost a language used to describe, model and analyse complex forms and structural interdependencies found in ecosystems. But while the elements of traditional mathematics, such as euclidean geometry, are basic visible forms, those of fractal geometry do not lend themselves to direct observation. They are algorithms translated into structures that, together with chaos theory, may alter the understanding of complex dynamic processes in ecosystems. The concept of fractal geometry questions our understanding of equilibria, and offers a holistic and integral model that can encompass aspects of ecosystem complexity. Structural complexity, which is governed by random processes, can be described by fractal properties and thus by simple rules. In other words, when considering fractal structures we include the dynamic process that created them. The fractal geometry of habitat surface structures, for instance, varies between different locations. The variation results from the stochastic processes, which structure and re-structure the habitat, resulting in habitats that exhibit statistical self-similarity.

The fractal concept may change the understanding of the link between various habitat-related phenomena, such as the surface complexity of both biotic and abiotic substrates, and body size distribution, abundances and species richness in benthic assemblages. Benthic sediments are the complex product of the interaction of parent material, hydrology, relief and biological activity acting over different spatial scales in time. Bed sediments display fractal properties because increasing the scale continues to reveal more and more detail. Similarly, sessile aquatic organisms such as algae and macrophytes, and biotic structures such as mussel beds and coral reefs, generate spatial heterogeneity and habitat for other organisms. These biotic structures may display fractal properties based on growth and branching processes following the L-system (Lindenmayer system). Moreover, particles and particle flocs may be deposited randomly between surface structures, such as aquatic moss patches or benthic sediments. The randomness of the deposition process apparently leads to an increase in surface complexity.

In addition, individuals of benthic populations may display a distribution pattern of fractal clustering, particularly in habitats that display fractal properties. Thus, the intertwining of the fractal properties results in a multifractal distribution pattern. Changes in fractal dimension when shifing between scales, as observed across transects in stream studies, may demarcate transition zones separating scale ranges dominated by different processes, where the environmental properties or constraints acting upon organisms are changing. Moreover, the fractal dimension of benthic species distribution patterns increases with the complexity of resource fragmentation, and the patterns of species abundance may relate to possible body size- and habitat-related fractal properties of food-web structures.

The fractal dimension appears to be a useful measure of spatial complexity, because spatial dependence frequently prohibits conservative statistical analysis and inference based on autocorrelated observations. As a measure of pattern, the fractal dimension describes variability at many scales jointly and fractal analysis provides information on the spatial configuration of process–pattern interactions.

Different from mathematical fractals, natural objects are characterized by the fact that their fractal features disappear if they are viewed at a sufficiently wide range of scales. Nevertheless, over a certain range of scales both biotic and abiotic habitat features display self-similarity or self-affinity and appear very much like mathematical fractals, and at such scales are regarded as random fractals. The distinction between random fractals and mathematical fractal sets that are used to describe them was emphasized by Mandelbrot (1983). However, natural phenomena such as brownian motion have been explained in terms of fractal mathematics. Movement paths of motile benthic organisms display properties analogous to fractal brownian motion, and these trajectories in turn appear to mirror the structural and resource properties of the environment.

Few studies have covered fractal aspects in aquatic ecosystems. Further investigations are necessary to understand the link between activity patterns and physical phenomena influencing patterns of dispersal and community structure. Perhaps the most exciting is the possibility that fractal geometry could be increasingly useful in establishing links between structure and function in benthic ecosystems.

ACKNOWLEDGEMENTS

I am very grateful to Dr J. Schmid-Araya for contributing many ideas to this review. I thank Drs J. Schmid-Araya and S. Wiedenbrüg for providing unpublished data and Dr B. Robson for a summary of her riverine investigation on fractal habitat structures. I also thank Dr C. de Jong for providing me with information on her work related to river bed geometry. I am especially thankful to Dr D. Raffaelli for helpful comments and a constructive review of an earlier draft. Part of the funding for this work was provided by National Environmental Research Council (NERC) Grant Gr3/09844.

REFERENCES

Adler, P.M. (1996). Transport in fractal porous media. *J. Hydrol.* **187**, 195–213.
Barabási, A.-L. and Stanley, H.E. (1995). *Fractal Concepts in Surface Growth.* Cambridge University Press, Cambridge.
Blackburn, T.M., Brown, V.K., Doube, B.M., Greenwood, J.J.D., Lawton, J.H. and Stork, N.E. (1993). The relationship between abundance and body size in natural animal assemblages. *J. Anim. Ecol.* **62**, 519–528.
Boyd, D.W. (1973). Improved bounds for the disc packing constant. *Aequat. Math.* **20**, 99–106.
Bradbury, R.H., Reichelt, R.E. and Green, D.G. (1984). Fractals in ecology: methods and interpretation. *Mar. Ecol. Prog. Ser.* **14**, 295–296.
Bretschko, G. and Leichtfried, M. (1988). Distribution of organic matter and fauna in a second order, alpine gravel stream (Ritrodat-Lunz study area, Austria). *Verh. Int. Verein. Limnol.* **23**, 1333–1339.
Brunauer, S., Emmett, P.H. and Teller, E. (1938). Absorption of gases in multimolecular layers. *J. Am. Chem. Soc.* **60**, 309–319.
Burrough, P.A. (1981). Fractal dimensions of landscapes and other environmental data. *Nature* **294**, 240–242.
Burrough, P.A. (1983). Multiscale sources of spatial variation in soil. 1. The application of fractal concepts to nested levels of soil variation. *J. Soil Sci.* **34**, 577–597.
Clarke, M.R.B. (1980). The reduced major axis of a bivariate sample. *Biometrika* **67**, 441–446.
Coxeter, H.S.M. (1961). *Introduction to Geometry.* Wiley, New York.
Cressie, N.A.C. (1991). *Statistics for Spatial Data.* John Wiley, New York.
Danielsson, P.E. (1980). Euclidean distance mapping. *Comput. Graphics Image Proc.* **14**, 227–248.
De Jong, C. and Ergenzinger, P. (1995). The interrelations between mountain valley form and river-bed arrangement. In: *River Geomorphology* (Ed. by E.J. Hickin), p. 56–91. John Wiley, New York.
Falconer, K.J. (1985). *The Geometry of Fractal Sets.* Cambridge University Press, Cambridge.
Falconer, K.J. (1990). *Fractal Geometry. Mathematical Foundations and Applications.* John Wiley, Chichester, UK.
Falconer, K.J. (1992). The dimension of self-affine fractals II. *Math. Proc. Camb. Phil. Soc.* **111**, 169–179.
Feder, J. (1988). *Fractals.* Plenum Press, New York.
Flook, A.G. (1978). The use of dilation logic on the quantimet to achieve fractal dimension characterisation of textured and structured profiles. *Powder Technol.* **21**, 295–298.
Frontier, S. (1985). Diversity and structure in aquatic ecosystems. *Oceanogr. Mar. Biol. Ann. Rev.* **23**, 253–312.
Frontier, S. (1987). Application of fractal theory to ecology. In: *Developments in Numerical Ecology* (Ed. by P. Legendre and L. Legendre), pp. 336–378. Springer, Berlin.
Goodarz-Nia, I. (1975). Floc simulation: effect of particle size distribution. *J. Colloid Interface Sci.* **52**, 29–40.
Gordon, N.D., McMahon, T.A. and Finlayson, B.L. (1992). *Stream Hydrology. An Introduction for Ecologists.* John Wiley, Chichester, UK.
Gore, J.A. (1978). A technique for predicting in stream flow requirements of benthic macroinvertebrates. *Freshw. Biol.* **8**, 141–151.

Griffiths, D. (1998). Sampling effort, regression method, and the shape and slope of size–abundance relations. *J. Anim. Ecol.* **67**, 795–804.

Gunnarsson, B. (1992) Fractal dimension of plants and body size distribution in spiders. *Funct. Ecol.* **6**, 636–641.

Hack, J.T. (1957). Studies of the longitudinal streams in Virginia and Maryland. *U.S. Geol. Surv. Prof. Papers* 294B.

Hastings, H.M. and Sugihara, G. (1993). *Fractals. A User's Guide for the Natural Sciences.* Oxford University Press, Oxford.

Hildgen, P., Nekka, F., Hildgen, F. and McMullen, J.N. (1997). Macroporosity measurement by fractal analysis. *Physica A* **234**, 593–603.

Holden, A.V. and Muhamad, M.A. (1987). A graphical zoo of strange and peculiar attractors. In: *Chaos* (Ed. by A.V. Holden), pp. 15–35. Manchester University Press, Manchester.

Jeffries, M. (1993). Invertebrate colonization of artificial pondweeds of differing fractal dimension. *Oikos* **67**, 142–148.

Kaye, B.H. (1978). *Sequential Mosaic Amalgamation as a Strategy for Evaluating Fractal Dimensions of a Fineparticle Profile.* Institute of Fineparticle Research, Laurentian University, Sudbury, Ontario.

Kaye, B.H. (1989). *A Random Walk Through Fractal Dimensions.* VCH Publisher, New York.

Kotliar, N.B. and Wiens, J.A. (1990). Multiple scales of patchiness and patch structure: a hierarchical framework for the study of heterogeneity. *Oikos* **59**, 253–260.

Krohn, C.E. (1988). Fractal measurements of sandstone, shales and carbonates. *J. Geophys. Res.* **93**, 3297–3305.

Lawton, J.H. (1986). Surface availability and insect community structure: the effects of architecture and fractal dimension of plants. In: *Insects and the Plant Surface* (Ed. by B.E. Juniper and T.R.E. Southwood), pp. 317–331. Edward Arnold, London.

Lawton, J.H. (1989). What is the relationship between population density and body size in animals? *Oikos* **55**, 429–434.

Leichtfried, M. (1985). Organic matter in gravel streams (Project Ritrodat Lunz). *Verh. Int. Verein. Limnol.* **22**, 2058–2062.

Leichtfried, M. (1991). POM in bed sediments of a gravel stream (Ritrodat-Lunz study area, Austria). *Verh. Int. Verein. Limnol.* **24**, 1921–1925.

Leopold, L.B., Wolman, M.G. and Miller, J.P. (1964*). Fluvial Processes in Geomorphology.* W.H. Freeman, San Francisco.

Mackey, A.P. (1977). Growth and development of larval Chironomidae. *Oikos* **28**, 270–275.

Macosko, C.W. (1994). *Reology. Principles, measurements, and applications.* Wiley–VCH, New York.

Maina, J.N. (1998). *The Gas Exchangers. Structure, Function, and Evolution of the Respiratory Process.* Zoophysiology, Vol. 37. Springer, Berlin.

Mandelbrot, B. (1983). *The Fractal Geometry of Nature.* Freeman, San Francisco.

Mandelbrot, B. (1984). Fractals in physics: squig cluster, diffusions, fractal measures and the unicity of fractal dimensionality. *J. Stat. Phys.* **36**, 895.

May, R.M. (1978). The dynamics and diversity of insect faunas. In: *Diversity of Insect Faunas* (Ed. by M. Cody and J.M. Diamond), pp. 81–120. Harvard University Press, Cambridge, Massachusetts.

McArdle, B.H. (1988). The structural relationship: regression in biology. *Can. J. Zool.* **66**, 2329–2339.

Meakin, P. (1987). Scaling properties for the growth probability measure and harmonic measure of fractal structures. *Phys. Rev. A.* **35**, 2234–2245.

Minkowski, H. (1901) Über die Begriffe Länge, Oberfläche und Volumen. *Jahresber. Deut. Math.* **9**, 115–121.
Morse, D.R., Lawton, J.H., Dodson, M.M. and Williamson, M.H. (1985). Fractal dimension of vegetation and the distribution of arthropod body lengths. *Nature* **314**, 731–733.
Nikora, V.I. (1991). Fractal structures of river plan forms. *Water Resour. Res.* **27**, 1327–1333.
Nolte, U. (1991). Seasonal dynamics of moss-dwelling chironomid communities. *Hydrobiologia* **222**, 197–211.
Palmer, M.W. (1988). Fractal geometry: a tool for describing spatial patterns of plant communities. *Vegetatio* **75**, 91–102.
Peters, R.H. (1983). *The Ecological Implications of Body Size*. Cambridge University Press, Cambridge.
Rammal, R. (1984). Random walk statistics on fractal structures. *J. Stat. Phys.* **36**, 547.
Richardson, L.F. (1961). The problem of contiguity: an appendix of statistics of deadly quarrels. *General Systems Yearbook* **6**, 139–187.
Robert, A. and Richards, K.S. (1988). On the modelling of sand bedforms using the semivariogram. *Earth Surf. Proc. Landf.* **13**, 459–473.
Robert, A. and Roy, A.G. (1990). On the fractal interpretation of the mainstream length–area relationship. *Water Resour. Res.* **26**, 839–842.
Rooke, J.B. (1986). Macroinvertebrates associated with macrophytes and plastic imitations in the Erasoma River, Ontario, Canada. *Arch. Hydrobiol.* **106**, 307–325.
Russ, J.C. (1992). *Image Processing Handbook*. CRC Press, Boca Raton, Florida.
Russ, J.C. (1994). *Fractal Surfaces*. Plenum Press, New York.
Schmid, P.E. (1992). Habitat preferences as patch selection of larval and emerging chironomids (Diptera) in a gravel brook. *Neth. J. Aqua. Ecol.* **26**, 419–429.
Schmid, P.E. (1993). Random patch dynamics of larval Chironomidae (Diptera) in the bed sediments of a gravel stream. *Freshw. Biol.* **30**, 239–255.
Schmid, P.E. (1997). Stochasticity in resource utilization by a larval Chironomidae (Diptera) community in the bed sediments of a gravel stream. In: *Groundwater/Surface Water Ecotones: Biological and Hydrological Interactions and Management Options* (Ed. by J. Gilbert, J. Mathieu and F. Fournier), pp. 21–28. Cambridge University Press, Cambridge.
Schmid, P.E. and Schmid-Araya, J.M. (1997). Predation on meiobenthic assemblages: resource use of a tanypod guild (Chironomidae, Diptera) in a gravel stream. *Freshw. Biol.* **38**, 67–91.
Schmid-Araya, J.M. (1993). Spatial distribution and population dynamics of a benthic rotifer, *Embata laticeps* (Murray) (Rotifera, Bdelloidea) in the bed sediments of a gravel brook. *Freshw. Biol.* **30**, 395–408.
Schwarz, H. and Exner, H.E. (1980). The implementation of the concept of fractal dimension on a semi-automatic image analyser. *Powder Technol.* **27**, 207–213.
Shorrocks, B., Marsters, J., Ward, I. and Evennett, P. J. (1991). The fractal dimension of lichens and the distribution of arthropod body lengths. *Funct. Ecol.* **5**, 457–460.
Sinsabaugh, R.L., Weiland, T. and Linkins, A.E. (1991). Epilithon patch structure in a boreal river. *J. N. Am. Benthol. Soc.* **10**, 419–429.
Smith, T.G., Jr., Marks, W.B., Lange, G.D., Sherriff, W.H., Jr. and Neale, E.A. (1989). A fractal analysis of cell images. *J. Neurosi. Methods* **27**, 173–180.
Snover, M.L. and Commito, J.A. (1998). The fractal geometry of *Mytilus edulis* L. spatial distribution in a soft-bottom system. *J. Exp. Mar. Biol. Ecol.* **223**, 53–64.
Sugihara, G. and May, R.M. (1990). Applications of fractals in ecology. *Trends Ecol. Evol.* **5**, 79–86.

Takayasu, H. (1990). *Fractals in the Physical Sciences*. Manchester University Press, Manchester.

Tarboton, D.G. (1996). Fractal river networks, Horton's laws and Tokunaga cyclicity. *J. Hydrol,* **187**, 105–117.

Tarboton, D.G., Bars, R.L. and Rodriguez-Iturbe, I. (1988). The fractal nature of river networks. *Water Resour. Res.* **24**, 1317–1322.

Tokeshi, M. (1990). Density–body size allometry does not exist in a chironomid community on *Myriophyllum. Freshw. Biol.* **24**, 613–618.

Tokeshi, M. (1993). Species abundance patterns and community structure. *Adv. Ecol. Res.* **24**, 111–186.

Tokeshi, M. (1999). *Species Coexistence. Ecological and Evolutionary Perspectives*. Blackwell Science, Oxford.

Turcotte, D.L. (1986). Fractals and fragmentation. *J. Geophys. Res.* **91**, 1921–1926.

Weiss, S.B. and Murphy, D.D. (1988). Fractal geometry and caterpillar dispersal: or how many inches can inchworms inch? *Funct. Ecol.* **2**, 116–118.

West, G.B., Brown, J.H. and Enquist, B.J. (1997). A general model for the origin of allometric scaling laws in biology. *Science* **276**, 122–126.

Wiens, J.A. (1976). Population response to patchy environments. *Ann. Rev. Ecol. Syst.* **7**, 81–129.

Wong, P., Howard, J. and Lin, J.-S. (1986). Surface roughening and the fractal nature of rocks. *Phys. Rev Lett.* **57**, 637–640.

Appendix A

Relationship between log length–log weight of ten larval chironomid species with different body-length ranges

Species	a (± 1 SE)	b (± 1 SE)	β	± 1 SE	d.f.	F	P	Length range (mm)
Subfamily: Tanypodinae								
Nilotanypus dubius M.	−6.123 ± 0.051	2.321 ± 0.056	0.989	0.024	1,40	1739.19	<0.001	0.45–4.80
Thienemannimyia geiskesi G.	−6.932 ± 0.061	2.703 ± 0.050	0.996	0.018	1,26	2907.37	<0.001	0.95–7.78
Subfamily: Diamesinae								
Diamesa cinerella M.	−6.301 ± 0.135	2.831 ± 0.096	0.978	0.033	1,39	864.99	<0.001	0.94–10.12
Subfamily: Orthocladiinae								
Corynoneura lobata E.	−6.593 ± 0.034	2.080 ± 0.059	0.985	0.028	1,38	1236.29	<0.001	0.36–2.25
Cricotopus annulator G.	−6.393 ± 0.068	2.298 ± 0.077	0.986	0.033	1,26	880.73	<0.001	0.66–4.40
Heleniella ornaticollis E.	−7.303 ± 0.154	2.512 ± 0.140	0.967	0.054	1,22	320.27	<0.001	0.98–5.30
Orthocladius (Eu.) rivulorum K.	−5.280 ± 0.045	2.374 ± 0.041	0.997	0.017	1,23	3351.06	<0.001	0.64–5.10
Synorthocladius semivirens K.	−5.989 ± 0.060	2.124 ± 0.080	0.982	0.037	1,27	710.88	<0.001	0.75–3.40
Subfamily: Chironominae								
Microspectra atrofasciata K.	−7.131 ± 0.085	2.602 ± 0.081	0.993	0.031	1,15	1032.28	<0.001	0.90–5.67
Rheotanytarsus nigricauda F.	−6.481 ± 0.075	2.218 ± 0.108	0.975	0.047	1,22	423.49	<0.001	0.80–3.50

a is the intercept and b the slope with ± 1 SE of the log–log relation. β is the standardized regression coefficient with ± 1 SE. d.f. are the degrees of freedom based on a number of groups of measured larvae. Depending on instar, each group represents mean length and dry-weight values of 5–15 larvae. F is the variance ratio.

Appendix B

Relationship between the maximum length at maturation of 10 different chironomid species and the exponent b of the length–weight relation for those species

Species	a (± 1 SE)	b (± 1 SE)	β	± 1 SE	d.f.	F	P	Maximum length (mm)
Chironomid species	1.748 ± 0.062	0.131 ± 0.012	0.969	0.087	1,8	125.11	<0.001	2.25–10.12

a is the intercept and b the slope with ± 1 SE. β is the standardized regression coefficient with ± 1 SE. d.f. are the degrees of freedom. F is the variance ratio.

Advances in Ecological Research
Volumes 1–30

Cumulative List of Titles

Aerial heavy metal pollution and terrestrial ecosystems, **11**, 218
Age-related decline in forest productivity: pattern and process, **27**, 213
Analysis of processes involved in the natural control of insects, **2**, 1
Ant-plant-homopteran interactions, **16**, 53
Biological strategies of nutrient cycling in soil systems, **13**, 1
Bray-Curtis ordination: an effective strategy for analysis of multivariate ecological data, **14**, 1
Can a general hypothesis explain population cycles of forest lepidoptera?, **18**, 179
Carbon allocation in trees: a review of concepts for modelling, **25**, 60
Catchment properties and the transport of major elements to estuaries, **29**, 1
Co-evolution of mycorrhizal symbionts and their hosts to metal-contaminated environments, **30**, 69
The cost of living: field metabolic rates of small mammals, **30**, 177
A century of evolution in *Spartina anglica*, **21**, 1
The climatic response to greenhouse gases, **22**, 1
Communities of parasitoids associated with leafhoppers and planthoppers in Europe, **17**, 282
Community structure and interaction webs in shallow marine hard-bottom communities: tests of an environmental stress model, **19**, 189
The decomposition of emergent macrophytes in fresh water, **14**, 115
Delays, demography and cycles: a forensic study, **28**, 127
Dendroecology: a tool for evaluating variations in past and present forest environments, **19**, 111
The development of regional climate scenarios and the ecological impact of greenhouse gas warming, **22**, 33
Developments in ecophysiological research on soil invertebrates, **16**, 175
The direct effects of increase in the global atmospheric CO_2 concentration on natural and commercial temperate trees and forests, **19**, 2
The distribution and abundance of lake-dwelling Triclads—towards a hypothesis, **3**, 1
The dynamics of aquatic ecosystems, **6**, 1
The dynamics of field population of the pine looper, *Bupalis piniarius* L. (Lep., Geom.), **3**, 207
Earthworm biotechnology and global biogeochemistry, **15**, 379
Ecological aspects of fishery research, **7**, 114

Ecological conditions affecting the production of wild herbivorous mammals on grasslands, **6**, 137
Ecological implications of dividing plants into groups with distinct photosynthetic production capabilities, **7**, 87
Ecological implications of specificity between plants and rhizosphere microorganisms, **21**, 122
Ecological studies at Lough Inc, **4**, 198
Ecological studies at Lough Hyne, **17**, 115
Ecology of mushroom-feeding Drosophilidae, **20**, 225
The ecology of the Cinnabar moth, **12**, 1
Ecology of coarse woody debris in temperate ecosystems, **15**, 133
Ecology of estuarine macrobenthos, **29**, 195
Ecology, evolution and energetics: a study in metabolic adaptation, **10**, 1
Ecology of fire in grasslands, **5**, 209
The ecology of pierid butterflies: dynamics and interactions, **15**, 51
The ecology of root lifespan, **27**, 1
The ecology of serpentine soils, **9**, 225
Ecology, systematics and evolution of Australian frogs, **5**, 37
Effects of climatic change on the population dynamics of crop pests, **22**, 117
The effects of modern agriculture, nest predation and game management on the population ecology of partridges (*Perdix perdix* and *Alectoris rufa*), **11**, 2
El Niño effects on Southern California kelp forest communities, **17**, 243
Energetics, terrestrial field studies and animal productivity, **3**, 73
Energy in animal ecology, **1**, 69
Estimates of the annual net carbon and water exchange of forests: the EUROFLUX methodology, **30**, 113
Estimating forest growth and efficiency in relation to canopy leaf area, **13**, 327
Evolutionary and ecophysiological responses of mountain plants to the growing season environment, **20**, 60
The evolutionary consequences of interspecific competition, **12**, 127
The exchange of ammonia between the atmosphere and plant communities, **26**, 302
Faunal activities and processes: adaptive strategies that determine ecosystem function, **27**
Fire frequency models, methods and interpretations, **25**, 239
Food webs: theory and reality, **26**, 187
Forty years of genecology, **2**, 159
Foraging in plants: the role of morphological plasticity in resource acquisition, **25**, 160
Fossil pollen analysis and the reconstruction of plant invasions, **26**, 67
Fractal properties of habitat and patch structure in benthic ecosystems, **30**, 339
Free-air carbon dioxide enrichment (FACE) in global change research: a review, **28**, 1
The general biology and thermal balance of penguins, **4**, 131
General ecological principles which are illustrated by population studies of Uropodid mites, **19**, 304
Generalist predators, interaction strength and food-web stability, **28**, 93
Genetic and phenotypic aspects of life-history evolution in animals, **21**, 63
Geochemical monitoring of atmospheric heavy metal pollution: theory and applications, **18**, 65
Heavy metal tolerance in plants, **7**, 2
Herbivores and plant tannins, **19**, 263
Human ecology as an interdisciplinary concept: a critical inquiry, **8**, 2
Industrial melanism and the urban environment, **11**, 373
Inherent variation in growth rate between higher plants: a search for physiological causes and ecological consequences, **23**, 188

Insect herbivory below ground, **20**, 1
Integrated coastal management: sustaining estuarine natural resources, **29**, 241
Integration, identity and stability in the plant association, **6**, 84
Isopods and their terrestrial environment, **17**, 188
Landscape ecology as an emerging branch of human ecosystem science, **12**, 189
Litter production in forests of the world, **2**, 101
Manipulative field experiments in animal ecology: do they promise more than they can deliver?, **30**, 299
Mathematical model building with an application to determine the distribution of Dursban® insecticide added to a simulated ecosystem, **9**, 133
Mechanisms of microarthropod-microbial interactions in soil, **23**, 1
Mechanisms of primary succession: insights resulting from the eruption of Mount St Helens, **26**, 1
The method of successive approximation in descriptive ecology, **1**, 35
The mineral nutrition of wild plants revisited: a re-evaluation of processes and patterns, **30**, 1
Modelling terrestrial carbon exchange and storage: evidence and implications of functional convergence in light-use efficiency, **28**, 57
Modelling the potential response of vegetation to global climate change, **22**, 93
Module and metamer dynamics and virtual plants, **25**, 105
Mutualistic interactions in freshwater modular systems with molluscan components, **20**, 126
Mycorrhizal links between plants: their functioning and ecological significances, **18**, 243
Mycorrhizas in natural ecosystems, **21**, 171
Nocturnal insect migration: effects of local winds, **27**, 61
Nutrient cycles and H$^+$ budgets of forest ecosystems, **16**, 1
Nutrients in estuaries, **29**, 43
On the evolutionary pathways resulting in C_4 photosynthesis and crassulacean acid metabolism (CAM), **19**, 58
Oxygen availability as an ecological limit to plant distribution, **23**, 93
The past as a key to the future: the use of palaeoenvironmental understanding to predict the effects of man on the biosphere, **22**, 257
Pattern and process of competition, **4**, 11
Phytophages of xylem and phloem: a comparison of animal and plant sap-feeders, **13**, 135
The population biology and turbellaria with special reference to the freshwater triclads of the British Isles, **13**, 235
Population cycles in small mammals, **8**, 268
Population regulation in animals with complex life-histories: formulation and analysis of damselfly model, **17**, 1
Positive-feedback switches in plant communities, **23**, 264
The potential effect of climatic changes on agriculture and land use, **22**, 63
Predation and population stability, **9**, 1
Predicting the responses of the coastal zone to global change, **22**, 212
The pressure chamber as an instrument for ecological research, **9**, 165
Primary production by phytoplankton and microphytobenthos in estuaries, **29**, 93
Principles of predator-prey interaction in theoretical experimental and natural population systems, **16**, 249
The production of marine plankton, **3**, 117
Production, turnover, and nutrient dynamics of above- and below-ground detritus of world forests, **15**, 303

Quantitative ecology and the woodland ecosystem concept, **1**, 103
Realistic models in population ecology, **8**, 200
The relationship between animal abundance and body size: a review of the mechanisms, **28**, 181
Relative risks of microbial rot for fleshy fruits: significance with respect to dispersal and selection for secondary defence, **23**, 35
Renewable energy from plants: bypassing fossilization, **14**, 57
Responses of soils to climate change, **22**, 163
Rodent long distance orientation ("homing"), **10**, 63
Secondary production in inland waters, **10**, 91
The self-thinning rule, **14**, 167
A simulation model of animal movement patterns, **6**, 185
Soil arthropod sampling, **1**, 1
Soil diversity in the Tropics, **21**, 316
Soil fertility and nature conservation in Europe: theoretical considerations and practical management solutions, **26**, 242
Spatial root segregation: are plants territorial?, **28**, 145
Species abundance patterns and community structure, **26**, 112
Stomatal control of transpiration: scaling up from leaf to regions, **15**, 1
Structure and function of microphytic soil crusts in wildland ecosystems of arid to semi-arid regions, **20**, 180
Studies on the cereal ecosystems, **8**, 108
Studies on grassland leafhoppers (Auchenorrhyncha, Homoptera) and their natural enemies, **11**, 82
Studies on the insect fauna on Scotch Broom *Sarothamnus scoparius* (L.) Wimmer, **5**, 88
Sunflecks and their importance to forest understorey plants, **18**, 1
A synopsis of the pesticide problem, **4**, 75
Temperature and organism size—a biological law for ecotherms?, **25**, 1
Terrestrial plant ecology and ^{15}N natural abundance: the present limits to interpretation for uncultivated systems with original data from a Scottish old field, **27**, 133
Theories dealing with the ecology of landbirds on islands, **11**, 329
A theory of gradient analysis, **18**, 271
Throughfall and stemflow in the forest nutrient cycle, **13**, 57
Towards understanding ecosystems, **5**, 1
The use of statistics in phytosociology, **2**, 59
Vegetation, fire and herbivore interactions in heathland, **16**, 87
Vegetational distribution, tree growth and crop success in relation to recent climate change, **7**, 177
Water flow, sediment dynamics and benthic biology, **29**, 155
The zonation of plants in freshwater lakes, **12**, 37

Index

Acaulospora 76, 80
Acinonyx jubatus 181
Acomys caharinus 225
Acomys russatus 225
Agrostis capillaris 73, 75, 84, 92
Alnus 14
Alpine tundra 32
Altitude, field metabolic rates (FMR) in small mammals 239–240
Aluminium-contaminated soils 73
   arbuscular mycorrhizas 78–79, 80
   ectomycorrhizas 101–102
   Ericaceae colonization 88, 89
Amanita muscaria 100, 101
Amazon river system 359
Ameriflux 167
Amino acids, root uptake 9–10
Ammonium absorption
   mycorrhizal uptake 12
   rhizosphere pH effects 15
   root uptake capacity 9, 10
Ammospermophilus 228, 233
Andropogon gerardii 83, 85, 86
Antechinus 239, 257
Apodemus sylvaticus 208, 240, 245
Apodidae (swifts) 189
Arbuscular mycorrhizas 73–88
   host-symbiont interactions 75–76
      colonization 80, 81
   metal resistance evolution 75, 78, 79, 80, 87, 105
   metal-contaminated sites 70, 71, 73–76, 105
      benefit from resistant strains colonization 87–88
      germ tube growth 79–80
      hyphal penetration 80
      morphological responses 81
      plant biomass production 85–86
      root colonization 81–82
      species diversity 76
      spore germination rates 76–79
      spore production 82
      toxic metal assimilation 83–85
   nutrient uptake 12
   metal micronutrients 82–83
Arctium tomentosum 20
Arsenic-contaminated soils, Ericaceae colonization 88, 89
Artificial pondweed colonization 388–389
Artimesia californica 32
Atmospheric nitrogen deposition 6
   heathland nutrient uptake 12–13

Basal metabolic rate (BMR) 206
   diet effects 241
   interspecific variability 209
   measurement criteria 206–208
   seasonal change 207
Bat migration 191
Benthic systems
   flocculated aggregates 348–349
   multifractals 358
   patch dynamics 373–380
   surface fractals 349–358
      direct measurement 351–352
      Fourier analysis 354–355, 356, 357
      indirect measurement 352–355
      SEM biofilm images 352, 353
Betula spp. 14, 98
Biomass allocation, plants 2, 53
   functional equilibrium concept 16
   leaves 16–17
   nutrient acquisition 15–19
   nutrient supply relationship 18–19

Biomass *continued*
  relative growth rate (RGR) relationship 16–17
  roots 15–18, 53
Biomass production, plants 28–33
  nutrient use efficiency (NUE) 28–31
Birds
  continuously flying, energy budgets 189
  migration 190–191
    energy expenditure 185, 188, 189
Body mass, small mammals
  field metabolic rates (FMR) relationship 178, 202–203, 221–229, 232, 233
  longevity relationship 260
  maximal sustainable energy expenditure levels 186
  resting metabolic rate (RMR) relationship 202–203, 243–244, 245
Bradypus tridactylus 222, 223
Branta leucopsis 185
Brillia 393
Brood number manipulation studies 257–258
Brunauer–Emmett–Teller (BET) method 351, 364, 365–366

Cadmium-contaminated soils 73
  arbuscular mycorrhizas 75, 76, 78, 79, 82
    host plant biomass production 85–86
    host plant phosphorus status 86–87
    metal assimilation 81, 82, 84
  ectomycorrhizas 96, 97, 101
  Ericaceae colonization 88
Calamagrostis epigejos 84
Calluna vulgaris 12, 28, 32, 89, 90, 91, 93, 94
Carbon cycle dynamics 114
  EUROFLUX project *see* EUROFLUX project
  global measurement networks 167
Carcinus maenas 322, 323
Carex 28, 46
Carex acutiformis 17

Carex diandra 17, 28
Carex lasiocarpa 17, 46
Carex rostrata 17
Carnivora 219, 297
Catchment area, fractal properties 358–361
Cavia porcellus 226
Ceanothus megacarpus 32
Cephalodella gibba 381
Chamaedaphne calyculata 28
Chironomids
  body size–abundance relationship 368–369, 372
  coexistence models 390, 392–393
  movement patterns 381–386
  sediment surface distribution 373–374, 375, 377–379
Chiroptera 226, 227
  energy budgeting 211–212, 219
  field metabolic rate (FMR) 238
  resting metabolic rate (RMR) 245
  body mass relationship 260
Chromium-contaminated soils, Ericaceae colonization 88, 89
Cobalt-contaminated soils, Ericaceae colonization 88
Coexistence, fractal resource use 390–395
  resource apportionment models 392–393
  resource fragmentation 394–395
Cold exposure 191
  maximal sustainable energy expenditure 195, 196, 199, 201, 202
  resting metabolic rate (RMR) acclimatization 245
Community press experiments *see* Manipulative field experiments
Conchapelopia pallidula 381, 385
Copper taxidermy mounts 210–211
Copper-contaminated soils 72, 73
  arbuscular mycorrhizas 75
    metal assimilation 83, 84
  ectomycorrhizas 96, 97
  Ericaceae colonization 88, 89
  role of mycorrhizas 70, 71, 89–93, 95, 105

INDEX

Coral reefs, fractal dimension 387–388
Corophium 322, 323
Corynoneura 374, 375, 378, 381, 393
Creatine phosphate metabolism 183
Cyphon 390

Daily energy expenditure (DEE), small mammals *see* Field metabolic rates (FMR), small mammals
Deciduous plants
  leaf lifespan 27–28
  leaf nutrient concentration 21–22, 23
  litter decomposition 50
Delichon urbica 259
Deschampsia flexuosa 12
Dietary habits, small mammals
  basal metabolic rate (BMR) 241
  field metabolic rate (FMR) 178, 240–241, 242, 261–262, 263
  resting metabolic rate (RMR) 246–247
Diomedeidae (albatrosses) 189
Dipodomys merriami (kangaroo rat) 224, 231, 232, 233
Doubly labelled water energy expenditure measurement 214–217

Ectomycorrhizas
  metal-contaminated soils 70, 71, 73, 95–104, 105
    aluminium 101–102
    cadmium 96, 97, 101
    copper 96, 97
    interspecific response variation 99–102
    lead 96, 97
    nickel 96, 97
    resistant/sensitive strain colonization studies 102–103
    role in metal resistance 95–98, 103–104
    zinc 96, 97, 99–101
  organic nitrogen absorption 12
Eddy covariance 114–115
  annual time-scale CO_2 flux above forests 115
  standardization of measurements 167
  system for EUROFLUX project *see* EUROFLUX project
Edentata 219
Empetrum nigrum 12
Energy expenditure
  basal metabolic rate (BMR) 206
  daily expenditure *see* Field metabolic rates (FMR), small mammals
  diet effects 241
  direct measurements 212–217
    comparative aspects 214
    doubly labelled water method 214–217
    heart rate telemetry 212–214, 217
    isotope elimintaion methods 214, 217
  indirect calorimetry 205–206, 283–289
  limits *see* Energy expenditure limitation
  maximal sustainable levels 184, 186–189
    body mass relationship 186
    cold exposure 195, 196, 199, 201, 202
    lactation studies 195–198, 199–200, 201
    stressor studies 195–197
  measurement criteria 206–209
  resting metabolic rate (RMR) 206–207
  running activities 181–182, 183, 184–185
  time and energy budget estimates 209–212, 217
    energy demand quantification 211–212
    standard operative temperature measurements 210–211
  trade-offs 188
Energy expenditure limitation 178, 181–189
  basal metabolic rate relationship 178
  bird migration 185, 186
  experimental studies 197–200
  extrinsic 178, 179, 189–192, 260–263
  optimal foraging theory 190

INDEX

Energy *continued*
  supplementation studies in summer 192
  survival strategies for winter 190–191
  intrinsic 178, 179, 192–197, 262–263
    central 195, 196, 197, 198–199, 200–202
    peripheral 195, 196, 199, 204
  running activities 181–183, 184–185, 186
  small mammals 185–186
Energy metabolism 183, 194
  lactate accumulation 183–184
  substrate utilization 183, 184–185
Energy requirements 180–181
  organ size phenotypic response 200
Eremitalpa namibensis 238
Erica tetralix 12, 28, 32
Ericaceae
  metal-contaminated sites colonization 88–89
  nutrient use efficiency (NUE) 32
Ericoid mycorrhizas 12, 13
  metal-contaminated sites 70, 73, 88–95, 105
    co-resistance 93–94
    copper 70, 71, 89–93, 95, 105
    iron 93
    role of mycorrhizas 95
    zinc 89, 90, 93–95, 105
  organic nitrogen absorption 12
Euclidean distance mapping (EDM) 347–348
EUROFLUX project 113–170
  data acquisition 127
    procedure 127–130
  data gap filling 158
    interpolation and parameterization 158
    neural networks 158–162
  data processing 130–144, 167–168
    flux computation 130–134
    frequency response losses correction 136–144
    software comparisons 134–136
    third axis rotation computation 170
    two-axis rotation matrix elements computation 168–170
  eddy covariance system 119–127
    air transport system 123–126
    analysis software 119
    infrared gas analyser 119, 120–123
    sonic anemometer 119–120
    temperature fluctuation measurements 120
    tower instrumentation 126–127
  error estimation 164, 166–167
  flux data quality control 144–154
    energy balance closure 147–154
    integral turbulence test 146–147
    raw data analysis 145
    stationarity test 145–146
  flux source footprint analysis 154–155
  measuring sites 116, 117
  meteorological measurements 127, 128
  night-time data corrections 162–164, 165, 166
  rationale 115
  summation procedure 156–157
  theoretical aspects 116, 118–119
Evergreens
  leaf lifespan 27
  leaf nutrient concentration 21–22, 23
  leaf nutrient resorption 24–25
  litter decomposition 49–51

Fabaceae 14
Factorial fertilizer experiments 7
Festuca arundina 83, 84, 85
Festuca rubra 84
Field metabolic rates (FMR), small mammals 177, 205–264
  altitude effects 239–240
  ambient temperature effects 178, 229–232, 236, 237, 238, 263
  ameliorating behavioural responses 231
  body mass relationship 178, 202–203, 221–229, 232, 233, 263
  daily energy expenditure measurement 205–206
  methods 206–217

database overview 218–221
  latitudinal distribution 219–220
  species distribution 219
database raw data 289–293
dietary factors 178, 240–241, 242, 261–262, 263
energy demand prediction 178
extrinsic limitation 260–263
instrinsic limitation 262–263
latitude effects 178, 235–239
life history trade-offs 179, 257–260
phylogenetic contrasts method 225–229
phylogeny 241–242, 294–297
resting metabolic rate (RMR) relationship 179, 200–202, 242–255
  dietary habit responses 261–262, 263
  interspecific review 202–204
  lactating mammals 256
  seasonal responses 232–235
  sustainable metabolic scope 255–264
Flocculated dispersion 348–349
Forest CO_2/water flux *see* EUROFLUX project
Fractal geometry 339–396
  artificial pondweed colonization 388–390
  boundary/profile lines 342–348
    dilation methods 346–348
    Euclidean distance mapping (EDM) 347–348
    Kolmogorov (box-counting) method 344–346, 355
    Minkowski logic 345, 346, 355
    mosaic amalgamation logic 347
    Richardson procedure 343–344, 355
  coexistence (species packing) models 390–395
  coral reefs 387–388
  deposition processes 348–349
  flocculated dispersion 348–349
  mussel beds 388
  random walks 380–386
  riverine ecosystems 358–380
  scale issues 339–340, 341–342

species abundance distribution 390–395
statistical self-similarity 342
surfaces 349–358
  multifractals 358
Freshwater field experiments 301
  control plots 312
  design weaknesses 318–319
  duration 314, 315
  experiments analysed 302, 303
  number of plots (replicates) 312
  plot sizes 303, 307

Gigaspora 76, 78
Gigaspora pellucida 80
Glomus 76, 79
Glomus manihotis 79, 80
Glomus mosseae 75, 76, 78, 79, 80, 81, 82, 84, 85, 86
Glycine, root uptake capacity 10

Habitat complexity 340, 369–370, 390
  fractal scaling 341
  see also Fractal geometry
Hausdorff dimension 355
Heart rate telemetry 212–214, 217
Heathland, mycorrhizal nutrient uptake 12
Hemibeldeus lemuroides 227
Herbivory defences 22–23
Hibernation 191
Hydrachnellidae, sediment surface distribution 373–374, 375, 376, 379
Hydrobia ulvae 322
Hymenoscyphus ericae 92, 94, 95

Indirect calorimetry 205–206, 283–289
Infrared gas analyser 119, 120–123
Insectivora 219, 227
  field metabolic rates (FMR) 238
  phylogeny 295
  resting metabolic rates (RMR) 245
Intertidal assemblage experiments 317, 319
  design weaknesses 319–320
  duration 324–325

Iron-contaminated soils 73
  Ericaceae colonization 88, 89
  role of mycorrhizas 93

Jasus 32

Kalmia polifolia 28
Keystone paradigm 319
Kolmogorov method 344–346, 355, 359–361, 363–364, 365
Kyoto Protocol 115

Lactation, small mammals
  energy expenditure 195–198, 199–200, 201
  field metabolic rates (FMR)/resting metabolic rate (RMR) relationship 256
Lagomorpha 219
  phylogeny 296
Lagopus scotticus 320
Latitude, small mammal influences
  field metabolic rate (FMR) 235–239
  resting metabolic rate (RMR) 244–245
Leaching nutrient losses, leaves 13
  leaf senescence 45–47
Lead-contaminated soils 73
  arbuscular mycorrhizas 84
  ectomycorrhizas 96, 97
  Ericaceae colonization 88, 89
Leaf biomass allocation 16–17
Leaf lifespan 26–28
  growth form influences 27–28
  leaf level nutrient-use efficiency 36–37
  nitrogen 38, 39, 43–44
  nutrient supply effects 28
Leaf nutrient levels 13–14
  leaf level nutrient-use efficiency 35
  nitrogen 38, 40
  N : P ratios 23
  stomatal conductance effect 13
Leaf senescence 19, 20
  leaf level nutrient-use efficiency 35–36
  nitrogen 40–41

nutrient leaching 45–47
nutrient resorption 24–26
  controls 25–26
  growth form influences 24–25
Leaf traits, nutritional 21–28
  leaf level nutrient-use efficiency 33, 34, 35–37, 54
  leaf lifespan 36–37, 43
  nitrogen 38–41
  nutrient concentration 35
  phosphorus 41, 42
  resorption efficiency 35–36
  lifespan 26–28
  nitrogen productivity (A_{NP}) correlates 29–30
  nutrient concentrations 21–24
    growth form influences 21–23
    nutrient supply effects 23–24
  nutrient resorption 24–26
    controls 25–26
Ledum groenlandicum 28
Ledum palustre 28
Life history trade-offs 257–260
Linum usitatissimum 28
Litter decomposition 2, 3, 24, 45–51
  climate effect 47–48
  growth form influences 49–51
  lignin : nutrient ratios 47, 48, 53
  litter chemistry 47–49
    secondary compounds 51–53, 54
  nutrient leaching 45–47
  nutrient use efficiency (NUE) relationships 51–53, 54
Longevity, small mammals 257, 259–260
  body mass relationship 260
Lotus corniculatus 73

MacArthur Fraction (MF) 392, 395
Macoma balthica 322
Macrobenthic organism movement patterns 380–381
Macrophyte colonization, fractal geometry 388–389
Macrotus californicus 238
Manganese-contaminated soils 73
  Ericaceae colonization 88, 89

Manipulative field experiments 299–330
  control plots 307, 308, 309, 310, 312, 314, 319
  design 300, 302–329
    reasons for weaknesses 312–320
    trends 305–312
  duration 307, 313, 314, 315, 320–324
    alternative stable states development 324
    short-term experiments 323–324
    time to final equilibrium (stopping rule) 321–323
  effect size 320, 328, 329
  experiments analysed 302, 303
  extent 325
  grain 324, 325
  interaction webs 322–323
  lag 325
  meta-analysis 327–328, 329–330
  number of plots (replicates) 304–305, 306, 309, 314, 319
  pilot experiments 318, 320, 322
  plot size 303–304, 305, 307, 325
  pseudoreplication 304–305
  publication 317–318, 327–328
  sampling unit independence 304, 305
  scale 314, 324–325, 326, 327, 329
    large 301, 305, 328–329
    spatial/temporal 300
  zero-effect outcome 320
  unpublished studies 327–328
Marine field experiments 301
  control plots 312
  design weaknesses 318–319
  duration 314, 315, 321–322
  experiments analysed 302, 303
  number of plots (replicates) 306, 307, 312
  plot size 307, 310, 311, 319, 325
Marmota flaviventris 219, 248
Marsupials 219
  field metabolic rates (FMR) 238
  phylogeny 294–295
  resting metabolic rates (RMR) 245, 247
Maximal sustainable metabolic rate 186–187, 189

Mean residence time of nitrogen (MRT) 29, 30, 54
  ericaceous species 32
Medeflux 167
Meiobenthic organism movement patterns 380–381
Metal smelting contamination 73, 75
Metal-contaminated soils
  mycorrhizal associations 71–72, 104–107
    benefits to plants 72, 73
    see also Arbuscular mycorrhizas; Ectomycorrhizas; Ericoid mycorrhizas
  plant resistance 70, 71
  tree species genetic adaptation 98–99
Microgale dobsoni 225, 227
Microgale tazalci 225, 227
Microtus agrestis 190, 201
Microtus pennsylvannicus 218
Migration 190–191
  energy expenditure 185, 188, 189
Mimulus guttatus 92
Mine spoil contamination 75, 79, 83, 85, 93
Minkowski method 345, 346, 347, 355, 363–364, 365, 384
Molina caerulea 12
Molybdenum-contaminated soils, Ericaceae colonization 88
Mosaic amalgamation logic 347
Multifractals 358
Mus spp. 227
Mussel beds, fractal dimension 388
Mustelids 255
Mycorrhizal associations
  litter decomposition pathway 24
  metal-contaminated soils 69–107
    colonization strategies 71–72
  nutrient uptake 11–13, 15
    nitrogen sources 12, 24
Mycorrhizal metal resistance 70, 71, 72
Mytilus californianus 319
Mytilus edulis 388

Nereis 323
NETFLUX 167

INDEX

Nickel-contaminated soils 72
  ectomycorrhizas 96, 97
  Ericaceae colonization 88, 89
Nile river system 359
Nitrate reductase 11
Nitrate, root uptake capacity 9–10
  low nutrient supply response 10–11
Nitrogen absorption
  forms of nitrogen 9–10
  leaves 13–14
  mass flow effects 14–15
  mycorrhizal uptake 12
  root exudate effects 15
  root uptake capacity 9–10
    low nutrient supply response 10–11
Nitrogen biogeochemical cycles 5–6
Nitrogen fixation 5
Nitrogen leaf levels
  growth form variation 21, 22–23
  senescent leaves
    leaching 45–47
    resorption 24, 25, 26
Nitrogen productivity (A_{NP}) 29–30, 54
  ericaceous species 32
Nitrogen-limited plant growth 4, 5
  atmospheric deposition influence 6
  biomass allocation to roots 18
  detection 7–9
  ecosystem management effects 6–7
  foliar N : P mass ratio 2, 7
  growth rate relationships 29
  reserve formation 19
  root hair formation 18
  symbiotic nitrogen fixation 14
Nitrogen-use efficiency (NUE_N) 3
  above-ground 32
  leaf level 38–41, 42–43, 54
    ecological consequences 44–45
    leaf lifespan 43–44
  litter decomposability relationship 51
Numenius arquata 323
Nutrient acquisition by plants 2, 9–15, 53
  biomass allocation 2, 15–19
  high nutrient supply conditions 13
  leaf uptake/loss 13–14
  mycorrhizal uptake 11–13
  organic nitrogen uptake 2, 9–10, 12, 15, 24
  rhizosphere effects 14–15
    root exudates 15
  root uptake 9–11
  specific root length (SRL) 2
  symbiotic nitrogen fixation 14
Nutrient resorption efficiency (r) 2, 3
Nutrient storage, plants 19–20
Nutrient-limited plant growth 2, 4–7, 53
  differential nutrient limitation 5–6
  leaf lifespan effects 28
  leaf nutrient concentrations 23–24
  N : P mass ratio as indicator 2, 5, 7–9
    species versus community biomass production 8–9
  nitrogen availability 5–6
  nutrient-use efficiency (NUE) 29, 30, 31, 33
  phosphorus sources 5
Nutrient-use efficiency (NUE) 2–3, 4, 53
  above-ground 32–33
  biomass production 28–31
  constraints 42–44
  ecological consequences 44–45
  leaf level 21, 33, 34–45
    estimation methods 34–35, 37–38
    leaf traits 33, 34, 35–37, 54
    nitrogen-use efficiency (NUE_N) 38–41, 42–43
    phosphorus-use efficiency (NUE_P) 41, 42
  litter decomposition rate relationships 51–53, 54
  low fertility soils 29, 30, 31, 33
  mean residence time of nitrogen (MRT) 29, 30, 32, 34, 54
  nitrogen productivity (A_{NP}) 29–30, 32, 54
  plant secondary compounds 51–53, 54
  trade-offs 54
  whole-plant level 31–32

Optimal foraging theory 190
Organic nitrogen uptake
  mycorrhizas 12

nutrient acquisition by plants/uptake capacity 2
plant absorption, litter decomposition pathway 24
rhizosphere pH effects 15
root uptake capacity 9–10
Organic phosphorus, root uptake capacity 10

Parorthocladius nudipennis 374, 375, 378
Patch structure 340
  fractal properties 341, 373–380
Patella vulgata 319
Paxillus involutus 100, 101, 102
Perognathus californicus 231
Perognathus formosus (pocket mouse) 231, 232, 233
Peromyscus maniculatus 239
Peromyscus spp. 227
Petauoides volans 248
Phascogale 257
Phosphatases, root 10, 15
Phosphorus absorption
  mass flow effects 14–15
  mycorrhizal uptake 12
    contaminated soils 86
  organic phosphorus 10, 15
  rhizosphere pH effects 15
  root uptake capacity 9, 10
    low nutrient supply response 10–11
Phosphorus biogeochemical cycles 5
Phosphorus leaf levels
  growth form variation 21, 22, 23
  senescent leaves
    leaching 45–47
    resorption 24, 25, 26
Phosphorus-limited plant growth 4, 5
  detection 7–9
  ecosystem management influences 6–7
  foliar N : P mass ratio 2, 7
  root hair formation 18
Phosphorus-use efficiency (NUE$_p$) 3
  above-ground 32
  leaf level 41, 42–43, 45, 54
  litter decomposability relationship 51

Phylogenetic relationships, small mammals 225–229, 294–297
Picea glauca 95, 97, 99
Pinus banksiana 95, 97, 99
Pinus strobus 41
Pinus sylvestris 97, 98, 99, 102, 103
Pisaster ochraceous 319
Pisolithus tinctorius 101, 102, 104
Plecotus auritus 238
Potassium, root uptake capacity 9
Pseudochirus spp. 227

Quercus rubra 97, 99

Random Assortment (RA) model 392, 393
Random Fraction (RF) model 392, 393, 395
Random Resource Allocation (RM) model 392, 393, 395
Respiratory quotient (RQ) 205, 206
Resting metabolic rate (RMR), small mammals 178–179, 206–207
  body mass relationship 202–203, 243–244, 245
  phylogenetic contrasts method 245–246
  dietary factors 246–247, 261–262, 263
  field metabolic rate (FMR) relationship 179, 200–202, 242–255
  dietary habit effects 261–262, 263
  interspecific review 202–204
  latitude relationship 244–245
  organ size relationship 200–201
  phylogeny effects 247, 255
Rhea americana 181
Rhododendron ponticum 90, 91, 94
Richardson procedure 343–344, 355
Riverine ecosystem fractal properties 358–380
  catchment area 358–361
  channel length 361–362
  mainstream length–basin relationship 361
  particle surfaces 364–367
    larval chironomid body-size relationship 367–372

Riverine *continued*
  pore–particle boundary 363–364, 365
  spatial scale 373–380
Rodents 219
  field metabolic rates (FMR) 238
  phylogeny 226, 227, 296
Root exudates 15
Root hair density 16
  nutrient supply effects 18
Root weight ratio (RWR) 16
Roots
  architecture 16
  biomass allocation 15–17, 53
    growth cabinet studies 17–18
    nitrogen limitation 18
  nitrate reductase 11
  nutrient efflux limitation 11
  nutrient uptake 9–11
    exudate effects 15
    forms of nitrogen 9–10
    low nutrient supply response 10–11
    mass flow effects 14–15
    mycorrhizal 11–13
    organic acids secretion 15
    uptake patterns 9–10
  phosphatases 10, 15
  root weight ratio (RWR) 16
  specific root length (SRL) 16, 17
Rotaria tardigrada 381, 385
Running activities 181–183
  energy expenditure 181–182, 183, 184–185
  maximal sustainable rates 188, 189
  substrate utilization 183, 184–185
  ultra-long distancees 188–189

Salvia leucophylla 32
Scaling issues
  field experiments 300, 301, 314, 324–325, 326, 327, 328–329
  fractal geometry 339–340, 341–342
Schizachyrium scoparium 46
Sclerocystis 76
Scutellospora 78
Seasonal effects, small mammals
  field metabolic rates (FMR) 231, 232–235
  winter adaptations 191, 231, 235
Sewage sludge contamination 73, 75, 79
Sigmodon hispidus 200
Sminthropsis crassicaudatus 248, 257
Sonic anemometer 119–120
Sorex areneus 223, 238
Specific root length (SRL) 2, 16, 17
  nutrient supply response 18
Sphagnum bogs 6
Standard operative temperature measurements 210–211
Stenostomum 384
Sternidae (terns) 189
Stomatal nutrient uptake/loss 13
Streambed sediment 340
  organism movement patterns 380–386
  particle surfaces 364–367
    larval chironomid body-size relationship 367–372
  pore–particle boundary 363–364, 365
Suillus bovinus 100, 101, 102, 103
Suillus luteus 96, 97, 100, 101, 102, 103
Surface absorption isotherms 351
Symbiotic nitrogen fixation 14

Tenrecidae 209, 255
Terrestrial field experiments 301
  control plots 312, 314
  design weaknesses 313–314
    acceptance 318–320
    authors' rationale 314, 316
    plot size selection critera 316
    repetition of experiments 317–318
    replicate numbers selection criteria 317
    zero-effect outcome 320
  duration 314, 315, 322
  experiments analysed 302, 303
  number of plots (replicates) 306, 307, 312, 314
  plot size 303, 307, 310, 311, 313–314
Thelophora terrestris 100
Thienemanniella 393
Thomomys bottae (pocket gopher) 228, 233
Time budgets 209–210
Torrenticola elliptica 375, 376

INDEX

Trade-offs
  plant nutrient-use efficiency (NUE) 54
  small mammal energy expenditure 188, 257–260

Vaccinium macrocarpon 90, 91, 94

Water vapour transfer
  EUROFLUX project *see* EUROFLUX project
  global measurement networks 167
Wetland ecosystems 7, 8
Winter survival strategies 190–191, 231, 235

Ybbs river system 359–361

Zinc-contaminated soils 72, 73
  arbuscular mycorrhizal associations 75, 76, 78, 79, 80, 81–82
  metal assimilation 82, 83, 84
  morphological responses 81
  plant biomass production 85–86
  plant phosphorus status 86–87
  ectomycorrhizas in metal resistance 96, 97, 99–101
  colonization studies 102–103
  Ericaceae colonization 88, 89
  role of mycorrhizas 89, 90, 93–95, 105

66.95!